Meteorology
Understanding the Atmosphere

Books in the Brooks/Cole Earth Sciences and Geography Series:

Meteorology
Meteorology: Understanding the Atmosphere, ACKERMAN/KNOX
Essentials of Meteorology, AHRENS
Meteorology Today, AHRENS
Meteorology for Scientists and Engineers, STULL

GIS Investigations in the Earth Sciences
Exploring the Dynamic Earth, HALL-WALLACE ET AL
Exploring Tropical Cyclones, HALL-WALLACE ET AL
Exploring Water Resources, HALL-WALLACE ET AL

Oceanography
Essentials of Oceanography, GARRISON
Oceanography: An Invitation to Marine Science, GARRISON

Physical Geology
Physical Geology: Exploring the Earth, MONROE/WICANDER
Introduction to Physical Geology, THOMPSON/TURK
Modern Physical Geology, THOMPSON/TURK
Essentials of Geology, WICANDER/MONROE

Historical Geology
Historical Geology: Evolution of Earth and Life through Time, WICANDER/MONROE

Combined Physical/Historical Geology
The Changing Earth, MONROE/WICANDER

Environmental Geology
Geology and the Environment, PIPKEN/TRENT

Geology/Earth Science Lab Manuals
Investigations into Physical Geology, MAZZULLO
Earth Lab: Exploring the Earth Sciences, OWEN/PIRIE/DRAPER

Earth Science
Earth Science, DUTCH/MONROE/MORAN
Earth Science Today, MURPHY/NANCE
Earth Science and the Environment, THOMPSON/TURK

Physical Geography
Essentials of Physical Geography, GABLER ET AL

World Regional Geography
Essentials of World Regional Geography, SALTER/HOBBS

Internet and Media Products

Blue Skies™ College Edition, CD-ROM, LIVING TEXT
Brooks/Cole Earth Science Resource Center:
http://earthscience.brookscole.com
CNN Today videos for Physical Geology, Meteorology, and Oceanography
World Regions Interactive CD-ROM, SALTER/DIBIASE/HOWARD

Meteorology
Understanding the Atmosphere

Steven A. Ackerman
University of Wisconsin, Madison

John A. Knox
University of Georgia

THOMSON

BROOKS/COLE

Australia • Canada • Mexico • Singapore • Spain • United Kingdom • United States

Earth Science Editor: Keith Dodson
Development Editors: Marie Carigma-Sambilay, Richard Morel
Assistant Editor: Carol Benedict
Editorial Assistant: Faith Riley
Technology Project Manager: Samuel Subity
Marketing Manager: Ann Caven
Marketing Assistant: Sandra Perin
Advertising Project Manager: Linda Yip
Project Manager, Editorial Production: Teri Hyde
Print/Media Buyer: Karen Hunt
Permissions Editor: Stephanie Keough-Hedges

Production Service: Graphic World, Inc.
Text Designer: Roy Neuhaus
Art Editor: Graphic World, Inc.
Photo Researcher: Terry Powell
Copy Editor: Graphic World Publishing Services
Illustrator: Graphic World, Inc.
Cover Designer: Roy Neuhaus
Cover Image: Taya Kashuba/Kalamazoo Gazette
Cover Printer: Transcontinental Printing Inc.
Compositor: Graphic World, Inc.
Printer: Transcontinental Printing Inc.

Printed in Canada
2 3 4 5 6 7 06 05 04 03 02

For more information about our products, contact us at:
Thomson Learning Academic Resource Center
1-800-423-0563

For permission to use material from this text, contact us by:
Phone: 1-800-730-2214
Fax: 1-800-730-2215
Web: http://www.thomsonrights.com

Library of Congress Cataloging-in-Publication Data

Ackerman, Steven A.
　Meteorology: understanding the atmosphere /
　Steven Ackerman, John Knox.
　　p. cm.
　Includes index.
　ISBN 0-534-37199-X
　　1. Meteorology. I. Knox, John, 1965- II. Title.

QC861.3 .A34 2002
551.5—dc21

2001058237

Brooks/Cole—Thomson Learning
511 Forest Lodge Road
Pacific Grove, CA 93950
USA

Asia
Thomson Learning
60 Albert Street, #15-01
Albert Complex
Singapore 189969

Australia
Nelson Thomson Learning
102 Dodds Street
South Melbourne, Victoria 3205
Australia

Canada
Nelson Thomson Learning
1120 Birchmount Road
Toronto, Ontario M1K 5G4
Canada

Europe/Middle East/Africa
Thomson Learning
Berkshire House
168-173 High Holborn
London WC1V 7AA
United Kingdom

Latin America
Thomson Learning
Seneca, 53
Colonia Polanco
11560 Mexico D.F.
Mexico

Spain
Paraninfo Thomson Learning
Calle/Magallanes, 25
28015 Madrid, Spain

About the Authors

Steven Ackerman is Professor of Atmospheric and Oceanic Sciences at the University of Wisconsin, Madison and is Director of the Cooperative Institute for Meteorological Satellite Studies (CIMSS). He received his B.S. degree in physics from the State University of New York-Oneonta, and his Ph.D. in atmospheric

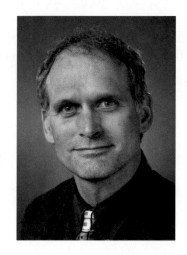

sciences at Colorado State University. Renowned for his ability to inspire active student participation in his classes, Ackerman has won numerous teaching and academic awards, including the Chancellor's Award for Distinguished Teaching (University of Wisconsin). Ackerman is also a member of the University of Wisconsin Teaching Academy. Ackerman's research interests center on understanding how changes in the radiation balance affect and are affected by changes in other climate variables such as clouds, aerosols, water vapor, and surface properties.

John Knox is Associate Research Scientist and Lecturer in the Department of Biological and Agricultural Engineering at the University of Georgia, where he has taught introductory weather and climate, weather analysis and forecasting, and aviation meteorology. He has also taught meteorology at Valparaiso University

and Barnard College of Columbia University. A National Science Foundation Graduate Research Fellow and Rhodes Scholar finalist, Knox received his B.S. in mathematics from the University of Alabama at Birmingham and his Ph.D. in atmospheric sciences from the University of Wisconsin, Madison. He was a post-doctoral fellow in climate systems at Columbia University in conjunction with the NASA/Goddard Institute for Space Studies (NASA/GISS) in New York City. At Wisconsin, Knox won the Eberhard Wahl Teaching Award and was a College of Letters and Science Teaching Fellow. He has published journal articles in geoscience education and served on the *Journal of College Science Teaching*'s Board of Advisors. His research in atmospheric dynamics focuses on Rossby waves, nonlinear balance, clear-air turbulence, and cyclone-induced windstorms, and is currently supported by NASA.

To Anne, Erin, and Alana, who are always lovingly patient with my meteorological distractions. I thank my parents and siblings for their good humor.

S.A.A.

To my family, most of all Pam and Evan; to my students; and to the two people who most inspired me to study and teach meteorology: the late Dr. Lyle Horn of the University of Wisconsin and J.B. Elliott of the National Weather Service (Birmingham, retired)

J.A.K.

Brief Contents

Contents

To The Student

Weather engulfs us. Its influence can be both dramatic and subtle. Weather tempers how we dress, how we live, the music we play, and the art we create. It can destroy our homes and threaten our lives. It affects our daily activities, leisure, holidays, transportation, commerce, agriculture, and nearly every other aspect of our lives. Our fascination with the weather has led to 24-hour weather networks, feature-length motion pictures, and an explosion of detailed weather data over the Internet.

Mark Twain said, "everyone talks about the weather, but nobody does anything about it." You may not be able to change the weather, but you can discover the processes that determine weather. We think that learning about meteorology can and should be both enjoyable and relevant to your everyday life. Knowledge gained from reading this book can help you better understand nightly television weather reports and interpret news articles on severe weather, impending climate change, greenhouse warming, the depletion of the ozone layer, and the causes and effects of El Niño. You will experience and be influenced by these events throughout your lifetime. *Meteorology: Understanding the Atmosphere* will help you to grasp the fundamentals and gain an appreciation of the complexities involved with these issues.

The World Wide Web has enhanced your opportunity to learn meteorology by applying textbook concepts to real-time weather conditions. A flood of sophisticated weather information that once was restricted to a few scientists is now just a click away on the World Wide Web. *Meteorology: Understanding the Atmosphere* will help you to make sense of the abundant weather information available to you on the Internet, serving as a reference as you investigate current conditions. In addition, dozens of Java applets on our text's Web site will help you understand the material in this book and allow you to explore topics in even greater detail.

Meteorology is a topic that easily generates interest in and an appreciation of a natural science. Our goal in writing this book was to provide you with a perspective on meteorology as a science in which observations play a key role. Thus, we provide many observations, both personal and from scientific instruments, throughout this book along with analysis of those observations. This approach of observing and then analyzing the atmosphere to gain an understanding is a scientific way of thinking, and it is how we, the authors, explore and understand the atmosphere ourselves. We hope that you find this approach exciting and that it inspires you to a lifetime of watching and understanding the weather.

To The Instructor

Meteorology: Understanding the Atmosphere is designed for use in a wide range of college and university introductory courses in meteorology and in weather and climate. This book is written in an interesting and clear manner that allows your students to immediately apply material to the world around them.

Our text emphasizes **observing the atmosphere and using those observations to explain atmospheric phenomena.** Just by paying attention to the weather outside, it is possible for a newcomer to the subject to observe key clues that explain how the atmosphere works. By learning how to interpret scientific observations of the atmosphere, students can deepen their understanding even more. The observations we examine in this text range from those made by students themselves to cutting-edge scientific measurements from radar and from space. Many of the images here have never appeared before in print. Throughout the text, weather phenomena come alive via conceptual models to explain their existence, visualization of their life cycles, and weather safety information to keep weather from becoming a killer.

This book focuses on understanding the basic concepts of meteorology. Most students have a better understanding of the world around them through observations and experiences than through mathematics. For this reason, throughout the text we begin by asking about a weather phenomenon, "what does it look like?" before delving into the theory behind it. Therefore, the book is accessible to those not majoring in geography or meteorology, while still providing detailed mathematics for more advanced students. We employ narratives and metaphors that in some chapters allow us to explore topics more deeply than is done in texts that use extensive mathematics.

Meteorology: Understanding the Atmosphere has several unique features. Weather maps and weather watches and warnings are introduced immediately in Chapter 1 to allow instructors the option of using current weather to explore topics discussed in each chapter. The physics of energy transfer in the atmosphere is presented in an accessible manner to students without a physics background in Chapter 2. These concepts are applied in Chapter 3 to describe observed temperature variations. Chapter 4 on the atmospheric water cycle combines clouds with other water phases. It is a concise and unified treatment of how water circulates through the atmosphere. Chapter 5 is an integrated approach that combines visual observations of the state of the atmosphere, including optics, with explanations of scientific measurements such as satellite imagery. Chapter 6 approaches the usually difficult discussion of forces via the simplifying idea of "balance," which is also how today's researchers make sense of this subject. These ideas are applied worldwide in Chapter 7 in the form of conceptual models of global winds.

The second half of our text examines topics in weather and climate in a variety of innovative ways. Tropical cyclones and El Niño are appropriately covered together in Chapter 8, a unified chapter on atmosphere–ocean interactions that reflects our growing appreciation of how the atmosphere and ocean interact to affect weather and climate. Chapter 9 enlivens the standard discussion of fronts and air masses with apt analogies to regional accents. Chapter 10 is a state-of-the art chapter on extratropical cyclones and anticyclones, presented in the compelling contexts of the stories of the *Edmund Fitzgerald* shipwreck and the John F. Kennedy, Jr., plane crash. Chapter 11 views the life cycles of severe weather from a variety of angles: the ground, the air, Doppler radar, and satellite. Chapter 12 is a comprehensive look at small-scale winds across the United States and the globe. Chapter 13 is the most complete weather forecasting chapter ever written for this level and is made accessible with three narratives illustrating the advances in forecasting during the past century. Chapters 14 and 15 address past climates and climate change as mysteries to be solved, rather than as cataclysmic scenarios. Finally, complex topics, such as global warming and adiabatic and diabatic temperature changes, are visited throughout the entire textbook.

The instructor-friendly structure of this book is based on our combined teaching experiences at five different universities. For example, Chapter 5 discusses how we observe the atmosphere using both our senses and scientific instruments. The modular structure of this chapter allows this material to be covered all at once or as a function of weather parameter. The tropical cyclone section in Chapter 8 comes early enough in the text that fall-semester instructors may easily cover it during hurricane season. Chapter 10 synthesizes and reinforces the material on forces, air masses and fronts from Chapters 6 and 9. Finally, Chapter 12's modular design allows instructors to cover as much, or as little, of small-scale winds as desired and to focus on a particular geographic region. Throughout the text, intertextual icons indicate related Java applets and Blue Skies exercises that expand your students' abilities to explore these topics beyond the confines of the lecture hall.

Our Web site—http://info.brookscole.com/ackerman—includes over three dozen unique Java applets that have been already acclaimed by the meteorology education community. These learning tools extend the textbook treatment of key topics such as weather map analysis, atmospheric circulation patterns, and numerical models.

Chapter Features

- Outlines and chapter goal lists at the beginning of each chapter
- Introductions focusing on observations of the atmosphere
- Intertextual icons that identify Java applets and Blue Skies exercises relating to material
- Extended boxes delving into advanced and unusual topics in each chapter
- End-of-chapter summaries to review the main ideas presented in the chapter
- A list of key terms at the end of each chapter
- Chapter-ending review questions that integrate chapter materials with the *Blue Skies* CD-ROM and the text's World Wide Web site

Web Features

The accompanying Web site includes over three dozen interactive Java applets that extend the textbook treatment of key topics such as weather map analysis, satellite interpretation, and numerical weather models. These applets are well tested and have received acclaim by the meteorology education community. The Web site also includes animations of weather phenomena and tools for assessing student learning.

Acknowledgments

This book could not have been possible without the efforts of many. Teri Hyde, Keith Dodson, and Nina Horne at Wadsworth and Brooks/Cole deserve special mention for their advice and captaining of the editorial process, which has resulted in a final product we are very proud of. Kim Leistner energized this project during its formative stages. Sam Subity has ably assisted in the development of this text's Web site, a key component of any 21st-century text. Tom Whittaker has created some of the world's best meteorological Java applets—try a few at info.brookscole.com/ackerman. Dick Morel became a special member of this project as a developmental editor and key advisor. Mike Ederer and John Denk at Graphic World Publishing Services shouldered much of the burden of editing and artwork. Terry Powell and her associates at The Photographer's Window dealt with myriad photo searches, and Stephanie Keough-Hedges handled a blizzard of permissions requests. Greg Thompson, Cynthia Johnson and Don Lloyd, among others, receive our heartfelt thanks for sharing their stunning images. Stino Iacopelli assisted with Chapter 10, which includes portions of his award-winning research. (We regret that permission was denied to reprint the lyrics of the meteorologically accurate song "The Wreck of the Edmund Fitzgerald" in that same chapter.) Pam Naber Knox read every word of page-proofs and considerably improved the content of the entire book. Anne Pryor and Erin Pryor-Ackerman provided editorial comments on several chapters. Finally, we thank the colleagues who took the time out of their busy schedules to review all or parts of the manuscript, including:

Mark R. Anderson,
University of Nebraska at Lincoln

John Arnfield,
Ohio State University

Leanne Avila,
University of Wisconsin

Mark Binkley,
Mississippi State University

Steven Businger,
University of Hawaii

Donna Charlevoix,
University of Illinois at Urbana-Champaign

William C. Culver,
St. Petersburg Junior College

Nancy Dignon,
Tallahassee Community College

John A. Ernst,
Embry-Riddle Aeronautical University

Terri Gregory,
University of Wisconsin

Vince Gutowski,
Eastern Illinois University

Leonard Hume, Jr.,
Cloud County Community College

Peter Jackson,
University of Northern British Columbia

Stephen Jascourt,
UCAR/COMET

Scott Jeffrey,
Catonsville Community College

Rudi Kiefer, Emeritus,
University of North Carolina at Wilmington

Steve Ladochy,
California State University

Frank Lombardo,
Daytona Beach Community College

Rich Miller,
Milwaukee Area Technical College

Scott Robeson,
Indiana University

Robert Rohli,
Louisiana State University

Paul Ruscher,
Florida State University

Catherine Souch,
Indiana University at Indianapolis

Harold Taylor,
Stockton College of New Jersey

Stanton E. Tuller,
University of Victoria

Steve Vavrus,
University of Wisconsin

Anthony J. Vega,
Clarion University

Charles Weidman,
University of Arizona

Wayne Wendland,
University of Illinois at Urbana-Champaign

Kay Williams,
Shippensburg University

Mark Wyman,
Cornell University

Mark Wysocki,
Cornell University

Douglas Yarger,
Iowa State University

Introduction to the Atmosphere

Introduction

It's a hot, muggy summer night at the baseball stadium. The Atlanta Braves are on their way to another winning season, and they are taking a night "off" to play an exhibition game against a minor league all-star team. The standing-room-only crowd, the largest in years, applauds as future Hall of Fame players take the field.

Midway through the game, however, the weather turns violent. High winds suddenly blow chairs off the stadium roof. Then the sky explodes with light and sound as lightning strikes an electric transformer on a pole out beyond center field. A fireball dances along the power lines and the stadium lights go dark.

Frightened, the baseball players run off the field into the dugouts and panicked fans shriek as the thunder crashes. One little boy dives under his stadium seat in terror, only to peek out and observe ominous purple and green clouds racing overhead. Reports of a funnel cloud—a tornado in the clouds above the stadium—spread among the crowd. Flooding rains descend, and sopping-wet spectators splash through puddles and duck lightning bolts as they flee to their cars.

Conversations on the way home focus on the rain, the wind, the lightning, and the possible tornado, not on the game everyone eagerly anticipated just a couple of hours before. The American pastime of baseball has been upstaged by the universal spectacle of the weather.

As this true story illustrates, weather affects every facet of our lives, even when we least expect it to. We all experience weather through our senses: the flash of lightning, the crack of thunder, the stickiness of a summer night, and the peculiar smell of rain. In some cases,

an event is so memorable that we experience long-ago weather in stories told at family gatherings. In this book we will pay close attention to how you sense the atmosphere, using sight, sound, touch, and even stories. We hope that years from now, you'll remember how observations of the atmosphere help to explain the weather around you.

Meteorology, the study of weather and climate, is a science—a young and exciting science. Meteorologists sense the atmosphere like everyone else. As we'll see, young scientists using only their eyes and brains have made some of the greatest discoveries in meteorology. Today, meteorologists also use a variety of specialized techniques and tools, including weather satellites, to augment their senses (Figure 1.1). In this book you will learn about the atmosphere in the context of how it is sensed by meteorologists. Our goal is to help you understand the science behind weather forecasts for tomorrow and predictions of global warming in future years.

We begin our exploration of meteorology in this chapter with the basics: what the atmosphere is made of and how our observations of it are turned into weather maps.

Weather and Climate

Weather is the condition of the atmosphere at a particular location and moment. Each day current weather conditions are given in local weather reports. These reports typically include current temperature, relative humidity, dew point, pressure, wind speed and direction, cloud cover, and precipitation. Such weather information is important to us because it influences our everyday activities and plans. Before going out for the day we want to know how cold or hot it will be and whether or

NOAA/NESDIS and SSEC, University of Wisconsin-Madison

■ Figure 1.1

This weather satellite image presents a view of North America as you would see it if you were above Earth looking down. Dark regions are areas of clear skies; gray and white regions are cloudy. The Western Great Plains are clear, while much of Wisconsin is cloud-covered. Notice the bright white thunderstorm clouds over Florida.

not it will rain or snow. **Meteorology** is the study of these weather variables, the processes that cause weather, and the interaction of the atmosphere with the Earth's surface, ocean, and life.

The fundamental cause of weather is the effect of the Sun on the Earth. At any time, only half of the Earth is warmed by the Sun, while the Earth's other side is shadowed. This causes uneven heating of the Earth's surface by the Sun every day, with some regions warmer than others. For reasons we'll explore in later chapters, these temperature differences cause weather: winds, clouds, and precipitation. Seasonal weather patterns result from variations in temperature caused by the Earth's tilt toward the Sun in summer and away from the Sun in winter. The distribution of water and land, and the topography of the land, contribute to the shaping of Earth's weather patterns on smaller scales. In Chapter 2 we explore the uneven heating of the Earth by the Sun and its effects on weather and climate in more detail.

The **climate** of a region, in contrast to the weather, is the condition of the atmosphere over many years. On average, Florida will have a mild climate all year long. Minnesota, on the other hand, will have a climate with warm, even hot, summers and very cold winters. The climate of a region is described by long-term averages of atmospheric conditions such as temperature, moisture, winds, pressure, clouds, visibility, and precipitation type and amount. The description of a region's climate must include extremes as well as averages, for example record high and low temperatures. In addition, the climate of a region may change over hundreds or thousands of years.

Climatology is the study of climate. Climatologists study the long-term averages and extremes of the atmosphere. Increasingly, climatologists also investigate the changes of climate in the past and possible climate changes in the future.

A close relationship exists between meteorology and climatology. Both fields study the atmosphere. However, climatology has been more concerned than meteorology with how oceans, landforms, and living organisms affect the atmosphere. The atmosphere, the thin ocean of air that we live in, is the main focus of meteorology. In this chapter we examine the basics of the atmosphere that are essential for both meteorology and climatology.

Northern Hemisphere

 ## The Earth's Major Surface Features

Our atmosphere receives energy from the Sun. The surface of the Earth exchanges energy and water with the atmosphere. The distribution of land and water therefore plays a major role in determining climatic conditions and weather patterns. Approximately 70% of Earth's surface is water. The four major water bodies are the Pacific, Atlantic, Indian, and Arctic Oceans.

Africa, Asia, Antarctica, Australia, Europe, North America, and South America are the seven continents. More than two thirds of these landmasses are located in the Northern Hemisphere (Figure 1.2). Differences in current climate and weather patterns between the northern and southern hemispheres can often be attributed to differences in the amount of land in each.

Surrounding Earth's surface is the atmosphere—a thin envelope of gases no taller than the distance of an hour's drive on the highway. The atmosphere protects us from the Sun's high-energy radiation and provides the air we breathe and the water we drink.

Southern Hemisphere

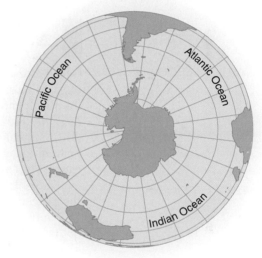

 Figure 1.2

The global distribution of land and water shown here strongly influences weather and climate patterns.

 ## Making an Atmosphere: Gases and Gravity

In our everyday lives we encounter matter in three forms: solid, liquid, and gas. For example, we are all familiar with water as solid ice, liquid water, or as a vapor. The atmosphere is made up primarily of a mixture of gases that includes liquid and solid particles suspended in air, such as water droplets, ice crystals, and dust particles.

The molecules of gases and liquids are in constant motion. They naturally spread out or diffuse from areas of high concentration to areas of low concentration. If someone peels an orange, its aroma will soon permeate the room. The aromatic molecules diffuse from an area of high concentration, right around the orange, to areas where there are few, if any, "orange aroma" molecules. But if our atmosphere is an area of concentrated air molecules, why don't they eventually diffuse into empty outer space?

Gravity, the mutual attraction between objects, is the force that holds the atmosphere in place. Gravity keeps the Moon orbiting the Earth, the planets orbiting the Sun, and you from floating out to space. The Earth's gravity exerts a pull on gas molecules in the atmosphere and keeps them from diffusing out to space.

Gravitational attraction between objects depends on the masses of the objects. If an object has a large mass, it will have a strong gravitational attraction and can attract objects that do not have much mass. For example, the Sun is large and is composed mostly of light gases such as hydrogen and helium. The Earth is much smaller than the Sun and its atmosphere contains little hydrogen and helium. The Moon is much smaller than the Earth and does not even have an atmosphere! Gravitational attraction also depends on the distance between the objects and weakens rapidly with distance. An object twice as far from Earth as another feels only one-quarter as much "pull" due to gravity.

By itself, gravity would turn the atmosphere into a layer cake with the heaviest molecules near the Earth's surface and the lightest molecules at the top of the atmosphere. However, weather processes "stir" the atmosphere, helping to keep it well-mixed almost to its outer edge. This is how heavier-than-air molecules, such as chlorofluorocarbons (see below), are able to reach the stratosphere.

Atmospheric Evolution and Composition

Gravitational attraction plays an important role in the evolution of the concentration of gases in our atmosphere. Since its formation approximately 4.5 billion years ago the Earth and its atmosphere have undergone extraordinary changes. In the beginning, the Earth's atmosphere was hot and consisted mostly of hydrogen (H), helium (He), methane (CH_4), and ammonia (NH_3). Only small amounts of these gases remain in today's atmosphere. The gases composing today's atmosphere are mostly nitrogen (N_2) and oxygen (O_2).

Table 1.1 lists the gases composing our current atmosphere. If you measured the percentage of the different gases in some fixed volume of air, you would find 78% nitrogen and approximately 21% oxygen. You would find only small amounts, or "traces," of other atmospheric gases. Water vapor (H_2O) is one of these trace gases. The amount of water vapor in the atmosphere varies from day to day and from place to place. This variability and the movement of water in all three phases underlie many aspects of weather, including changes in the weight of air (Box 1.1).

The gases in today's atmosphere are largely a result of emissions by volcanoes over billions of years. A volcanic eruption throws not only ash and rock, but also large amounts of gases, into the atmosphere. The major gases in a volcanic plume are water vapor, carbon dioxide (CO_2), and nitrogen. What happened to these gases after their release into the atmosphere over billions of years?

After its formation, the Earth began to cool. During the cooling process, the water vapor from volcanic eruptions condensed and formed clouds. Precipitation from the clouds eventually formed the oceans, glaciers, lakes, and rivers. The development of the oceans affected atmospheric concentrations of carbon dioxide. Some carbon dioxide from the atmosphere dissolved and accumulated in the oceans as they formed.

What happened to the nitrogen outgassed by volcanoes? Nitrogen is a chemically stable gas,

TABLE 1.1 Composition of the Atmosphere	
Gas	*% by volume*
Nitrogen	78.08
Oxygen	20.95
Argon	0.93
Trace Gases	
Carbon dioxide	0.037
Methane	0.00017
Ozone	0.000004
CFCs	0.00000002
Water vapor	Highly variable (0–4%)

which means it does not interact with other gases or the Earth's surface. Once nitrogen enters the atmosphere, it tends to stay there. This accounts for the high concentration in today's atmosphere; nitrogen has been accumulating over billions of years.

Volcanoes emit very little oxygen, so how did oxygen come to comprise such a large amount of today's atmosphere? Approximately 3 billion years ago tiny one-celled green-blue algae evolved in the ocean. Water protected the one-celled organisms from the Sun's lethal ultraviolet light. The algae produced oxygen as a by-product of photosynthesis, the process plants use to convert solar energy, water, and carbon dioxide into food. Today's oxygen levels are the result of billions of years of accumulation.

As the oxygen from plants slowly accumulated in the atmosphere, ozone (O_3) began to form. Ozone is both caused by and provides protection from damaging ultraviolet energy emitted by the Sun. The development of an atmospheric "ozone layer" allowed life to move out of the oceans and onto land. We cover ozone in more detail in Chapter 2.

Trace Gases and Aerosols

Nitrogen and oxygen make up 99% of the atmosphere. Yet the remaining 1% contains many of the atmospheric gases that are crucial to weather and climate. The three major trace gases in the atmosphere are carbon dioxide, water vapor, and ozone. These gases play important roles in the energy cycles of the atmosphere. Other important trace gases for atmospheric studies include methane and chlorofluorocarbons (CFCs). These gases are important because they interact with other gases and modify the energy balance of the atmosphere, a topic discussed in the next chapter. In addition to gases, small particles suspended in the atmosphere are also important in determining the quality of the air we breathe and the transfer of

BOX 1.1

Moist Air Is Lighter Than Dry Air

For now we will define moist air as a volume of air with many water vapor molecules and dry air as a volume of air that contains only a few water vapor molecules. To understand why moist air is lighter than dry air we have to define a few concepts.

- A molecule of water has the properties of water and is composed of two hydrogen atoms and one oxygen atom.
- The weight of an individual atom is represented by its atomic weight. The atomic weight of hydrogen (H) is 1, oxygen (O) is 16, nitrogen (N) 14, and carbon (C) has an atomic weight of 12. The weight of a molecule is determined by summing the atomic weights of its atoms. A water molecule (H_2O) has a molecular weight of 18 (1+1+16). Free nitrogen (N_2) has a molecular weight of 28 (14+14) and an oxygen molecule (O_2) has an atomic weight of 32.

- A fixed volume of a gas at constant pressure and temperature has the same number of molecules. It does not matter what the gas is, the same number of molecules will exist in that volume. This is known as Avogadro's Law.

To make a given volume of air more moist, we need to add water vapor molecules to the volume. To add water molecules to the volume, we must remove other molecules to conserve the total number of molecules in the volume (Avogadro's Law). Dry air consists mostly of nitrogen and oxygen molecules, which weigh more than water vapor molecules. To make a given volume of air moister we replace heavy molecules with ones that are lighter. Therefore, moist air is lighter than dry air at the same temperature. As we shall see later, an ingredient for the formation of severe thunderstorms is heavier dry air above lighter moist air, a condition we refer to as an *unstable atmosphere*.

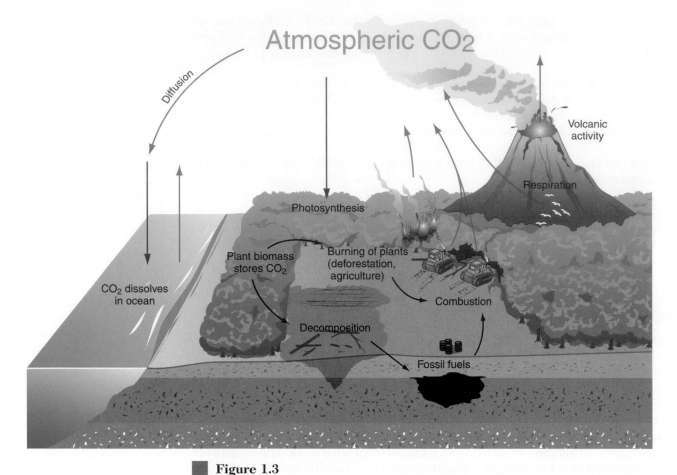

Figure 1.3

The carbon dioxide cycle. The blue lines represent processes by which carbon dioxide enters the atmosphere. Red lines represent the primary processes by which carbon dioxide is removed from the atmosphere. Black arrows represent processes that store carbon in the Earth.

energy in the atmosphere. For climatic predictions, it is important to know how and why the concentrations of these gases and particles change over time.

To know how the concentration of a gas changes, we have to know how it enters and departs the atmosphere. A **source** is a mechanism that supplies a gas to the atmosphere, and a **sink** removes a gas from the atmosphere. The routes by which a gas enters and leaves the atmosphere are known collectively as a cycle. In this section, we first consider the carbon dioxide and hydrologic (water) cycles before moving on to discuss methane, chlorofluorocarbons, and aerosols. (The formation and destruction of ozone is discussed in the next chapter.)

Carbon Dioxide Cycle

The atmospheric carbon dioxide cycle (Figure 1.3) describes how carbon dioxide moves between the atmosphere, ocean and the land. Nearly half the carbon dioxide that enters the atmosphere moves between the ocean and plants. Here we review how carbon dioxide enters and leaves the atmosphere.

As mentioned earlier, volcanoes inject carbon dioxide into the atmosphere and are therefore an atmospheric source of carbon dioxide. Plants, through the process of photosynthesis, use sunlight, water, and carbon dioxide to manufacture food. Plants remove carbon dioxide from the atmosphere during photosynthesis where it becomes incorporated into their tissues in the form of other chemicals such as

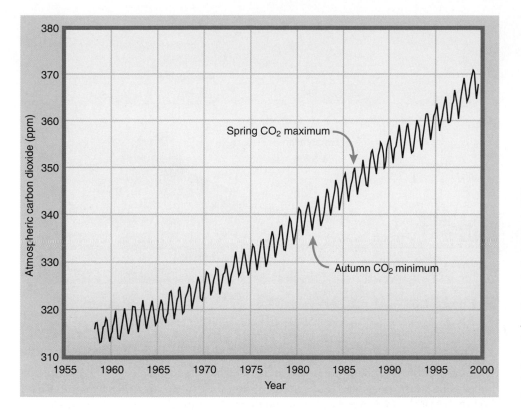

■ **Figure 1.4**

Carbon dioxide measurements made monthly since 1958 at Mauna Loa, Hawaii. Currently, human activity is causing a significant year-to-year change in the amount of carbon dioxide in the atmosphere. This accounts for the rise of the curve from left to right. The shorter-term seasonal oscillations in the curve are caused mostly by the worldwide effect of plants on the carbon dioxide cycle.

sugars. Photosynthesis is a process that allows plants to become a temporary sink of atmospheric carbon dioxide. When the plants die and decay, they release the stored carbon dioxide into the atmosphere. Dead plant tissue and other dead organisms are therefore a source of atmospheric carbon dioxide. Plant decomposition and geological forces over millions of years have generated coal and oil fields underground. The burning of these fuels returns carbon dioxide into the atmosphere. Through respiration animals inhale atmospheric oxygen and exhale carbon dioxide and are therefore another source of atmospheric carbon dioxide.

The atmospheric concentration of carbon dioxide is monitored throughout the world. The concentrations that have been carefully measured at Mauna Loa, Hawaii, since 1958 are shown in Figure 1.4. The steady increase in carbon dioxide concentration is attributed to the burning of fossil fuels and, to a lesser extent, deforestation. Imposed on this increasing trend is a repetitive cycle of peaks and valleys.

The life cycle of plants in the Northern Hemisphere drives this seasonal cycle of peaks and valleys of carbon dioxide. During winter, dormant plants stop removing carbon dioxide from the atmosphere. However, at the same time decaying plants continue to release the carbon dioxide stored in their tissues into the atmosphere. Since the source is greater than the sink, atmospheric concentrations of carbon dioxide increase throughout the winter until late spring. In summer, decomposition also occurs but photosynthesis is at a maximum and carbon dioxide is removed from the atmosphere in large quantities. The sink is now larger than the source. This causes a decrease in carbon dioxide concentrations throughout the summer, leading to minimum yearly values in early autumn.

The amount of carbon dioxide in the atmosphere is an important factor that influences atmospheric temperature. Warm periods in the Earth's long-term history are associated with high levels of atmospheric carbon dioxide. As we discuss in the next chapter and Chapter 15, the increase of atmospheric carbon dioxide caused by burning of fossil fuels plays a vital role in the planet's warming. When discussing predictions of global climate warming you should keep in mind that large quantities of carbon dioxide are dissolved or stored in the oceans. The ocean contains 50 times

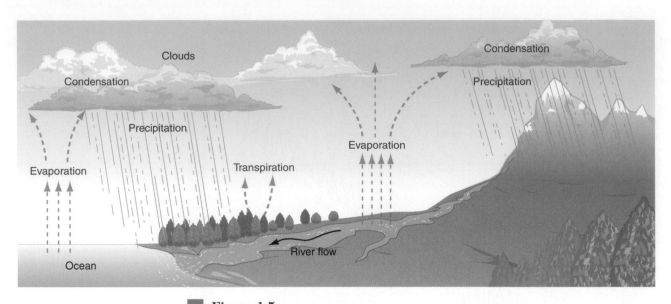

Figure 1.5

The hydrologic cycle describes how water enters and leaves the atmosphere.

more carbon dioxide than the atmosphere. Marine organisms use some of the carbon dioxide in the oceans to build shells. When they die, their shells accumulate on the bottom of the ocean and form carbonate rocks, removing carbon from the cycle. Scientists do not completely understand how much carbon dioxide the oceans will be able to absorb as the amount of carbon dioxide in the atmosphere changes.

Hydrologic Cycle

In the atmospheric sciences, water is very important because it couples, or connects, the atmosphere with the surface of the Earth. Water is also the only substance that exists naturally in the atmosphere in all three phases: gas, solid and liquid. Changing from one phase of water to another, such as from liquid to gas, is an important means of transferring energy in the atmosphere.

The hydrologic cycle (Figure 1.5) describes the circulation of water from the ocean and other watery surfaces to the atmosphere and to the land. A major source of atmospheric water vapor (i.e., water in the gas phase) is evaporation from the oceans. **Evaporation** is the change of phase of liquid water to water vapor. Evaporation from lakes and glaciers supplies a relatively small amount of water to the atmosphere. **Transpiration,** the process by which plants release water vapor into the atmosphere, is also a source of atmospheric water vapor. The surface of the Earth is the major source of atmospheric water vapor, so the amount of water vapor in the atmosphere is generally largest near the surface and rapidly decreases with distance from the surface.

Often atmospheric water vapor changes phase to form solid and liquid particles. The change of phase from water vapor to liquid is called **condensation.** You see condensation occurring whenever a cold drink glass becomes wet on a hot humid day. Similarly, clouds form in the atmosphere via condensation. Typically, 50% of the globe is covered with clouds. The occurrence of clouds is more frequent in some areas of the world and less frequent in others. Figure 1.6 depicts the cloud cover on a single day as measured from several different weather satellites. Global patterns in cloud cover are evident. A lack of clouds is observed at about 30° north and south latitude, particularly over the deserts of Africa. A band of clouds is observed in the vicinity of the equator, at approximately 5° north latitude. The reasons for the global cloud patterns shown in Figure 1.6 are discussed in Chapter 7.

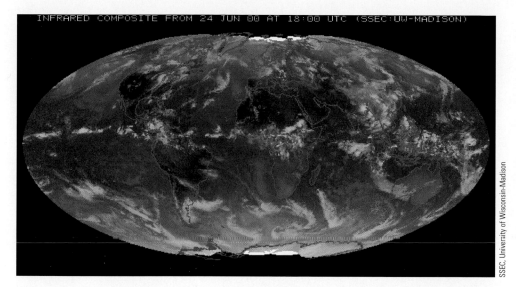

INFRARED COMPOSITE FROM 24 JUN 00 AT 18:00 UTC (SSEC:UW-MADISON)

SSEC, University of Wisconsin-Madison

Figure 1.6

Satellites are used to determine global distributions of cloud amounts. This is a view of Earth from a combination of weather satellites on the same day as shown in Figure 1.1. The global cloud patterns provide insight regarding atmospheric wind patterns. Can you see the swirl of clouds over the upper midwestern United States that stood out prominently in Figure 1.1?

Precipitation, such as rain, snow, sleet, freezing rain, and hail, falls from clouds. Precipitation is a sink of atmospheric water as it removes water from the atmosphere. Precipitation returns water to the Earth's surface after it has evaporated and completes the hydrologic cycle. Precipitation on land may collect in lakes, run in rivers directly back to the sea, or percolate into the soil.

Precipitation may fall on glaciers, which then store the water. The occurrence and extent of glaciers are functions of the temperature of the atmosphere and the amount of precipitation. The presence of glaciers will also affect the atmospheric temperature. They reflect more solar energy back to space, reducing the amount that can be absorbed to heat the Earth. Like rivers, glaciers flow, but usually extremely slowly. Evaporation from lakes returns water to the atmosphere. Lake water and precipitation seep below the surface of the Earth and are stored there. Water stored in the soil ultimately flows back to the oceans, as does precipitation and river and glacier flows, completing the cycle.

Since water plays a major role in weather and climate, it is important to understand the hydrologic cycle. A change in one component of the hydrologic cycle can affect weather. For example, a decrease in the amount of cloud cover over land during the day will allow more solar energy to reach the surface and warm the ground and the atmosphere above.

Methane

In addition to adding carbon dioxide to the atmosphere, human activities are changing the atmospheric concentration of other trace gases, such as methane. Figure 1.7 shows that the concentration of methane has doubled since the beginning of the Industrial Revolution! We do not fully understand the observed increases in atmospheric methane. Human activities that contribute to the increased methane concentration include the decay of organic substances in rice paddies (the cultivation of rice has doubled since 1940), the burning of forests, coal mining, and cattle-raising (methane is a by-product of their digestive process).

Methane and other trace gases play a role in the global warming debate. An understanding of how trace gases increase the global average temperature requires us to revisit so-called *greenhouse warming* throughout this book. While carbon dioxide is an important greenhouse gas, water vapor is the most important gas in the greenhouse warming scenario. Other important gases that contribute to greenhouse warming are chlorofluorocarbons.

http://info.brookscole.com/ackerman

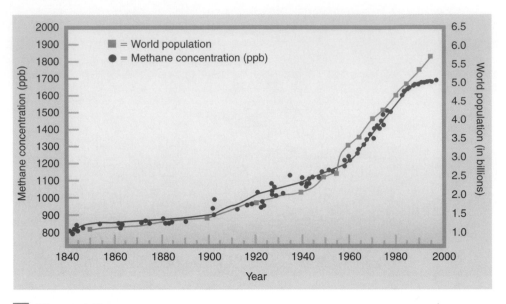

■ Figure 1.7

Methane concentrations since 1840 (shown as the number of molecules of methane per billion molecules in a given volume of air) compared to world population growth. Coal mining, rice production, leaks in natural gas pipelines, and fermentation in the digestive tracts of domestic animals such as cattle all may have contributed to the increase in the concentration of methane in the past several decades. (Data from: http://www.epa.gov/ghginfo/topics/table1-3.htm)

■ Chlorofluorocarbons

Chlorofluorocarbons (CFCs) do not occur naturally. CFCs were invented by chemists in 1928 and were used as propellants in spray cans, Styrofoam™ puffing agents, and as coolants for refrigerators and air conditioners. The use of two CFC compounds (CFC-11 and CFC-12) is plotted in Figure 1.8. Notice that usage decreased in the middle of the 1970s. When these human-made, or **anthropogenic,** gases were found to destroy ozone (Chapter 2), the United States banned the use of CFCs, leading to the initial decrease shown in Figure 1.8.

■ Figure 1.8

Global use of the key chlorofluorocarbons CFC-11 and CFC-12 from the time of their invention through the early 1990s. The discovery of the danger posed by CFCs to the ozone layer led to successful international agreements to limit their use.

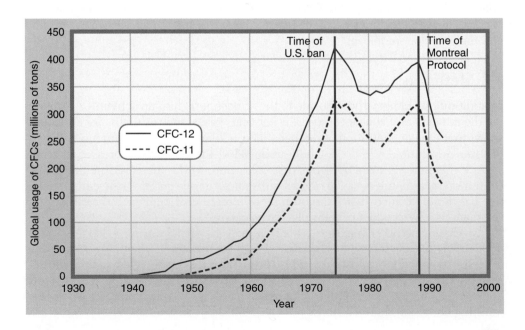

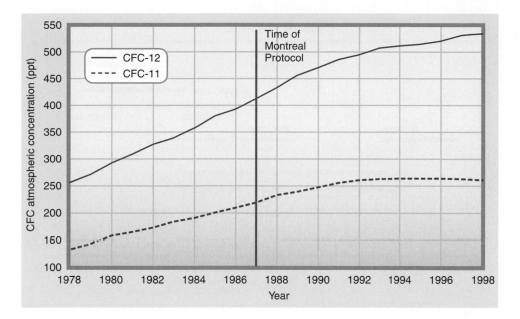

■ **Figure 1.9**

Concentrations of CFC-11 and CFC-12 in parts per billion per volume as a function of year. While the usage of CFCs has declined in recent years (see Figure 1.8), the concentration in the atmosphere has not declined as quickly. This is because CFCs are very chemically stable, so once these molecules enter the atmosphere they remain there for approximately a century.

In the 1980s other countries increased their use of CFCs. In response to the discovery of CFCs role in destroying ozone, representatives from 23 nations met in Montreal, Canada, in 1987 to address concerns of ozone depletion by CFCs. The resulting Montreal Protocol called for a 50% reduction in the usage and production of CFCs by the year 1999. This and subsequent international agreements explain the decline shown in Figure 1.8 in the late 1980s. While the use of these chemicals has declined, their concentrations in the atmosphere have not responded as quickly, as shown in Figure 1.9. This is because CFCs are very stable molecules and will stay in the atmosphere for nearly 100 years before they decompose.

■ Aerosols

Clouds are not the only liquid and solid particles present in the atmosphere. Smoke, salt, ash, smog, and dust are examples of particles suspended in the atmosphere. These particles are collectively known as **aerosols.**

The size of an aerosol particle is measured in **microns** (one millionth of a meter) and varies with the type of aerosol (Figure 1.10). The amount and type of aerosol can influence the climate of a region by modifying the amount of solar energy that reaches the surface. The largest airborne particles are those thrown into the atmosphere during volcanic eruptions, forest fires, and tornadoes. Most atmospheric aerosols are too small to be visible to the naked eye, but they serve an important purpose discussed in Chapter 3—they are needed to form most clouds.

Aerosols are more prevalent in certain regions of the world than others. For example, dust derived from soil is very common over deserts. Atmospheric winds can transport this dust to other regions of the globe where it may be deposited. Dust deposits are found on the ice sheets of Greenland. Dust is not generated in large quantities over Greenland, and so these deposits must have originated in another location and then been transported towards the Arctic. Analysis of the Greenland ice sheet indicates an increased amount of dust present in the atmosphere at the end of the last glacial period, suggesting persistent dry, windy conditions. Dust deposits over glaciers are therefore a window into the weather of centuries past.

Sea salt is an aerosol that commonly occurs over the oceans and along shorelines. Sea salt enters the atmosphere from the oceans when spray from waves evaporates,

Mexican Smoke over the Midwestern United States

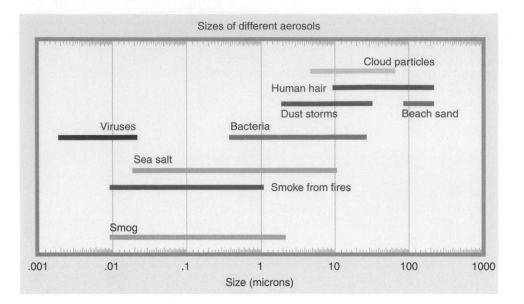

Figure 1.10

Aerosols are liquid and solid particles that are suspended in the atmosphere. Typical sizes of different aerosols are shown. For reference, the diameter of human hair, beach sand, viruses, and bacteria are also shown.

The Mount St. Helens Eruption: A Satellite View

leaving behind tiny salt particles in the atmosphere. Sea salt particles are important in the formation of clouds.

Other primary sources of aerosols are wind erosion, fires, volcanoes, and human activity. Wind erosion is a natural way of getting aerosols into the atmosphere. Winds lift soil particles off the bare ground and transport them into the atmosphere. Fires produce large amounts of aerosols. Air heated by fire can lift particles several thousand feet above the ground, where winds can carry the aerosols far away from their original source (Figure 1.11).

Volcanic eruptions spew huge amounts of particles into the atmosphere. The eruption of Mount St. Helens in Washington State on May 18, 1980, injected approximately 472 million metric tons (520 million tons) of ash into the atmosphere! Chapter 3 presents a discussion of how aerosols generated in the largest volcanic eruptions can lower Earth's temperature.

Airborne particles generated by human activity are referred to as *anthropogenic aerosols*. Sources of these aerosols include fuel combustion, construction, crop spraying, and industrial processes. Large anthropogenic aerosols are generated in mechanical processes such as grinding and spraying. The smaller particles are generated in processes where combustion is incomplete, such as in automobiles.

Atmospheric Pressure and Density

Basic Concepts

Pressure is the force exerted on a given area. The atmosphere is made up of gas molecules that are constantly in motion. These molecules exert a pressure when they strike an object. The air molecules hitting one square inch of your skin exert a pressure of about fifteen pounds. The pressure exerted by the molecules hitting you is a function of the speed, number, and mass of the molecules.

Molecules that comprise the air are moving in all directions. These molecules exert a pressure on objects as they collide with them. Since the molecules are moving in all directions, the pressure is exerted in all directions.

The pressure exerted by the atmosphere can be illustrated using the everyday example of a balloon. If we increase the number of molecules on the inside of a balloon by pumping air into it (but not allowing the temperature to change), we increase the number of molecules colliding and therefore exerting a pressure on the

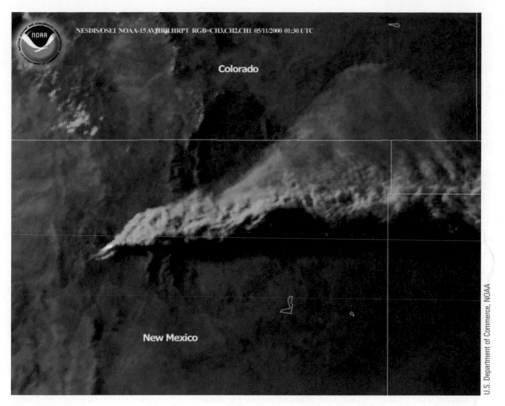

U.S. Department of Commerce, NOAA

Figure 1.11

Smoke from the Los Alamos, New Mexico, forest fire on the evening of May 10, 2000. In this picture, taken by a weather satellite, the smoke plume extends upward for thousands of feet from the fire (red region at left). Strong winds have blown the smoke eastward (right). This fire was intentionally set by the National Park Service to clear away dry underbrush in a park on May 4, but meteorological conditions were not fully taken into account. Shortly after the fire was set, high winds fanned the flames out of control and toward Los Alamos. The fire eventually caused the evacuation of 25,000 people and damaged or destroyed 115 buildings at the Los Alamos National Laboratory, the birthplace of the atomic bomb and current home to radioactive materials. No leaks of radioactivity were reported to have been caused by the fire.

walls of the balloon. If we pump enough molecules in, the balloon will bulge outward because of the pressure differences between air outside and inside the balloon. Atmospheric pressure differences not only allow us to blow up balloons, but also allow sounds to propagate through the air.

The concept of pressure also helps explain how to pop a balloon. Everyone knows that the quickest way to pop a balloon is to stick something small and pointed into it, such as a needle. Why not use a large object instead, such as a baseball bat? The reason is that the fine point of the needle comes into contact with a much smaller area of the balloon's surface than the bat does. Therefore, if you apply the same amount of force when you poke a balloon with a needle versus a bat, the needle will exert a much greater pressure—force *per area*—than the bat. The high pressure of the needle creates an opening in the balloon's surface. Then we can use the concept of *atmospheric* pressure to explain the rest. The air inside the bursting balloon rushes out because the atmospheric pressure inside the balloon is higher than the atmospheric pressure outside it.

The concentration of molecules is measured in terms of density, or mass per unit volume. The atmospheric density at sea level for a standard temperature of 15° C (59° F) is 1.225 kilograms per cubic meter. By comparison, the density of liquid water is much greater, approximately 1025 kilograms per cubic meter. The pressure, temperature, and density of a gas are related to one another through a mathematical formula known as the ideal gas law (Box 1.2). Changing one of these variables will cause a change in one or both of the others.

Air has mass and because of gravity has weight. Because air is compressible, the air near the surface is compressed by the weight of the air above. Figure 1.12 depicts how density decreases with distance from the surface. As you move away from the surface you are decreasing the amount of air above you, and therefore its weight. As a result, as the distance from the surface increases, the density of the air decreases. A decrease in density results in a decrease in pressure, because there are fewer molecules in the same volume of air.

■ Pressure and Altitude

Television weather reports show atmospheric pressure in terms of inches or millimeters of mercury. In meteorology it is more common to report atmospheric pressure in millibars or Pascals rather than inches of mercury. One millibar (mb) equals 100 Pascals and 0.76 millimeters (0.03 inches) of mercury. The average surface pressure at sea level is 1013.25 mb or 29.92 inches of mercury. Pressure is also expressed in terms of pounds per square inch (psi); air pressure in tires is frequently measured in psi. One millibar is equivalent to 0.0145 psi. A common pressure in automobile tires is 32 psi, which converts to 2206 mb or 65.3 inches of mercury—more than twice the atmospheric pressure at sea level! But, as we will see, atmospheric pressure, like density, decreases rapidly with altitude.

But first we need to explain the science behind the terminology of atmospheric pressure. What does atmospheric pressure in "inches of mercury" really mean? The TV meteorologist is reporting atmospheric pressure in terms of the height of a column of mercury. Consider Figure 1.13, in which a tube with no air in it is placed upside down in a liquid. Because gravity pulls objects towards the center of the Earth, the weight of an object exerts a downward pressure on a surface it rests on. Air has mass and because of gravity exerts a downward pressure on objects. The molecules

BOX 1.2

The Ideal Gas Law

Mathematics is a convenient and powerful tool for expressing physical ideas. Most atmospheric processes can be explained in descriptive terms; however, mathematical equations provide further insight. The ideal gas law describes the relationship between pressure, temperature, and volume of a gas. It is written as:

$$\text{Pressure} \times \text{Volume} = k \times \text{Temperature}$$

In this equation, k is a constant of proportionality that depends on the size of the gas sample. Although this law is an approximation of how a gas behaves, it is an excellent one for atmospheric studies. The ideal gas law can also be written as

$$\text{Pressure} = R \times \text{Density} \times \text{Temperature}$$

where R is a constant that depends on the gases that compose the atmosphere. For dry air, the value of R is 287.05 Joules per kilogram per degree Kelvin.

We will use the ideal gas law to understand the relationship between pressure, volume, and temperature of a gas. Changing one of these properties changes the others. Let us start by considering a fixed volume of a gas. You can envision this as a gas stored in a thick metal container. If we double the temperature of the gas, then, to make the left hand side of the equation equal to the right hand side, the pressure inside the container must double. If we fix pressure at a constant value, then doubling the temperature of the gas would also double the volume of the gas.

It is worthwhile to note that the study of how one variable changes with respect to another variable is a fundamental concept of the college mathematics subject known as *calculus*. Because weather variables change with respect to each other and across time and space, meteorologists must study calculus.

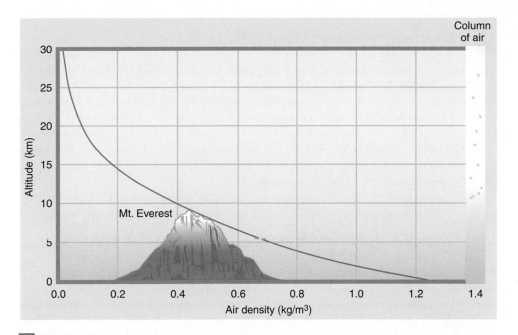

Figure 1.12

The density of the air decreases rapidly with increasing distance from the Earth's surface.

of air collide with and exert a pressure on the surface of the fluid. This pressure pushes the fluid up the tube. The greater the pressure the higher the column of fluid. As we go higher in the atmosphere, the number of molecules colliding with the fluid surface decreases and so the pressure exerted on the fluid is less. This causes the height of the fluid column to drop as the measuring device, known as a barometer, is carried to a higher altitude. You may find it useful to think of atmospheric pressure as the weight of a column of air above you. This will help you to understand the rule: *atmospheric pressure always decreases with increasing altitude.* This property of the atmosphere helps explain how the human body reacts to airplane travel and mountain-climbing (Box 1.3).

Denver is approximately 1.6 kilometers (one mile) above sea level, while San Francisco is near sea level. Because pressure always decreases with altitude, the pressure measured at Denver will always be lower than that measured at San Francisco. When comparing the pressure of different locations, meteorologists always adjust the pressure measurements to a single altitude: sea level. This removes the effect of altitude on pressure and allows the meteorologist to focus on the smaller, but important, pressure differences due to weather systems.

For example, let's say that the atmospheric pressure measured by a barometer in Denver is 850 mb and a barometer at sea level in San Francisco reads 1013 mb, which is the typical pressure at sea level. Denver's pressure sounds very low compared with San Francisco's. However, until we take into account Denver's high altitude we are comparing apples with oranges.

How can we adjust Denver's pressure to sea level? A city at sea level, such as San Francisco, has one more mile of air above it than Denver. To adjust the Denver pressure for sea level we have to add the pressure caused by the difference resulting from that extra mile of air. The lowest mile of air has a pressure of approximately 170 mb. The pressure in Denver *adjusted to sea level* is then 850 mb + 170 mb = 1020 mb. (This assumes a temperature of 25° C; see the text's website for details.)

This calculation shows that in our example Denver's weather is being affected by higher-than-average pressure, despite our initial impressions to the contrary. In a similar way, atmospheric pressure readings across the world are standardized by calculating what the pressure would be if the weather station were located at sea level.

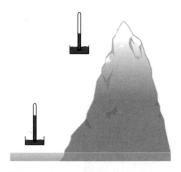

Figure 1.13

Blaise Pascal demonstrated that atmospheric pressure decreases with increasing altitude using a barometer to measure pressure as he climbed a mountain. For his work with pressure, Pascal was rewarded by having a unit of pressure named after him. One millibar is equal to 100 pascals.

Sea-Level Pressure Conversion

Dividing up the Atmosphere

Meteorologists find it useful to divide the atmosphere into layers. We can divide the atmosphere vertically according to pressure. For example, we often will discuss the winds at an altitude where the atmospheric pressure equals 500 mb. Why are we interested in 500 mb? Since the average atmospheric pressure at the Earth's surface is 1013 mb, if we were at an altitude of 500 mb, approximately half the atmosphere would lie above us and half below us. We will learn in later chapters that the winds at about 500 mb help "steer" atmospheric storms. This makes the 500-mb level especially important for weather forecasting.

We can also divide the atmosphere according to temperature. Unlike pressure, temperature does not always decrease with increasing distance from the surface. If

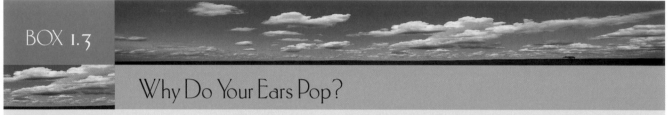

BOX 1.3

Why Do Your Ears Pop?

Have you ever felt pressure in your ears when you were on an airplane flight or climbing a mountain? Have you noticed that the discomfort is worse if you have a cold, and that chewing gum or yawning helps alleviate it? Have your ever had trouble hearing an airline crew member make an announcement shortly after takeoff or shortly before landing, because of loudly crying babies? A combination of meteorology and biology explains all of these observations.

As discussed in this chapter, atmospheric pressure decreases rapidly as you go up. In most of the western and eastern United States, it is possible to drive from sea level to 1 kilometer (3300 feet) elevation or higher in a few hours. This is equivalent to a pressure decrease of roughly 100 mb.

Pressure changes in an airplane are even quicker. Commercial airplanes cruise at an altitude of about 10 kilometers (33,000 feet) where atmospheric pressure is so low the interior of the plane must be pressurized to keep the crew members and the passengers conscious! Although the plane's cabin is pressurized, during flight it is not kept at the same pressure as at the ground. To do so would place too much stress on the plane's structure when the far less dense air at cruising altitude surrounds it. Because of this, cabin pressure decreases some during takeoff and increases some during landing.

If you chart the changes in surface atmospheric pressure throughout the day, you will observe that pressure does not usually change more than a few millibars a day. Because pressure varies slowly with time, our bodies have had no need to adapt to rapid changes in pressure. Yet, this is precisely what happens in an airplane or when climbing a mountain. If there is a rapid change in pressure our bodies are slow to react to it. To understand how a rapid pressure change hurts our ears we have to review some biology.

The pinna (the part of the ear that is visible) collects sounds and directs the sound waves into the auditory canal to the eardrum (tympanic membrane). Sound waves traveling

through air are variations in pressure. When these pressure variations strike the eardrum they cause the eardrum to vibrate. Tiny bones (the hammer, anvil, and stirrup) sense these vibrations and pass them to the inner ear. To hear sounds clearly requires the tympanic membrane to vibrate freely. Air on either side of the membrane exerts a pressure that maintains the proper tension for the eardrum to vibrate. The ear is connected to the nasal passage and throat by the Eustachian tube. This tube is a passageway for air to enter and leave the middle ear to balance the pressure exerted by air on the auditory canal side of the eardrum.

A rapid change in altitude results in a pressure imbalance on the eardrum. When the air pressure rapidly drops, the eardrum bulges outward, causing discomfort. To balance this pressure difference at the eardrum, air molecules must escape through the Eustachian tube. When the pressure inside your middle ear adjusts to the outside pressure, the eardrum returns to its normal tension. This movement of the eardrum makes a popping sound, and we say that your ear "pops."

Now we can explain the usefulness of yawning and chewing gum on an airplane. When you yawn or swallow you are allowing the Eustachian tube to open wider and equalize this pressure difference more rapidly. A cold may block the Eustachian tube, making it more difficult to adjust to changes in air pressure. This increases the distortion of the eardrum and causes pain.

Babies on airplanes are too young to understand that they need to yawn or swallow. Furthermore, their Eustachian tubes are tiny and are easily blocked during colds and ear infections, which babies frequently endure. Therefore, during airplane takeoffs and landings babies feel intense pressure and pain in their ears, and they respond by crying. Remember this with some sympathy the next time you fly on an airplane to the serenade of squalling infants!

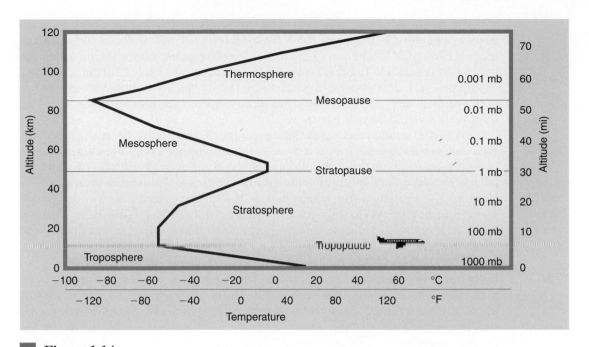

■ **Figure 1.14**

The atmosphere is subdivided into layers based on how air temperature changes with altitude.

we were to average temperature over many years at many different locations across the globe for many different altitudes, however, a distinct pattern would emerge between temperature and distance from the surface. Figure 1.14 presents an example of this pattern. Based on this temperature profile, the atmosphere can be divided into four main layers: the troposphere, stratosphere, mesosphere, and thermosphere.

From the surface up to approximately 10 to 16 kilometers (6 to 10 miles), temperature generally decreases with altitude. This is because the atmosphere is nearly transparent to the Sun's energy and is instead heated from below by the Earth's surface, like a pot of water on a stove. The region of the atmosphere closest to Earth, where the temperature decreases as you go up, is called the **troposphere.** The word troposphere is derived from the Greek word *tropein* meaning "to change." Almost all of what we normally call "weather" occurs in the troposphere, and about 80% of the atmosphere's mass is located in the troposphere. For these reasons, the troposphere is the focus of this book.

The top of the troposphere is referred to as the **tropopause.** It acts as an upper lid on most weather patterns, just as a lid on a pot of water on a stove keeps the water from escaping. The height of the tropopause is a function of latitude. It is higher in the equatorial regions, where the air is warmer and expands upward, than in the cold polar regions.

Above the tropopause lies the **stratosphere,** where temperature increases with altitude. Temperature is increasing because ozone molecules in the stratospheric "ozone layer" (see Chapter 2) are absorbing solar energy near the top of the stratosphere. Air flow in the stratosphere is much less turbulent than in the troposphere. For this reason jet aircraft like to cruise at stratospheric altitudes, where the flight is less "bumpy." The increasing temperature with altitude of the stratosphere makes it difficult for tropospheric air to mix into the stratospheric air, as explained in later chapters. The lack of mixing and turbulence makes the stratosphere very layered or "stratified"—hence its name. In the vicinity of storms with strong up-and-down wind motions, such as thunderstorms and low-pressure systems, tropospheric air can penetrate into the stratosphere and stratospheric air can descend into the troposphere.

Blue Skies

Atmospheric Basics/Layers of the Atmosphere

http://info.brookscole.com/ackerman

The **stratopause** marks the top of the stratosphere. On average the stratopause occurs at an altitude of approximately 50 kilometers (31 miles). Its average temperature is close to 0° C and during the winter the stratopause can be warmer than the ground far beneath it. But don't consider the stratopause a likely Christmas vacation destination. The atmospheric pressure there is only one one-thousandth as great as at the surface. Human beings could not survive at such low pressures; the liquids in their bodies would literally boil away!

Above the stratopause lies the **mesosphere.** Temperature decreases with altitude in the mesosphere, just as it does in the troposphere. The **mesopause** separates the mesosphere from the thermosphere at an average altitude of about 85 kilometers (53 miles). In the **thermosphere** the temperature again increases with altitude. The **density** of the atmosphere in the thermosphere is so low that above about 120 kilometers (75 miles) the atmosphere gradually blends into interplanetary space. The entire Earth's atmosphere is extremely thin, less than 2% as thick as the Earth itself.

From the mesosphere on up, the atmosphere becomes more and more affected by high-energy particles from the Sun. These particles break apart atmospheric molecules, which then form ions. For this reason, the region of the upper mesosphere and thermosphere is sometimes called the *ionosphere.* Some interesting visual phenomena occur in the mesosphere and thermosphere due to the interaction of solar particles and the atmosphere, such as the aurora. In addition, the ionosphere reflects radio waves, permitting radio stations to be heard far beyond the horizon.

An Introduction to Weather Maps

Basic Concepts

An important part of studying meteorology is simply paying attention to weather conditions and applying your knowledge to what you observe. However, many different variables including temperature, moisture, cloudiness, precipitation, and others, are used to describe weather. All of these must be considered and analyzed numerically. Furthermore, the weather at one location is often caused by larger weather patterns. So it is not enough to consider the weather locally. We must also analyze the weather for many other locations. This means even more numbers.

To identify key weather-making patterns, we need to see all the numbers describing weather at many locations, all at once. It is also vital to see how the observations relate to each other geographically. A list of numbers doesn't help much; we need a picture. So, we plot weather variables on a weather map to understand the relationships among them. To paraphrase the saying, "A picture is worth a thousand words," a weather map is worth a thousand numbers—or more. In this section, we learn how to interpret weather maps.

Perhaps the most obvious features on weather maps are fronts. A **front** is a boundary between two regions of air that have different meteorological properties, such as temperature and humidity. A **cold front** denotes a region where cold air is replacing warmer air. A **warm front** indicates that warm air is replacing cooler air. We will discuss fronts in detail in Chapters 9 and 10 of this book.

All weather maps today depict frontal locations because they are regions of rapidly changing and sometimes dangerous weather conditions. The frontal lines drawn on a weather map represent the locations where the fronts meet the Earth's surface (Figure 1.15). On a weather map, a blue line with blue triangles indicates a cold front. The triangles point in the direction the front is moving. A warm front is shown as a red line with red semi-circles pointing in the direction of frontal movement. A **stationary front** is a front that is not moving and is represented as shown in the lower-left corner of Figure 1.15. The **occluded front,** represented as a purple line with alternating triangles and semi-circles, has characteristics of both cold and warm fronts and is discussed in more detail in Chapters 9 and 10.

To locate the position and type of a front on any given day requires the analysis of the weather conditions from locations over a wide geographic area. To help in this analysis we often draw lines on weather maps connecting locations that have the same temperature. Lines of constant temperature are called **isotherms** (from *iso,* "same," and *therm,* "temperature"). Similarly, **isobars** connect locations with the same sea-level atmospheric pressure. Both isotherms and isobars are often shown on television and newspaper weather maps. **Isotachs** connect locations with the same wind speed. **Isopleth** is a more general term describing contours along which any particular variable is constant.

Learning to Contour Weather Maps

The Station Model

Television weather reports sometimes represent local weather conditions with smiling suns, rainy clouds or flashing bolts of lightning. However, one smiling sun can cover several states and gives you no information about temperature, wind, and pressure. Meteorologists need a way to condense *all* the numbers describing the current weather at a location into a compact diagram that takes up as little space as possible on a weather map. This compressed graphical weather report is called a **station model.**

A simplified example of a station model plot used to represent meteorological conditions near the surface for a specific location is shown in Figure 1.16. The station model depicts weather conditions at a particular time in a specific location, plus cloud cover, wind speed, wind direction, temperature, dew point temperature, atmospheric pressure adjusted to sea level, and the change in pressure over the last three hours. In Figure 1.16, nine weather variables commonly reported on the evening news are plotted—much more information than a smiling sun can convey, in much less space.

Let's decode the station model at the top of Figure 1.16. First, cloud cover is depicted graphically by the type of shading inside the circle, which represents the location of the weather station. The circle is completely shaded, so this means 100% cloud cover or "overcast" conditions.

Next, the temperature reading is located at the ten o'clock position with respect to the circle (i.e. just above and to the left of it). In the United States, temperature is reported on weather maps in degrees Fahrenheit. Therefore, in our example the temperature is 76° F. The dew point temperature, which is a measure of the amount of moisture in the atmosphere, is located at the eight o'clock position. In our example it is 55° F.

Current weather conditions are symbolically represented in between the temperature and dew point temperature. Some symbols for common weather conditions are shown in the lower part of Figure 1.16. In our example, the station is currently reporting light rain. If there is no precipitation, haze, or fog occurring at the time of observation, the current weather condition location is left blank.

Wind speed and direction are also depicted graphically on a station model by the position of the "flagpole" extending out from the circle at the center. The pole points to the direction *from* which the wind is coming. So, the wind blows from the "flagged" end toward the "pointed" end. The "flags" are long and short lines that indicate wind speed, as explained in Figure 1.16.

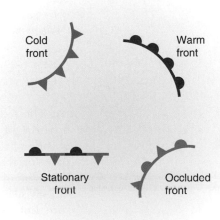

Figure 1.15

The colors and symbols for the four types of fronts: cold, warm, stationary, and occluded.

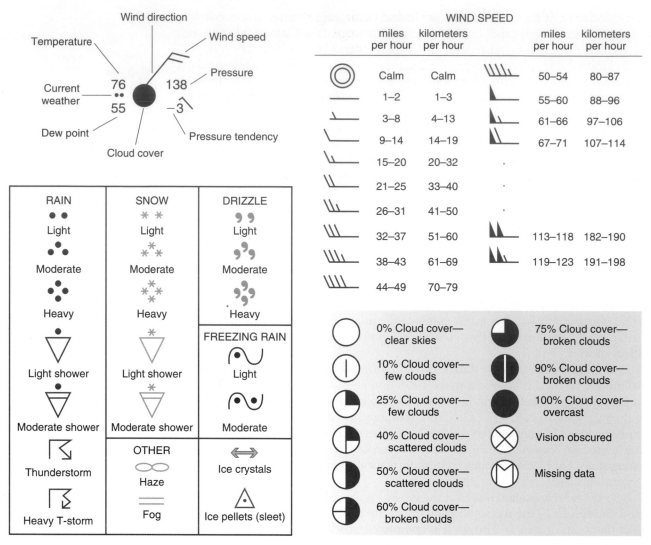

Figure 1.16

The station model for a typical weather situation (top left) and selected symbols used in the station model and their meaning. See the text and the Web site for an explanation of how to decode a station model.

The measured atmospheric pressure is adjusted to sea level and printed in the upper right of the station model. The units used are millibars. Sea-level pressures outside of tornadoes and hurricanes are always between 900 and 1099 mb. To save space on the map, the leading 9 or 10 is dropped, as is the decimal point. To decode the value of pressure on the station model, add a 9 if the first number is a 7, 8, or 9; otherwise add a 10. (This rule does not work in situations with very low or high pressure, but in those rare cases the pattern of isobars helps you know whether the leading digit(s) should be a 9 or a 10.) In our example, the "138" means that the station's sea-level atmospheric pressure is 1013.8 mb. If it had been "831" instead, the sea-level pressure would be 983.1 mb.

The *change* in surface pressure during the last three hours is plotted numerically and graphically on the lower right of the station model. The change in pressure is represented by a value (in tenths of a millibar) and a line that describes how the pressure was changing over time from left to right. In our example, the line goes up and then goes down, indicating that the pressure rose and then fell over the past three hours, a total change of 0.3 mb. A more complete explanation of how to decode pressure tendency symbols is provided on the text's Web site.

Many other weather variables can be depicted on the station model: visibility (next to current weather conditions), weather conditions during the past hour (lower right), precipitation amounts (near pressure tendency), and peak wind gusts during the past hour (number in knots next to the "flags" on the "flagpole"). In this text, we will not look at all station model variables at once, but instead will focus on different variables in different chapters. This text's Web site contains a complete guide to the station model and examples of current weather maps that use surface station models.

An example of a surface weather map is given in Figure 1.17. It includes station model plots and a cold front attached to a low-pressure area over the upper Midwest United States. This map corresponds to the satellite image shown in Figure 1.1. Notice that the clouds in the center of Figure 1.1 correspond closely to the location of the low pressure and the cold front. The surface weather map also shows western Nebraska to be cloud-free, while much of the Southeast is covered with clouds, in agreement with Figure 1.1.

Focusing on the station model for Tallahassee, Florida, in the inset in Figure 1.17, we see that it is mostly cloudy there, the temperature is 77° F, the dew point temperature is 74° F, the sea-level pressure is 1018.4 mb, the winds are light and from the north, and Tallahassee is experiencing a thunderstorm.

The weather observations in Figure 1.17 were all made at the same time. This time is printed in the bottom left corner of each weather map. Unlike television programming, Eastern Standard Time is not always used for these maps! Below we learn how to decode meteorological time.

Decoding the Surface Station Model

Time Zones

To depict current weather patterns using a weather map and to predict future weather, it is important to coordinate the time of global weather observations. To aid in this coordination weather organizations throughout the world have adopted the Coordinated Universal Time (UTC, for Universel Temps Coordonné) as the reference clock. UTC is also denoted by the abbreviations GMT (Greenwich Meridian Time) or, often with the last two zeroes omitted, Z (Zulu).

The reference **time zone** for UTC is centered on Greenwich, England. The International Date Line (180° longitude, halfway around the world from Greenwich) separates one day from the next. Just to the east of the date line is 24 hours earlier than just to the west of the date line.

Meteorology also uses the 24-hour military-style clock. For example, 1:30 PM UTC is 1330 (1200 + 130) UTC, and 0130 UTC is 1:30 AM. Observations of the upper atmosphere are coordinated internationally to be made at 0000 UTC (midnight at Greenwich) and 1200 UTC (noon at Greenwich). The Eastern Standard Time zone of the United States is 5 hours earlier than Greenwich time. So, an observation made at 1300 in New York City would be recorded as 1800 UTC time.

Why is Eastern Standard Time 5 hours earlier than Greenwich time? Since the Earth has 360 degrees of longitude and rotates around its axis once every 24 hours, this means that local time should be one hour earlier for every (360°/24) = 15 degrees of longitude west of Greenwich. (In practice the exact lines dividing time zones are sometimes drawn along rivers and state and province boundaries, not along longitude lines.) New York City is at approximately 74° west longitude, or 74°W. This is almost 5 × 15 degrees of longitude west of Greenwich. Using the 15-degree rule above, this explains why Eastern Standard Time is 5 hours earlier than Greenwich or UTC time.

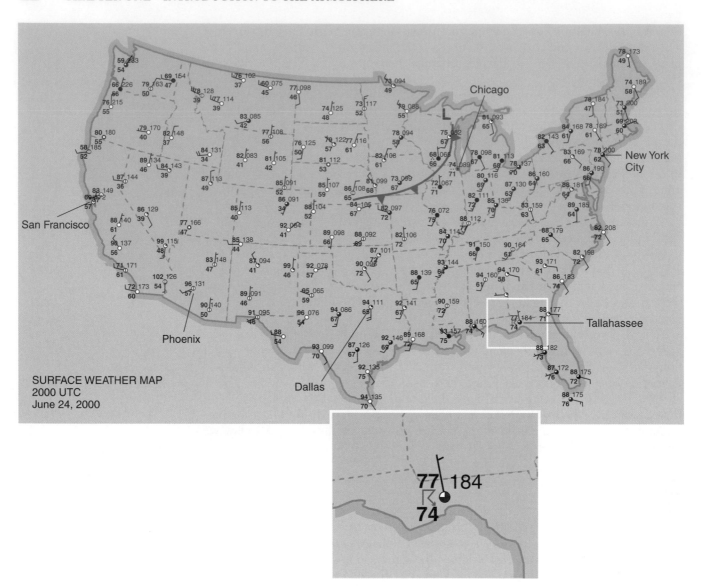

Figure 1.17

A surface weather map employing the station model at 2000 UTC on June 24, 2000. See the text for an explanation of the weather at Tallahassee, Florida (inset). Compare this map and the weather described by it to the satellite image in Figure 1.1.

Using the same reasoning, New Orleans (90°W) and the rest of the Central Standard Time zone is 6 hours earlier; Denver (105°W) and the rest of Mountain Standard Time is 7 hours earlier; Los Angeles (118°W) and the rest of Pacific Standard Time is 8 hours earlier; Juneau (134°W) and the rest of Alaskan Standard Time is 9 hours earlier; and Honolulu (158°W) and the rest of Hawaiian Standard Time is 10 hours earlier than Greenwich.

In our example in Figure 1.17, the weather observations were taken at 20Z or 2000 UTC on June 24, 2000. (This was during daylight saving time, which adds one hour to standard time.) Although the observations were all taken at about the same moment, the clock read 4:00 PM in cloudy New York City, 3:00 PM in rainy Chicago and windy Dallas, 2:00 PM in hot Phoenix and 1:00 PM in cool San Francisco.

Weather Watches, Warnings, and Advisories

As you follow the weather you will notice that certain meteorological conditions may exist that pose a threat to life and property. Under these conditions the National Weather Service issues weather watches and weather warnings. A weather watch informs us that current atmospheric conditions are favorable for hazardous weather. When the hazardous weather will soon occur in an area, then a warning is issued. Weather watches and warnings are issued for tornadoes, hurricanes, severe thunderstorms, winter storms, and flooding. Figure 1.18 depicts the weather warnings and watches issued on the day corresponding to Figure 1.17. Notice that many of the watches and warnings are in the vicinity of the cold front.

The National Weather Service issues weather watches and warnings under specific weather conditions (Table 1.2). It is important to understand the difference between a weather watch, a weather warning, and a weather advisory. The term **watch** implies that you should be aware that a weather hazard *may* develop in your area. A **warning** is issued when the hazard *is* developing in your area. You should take immediate action in the event of a warning. An advisory is a less urgent statement issued to bring to the public's attention a situation that may cause some inconvenience or difficulty for travelers or people who have to be outdoors.

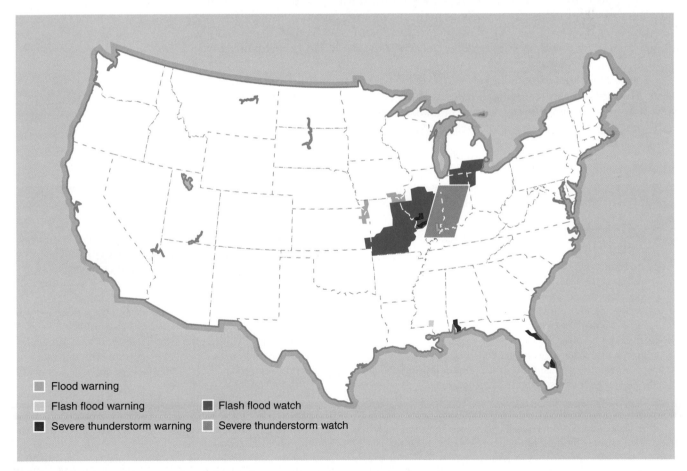

■ **Figure 1.18**

The National Weather Service's weather watches and warnings for June 24, 2000 (compare to Figures 1.1, 1.6, and 1.17). A severe thunderstorm watch has been issued for most of Indiana, southeast Illinois, and western Kentucky.

TABLE 1.2   National Weather Service Criteria for Issuing Selected Weather Watches and Warnings

Weather Hazard	Watch	Warning
Hurricane	Hurricane conditions are expected in a given region within 36 hours.	The arrival of a hurricane is expected within the next 24 hours.
Severe thunderstorm	Conditions are favorable for the development of severe thunderstorms in and close to the watch area.	A severe thunderstorm has been sighted visually or indicated by radar. Issued when a thunderstorm produces hail 0.75 inch or larger in diameter and/or winds equal to or exceeding 58 mph.
Tornado	Conditions are favorable for the development of tornadoes in or close to the watch area.	A tornado has been sighted visually or indicated by radar.
Flash flood	Conditions are detected that could result in flash flooding in or close to the watch area	Flash flood will occur within 6 hours and will threaten life and/or property. Dam breaks or ice jams can also create flash flooding.
Winter storm	Conditions are favorable for the development of hazardous weather such as heavy snow, sleet, or freezing rain.	Hazardous winter weather is imminent or very likely.

Pinpointing the location of hazardous weather in advance is extremely difficult. For this reason, watches are usually issued for large regions, sometimes covering several states (see Figure 1.18). Warnings are issued for much smaller areas, often only a county or two as in Figure 1.18, because they are based on actual observations of hazardous weather.

You should know in advance what to do if hazardous weather threatens your area. Throughout this text we will discuss many types of weather hazards, the ways in which they kill, and the steps you can take to protect yourself from the awesome, and sometimes awful, pageant of weather.

PUTTING IT ALL TOGETHER

SUMMARY

Weather is the state of the atmosphere at a particular time, and meteorology is the study of weather. The state of the atmosphere on longer time scales, and its interactions with oceans, land, and living things, is called climate. As with daily weather, the climate of a region changes with time, although on much longer time scales than the weather.

Earth's atmosphere has been shaped by billions of years of volcanic emissions and plant life. It is composed primarily of nitrogen and oxygen gases. Atmospheric concentrations of water vapor, carbon dioxide, ozone, methane, chlorofluorocarbons, and aerosols are very small but also very important to the study of weather and climate. Understanding the concentrations of water vapor and carbon dioxide, among other gases, requires taking into account the cycles of water and carbon from the air to the land and ocean and back to the air. The concentrations of most of these trace gases are increasing as a result of human activities.

The different molecules of the atmosphere are continually moving, exerting pressure in all directions. Atmospheric pressure is related to the weight of the column of air above you. As your altitude increases, the number of molecules above you decreases.

For this reason atmospheric pressure always decreases with distance from the surface. This simple concept is important in understanding the formation of clouds and the transfer of energy in the atmosphere. To compare measurements of atmospheric pressure at locations that are at different altitudes, the measurements must be adjusted to sea level.

Unlike pressure, temperature does not always decrease with increasing distance from the surface. The vertical changes in temperature neatly divide the atmosphere into four distinct layers: the troposphere, stratosphere, mesosphere, and thermosphere. Most weather occurs in the troposphere, which is the focus of this text.

Weather maps show the weather conditions for many locations at the same time. Many variables describe weather at a specific location, including temperature, pressure, and wind. The station model is a graphic way to compress large amounts of weather data for one location into a very small area on a weather map. Weather observations are taken at the same time at locations all over the world. This requires a system of time zones and a universal standard of time, known as UTC and referenced to Greenwich, England.

The National Weather Service issues watches when hazardous weather may occur in a region, and issues warnings for more localized areas when hazardous weather has been observed nearby. Warnings require immediate action on your part to protect life and property.

 KEY TERMS

You should understand all of the following terms. Use the glossary and this chapter to improve your understanding of these terms.

Aerosols	Micron
Anthropogenic	Millibars
Barometer	Molecule
Barometric pressure	Occluded front
Carbon dioxide	Ozone
Chlorofluorocarbons (CFCs)	Pressure
Climate	Sink
Climatology	Source
Cold front	Stationary front
Condensation	Station model
Density	Stratopause
Diffuse	Stratosphere
Evaporation	Thermosphere
Front	Time zones
Gravity	Trace gas
Hydrologic cycle	Transpiration
Isobar	Tropopause
Isopleth	Troposphere
Isotach	UTC
Isotherm	Warm front
Mesopause	Warning
Mesosphere	Watch
Meteorology	Water vapor
Methane	Weather

 REVIEW QUESTIONS

1. What is one difference between weather and climate?
2. Explain why the meteorology of the Northern Hemisphere differs from that of the Southern Hemisphere, with reference to Figure 1.2.
3. Although the production of chlorofluorocarbons has been drastically reduced, their atmospheric concentrations are not yet decreasing significantly. Why?
4. If pressure is force per area, then the atmospheric pressure over a large area such as the United States must be tiny compared to the atmospheric pressure over a city. Find the flaw in this reasoning.

5. Place a balloon over the open end of an empty glass bottle. Run hot water over the bottle (not the balloon). What happens to the balloon? Explain what you observe.
6. Fill a balloon with air and place it in a refrigerator. Remove it from the refrigerator after twenty minutes and describe its appearance. Explain what you observed.
7. What aspect of meteorology might make it easier to hit a home run at Coors Field in Denver than at Pro Player Stadium in Miami?
8. Why is there no such field as lunar (Moon) meteorology?
9. The pressure at a weather station 1.6 kilometers above sea-level is 900 mb. What is the sea level pressure of this station? Is this station experiencing unusually high, low, or normal atmospheric pressure?
10. Why does atmospheric pressure decrease with altitude?
11. If a raft filled with air is thrown into a cold lake it appears to deflate, even though no air escapes. Explain.
12. Why does air that is rapidly compressed into a small volume get hot?
13. You are halfway between two layers of the atmosphere in which temperature increases with altitude. What is your approximate pressure and what layer of the atmosphere are you in?
14. Using the station model decoding information in Figure 1.16, completely decode the station model for New York City in Figure 1.17.
15. You are asked to investigate the weather conditions at Boston at 7:00 PM on Monday, February 28, 2000. To get this information, you have to know the right day and time for this situation in coordinated universal time (UTC). What is the correct UTC day, date and time?
16. Using the *Blue Skies* CD-ROM exercise "Atmospheric Basics/Layers of the Atmosphere," explore the troposphere and stratosphere using the sample data provided.

 WEB ACTIVITIES

Choose Chapter 1 on the textbook's Web site:
http://info.brookscole.com/ackerman
and select from the following resources:
- Interactive Modules that illustrate and extend your understanding of key topics in this chapter
- Tutorial Quizzes to test your mastery of terms and concepts

 For additional readings, go to the InfoTrac College Edition, your online library, at: http://www.infotrac-college.com

The Energy Cycle

After completing this chapter, you should be able to:

■ Describe the different ways in which energy is transferred in the atmosphere
■ Calculate the temperature changes of rising air parcels, both dry and moist
■ Determine why very hot objects glow, but room-temperature objects don't
■ Explain why the Earth has seasons
■ Relate the greenhouse effect to the presence of trace gases in the atmosphere

 Introduction

Whether something warms or cools is related to its energy gains and losses. So, as you stand facing an evening bonfire, your front warms because it gains more energy than it loses, while your back cools as it loses more energy to the cooler night air than it gains. If the night is chilly and you are too close to the fire you become uncomfortable; your front is too hot and your back too cold. You can modify your energy imbalance in several ways. You can turn around and place your back to the fire, or step further away and put a blanket over your back. In both cases you have changed your energy budget.

The same thing happens to the Earth. At any particular moment half of the Earth is facing the Sun and half isn't. Over a year, the tropical regions of the planet experience a net energy gain while the poles have a net energy loss. This global imbalance of energy is largely due to differences in how high the Sun gets in the sky and the length of day. The atmosphere and oceans respond to this energy imbalance by transporting heat from the equatorial regions toward the poles. Transport of heat is the reason we have weather.

This chapter introduces the methods of heat transfer important for understanding weather and climate. The chapter contains definitions of terms and concepts used throughout the book. Learn them now and the next chapters will be easier to follow.

 Force, Work, and Heat

The movement of air is important in defining the weather and climate of a given region. What causes this movement? To explain it, we have to understand a little about force, work, and heat.

Put your book down and give it a push. When you push your book, you exert a force on the book that causes it to move. But your book doesn't keep moving. Why not? How fast and how far the book moves depends on how hard you push it and the force of friction acting between the book and the surface.

The same forces act in the atmosphere. In mathematical terms, the **force** exerted on an object is the mass of the body multiplied by the acceleration the force causes in the body. **Acceleration** is a change in speed or a change in direction of an object's movement. When you push your book, its speed changes. Similarly, forces acting on the atmosphere cause the wind to speed up or slow down or change direction. (We explore atmospheric forces in detail in Chapter 6.)

When you push your book, you use energy to do work on the book. **Work** is done on an object, whether it is your book or the air, when a force moves it. The amount of work done on your book is the distance traveled times the force in the direction of that displacement. Wind is air in motion; to move air requires work.

Doing work requires energy. **Energy** is the capacity to do work. The amount of energy needed depends on the amount of work to be done. It does not take much energy to push a book or lift a glass of water. The energy required to carry a bucket of water up to the top of the Empire State Building is a different matter!

We will study four different forms of energy: heat energy, electrical energy, kinetic energy, and potential energy. Energy can be converted from one form to another, but the total energy is always conserved.

In meteorology and climatology we are primarily concerned with kinetic and potential energy. **Kinetic energy** is the work that a body can do by virtue of its motion. A major part of this energy is related to how fast a mass is moving. The kinetic energy of a moving object is also related directly to its mass. A freight train moving at only 16 km/hr (10 mph) has 36 times less kinetic energy than the same train moving at 96 km/hr (60 mph). A sport–utility vehicle on a highway has more kinetic energy than a car moving at the same speed.

Potential energy is the work an object can do as a result of its relative position. You do work as you lift a book off the desk. Once off the desk, the book has potential to do work because of gravity. The higher you lift the book the greater the potential energy. When you let go of the book it falls and potential energy is converted to kinetic energy. If you drop the book from the tenth floor of a building, it becomes a dangerous missile. If you drop the same book from a height of an inch it can't do much damage. So, potential energy represents stored energy that can be converted to other forms of energy, such as kinetic energy. The potential energy of the book is represented by its height from the surface of the desk.

In studies of the atmosphere, meteorologists find it useful to refer to an isolated **parcel of air.** A parcel of air is a hypothetical balloon-like bubble of air, flexible but impermeable, and perhaps as large as a parking lot. Inside this bubble, basic weather variables such as temperature and moisture are the same. As this parcel of air moves around in the atmosphere, mass and energy do not cross its imaginary boundary. The air around the parcel is the "environment" or the "surroundings." Even though the real atmosphere is not composed of parcels, meteorologists can make sense of the atmosphere by looking at it in this way. If we imagine such a parcel of air, we can move it through the atmosphere as an intact unit and compare its temperature to the temperature of its environment.

Temperature relates to energy, because temperature is a measure of the average kinetic energy of a substance. A thermometer in a glass of ice water records a lower temperature than a thermometer in a pan of boiling water, because the molecules of H_2O in the boiling water are much more energetic. That energy is imparted to the thermometer and shows up as a higher temperature.

There are three commonly used scales for measuring the temperature of an object. **Fahrenheit** (named after the German instrument maker G.D. Fahrenheit) is commonly used in the United States to report temperatures near the surface (Figure 2.1). At sea level ice melts at 32° F and water boils at 212° F. (It is common to say that water freezes at 32° F, but we will see in Chapter 4 that water droplets can stay unfrozen at temperatures as low as −40° F!) The **Celsius** (or *centigrade,* named after the Swedish astronomer A. Celsius) temperature scale is based on the melting and boiling points of water—ice melts at 0° C and water boils at 100° C. This temperature scale is used every day throughout the world to report the air temperatures near the ground. The **Kelvin** scale (named after one of Britain's foremost scientists, William Thomson, who in 1892 became Lord Kelvin) is an absolute scale in which zero is the lowest possible temperature. The Celsius and Kelvin scales are the same, except that the Kelvin temperature is a little more than 273 degrees higher than the Celsius temperature. All three temperature scales are summarized in Figure 2.1.

With this definition of temperature, we can now define the **calorie** (abbreviated *cal*) as the unit used to measure amounts of energy. A calorie is the energy needed to raise the temperature of 1 gram of water 1 degree Celsius (from 14.5° C to 15.5° C.) The dietary 'Calorie' (with a capital "C") used in quantifying the energy content of foods is actually a kilocalorie or 1000 calories (with a small "c"). A **Joule** is another unit used to measure amounts of energy. One Joule equals 0.2389 calories.

The term **power** refers to the rate at which energy is transferred, received, or released. The **watt** (W) is a unit of power that represents the transfer of one joule of energy per second. You are probably familiar with the term watt from light bulbs. A 100-watt light bulb indicates the rate at which electric energy (from your outlet) is consumed by the bulb. A 100-watt bulb consumes more electrical energy than a 60-watt bulb and consequently is brighter. In weather and climate the term *watt* is also used to indicate the rate of energy change. We are also interested in how much energy flows across an area. This energy flow is expressed in units of watts per

Fun with Temperature Conversions

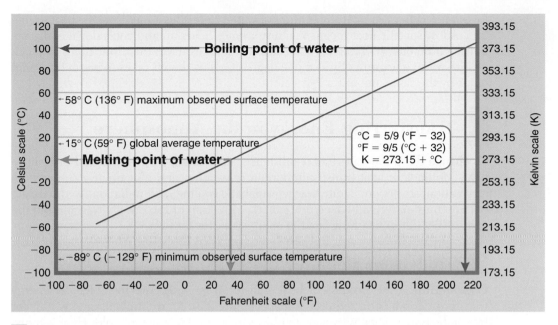

Figure 2.1

The three scales of temperature. Fahrenheit (°F), Celsius (°C), and Kelvin (K) all represent the temperature of matter. You can convert between scales using the graph or by the arithmetic formulas at right.

square meter of area (W/m²). For example, the average amount of solar energy hitting the top of the atmosphere is 1368 W/m².

Heat is energy produced by the random motions of molecules and atoms; it is the total kinetic energy of a sample of a substance. Both heat and temperature are related to kinetic energy and therefore to one another. Consider the heat of a freshly brewed cup of coffee, and that of Lake Erie. The temperature of the coffee is greater than that of Lake Erie. However, the total kinetic energy—the heat contained in Lake Erie—is much greater than the heat contained in a cup of brewed coffee, as there are many more molecules that are moving. If the cup of coffee is gently placed into Lake Erie without spilling any coffee, the temperature of the coffee will eventually be the same as that of the lake. For the temperature of the coffee to decrease, the kinetic energy of the molecules must decrease, and since energy cannot be created or destroyed it must be converted to another form or transferred to the environment. In this case energy, or heat, was transferred from the cup of coffee to Lake Erie. It may help to think of heat as *the energy transferred between objects as a result of the temperature difference between them.* Absorbed heat may increase a system's internal energy or it may be used by the system to do work.

Though the cup of coffee transferred heat to Lake Erie, the temperature of the lake did not increase perceptibly. This is because the temperature change of an object depends on:

1. How much heat is being added—a single cup of coffee, though hot, does not contain much heat compared to Lake Erie, because the coffee does not have much mass relative to the lake.
2. The amount of matter—the more matter, the more heat is required to change its temperature. Lake Erie contains a lot of water molecules and therefore a lot of matter!
3. The specific heat of the substance—but what is specific heat?

The **specific heat** of a substance is the amount of heat required to increase the temperature of one gram of that substance one degree Celsius. Because it takes a lot of energy to raise the temperature of water, water has a high specific heat (Table 2.1). A low specific heat means that a substance heats up and cools down easily, requiring little energy. Table 2.1 indicates that it takes more than four times as much heat to warm one gram of water one degree Celsius than it takes to warm one gram of air by the same amount.

 ## Transferring Energy in the Atmosphere

To change the temperature of a substance, such as air, we need to add or remove heat. Methods of heat transfer important to weather and climate are: conduction, convection, advection, latent heating, adiabatic cooling, and radiation. In discussing energy transfer, we will concentrate on the direction of the energy transfer and the factors that determine how fast the energy transfer occurs.

Conduction: Requires Touching

Conduction is the process of heat transfer from molecule to molecule; energy transfer by conduction requires contact (Figure 2.2). An example of energy transfer by conduction is when we touch an object to feel if it is warm or cold. Heat is transferred from the warmer object to the colder one. The amount of heat transferred by conduction depends on the temperature difference between the two objects and their **thermal conductivity.** The ability of a substance to conduct heat by molecular motions is defined by its thermal conductivity.

If you walk into a cool room and touch a piece of wood and a piece of metal, the metal "feels" colder. The two objects are actually at the same temperature but the metal feels colder because metal has a much higher thermal conductivity than wood. Heat is rapidly conducted from your warm finger to the cooler metal, making your finger cool. Wood has a low heat conductivity and the amount of heat transferred is smaller; it does not feel as cold as metal.

TABLE 2.1	The Specific Heat of a Substance is the Amount of Heat Required to Increase the Temperature of One Gram of the Substance 1° C	
	Specific Heat	
Substance	(cal/g/°C)	(J/kg/°C)
Water	1.0	4186
Ice	0.50	2093
Air	0.24	1005
Sand	0.19	795

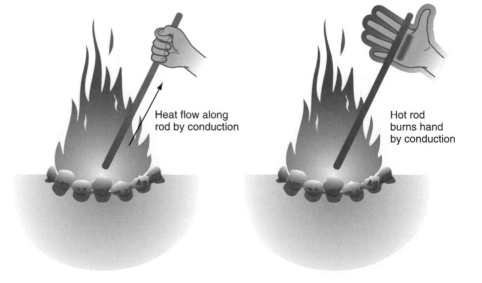

Heat flow along rod by conduction

Hot rod burns hand by conduction

■ **Figure 2.2**

Touching objects is an example of heat transfer by conduction. When something you touch feels hot, heat is being conducted from the object to your body.

Water is a good conductor of heat, while air is a poor heat conductor when it is not moving. This is why air between two pieces of glass in a storm window keeps a room insulated from the cold in winter. Dry sand is a poor conductor because of the air between the sand grains. Wet sand is a better conductor than dry sand because the air spaces are filled with water, which is a good conductor, as are the individual grains of sand. This is why running barefoot on a dry, sandy beach on a hot, sunny day can be a painful experience!

Since air is a poor conductor, conduction is not an efficient mechanism for transferring heat in the atmosphere on a global scale. But conduction is good for transferring energy over small distances, and is an important form of heat transfer near the ground.

▮ Convection: Hot Air Rises

If the ground is hot, heat is transferred to air molecules in contact with the surface via conduction. The heated parcel of air rises and cooler air sinks to replace the rising warm air. This results in a net transfer of heat upward, away from the surface. Warmth moves upward, far away from the surface, because of the movement of the fluid (air in this case), not because the air high up is in direct contact with the ground. This process of transferring energy vertically is called **convection** (Figure 2.3).

In the atmosphere the rate of energy transfer by convection depends on how hot the rising air parcel is and the vertical temperature pattern of the surrounding atmosphere. In certain regions of the globe where the ground is much warmer than the air above it, convection is an important process for moving heat vertically. Convection is strong over deserts during the summer, where energy from the Sun rapidly heats up the sand. Convection is an inefficient mode of heat transfer in polar regions, where the surface air is in contact with a surface that is often cooler than is the air above it.

▮ Heat Advection: Horizontal Movement of Air

The horizontal transport of heat in the atmosphere is referred to as **heat advection** (Figure 2.4). Warm air advection occurs when warm air replaces cooler air. On weather maps (Chapter 1), temperature advection is occurring wherever the wind "flagpoles" are pointed across the isotherms. In winter snowstorms, warm air advection moves warm air poleward while cold air advection brings cold air towards the tropical regions. Advection is an important process throughout the troposphere.

▮ Figure 2.3

The shape of a flame results from convection. Air near the flame heats by conduction, becomes less dense than the surrounding air, and rises. Because of convection, the rising air creates a flow of fresh air to the flame.

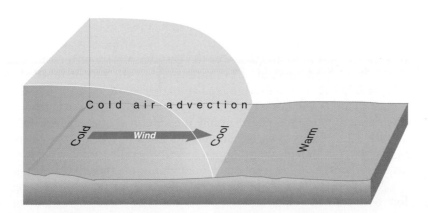

▮ Figure 2.4

Cold air advection occurs when colder air replaces warmer air. Warm air advection occurs when warmer air replaces colder air.

■ Latent Heating: Changing the Phase of Water

In the atmosphere, only water exists in all three phases: solid, liquid, and gas. Ice is the solid form of water and water vapor is the gas phase. In everyday language, the liquid form of water is generally referred to as "water." In this section, however, using this common term would lead to some confusion. So, in this section only, water in the liquid phase is referred to explicitly as *liquid water.*

Changing the phase of a substance either requires or releases energy. Changing the phase of water adds or removes energy from its surroundings. Since the Earth's surface is 70% liquid water, understanding the phase changes of water is important for understanding atmospheric energy. How does a change of phase occur?

Ice cubes melt and puddles evaporate because heat is added or removed from water, causing a change of phase. **Latent heat** is the heat absorbed or released per unit mass when water changes phase. This change of phase does not necessarily result in an increase in the water temperature. For example, adding heat to an ice cube may result in a change of phase of the water, from a solid to a liquid. However, the temperature of a liquid water–ice mixture will not increase until all the ice is melted. Energy cannot be destroyed, so what happens to the heat energy if it isn't used to change the temperature of the water?

Water molecules of ice and liquid water are bound together by molecular forces. In the ice phase, the water molecules have low enough kinetic energies that intermolecular attractions bind them into a highly ordered, crystalline form. To break these intermolecular attractions and transform ice into liquid water, energy must be added. **Latent heat of melting** is the amount of energy absorbed by water to change one gram of ice into liquid water, and is equal to 80 cal for each gram of ice.

If heat must be added to melt ice, the opposite occurs when water freezes. The amount of energy *released* into the environment when water freezes is also 80 cal per gram of ice, and is referred to as the **latent heat of fusion.**

The transition of water from the liquid phase to the gas phase is called vaporization or *evaporation.* The molecules of a gas are essentially free of one another, having no bonds between them. To convert liquid water to a vapor requires the addition of a sizable amount of energy (more than seven times the energy required to melt ice) to break the binding forces that keep the molecules in a fluid state. The amount of heat required to evaporate one gram of liquid water is referred to as the **latent heat of vaporization.** The latent heat of vaporization is a function of water temperature, ranging from 540 cal per gram of water at 100° C to 600 cal per gram at 0° C. That is, it takes slightly more energy to evaporate cold water than to evaporate the same amount of warmer water. (In addition, to raise the temperature of liquid water from 0° C to 100° C also requires an input of energy; this is where the specific heat of water comes in.)

Water vapor condenses to form liquid water. *Condensation* is the opposite of evaporation. **Latent heat of condensation** represents the amount of energy released when water vapor condenses to a liquid form. The latent heat of condensation is a function of temperature and has the same range as the latent heat of vaporization.

Water vapor may change directly to ice in a process known as **deposition.** Conversely, ice may also directly enter the gas phase without melting (called **sublimation**). The **latent heat of sublimation** equals the **latent heat of deposition.** A total of 680 cal are required to change one gram of ice at 0° C (32° F) into vapor. Notice that this number is equal to 600 cal + 80 cal. This means that the latent heats of sublimation/deposition are simply the sum of the latent heats of melting/fusion and the latent heats of vaporization/condensation. However, on Earth the sublimation and deposition of water happen less frequently, and on a far smaller scale, than do the processes of evaporation, condensation, melting, and freezing.

Where do we see phase changes of water in everyday life? Changes of phase affect both human beings and the atmosphere by removing or adding energy to the air in which phase changes of water are occurring. For example, when we physically exert ourselves, we sweat. Perspiration is a method for maintaining our body temperature. To evaporate the liquid water on our skin requires energy; some of this energy is taken from our skin. This energy transfer from our skin to the sweat cools the skin. So, evaporation is a cooling process that removes energy from the physical environment—in this case, our bodies. On the other hand, condensation, the opposite of evaporation, is a heating process that supplies energy to the environment. When water vapor changes into the liquid water or ice phase to form clouds, energy is released into the atmosphere. Changes of phase are also important in the energy gains and losses at ground level. The formation of dew (condensation) or frost (deposition) releases heat.

Changing the phase of water is an efficient method of transferring energy globally and provides an energy source for much of our weather. However, we are interested in some phase changes more than others. The reason is simple: 600 cal per gram is a lot more than 80 cal per gram! In other words, the phase changes of evaporation and condensation have by far the largest latent heats of the common phase changes of water. This obvious fact means that throughout this text we will be very interested in where, when, and how much evaporation and condensation occur. Wherever these two processes occur, there is a large transfer of energy going on.

To appreciate the amount of energy involved in condensation and evaporation, consider how much energy you would use to boil away a pot of water. Condensing an equivalent amount of water vapor would release a similar amount of energy. Now remember the heaviest rainstorm that you have experienced in your life. Imagine how much time and energy that was required to evaporate all the water that eventually condensed to form the rains. All of that rainwater was at one time water vapor and later condensed, thus releasing energy into the atmosphere and supplying fuel for the storm. If just 2 kilograms (4.4 pounds) of atmospheric water vapor are condensed into a liquid, 1,194,230 cal of latent heat are released. If all this energy were used to heat 2 kilograms of air, the air would warm more than 2000° C! The air does not get this hot when storm clouds form, however, because much of the released latent heat energizes the movement of air within the storm. This is a major, yet invisible, energy source for the winds of the atmosphere, which we will explore later in this text.

Figure 2.5 summarizes the phase changes of water vapor and the associated energy changes. To understand much of meteorology, it is very important that you know which processes in Figure 2.5 absorb energy and which ones release energy—so this is a figure to study now and refer to later.

Adiabatic Cooling and Warming: Expanding and Compressing Air

Air that moves up and down in the atmosphere undergoes temperature changes. To illustrate this concept we follow a parcel of air as it ascends in the atmosphere.

We start with our parcel of air near the ground. As we learned in Chapter 1, pressure always decreases with altitude. Therefore, as the parcel rises (Figure 2.6) the pressure exerted on the outside of the parcel decreases. The molecules inside the rising parcel exert a pressure that is greater than the pressure exerted by the molecules outside the parcel. The internal pressure increases the parcel's volume until the internal pressure equals the external pressure of the environment. In other words, *a rising parcel always expands because atmospheric pressure always decreases with altitude.*

Increasing the volume of the parcel requires work, which means energy must be involved. The air molecules are expending energy—in this case kinetic energy—to do the work required for expansion. Where does this kinetic energy go? Remember that energy cannot be destroyed but it can be converted to another form.

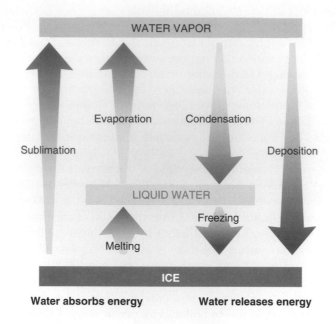

■ **Figure 2.5**

The phase changes of water. Blue lines represent a phase change that removes energy from the atmosphere. Phase changes that add energy to the environment are represented by red lines. When water vapor changes to liquid water or ice, clouds are formed and large amounts of energy are released into the atmosphere.

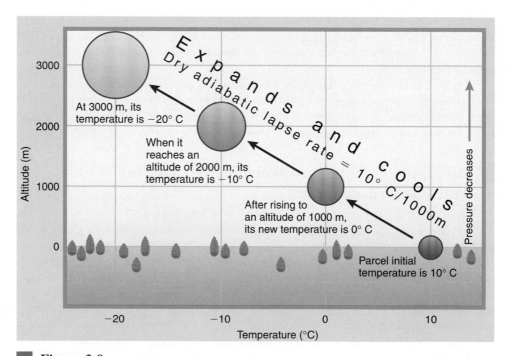

■ **Figure 2.6**

Adiabatic cooling and warming occurs as a parcel of air moves up and down in the atmosphere. As an air parcel rises it expands because the surrounding pressure decreases with altitude. Energy is required to do the work of expansion. Kinetic energy is converted to potential energy. Temperature is proportional to kinetic energy and so the parcel temperature decreases at a rate of 10° C per kilometer. The cooling of ascending air is important in the formation of clouds and precipitation.

As the parcel rises away from the ground, its potential energy increases. So the air molecules' kinetic energy is being converted to potential energy.

The change in kinetic energy of the parcel affects the parcel's temperature. Recall that a parcel's temperature is the average kinetic energy of its molecules. Since the average kinetic energy is decreasing, the temperature of the parcel must also be decreasing. This leads to a hard-and-fast rule of meteorology: *A rising parcel of air always cools.* How fast the parcel cools depends on gravity, the amount of moisture in the parcel (see below), and the specific heat of the air in the parcel. A dry parcel's temperature will decrease by approximately 10° C for each kilometer it rises.

Now let's consider what would happen if the parcel returned to its original position. As it descended it would be compressed, because atmospheric pressure increases closer to the Earth's surface. This compression increases the average kinetic energy of the molecules as they collide within the collapsing imaginary boundary of the parcel. As the parcel descended, the potential energy of the molecules would be converted back to kinetic energy. When the parcel returned to its original level its temperature would have returned to its initial temperature. *A descending parcel of air always warms.*

In following our air parcel, we've been tracking an **adiabatic process.** An adiabatic process is a process in which no heat energy is gained or lost by the system in question (in our case a parcel of air). The rate of cooling due to expansion (which is also the rate of warming due to compression) is referred to as the **dry adiabatic lapse rate.** Although water molecules exist in the parcel in the vapor phase, the parcel is dry because no clouds formed as it rose. A dry rising parcel of air cools adiabatically at a rate of almost 10° C per 1000 meters (or approximately 5.5° F per 1000 feet). A dry sinking parcel of air warms adiabatically at the exact same rate.

■ Diabatic Cooling and Warming: Adding and Subtracting Heat

A quick glance out the window during a thunderstorm tells you that the atmosphere is not dry. Therefore, we have to modify our understanding of temperature changes to take water into account.

A **moist parcel of air** is a parcel in which water molecules are changing phase from vapor to liquid or ice. A phase change of water vapor to liquid water or ice releases energy, warming the parcel through latent heating. For an ascending moist parcel of air, two processes are going on at once. Expansion is cooling the parcel while condensation (or deposition) is warming the parcel by latent heating. The cooling process from expansion is always larger than the latent heating, so the parcel temperature decreases. The rate that the rising moist parcel cools is called the **moist adiabatic lapse rate.**

Because heat is being added by the phase change of water vapor, the cooling rate of a rising moist parcel is less than the dry adiabatic lapse rate. The specific moist adiabatic rate for a given parcel depends on whether liquid or ice particles form and how much water vapor changes phase, and as a result differs from one place and time to another. To simplify our discussion, we will use a moist adiabatic lapse rate of 6° C per 1000 meters.

Consider a rising parcel at the ground with an initial temperature of 10° C (50° F) (Figure 2.7). The parcel rises to an altitude of 1000 meters as a dry parcel. Therefore, it cools at the dry adiabatic lapse rate. When it reaches 1000 meters it has a temperature of 0° C (32° F). At this point the figure indicates that the water vapor in the parcel is condensing (where and why this happens is covered more fully in Chapter 4). This means that the parcel will continue to cool as it rises, but now at the *moist* adiabatic lapse rate. The parcel then rises to 2000 meters, where it has a temperature of −6° C (21° F). If the parcel continues at the moist adiabatic lapse rate, what will its temperature be at an altitude of 3000 meters? Because the parcel is still saturated, it will cool to six degrees less than −6° C, or −12° C (10° F).

Blue Skies

Moisture and Stability/ Adiabatic

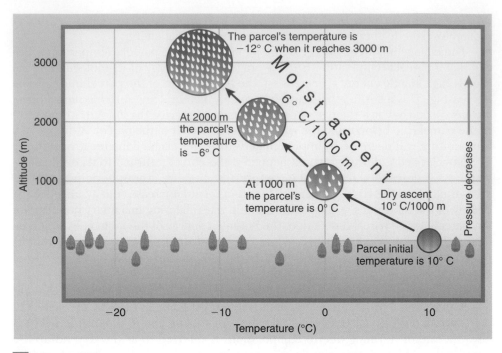

Figure 2.7

Moist adiabatic cooling and warming occurs when condensation (or deposition) is occurring inside a parcel of air. The rate of cooling during ascent is less than the dry adiabatic lapse rate because the cooling caused by expansion is partly offset by the warming resulting from the latent heat of condensation (or deposition). A typical moist adiabatic lapse rate is 6° C per kilometer, but it varies widely.

Comparing Figures 2.6 and 2.7, the main point is that the presence of moisture has caused the parcel to be much warmer at 3000 meters than it would have been otherwise. Where does this warmth come from? It comes from the latent heat of condensation of water.

In later chapters, we will determine the potential for thunderstorm formation by lifting a parcel of air and comparing the parcel temperature to the temperature of its surroundings. If the parcel is colder than its surroundings, it will tend to sink back to its original level. This condition is not favorable for storm development, because too much energy is required to lift the parcel and overcome the force of gravity. When the parcel temperature is much warmer than its surroundings, severe storms are possible. This is because air that is warmer than its environment keeps rising, creating tall thunderstorm clouds. Therefore, the presence of moist air greatly assists in the formation of thunderstorms and severe weather. Solar heating, a topic of the next section, is also an ingredient in severe storm development.

Radiative Heat Transfer: Exchanging Energy with Space

Earth receives energy from the Sun. This solar energy moves through empty space from the Sun to the Earth and is the original energy source for our weather and climate. Solar radiation is one form of **radiant energy,** or energy in the form of waves that are not composed of matter. Radiant energy is also called **radiation** and **electromagnetic energy.** In meteorology, "radiation" refers to a broad range of energy types, not just the type associated with nuclear energy. All electromagnetic radiation propagates through matter or empty space in the form of a wave with electric and magnetic fields.

Waves are characterized by two properties: **wavelength,** the distance between wave crests, and **amplitude,** one half the height from the peak of the crest to the

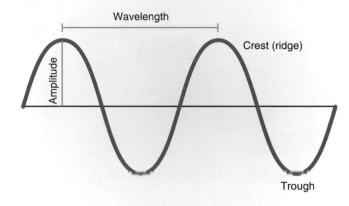

Figure 2.8

All waves, no matter what kind, are defined by the distance between crests (the wavelength) and their amplitude.

lowest point of the wave (Figure 2.8). The size of the wavelengths of radiation range from ultra-long radio waves to high-energy gamma rays (Figure 2.9). The amount of energy in the wave increases for smaller wavelengths. The distance between wave crests is measured in terms of a micrometer or micron (Chapter 1; abbreviated μm), which is approximately $\frac{1}{100}$ the diameter of a human hair. In meteorology, we are concerned with radiant energy with wavelengths between 0.1 and 100 μm.

We can make a useful distinction between the types of radiation in this wavelength range. The Sun emits most of its radiant energy with a wavelength between 0.2 to 4 μm, on the short end of the 0.1 to 100 μm range. For this reason, solar radiation is sometimes referred to as **shortwave radiation.** Solar energy includes ultraviolet, visible, and near-infrared radiation.

Within the wavelength range of solar energy lies **ultraviolet (UV) light** where wavelengths range from 0.2 to 0.4 μm. UV radiation is responsible for the tanning of our skin and skin cancer. Our eyes are sensitive to visible radiation, meaning that they detect electromagnetic energy characterized by wavelengths ranging from 0.4 to 0.7 μm. The color blue has a wavelength of approximately 0.45 μm, and red has a wavelength of 0.7 μm. Because visible light has wavelengths that are longer than UV light, they are not as energetic.

Longwave radiation (or **terrestrial radiation**) emitted by the Earth is less energetic than solar radiation and is characterized by much longer wavelengths, primarily between 4 and 100 μm with a maximum near 10 μm.

Why is it that the Earth and the Sun emit energy in such different wavelength bands? To understand, we must learn a few fundamental laws of radiation.

Radiation Laws

The first law of radiation we need to know is that *all objects with a temperature above absolute zero emit radiation.* You, this book, a hot cup of coffee, a roaring fire, and a cold cup of coffee all emit radiation. Even cold, nearly empty outer space emits some radiation from the original "Big Bang" billions of years ago.

Some objects emit more radiation than others, however. The amount of radiant energy that is emitted by an object depends on its temperature. The **Stefan-Boltzmann Law** makes our first law more precise by stating that the amount of energy per square meter per second that is emitted by an object is

http://info.brookscole.com/ackerman

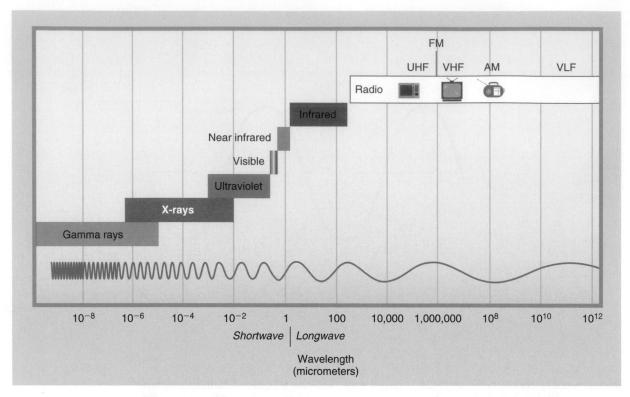

■ **Figure 2.9**

Electromagnetic energy spans a large spectrum of wavelengths. In this course we are interested primarily in solar (or shortwave) and infrared (or longwave) radiation. (Source: Adapted from Aguado, E. and Burt, J., Understanding Weather and Climate, *2nd ed., Prentice-Hall, 2001, p. 34.)*

related to the fourth power of its Kelvin temperature. Therefore, *a warmer object emits much more radiation than a cooler object.* For example, the Sun emits about 160,000 times as much energy as the Earth because it is about 20 times hotter.

The wavelength of radiant energy emitted depends on the temperature of the emitting body. German physicist Wilhelm Wien (pronounced "ween") won the 1911 Nobel Prize in physics for this discovery, which is now called **Wien's Law.** His law is simple division:

$$\frac{\text{Wavelength } (\mu m) \text{ of maximum}}{\text{radiation emitted by object}} = \frac{2900}{\text{Object's temperature in Kelvins}}$$

Let's use this equation to answer our questions about solar and terrestrial radiation. Using Wien's Law, an object with a near-room temperature of 290 K (17° C or 62° F) maximizes its radiant energy output at 10 μm. From Figure 2.9, we see that this maximum occurs in the infrared region. Therefore, room-temperature objects emit longwave heat energy, not visible light or X-rays. The Earth's average temperature is 288 K (15° C or 59° F), and so this explains why the Earth emits longwave radiation.

Now consider a very hot object with a temperature of about 2900 K, such as the filament in a halogen lamp. By Wien's Law it has a wavelength of maximum energy output of about 1 μm, which is much shorter than the 10-μm wavelength that characterizes the energy coming from the room the lamp is in. The Sun is about twice as hot as a halogen lamp filament, with a temperature around 5800 K. By Wien's Law, its energy output therefore maximizes at wavelengths that are half as long as in the halogen lamp case, in the shortwave region around 0.5 μm. This explains why halogen lamps and the Sun emit light—but you, this text, and the room around you

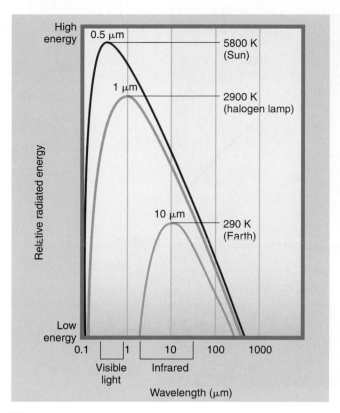

██ **Figure 2.10**

Energy curves for three objects having temperatures of 5800 K, 2900 K, and 290 K. These temperatures correspond roughly to the temperature of the Sun, the temperature of a halogen lamp filament, and the globally averaged temperature of the Earth. Notice that the hotter the object, the larger the curve is (Stefan-Boltzmann Law). The figure also shows that the hotter the object, the shorter the wavelength of the peak of the curve (Wien's Law).

emit only heat, not light. Wien's Law can be therefore be summarized as, *the hotter the object, the shorter the wavelength of maximum emission of radiation.*

Figure 2.10 illustrates both the Stefan-Boltzmann Law and Wien's Law. The curves shown are the amount of energy as a function of wavelength for the three different temperatures we have discussed: 290 K, 2900 K and 5800 K. Notice that the 2900 K energy curve is higher than the 290 K energy curve at every point, and the 5800 K curve is higher than the 2900 K curve. This is a consequence of the Stefan-Boltzmann Law. Notice also that the 5800 K curve peaks at the shortest wavelengths, followed closely by the 2900 K curve. The 290 K curve, in contrast, peaks farther to the right in the infrared heat region. This shift in peak energy emission is the main point of Wien's Law.

For those who don't like arithmetic or energy curves, Figure 2.11 tells the same story in a single photograph. It is a photograph of a person holding a lighted match, but the camera is seeing infrared heat energy, not light. The burning match is the hottest object in the picture and is emitting the most infrared energy, just as it should by the Stefan-Boltzmann Law. It is color-coded white. The person's eyeglasses and necktie are emitting the least energy and are color-coded blue. Wien's Law explains why you can see a hot match glow, but you can't see a necktie glow.

Energy Curves

The ART of Radiative Heat Transfers

If rooms and neckties don't glow, how can we see them? To explain this, we need to learn what happens when emitted radiation interacts with an object.

http://info.brookscole.com/ackerman

NASA/JPL/Caltech

■ **Figure 2.11**

An infrared photograph show-ing the heat coming from a man holding a burning match. White indicates the most infrared energy emission, and blue indicates the least in-frared energy emission. (Cour-tesy of NASA/JPL/Caltech)

Planetary Effective Temperatures

As with all forms of energy, radiation can change form, but it must be con-served. When radiation interacts with an object, it can be:

1. *Absorbed:* Absorption of radiation by a molecule increases the energy of the mol-ecule. If this energy is in the form of kinetic energy, then the temperature of the object or gas increases. The atmosphere is an effective absorber of heat energy.
2. *Reflected:* Energy reflected by an object is sent back. Mirrors are good reflectors of visible light, as are clouds.
3. *Transmitted:* Energy transmitted through an object passes through the object, though it may change direction. The atmosphere is almost completely transpar-ent to visible radiation.

The initials of these three processes are *ART,* making them easy to remember. (A fourth process, scattering, will be discussed in Chapter 5.)

Most of what we see comes from the reflection of visible light off objects. The **albedo** of an object describes the percentage of light that it reflects. The higher the albedo, the whiter the object. Freshly fallen snow has an albedo of 90%, whereas green grass has an albedo of 25%.

Albedo is a key determiner, along with distance, of the temperatures of the plan-ets in the solar system. Whiter planets are able to absorb less solar radiation than darker planets. Figure 2.12 shows visible photographs of Venus, Earth, and Mars from U.S. space missions. Bright white Venus has the highest albedo, at about 80%; Earth's albedo is around 30%; and Mars, the darkest of the three, has an albedo close to 20%. Mars is darker than Earth and farther away from the Sun, so it receives much less energy than the Earth does but absorbs a higher percentage of what reaches it. The temperature of Mars is therefore lower than Earth's, on average 40° C cooler. (Venus is much whiter than Earth, which should keep it cooler than Earth despite Venus's proximity to the Sun. However, a thick carbon dioxide atmosphere causes Venus to be much hotter than Earth, an effect that we will discuss shortly.)

The percentage of energy that is not reflected is either absorbed or transmitted. The structure of a molecule can be altered if it absorbs high-energy radiation.

A   B   C

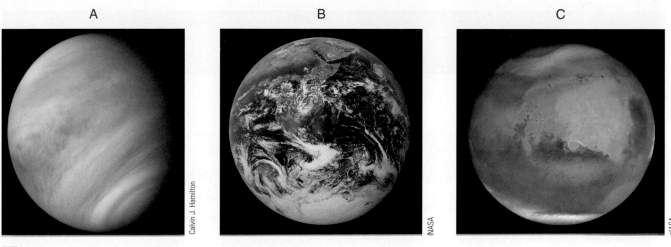

Calvin J. Hamilton   NASA   NASA

Figure 2.12

Photographs of: **A,** *Venus on February 5, 1974.* **B,** *Earth on December 7, 1972.* **C,** *Mars on June 26, 2001. These were taken from the Mariner 10, Apollo 17, and Hubble Space Telescope missions, respectively. Venus is whitish, Earth is blue and white, and Mars is red. Therefore, the albedo of Venus is high, Earth's albedo is relatively low, and Mars's albedo is even lower than Earth's.*

Photodissociation occurs when absorption of ultraviolet radiation results in breaking of chemical bonds. Photodissociation is an important process in the formation of ozone, O_3 (Box 2.1).

Absorption of electromagnetic energy at wavelengths longer than UV does not disrupt the molecule's structure. Instead, the molecule vibrates or spins after absorbing the radiation and thus increases its kinetic energy. Absorption of radiation with wavelengths greater than 0.2 μm warms the atmosphere and surface.

How much radiation energy the atmosphere or an object absorbs depends on:

1. *The radiative properties of the material:* Most objects are selective in the wavelengths that they absorb.
2. *The amount of time the object is exposed to the emitted energy:* The longer it is exposed to radiation, the more energy it can absorb.
3. *The amount of material:* Very thin objects may transmit and not absorb all the energy reaching them. Increasing its thickness can increase the amount of energy an object absorbs.
4. *How close the object is to the source of energy:* The farther an object is from a source of radiation, the less energy that reaches it, and it can absorb less energy.
5. *The angle at which the radiation is striking the object:* Radiation shining directly onto an object in a concentrated beam is absorbed more effectively than is a spread-out beam hitting the object at an angle.

Just as some substances are better conductors of heat than others, some objects emit and absorb radiation better than others. A **blackbody** is an object that absorbs all the electromagnetic energy that falls on the object, no matter what the wavelength of the radiation. A perfect blackbody does not exist, but it is a useful reference for determining how good a body is at emitting and absorbing radiation. (While an object may visually appear black, it does not mean it is a blackbody.) The energy curves drawn in Figure 2.10 are based on the assumption that the objects are perfect blackbodies.

If an object absorbs electromagnetic energy of a certain wavelength, it will also emit energy at that wavelength. This is **Kirchhoff's law,** our last law of radiation: *A good absorber of radiation is also a good emitter of radiation at that same wavelength.* Remember, how much radiation an object emits also depends on its temperature.

http://info.brookscole.com/ackerman

Ozone

Ozone is a trace gas that plays a role in myriad radiative processes in the Earth's atmosphere. Ozone is formed by interacting with solar radiation, it is destroyed by solar radiation, it absorbs terrestrial radiation, and recent losses in stratospheric ozone are firmly linked to the presence of chlorofluorocarbons (CFCs), another trace gas that also absorbs terrestrial radiation.

Stratospheric ozone (O_3) represents about 90% of the total amount of ozone in the atmosphere. It is produced in the stratosphere by the combination of ultraviolet radiation and an oxygen molecule (O_2). The basic steps in its formation are:

1. UV radiation strikes an oxygen (O_2) molecule, splitting it into two O atoms.

2. Two O atoms collide with 2 O_2 to produce 2 O_3 molecules.

This reaction is written in chemical equations as

$$O_2 + UV \rightarrow O + O$$

$$2O + 2O_2 \rightarrow 2O_3$$

Net Reaction: $3O_2 + UV \rightarrow 2O_3$

UV radiation is also involved in the destruction of O_3.

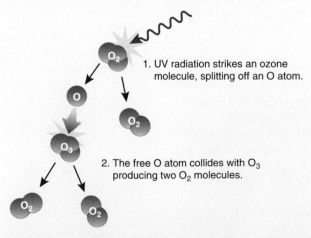

1. UV radiation strikes an ozone molecule, splitting off an O atom.

2. The free O atom collides with O_3 producing two O_2 molecules.

This destruction is expressed as

$$O_3 + UV \rightarrow O + O_2$$

$$O + O_3 \rightarrow 2\ O_2$$

Net Reaction: $2O_3 + UV \rightarrow 3O_2$

This cycle of creation and destruction of ozone does not cause any long-term loss of ozone. However, in 1970, meteorologist Paul Crutzen proposed the following catalytic reaction that results in the chronic destruction of ozone by molecules like chlorine.

$$X + O_3 \rightarrow XO + O_2$$

$$O_3 + UV \rightarrow O_2 + O$$

$$O + XO \rightarrow X + O_2$$

Net Reaction: $2O_3 + UV \rightarrow 3O_2$

In this sequence of reactions, X is an atom or molecule that acts as a catalyst to convert O_3 to O_2. Note that X does not change in the net reaction, so it can continue to destroy O_3 molecules.

There is a delicate balance between the production and destruction of ozone, resulting in what is referred to as an ozone shield that protects us from high-energy UV radiation. This natural balance has recently been disrupted by human activities causing *ozone depletion* (meaning that the destruction of O_3 exceeds the creation of O_3). How has this happened?

One molecule that can serve as the catalyst molecule X is chlorine (Cl). But how does chlorine get into the stratosphere? In the 1930s CFCs were first produced for use in refrigeration, air conditioning, solvents, aerosol spray cans, and Styrofoam™ puffing agents. They leaked into the atmosphere as a result of their use. CFCs are very stable in the troposphere with lifetimes of approximately 100 years. This long lifetime allows CFCs to eventually make it into the stratosphere where they are split, or dissociated, by UV light to produce chlorine atoms. The destruction of ozone then occurs with the following chemical reactions:

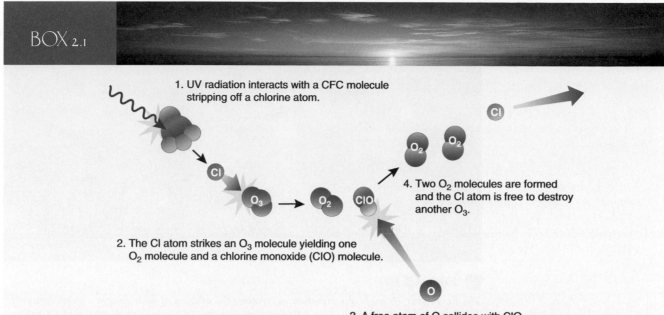

BOX 2.1

1. UV radiation interacts with a CFC molecule stripping off a chlorine atom.

2. The Cl atom strikes an O_3 molecule yielding one O_2 molecule and a chlorine monoxide (ClO) molecule.

3. A free atom of O collides with ClO.

4. Two O_2 molecules are formed and the Cl atom is free to destroy another O_3.

The chemical reactions shown above predicted a slow but worrisome decrease in stratospheric ozone. Observations during the 1980s indicated, to the contrary, a very rapid and alarming decrease in ozone over Antarctica. Why?

Extremely cold stratospheric clouds composed of ice particles exist over Antarctica, but not over the midlatitudes or tropical regions. Complex chemical reactions involving chlorine and other ozone depleting molecules occur on the surface of these ice particles that greatly accelerate the depletion of ozone. When the Sun rises in the Southern Hemisphere springtime (e.g., late September), this adds the crucial ingredient of UV radiation to the mix of clouds and chemicals. Ozone is then destroyed rapidly. The depletion of ozone over Antarctica is so pronounced that it has been called the "ozone hole" (the blue region in the satellite image of ozone amounts below). A smaller but noticeable Arctic "ozone hole" has also developed at times in the past decade.

Fortunately, chlorine alone does not remain for centuries in the stratosphere. Thanks to the cooperation of scientists, governments, and industry, the use of CFCs has declined drastically in just the past two decades. As a result, stratospheric ozone may eventually return to pre-1930 levels—but it will take decades, if not a century, to do so.

In 1995 Drs. Crutzen, Mario J. Molina and F. Sherwood Rowland won the Nobel Prize in chemistry for their work concerning the formation and decomposition of ozone. Crutzen became the first meteorologist to win the Nobel Prize.

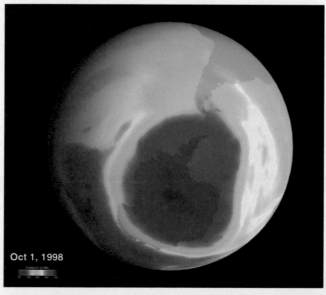

Oct 1, 1998

Dobson Units

NASA

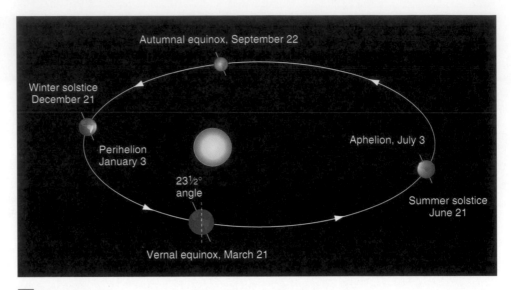

Figure 2.13

The Earth's tilt, or angle of inclination, determines the amount of solar energy a given region of Earth receives as it orbits the Sun. The northern hemisphere summer occurs when the Earth is tilted towards the Sun. Earth is about 3% further away from the Sun in June than in December. (The ellipse shape of the orbit is greatly exaggerated in this figure.)

Now that we know how radiation behaves, we need to examine more closely how solar radiation reaches Earth, how it is distributed across the Earth, and how the atmosphere affects the transfer of radiation.

The Sun Supplies Energy to Earth

The amount of solar energy reaching Earth at any particular latitude is defined by how the Earth orbits the Sun. Therefore, it is a function of the time of year and the tilt of the Earth with respect to the Sun. In the 17th century, Johannes Kepler discovered that Earth, along with the other planets, orbits the Sun in a path that traces out an ellipse. It takes one year (365.24 days) for Earth to make one complete revolution around the Sun. Because of its elliptic path, Earth's distance from the Sun varies with the time of year. Currently, Earth is farthest from the Sun on July 3 (**aphelion**), and is closest to the Sun on January 3 (**perihelion**) (Figure 2.13). The difference in the Earth–Sun distance between the time of aphelion and perihelion is greatly exaggerated in Figure 2.13; it is only about 5 million kilometers (3 million miles), or 3% of the total Earth–Sun distance.

If Earth is closer to the Sun in January than July, then why is it colder over North America in January than July? It is because the tilt of the Earth, not its distance from the Sun, is the primary orbital parameter that affects the Earth's temperature. As the Earth orbits the Sun, its axis of rotation is tilted at an angle of 23.5° from its orbital plane (see Figure 2.13). This tilt is referred to as the **angle of inclination.** Since Earth's axis of spin always points in the same direction—toward the North Star—the orientation of Earth's axis to the Sun is always changing as the Earth orbits around the Sun (see Figure 2.13). As this orientation changes, so does the distribution of sunlight on Earth's surface at any given latitude.

At high noon on the **solstices** (*sol*, "Sun," and *stice*, "come to a stop"), the Sun's rays strike the equator at an angle of 23.5°. On these days, the Sun is highest in the midlatitude summer sky. On approximately June 21 the northern spin axis is tilted

23.5° towards the Sun. On this day, the Northern Hemisphere summer **solstice,** latitudes south of approximately 66.5°S remain in complete darkness. This latitude is referred to as the *Antarctic Circle.* Around December 21 on the Northern Hemisphere winter solstice, the northern spin axis is pointed away from the Sun and latitudes north of the Arctic Circle (66.5°N) have 24 hours of darkness.

The **equinoxes** (*equi,* "equal," and *nox,* "night") occur when the Sun's rays strike the equator at noon at an angle of 90°. In the Northern Hemisphere, the **vernal** or **spring equinox** occurs on March 21 or 22 and the **autumnal** or **fall equinox** on September 22 or 23. During the equinoxes all locations on Earth experience 12 hours of daylight and 12 hours of darkness. On June 21 the Sun is directly overhead the Tropic of Cancer (23°27'N) at noon. The Tropic of Capricorn (23°27'S) is the latitude at which the noon Sun is overhead on December 21.

The angle of inclination is responsible for the seasonal variation in the amount of solar energy distributed at the top of the atmosphere, and plays a key role in determining the seasonal variation in surface temperature. The variation of solar energy by latitude is caused by: changes in the angle that the Sun's rays hit the Earth, the amount of atmosphere the Sun's rays have to pass through, the number of daylight hours, and changing solar cycles.

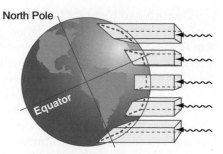

Northern Hemisphere Winter

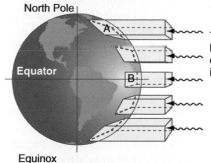

Equinox

The same amount of light is distributed over a larger area in A than in B

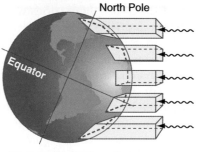

Northern Hemisphere Summer

Figure 2.14

The distribution of the Sun's energy on Earth's surface changes throughout the year. The Sun's energy is always more concentrated in the equatorial regions than the polar regions.

If you shine a flashlight at the ceiling, the region that is illuminated shrinks or grows depending on whether you point it directly at the ceiling or at an angle. Similarly, the Sun's energy spreads out over differing geographic areas when it reaches Earth's spherical surface. Solar energy is most concentrated in the area where the Sun is overhead at noon, the location of which changes with the season (Figure 2.14).

The angle at which the Sun's energy strikes a particular location on Earth is called the **solar zenith angle** (Figure 2.15). This angle is equal to 0° when the Sun is directly overhead and increases as the Sun sets, until the Sun is on the horizon and the solar zenith angle equals 90°. The solar zenith angle indicates the concentration of the Sun's rays at a given instant. Maximum intensity of the Sun's radiation occurs when the solar zenith angle is 0°, or directly overhead.

The solar zenith angle is a function of time of day, time of year, and latitude. Figure 2.16 demonstrates typical paths that the Sun traces from sunrise to sunset at two locations in the Northern Hemisphere. At local noon the Sun reaches the highest point on its path. Poleward of 23.5°N and 23.5°S, the Sun can never be seen directly overhead. Poleward of the Arctic and Antarctic Circles and between the spring and autumnal equinoxes, the Sun never sets. Instead the solar zenith angle remains approximately fixed throughout the day as the Sun circles the horizon.

Incoming Solar Radiation

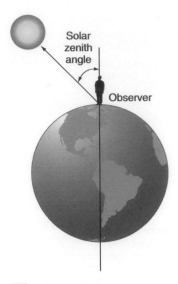

Figure 2.15

The solar zenith angle is measured from directly overhead to the position of the Sun. A solar zenith angle of 90° means the Sun is on the horizon.

Referring to Figure 2.14, we see that the Sun's energy spreads out over a larger area when the zenith angle is large. In addition, large zenith angles also cause the Sun's energy to pass through more atmosphere because they intersect a larger region of the atmosphere. This means that there is more chance for atmospheric absorption of solar energy. The surface, then, receives less energy.

The angle of inclination also defines the length of daylight for a given latitude. As Earth orbits the Sun it spins about its axis approximately once every 24 hours. This spinning explains our daily cycle of night and day and the resulting daily, or **diurnal,** variations in the amount of solar energy and in temperature. (Diurnal temperature changes are discussed in detail in the next chapter.) On June 21, the North Pole, because it is facing the Sun, experiences 24 hours of daylight, while the South Pole is in complete darkness. As we noted, on the equinoxes each region of the globe has 12 hours of daylight. The equator *always* has 12 hours of daylight.

The average amount of solar energy that reaches the outer limits of our atmosphere on a surface that is perpendicular to the solar rays is referred to as the **solar constant.** Its value is about 1368 watts per square meter. This "constant" actually can fluctuate by as much as 0.4% in a week, and it regularly changes by about 0.1% over a regular period of eleven years. These are small differences; however, it turns out that a change of just a few watts per square meter can make an enormous difference in climate.

Let's put all these ideas together now as we look at the changing amounts of incoming solar energy striking the top of the atmosphere at different latitudes over the course of a year. Figure 2.17 shows how the amount of energy incident at the top of the atmosphere varies with time of year for four latitudes: 70°N, 30°N, the equator, and 70°S.

During the late winter, 70°N receives no solar energy because the Sun is below the horizon. The Sun finally appears above the horizon in late January and the energy input continues to increase until June 21, when the Sun reaches its smallest solar zenith angle. This is also the day that the Sun appears directly overhead the Tropic of Cancer (23.5°N) at noon. At the equator, the amount of solar energy at the top of the atmosphere is greatest at the equinoxes, when the Sun appears directly overhead at noon at the equator.

The shape of the energy distribution at 70°S is opposite to the distribution at 70°N: a minimum occurs in June and a maximum in December. Notice that during their respective summers, the incident energy at 70°S is a little greater than the

Figure 2.16

The Sun rises in the east and sets in the west, tracing out a path in the sky similar to that indicated above. The top case represents conditions at the Tropic of Cancer and the bottom portion represents the Sun's path in the Northern Hemisphere midlatitudes. The more northern path represents summertime conditions.

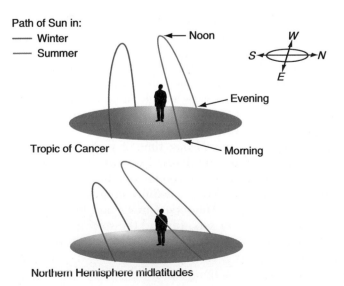

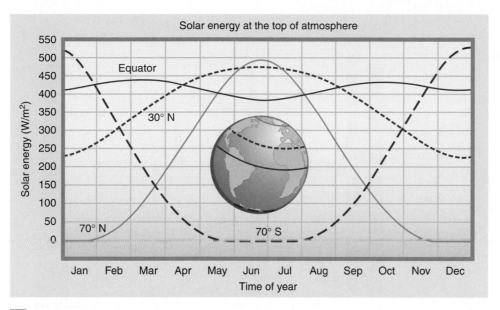

 Figure 2.17

The amount of solar energy striking the top of the atmosphere for four different latitudes (70°N, 30°N, the equator, and 70°S) as a function of time of year. Zero solar energy means that the Sun never rises.

amount at 70°N. This is because of the Earth's elliptical path around the Sun, which brings the Earth a little closer to the Sun in December than in June.

The energy distribution at 30°N has the single summer peak characteristic of locations poleward of the Tropics of Cancer and Capricorn, but its summertime maximum and especially its wintertime minimum are less drastic than at 70°N.

Radiative Properties of the Atmosphere

Before solar rays can reach the surface of Earth, they have to pass through the atmosphere. The trek through the Earth's thin atmosphere represents less than 0.0001% of the distance the rays travel from the Sun to the Earth. But what happens there makes all the difference to us, because it permits human life to survive on Earth.

Atmospheric gases are selective in the solar wavelengths that they absorb. Of the atmospheric gases, ozone absorbs most of the shortwave radiation. Ozone absorbs UV energy and a small amount of visible energy in the 0.4- to 0.56-μm spectral region. Without this absorption, human and animal life would perish because UV light destroys the genetic code of life. Water vapor weakly absorbs radiation at several wavelengths between 0.7 and 4.0 μm. Carbon dioxide is a very weak absorber of solar energy as are methane and chlorofluorocarbons (or CFCs). The left-hand side of Figure 2.18 summarizes the wavelength dependence of atmospheric absorption of solar radiation. Note that nearly all of the visible energy from the Sun is transmitted to the surface, while radiation with wavelengths less than 0.28 μm does not reach the surface. The percentage of solar energy absorbed by the atmosphere basically depends on how much ozone and water vapor are present. A 1% decrease in ozone, for example, leads to a 2% increase in the amount of UV light at the surface.

Atmospheric gases are also selective in the wavelengths at which they absorb and emit terrestrial radiation (right-hand side of Figure 2.18). In addition to absorbing solar radiation, water vapor absorbs (and therefore emits) terrestrial energy

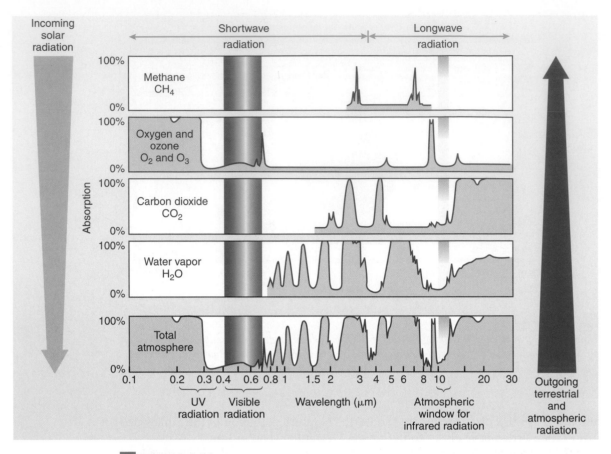

Incoming solar radiation (left vertical arrow)

Shortwave radiation · **Longwave radiation** (top)

Methane CH$_4$
Oxygen and ozone O$_2$ and O$_3$
Carbon dioxide CO$_2$
Water vapor H$_2$O
Total atmosphere

Absorption (vertical axis, 0%–100%)

Wavelength (μm): 0.1 0.2 0.3 0.4 0.6 0.8 1 1.5 2 3 4 5 6 8 10 20 30

UV radiation · Visible radiation · Atmospheric window for infrared radiation

Outgoing terrestrial and atmospheric radiation (right vertical arrow)

■ **Figure 2.18**

Absorption of shortwave (left) and longwave (right) radiation by the atmosphere. Low amounts of total absorption (white areas in bottom row) indicate wavelength regions in which solar radiation can reach the Earth's surface and terrestrial radiation can escape to outer space. The visible light spectrum and the infrared atmospheric window stand out in this regard. Notice the prominent roles of water vapor and carbon dioxide in absorbing Earth's longwave radiation.

at wavelengths between 5 and 8 μm, and beyond 12 μm. Carbon dioxide and ozone emit and absorb energy at wavelengths near 15 μm and 9.6 μm, respectively.

Atmospheric gases only weakly emit and absorb energy in the 10 to 12 μm region. This spectral region is referred to as the infrared **atmospheric window** because the atmosphere is relatively transparent to infrared radiation emitted by the surface at these wavelengths. Another window exists around 4 μm, but the other atmospheric window is more important for weather and climate because the Earth's peak in emitted energy happens to occur right in this window, at 10 μm. The atmospheric window thus allows the Earth to cool off and emit its energy into outer space.

Clouds are good reflectors of solar energy, and they are good emitters and absorbers of longwave energy. Clouds also emit and absorb radiation in the 10 to 12 μm IR atmospheric window. So, when clouds are present, the window is effectively "shut."

Both the infrared atmospheric window and the window-shutting effects of clouds are "visible" from satellite. Figure 2.19 shows the radiation observed by all the different channels of the GOES-8 weather satellite for a North American August evening. The redder the colors, the more radiation is being observed that is being emitted or reflected by the Earth and atmosphere below. Channels 6, 7, and 8 in the infrared atmospheric window are the "warmest," followed by Channels 16, 17, and 18 in the window near 4 μm (compare to Figure 2.18). This means the Earth is cooling off to space in the red regions.

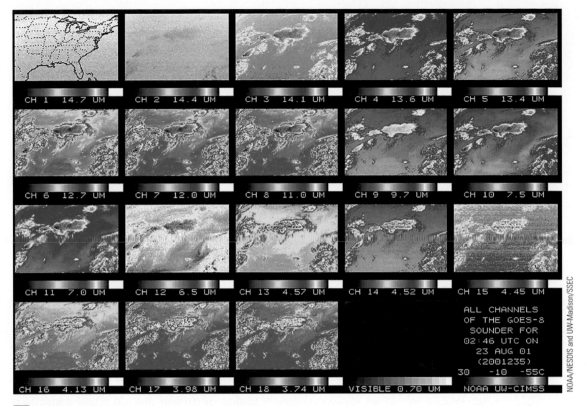

■ **Figure 2.19**

The 19 channels of viewing of the GOES-8 weather satellite. This image, from 0246 UTC on August 23, 2001 depicts the state of atmospheric radiation over North America (see map at top left). The channels decrease in wavelength from Channel 1 (14.7 μm) to Channel 19 (0.7 μm), the only visible channel. Red areas indicate regions of high infrared radiation emission; blue indicates regions of lower infrared emission; and black indicates zero visible radiation emission or reflection.

However, in Figure 2.19 clouds are causing the very low (white) regions of radiation over the Great Lakes in the center portion of most of the channels. The reason we know that clouds are shutting the window is that the water vapor in these clouds absorbs and emits strongly in the wavelength range of Channel 11, and the clouds stand out in that picture. Almost no radiation at all is seen in Channel 19 at 0.7 μm. There is a simple reason why: this picture was taken at night, and 0.7 μm is in the wavelength range of visible light. (This figure also illustrates how weather satellites make use of the science of radiation, a topic we'll return to in Chapter 5.)

■ **The Greenhouse Effect**

The selective nature of radiation absorption by atmospheric gases is the fundamental cause of the **greenhouse effect.** Much of the shortwave, or solar, energy passes through the atmosphere and warms the surface. While the atmosphere is nearly transparent to shortwave radiation, it efficiently absorbs terrestrial or longwave radiation emitted upward by the surface. So, while carbon dioxide and water vapor make up only a very small percentage of the atmospheric gases, they are extremely important because they absorb this longwave radiation and emit it throughout the atmosphere.

What happens when the atmosphere absorbs longwave radiation emitted by the surface? The atmosphere gains energy through absorption, but does not accumulate this energy continually. Instead, the atmosphere loses energy by emitting longwave radiation in all directions. Some of this longwave energy is emitted towards the Earth and absorbed by the surface. The Earth's surface is heated by shortwave and longwave absorption emitted by gases in the atmosphere. If the atmosphere did not absorb and emit longwave radiation, the surface of Earth would be approximately 33° C (60° F) cooler than it is today!

Because the greenhouse effect keeps the Earth from freezing into a subzero slab of ice, it is a good thing. A separate issue is whether or not humans have added to the greenhouse effect and made too much of a good thing.

Greenhouse Warming

■ Greenhouse Warming: The Basics

Greenhouse warming and the **enhanced greenhouse effect** are the terms used to explain the relationship between the observed rise in global temperatures and an increase in atmospheric carbon dioxide. In this chapter we have considered one aspect of the enhanced greenhouse effect: absorption and emission of radiation by certain atmospheric gases. How does this play a role in the enhanced greenhouse effect? Let's use carbon dioxide as an example.

Increasing the carbon dioxide concentrations in the atmosphere does not appreciably affect the amount of solar energy that reaches the Earth's surface. However, since carbon dioxide absorbs longwave radiation, then the more carbon dioxide, the more heat is absorbed. This increases the temperature of the atmosphere. Then, by the Stefan-Boltzmann Law, the now-warmer atmosphere will emit more energy than it did when there was less carbon dioxide in it. This, in turn, increases the amount of longwave energy striking the Earth's surface, and it too warms up. By this simple argument, increased concentration of carbon dioxide could result in a warming of Earth's atmosphere and surface.

Gases that are transparent to solar energy while absorbing terrestrial energy will warm the atmosphere because they allow solar energy to reach the surface and inhibit longwave radiation from reaching outer space. These radiatively active gases are called **greenhouse gases.** This nickname makes an analogy with a greenhouse, which lets solar radiation come through its glass ceiling and walls but traps heat energy inside. In addition to carbon dioxide, other important greenhouse gases are water vapor, ozone, methane, and CFCs. Methane and CFCs are important despite their very small concentrations because they absorb terrestrial radiation in the 10 to 12 μm infrared atmospheric window.

Water vapor is the most important greenhouse gas because of its relative abundance and its ability to absorb a lot of longwave energy in many different wavelengths. A warmer atmosphere can mean more water vapor in the atmosphere and possibly more clouds. However, clouds affect both shortwave and longwave radiation, complicating the situation considerably. We will revisit the enhanced greenhouse effect in Chapter 4, where we cover water vapor and clouds in more detail.

■ The Global Average Energy Budget: Heat Is Transferred from the Surface to the Atmosphere

 Blue Skies

Atmospheric Basics/Energy Balance

When you balance your checkbook you are concerned with deposits and withdrawals. Deposits represent a gain to the account and withdrawals a loss. Gains are positive and losses are negative. The net result of adding together all of the gains and losses tells you if you are getting richer or poorer.

Similarly, when we study energy it is helpful to make a "budget" of the gains and losses of all types of energy over long periods of time, such as a year. Net gains in energy lead to warmer temperatures, and losses lead to cooler temperatures. This approach works on many different scales, from the entire planet to a single location. The budget for each square meter of Earth is shown in Figure 2.20, and now we will step through this "bookkeeping."

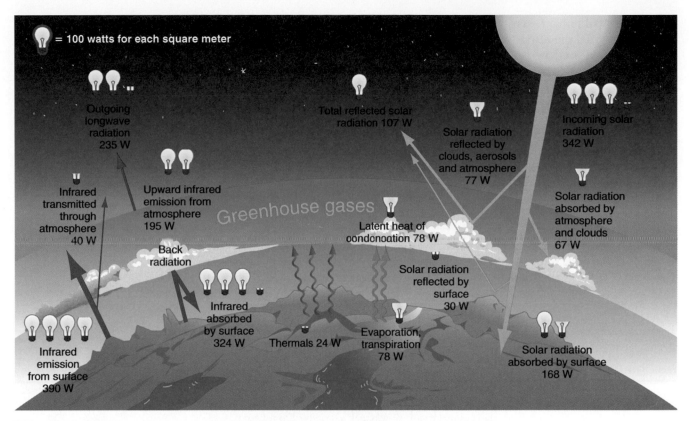

Figure 2.20

The annual average energy balance of Earth. The flow of solar radiative energy is depicted with yellow (solar) and red (infrared) arrows with widths proportional to their amounts for a square meter. Each light bulb represents 100 W for each square meter of surface area. On average, the atmosphere is losing energy by radiative processes while the surface of the Earth has a surplus of radiation energy. Energy transfer from the surface to the atmosphere is one reason why the average temperature in the troposphere decreases with increasing distance from the surface.

The globally averaged, total amount of solar energy incident on each square meter of area at the top of the atmosphere is 342 W (top right of Figure 2.20). This is an energy gain for the planet. The annual average albedo of the planet is 30%. So approximately 30% or 107 W per square meter of the incident solar energy at the top of the atmosphere is sent back out into space. The solar energy reflected back to space is considered a loss of energy. Seventy percent (235 W for each square meter of area) is absorbed by the atmosphere and surface. Solar radiation absorbed by each square meter of the atmosphere is 67 W. Each square meter of the surface absorbs 168 W. In the atmosphere, water vapor, clouds, aerosols, and ozone absorb solar energy.

The Earth's surface also gains energy as a result of atmospheric emission of longwave energy (left-hand side of Figure 2.20). At the same time it loses its own longwave energy by emission to the atmosphere. On average, each square meter of the Earth's surface emits 390 W. Some of the surface-emitted energy escapes to space, and the atmosphere absorbs the rest. Only 40 W per square meter area of the Earth's surface is transmitted through the atmosphere directly to space, much of it through the infrared atmospheric window. The atmosphere absorbs 350 W. The atmosphere emits radiant energy out to space and toward the surface. On average, each square meter area of the atmosphere emits 195 W to space and 324 W to the surface.

Now it's time to add up the numbers of the energy budget for the Earth and the atmosphere's "joint bank account." The Earth's surface has a net gain of 102 W of radiant energy (168 W – 390 W + 324 W) and the atmosphere experiences a net

The Energy Budget

loss of 102 W of radiant energy (67 W + 390 W − 40 W − 324 W − 195 W). The surface's gains equal the atmosphere's losses. This means that the Earth and its atmosphere, when viewed as one system, are in radiative balance.

A loss of 102 W for each square meter area of the atmosphere is equivalent to the atmosphere cooling more than 200° C over the course of a year! We do not observe this large cooling because energy is transferred from the surface to the atmosphere through ways other than radiation. The transfer of 102 W is accomplished by two kinds of heat transfer: sensible and latent heat transfers (center of Figure 2.20). **Sensible heating** represents the combined processes of conduction and convection, and amounts to a total of 24 W transferred from the surface to the atmosphere. Latent heating transfers 78 W from the surface to the atmosphere. Evaporation from oceans and lakes and sublimation from glaciers cools the surface. Some of the water that evaporates into the atmosphere condenses to form clouds and precipitation, releasing latent heat.

When globally averaged over a year, Earth's net energy gains balance energy losses, or nearly so. But this is not the case when the radiation gains and losses are averaged in relation to latitude (Figure 2.21). As we learned earlier, any object with a temperature above absolute zero emits energy. This means that all parts of the

BOX 2.2

Monitoring the Earth's Energy Budget

NASA's Earth Radiation Budget Experiment or ERBE (pronounced *er-bee*) is a multisatellite system designed to measure the radiation budget—how much heat is lost and gained—between space and Earth. On each of three satellites are instruments that measure how much solar energy the Earth absorbs and how much energy our planet emits. Observations from ERBE have improved our understanding of how clouds influence the energy budget of our planet.

Just because a given region of Earth is receiving more radiative energy than it is losing to space does not mean that the region's temperature is increasing. The atmosphere and ocean may move this excess energy to a different region of the globe, perhaps to a region that is losing more energy than it is receiving.

The determination of Earth's radiation budget is essential for atmospheric modeling and climate studies. A major research problem addressed with the ERBE program is how clouds affect the radiative energy budget of the planet, and thereby influence climate. Clouds generally reduce the radiation emitted to space and thus result in heating the planet. At the same time, clouds also reduce the absorbed solar radiation, because of a generally higher albedo (or brightness) than the under-lying surface. A higher albedo helps cool the planet. The effect clouds have on solar and terrestrial energy offset each other in terms of the energy budget of the planet. The latest results from ERBE indicate that, on average globally, clouds tend to cool the planet.

The picture below depicts Earth net radiation budget for July as determined by ERBE. Notice the net energy gains over the oceans and the net energy loss over Greenland. The ERBE data can be explored on the text's Web site.

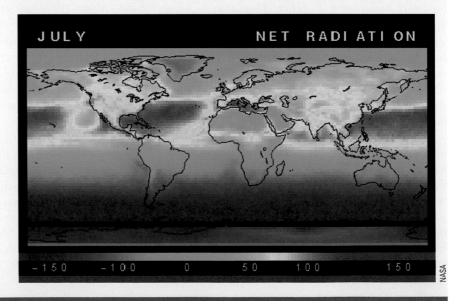

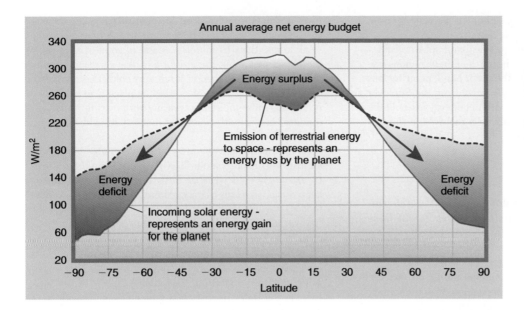

Annual average net energy budget

Energy surplus

Emission of terrestrial energy to space - represents an energy loss by the planet

Energy deficit

Energy deficit

Incoming solar energy - represents an energy gain for the planet

Figure 2.21

The radiation balance of the planet as a function of latitude. Atmospheric and oceanic circulations move energy from the latitudes with a surplus of radiation to regions with a deficit (purple arrows). Because tropical latitudes cover more surface area than polar latitudes on a sphere, the apparently small area of surplus does equal the deficit. For this reason, the Earth's temperature remains more or less constant.

Earth's Radiation Budget from ERBE

globe, including the poles, emit longwave energy at all times. In budget terms, all parts of the globe are always "paying out" energy. However, only the tropics receive enough sunlight all year long to counterbalance these losses and "turn a profit." The poles run deep into energy debt during the dark winter and are not able to make it up during the summer because even 24-hour sunlight is at a perpetually high zenith angle. The end result is that the tropics gain and the poles lose radiant energy on a yearly basis. Box 2.2 discusses the NASA program that strives to understand Earth's radiation budget.

Since the tropics are not observed to be continually heating and the polar regions continually cooling, energy must be transported from the tropics to the poles. This transport is accomplished by the atmosphere and ocean currents and is the reason we have weather!

PUTTING IT ALL TOGETHER

SUMMARY

Conduction, convection, advection, latent heating, and radiation are important processes that transfer energy in the atmosphere. Conduction moves energy by physical contact. Convection and latent heating transfer heat over great distances through vertical motions and phase changes of water. Advection by the wind moves heat horizontally. Radiative processes transfer heat throughout the entire atmosphere and into space. Radiation can be absorbed, reflected, or transmitted.

The Sun powers Earth's weather and climate. Solar energy streams through space as electromagnetic waves. Most of this solar radiation has wavelengths between 0.2 and 4 μm. The reason the Sun emits radiation with such short wavelengths is because, according to Wien's Law, the Sun is very hot.

Earth intercepts a portion of this solar energy. The total amount intercepted is determined by the tilt of the Earth's axis and how far the Earth is from the Sun. While Earth is a little closer to the Sun in January than in July, Northern Hemisphere temperatures are warmer in July because Earth's axis of rotation

points towards the Sun. The Earth's tilt, combined with its orbit around the Sun, causes seasons.

Approximately 50% of the solar energy reaching the planet passes through the atmosphere and is absorbed by the surface of the Earth. The surface albedo is about 30%, meaning that the Earth reflects 30% of the incoming sunlight.

Since any object that has a temperature emits radiation, the Earth emits energy and is constantly losing energy to space as longwave or terrestrial radiation. This energy emitted to space has wavelengths of 4 to 100 μm. Both the surface and atmosphere emit longwave radiation, not shortwave radiation like the Sun, because they are much cooler than the Sun.

Much of the energy that escapes from the Earth to space is in the narrow band of 10 to 12 μm called the *infrared atmospheric window*. At other wavelengths, the Earth's atmosphere absorbs much of the longwave radiation emitted by the Earth and warms the Earth's surface in return. This is known as the greenhouse effect and it keeps the Earth from freezing.

The atmosphere is always losing radiant energy, and the Earth's surface has a surplus of radiation energy. Conduction, convection, evaporation, and condensation transfer energy from the surface of the Earth to the atmosphere.

Averaged over many years, the tropical and subtropical regions of the planet gain more solar energy than these regions lose to space by longwave radiation. On the other hand, polar regions lose more longwave energy than they receive from the Sun over a given year. This energy imbalance, with net energy losses at the pole and net energy gains in the tropics, is the driving force of weather and climate.

 KEY TERMS

You should understand all of the following terms. Use the glossary and this chapter to improve your understanding of these terms.

Acceleration	Latent heats of:
Adiabatic process	condensation,
Advection	deposition, fusion, melting,
Albedo	sublimation, and
Amplitude	vaporization
Angle of inclination	Longwave radiation
Aphelion	Moist adiabatic lapse rate
Atmospheric window	Moist parcel of air
Blackbody	Parcel of air
Calorie	Perihelion
Celsius	Photodissociation
Conduction	Potential energy
Convection	Power
Deposition	Radiation/radiant energy
Diurnal	Sensible heat
Dry adiabatic lapse rate	Shortwave radiation
Electromagnetic energy	Solar constant
Energy	Solar radiation
Enhanced greenhouse effect	Solar zenith angle
Equinoxes	Solstices
Fahrenheit	Specific heat
Force	Stefan-Boltzmann Law
Greenhouse effect	Sublimation
Greenhouse gases	Temperature
Greenhouse warming	Terrestrial radiation
Heat	Thermal conductivity
Heat advection	Ultraviolet light
Joule	Watt
Kelvin	Wavelength
Kinetic energy	Wien's Law
Kirchhoff's Law	Work
Latent heat	

 REVIEW QUESTIONS

1. What are the major mechanisms for transferring energy in the atmosphere?

2. A documentary filmmaker once interviewed graduating Harvard University seniors on-camera by asking them the question, "Why are we warmer in summer than in winter?" The consensus answer from the Harvard graduates was, "Because we're closer to the Sun in summer than in winter." Is this answer correct? If not, what is the correct answer?

3. I was walking through a store one day and came across an advertisement of an object that claimed to defrost a one-inch steak in less than 10 minutes. Made of some sort of metal, the object was about the size of a cookie sheet and was not electric. When I touched it, it felt very cold. The salesperson demonstrated the 'super defroster' by placing an ice cube on it. The ice cube melted before my very eyes in a matter of minutes. How could something that felt so cold melt ice so quickly?

4. To keep warm when sleeping on a cold winter night, some animals burrow into snow. How does this keep them warm?

5. In extremely cold weather birds fluff up their feathers to keep warm. In terms of energy losses, what is the bird doing to keep warm?

6. Explain why "a watched pot of water never boils" in terms of specific and latent heats.

7. Consider a puddle of water that is evaporating. The faster molecules escape, leaving behind those with less energy. Thus, during evaporation, the average energy of the molecules in the water decreases as molecules continually leave. The water temperature should be lowered; however, the temperature of a puddle often isn't much different that the surrounding air. Explain why the temperature of the puddle does not continually decrease.

8. You pump up a bicycle tire with a hand pump in which air is compressed in the barrel of the pump and forced into the tire. You observe that the barrel of the pump gets hot. Why? Was the process adiabatic or diabatic?

9. If a dry parcel of air at the surface has a temperature of 20° C and rises to 3 kilometers above the surface, what will its temperature be? If it sinks back to the ground, what will its final temperature be? Repeat your calculations, but this time assume that the parcel of air is moist throughout its trip from the surface to 3 kilometers and back to the surface.

10. If an object in a fire is heated until it glows "red-hot," what is its approximate temperature? Assume that its wavelength of peak energy emission is 3 μm, so that the front part of its energy curve will include the red part of the visible spectrum.

11. The leftover radiant energy from the "Big Bang" is measured to be about 2.9 K. What is the wavelength of maximum emission of this energy? What kind of equipment would you need to detect this energy? (Hint: it would be a specialized version of a common electronic device. See Figure 2.9.)

12. Why will covering a soil with a white powder, such as lime, reduce the soil's daytime temperature?

13. The outer planet Pluto and Earth have the same albedo. Why is Pluto much colder than Earth?

14. Using the *Blue Skies* CD-ROM exercise "Atmospheric Basics/Energy Balance," compare the energy budget of Boulder, Colorado, versus Goodwin Creek, Mississippi, at noon on June 3, 1998. What are two reasons why Goodwin Creek is absorbing more solar radiation at the surface than is Boulder? (Hint: Look at all the budget numbers, and a map.) Which location is likely to have a greater temperature rise from noon to 1:00 PM on that day? What two numbers do you need to compare for each location to answer this question?

15. How would the length of day vary with latitude if the Earth's angle of inclination were 0°?

16. Why is the greenhouse effect beneficial? If enhanced, how can it become a problem?

■ WEB ACTIVITIES

Choose Chapter 2 on the textbook's Web site:
http://info.brookscole.com/ackerman
and select from the following resources:

• Interactive Modules that illustrate and extend your understanding of key topics in this chapter

• Tutorial Quizzes to test your mastery of terms and concepts

 For additional readings, go to the InfoTrac College Edition, your online library, at:
http://www.infotrac-college.com

Temperature

After completing this chapter, you should be able to:

■ Describe the various surface temperature cycles

■ Interpret temperature cycles in terms of the surface energy budget

■ Explain differences in temperature cycles due to the effects of location, altitude, and cloud cover

■ Define the concept of stability

■ Relate temperature inversions to the surface energy budget and to stability

 Introduction

Can the weather determine the outcome of wars? The experiences of Napoleon I and Hitler suggest it can. Their armies encountered two bitter Russian winters over a century apart. The rest, as they say, is history.

In the fall of 1812, Napoleon's army—a half-million strong, the largest the world had ever seen—entered the outskirts of Moscow, but was forced to retreat. The premature onset of winter turned the retreat into a complete disaster. Figure 3.1 depicts Napoleon's army as a dwindling river of soldiers advancing from left to right, and then retreating from right to left as temperatures dropped as low as −34° C (−30° F). One of Napoleon's generals reported that it was so cold that birds fell dead to the Earth. So did Napoleon's soldiers. Fewer than 10,000 men fit for combat remained with Napoleon's main force after the retreat. (Napoleon met his Waterloo two and a half years later, where once again weather was his nemesis.)

In October 1941, the German Nazis' Operation Barbarossa blitzkrieg on Moscow bogged down as the ground turned into a quagmire because of heavy rains. The rainy season was followed by one of the coldest winters in two centuries. Adolf Hitler's invading force had expected such a quick victory that they did not even bring winter gear with them. Cold outbreaks that winter plummeted the temperature just west of Moscow to a mind-numbing −52° C (−63° F). The Germans' equipment could not function properly at such bitter temperatures. Desperately needed supplies failed to reach the army because of breakdowns in the rail system brought on by the cold weather. Bombs were muffled by deep snow and railway bridges had to be constructed out of blocks of ice.

Stymied by the weather, the German troops abandoned the attack on December 6, 1941 (the day before Japan bombed Pearl Harbor). Then the Soviet army launched a counteroffensive against the Nazis. By the end of December more German troops died because of exposure to the cold than through the more conventional horrors of war.

Like Napoleon, Hitler was haunted by weather woes. Two and a half years later the weather forecast for D-Day (Chapter 13) would help put a permanent end to the Nazi quest for world domination.

Historians can provide many examples of how weather, especially temperature, has controlled battles or even entire wars. In this chapter, we provide examples of temperature observations at different locations to explain what controls the temperature.

 Surface Temperature

Temperature, as we learned in the last chapter, represents the average kinetic energy of the air molecules. A change in the air's temperature depends on its net energy budget, its specific heat, and whether or not phase changes of water have occurred. But before we explore temperature, we must be clear on what a "surface temperature" really is.

For consistency's sake, throughout the world meteorologists measure temperature at the same reference height, 1.5 meters (about 5 feet) above the ground, usually on a grass-covered surface. To avoid solar heating of the thermometer, temperature measurements are made in the shade. Air temperatures at this 1.5-meter height are called **surface temperatures.** The surface temperature is the temperature of the air *near* the ground, not

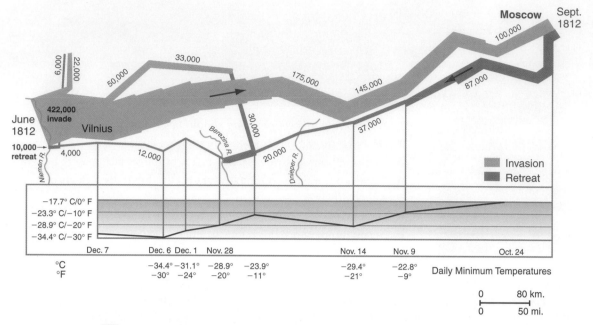

the temperature of the ground itself. (Further details on meteorological measurements will be covered in Chapter 5.)

Today's average global surface temperature is approximately 15° C (or about 59° F). The coldest temperature ever recorded at the surface is −89° C (−129° F) at Vostok, Antarctica on July 21, 1983. The record maximum surface temperature is 58° C (136.4° F) observed at El Azizia, Libya on September 13, 1922.

How can the temperature vary so widely from one day and location to another? This chapter examines the reasons for the observed temperature variations across the globe, at different latitudes, throughout the year, and at different times of the day.

Surface Energy Budget

In the previous chapter we discussed the transfer of energy and how the balance of energy gains and losses determine an object's temperature. The simple diagram in Figure 3.2 summarizes the flow of energy through the Earth–atmosphere system. Any system, from Earth and its atmosphere all the way down to a volume of air 1.5 m above the ground, is in energy balance if its energy gains equal its energy losses. The Earth–atmosphere system, averaged over a year, is in energy balance. And, as we discussed in the last chapter, if energy gains and losses are equal, the temperature of the system does not change.

But over short periods of time and in localized regions, energy gains do not always balance energy losses. When an energy imbalance exists, energy is stored within or removed from the system. When energy gains exceed losses, the temperature increases. Conversely, when energy losses exceed energy gains, the system will cool.

An energy imbalance is common for many systems we are familiar with. For example, imagine running barefoot on the blacktop in a parking lot on a hot sunny summer day. You will immediately notice that the blacktop is very hot. Why? The black-

The Energy Budget

top is exchanging energy with the bottom of your feet. Your feet are naturally warm and therefore emit energy. However, the blacktop is even warmer and emits much more energy. Enough heat is transferred from the blacktop to your feet to overcome their energy loss. As a result, each foot's energy gains are greater than its losses, and the soles of your feet get hot. At the same time, heat is also exchanged with the air above the blacktop, affecting the air temperature.

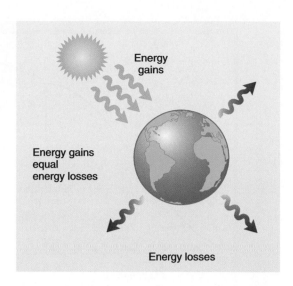

■ **Figure 3.2**
Earth gains energy from the Sun and loses terrestrial radiation to space, resulting in a global balance of energy.

The temperature of the air near the ground is determined by energy exchanges with the surface. Convection, conduction, latent heating, and radiation all play roles in transferring energy between the air and the ground. In addition, irregular air motions, referred to as **turbulence,** mix heat and moisture from the surface higher up in the atmosphere. We examine the complicated topic of turbulence in depth in Chapter 12.

Conduction, convection, radiation, sensible and latent heat transfers, and turbulence all act at the same time to transfer energy between the atmosphere and its surroundings. This makes it surprisingly difficult to express the relationship between air temperature and surface conditions in a simple formula. Instead, let's turn our attention to observed variations in the air temperature across the world. By comparing the observed temperature patterns of different cities, we will reveal the processes that govern temperature across the globe.

■ Temperature Cycles

In Chapter 1, we studied the cycles of water and carbon dioxide that described the ways that those molecules repeatedly moved from the atmosphere to the land and back again. Repetitive patterns happen in weather and are also called *cycles.* For example, you have probably noticed that temperatures are usually warmer in the afternoon than during the night, day after day. This repeating pattern of daily temperatures is the **diurnal temperature cycle.** This cycle includes the maximum and minimum daily temperatures and the times of day they usually occur (Figure 3.3). In the typical diurnal temperature cycle, the maximum temperature occurs during mid- to late afternoon. The minimum temperature is reached around dawn.

The **diurnal temperature range** is the difference between the maximum and minimum temperatures of any given day. The **daily mean temperature** is usually determined by averaging the maximum and minimum temperature for a 24-hour period, or sometimes by averaging all 24 hourly temperature measurements.

Everyone knows that temperatures are usually colder in winter and warmer in summer. This very regular cycle of temperature throughout the year is the **seasonal** or **annual temperature cycle.** Plotting the monthly mean temperature as a function of month represents the annual temperature cycle. The **monthly mean** (or monthly average) **temperature** is calculated by adding together the daily mean temperature for each day of the month and dividing by the number of days in the month.

Similar statistics can be compiled for annual temperatures. The **annual temperature range** is the difference between the warmest and coldest monthly mean temperatures of a given geographic location. It can also be defined as the difference between the highest and lowest daily temperatures observed in a given year or

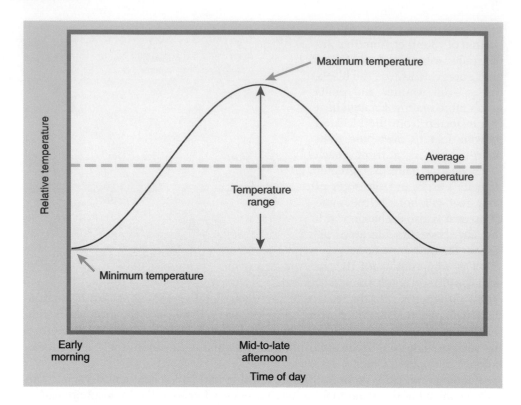

Figure 3.3

Temperature cycles are characterized by a maximum, minimum, the range between maximum and minimum, and the average temperature.

years. The **annual average temperature** is simply the sum of the monthly mean temperatures divided by twelve.

The seasonal and diurnal temperature cycles of a given region reflect the net energy gains and losses. When energy gains exceed losses, the temperature warms, and the temperature cools when energy losses surpass energy gains.

We know that temperatures usually rise and fall day after day and from season to season. This means that energy imbalances occur locally on daily, seasonal, and yearly time scales. What causes these energy imbalances? The five major factors are: latitude, surface type, elevation and aspect, relation to large bodies of water, and cloud cover. First we will discuss these factors in terms of how they affect the annual temperature cycle.

Annual Temperature Cycle

The geographic setting influences the temperatures at a given location. Comparing the annual temperature cycles of different locations allows us to observe how latitude, surface type, elevation and aspect, relation to large bodies of water, and cloud cover all influence temperature cycles.

Latitude

Annual Temperature Cycles

As discussed in Chapter 2, the tilt of Earth's axis—the angle of inclination—affects the amount of incoming solar energy, and is the reason for the seasonal cycle in temperature. The amount of incident solar energy at the top of the atmosphere, or **insolation,** is a function of time of the year, time of day, and latitude.

Figure 3.4 shows the monthly mean temperatures for each month of the year (which together make up the annual temperature cycle) of New York City and Miami. Both are large cities on the East Coast, both are near a large body of water, and both are at nearly the same low altitudes above sea level. Miami's annual average temperature is 24° C (75° F) while New York City's is 11.5° C (53° F). The main reason for this temperature difference is that Miami is closer to the equator. Therefore, the Sun is higher in the sky all year at Miami and Miami receives a greater amount of insolation than New York City throughout most of the year (Figure 3.5).

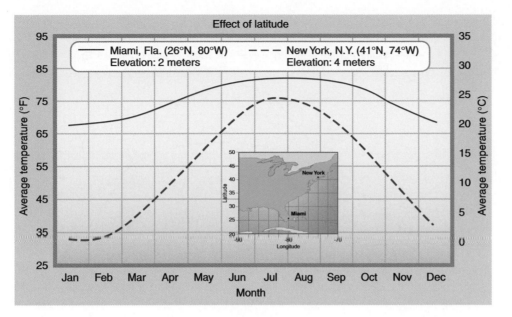

■ Figure 3.4

Comparison of the annual temperature cycle observed at Miami, Florida, and New York City. The difference in latitude between Miami and New York creates large differences in temperature between these two cities, which are otherwise similar in location and altitude.

The maximum monthly mean temperatures of both New York City and Miami occur in July, and the minimum in each city occurs in January. The maximum temperature occurs after, or **lags,** the time of maximum solar input, which occurs in June. Maximum solar radiation is received on the summer solstice in late June. However, after the summer solstice, energy gains still exceed energy losses. This means that the atmosphere continues to warm. Not until late July are the energy losses (such as emission of radiation) larger than the energy gains, causing the mean temperature to begin to decrease. Energy losses are greater than the energy gains until late January or February, even though the minimum solar energy received occurs at the winter solstice. This explains why the coldest temperatures come in January, not December. These lags exist largely because of the Earth's oceans.

New York's annual range of temperature is much larger than Miami's range, 24° C (44° F) versus 8° C (15° F). Latitude influences the annual temperature range because it affects: (1) the seasonal variation of the insolation, (2) the solar zenith angle, and (3) the length of day. Throughout the year these three factors vary less in Miami than in New York, causing the temperature range to be smaller for Miami.

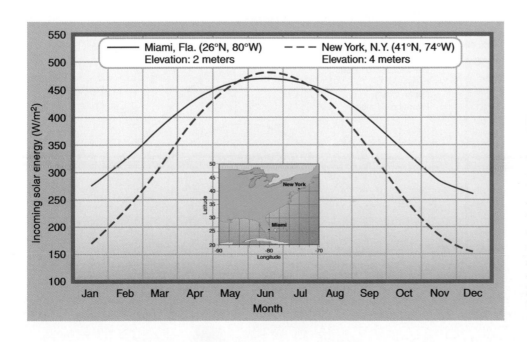

■ Figure 3.5

The incoming solar energy at the top of the atmosphere over Miami, Florida, and New York City. New York receives more solar energy than Miami near the summer solstice because the length of day is greater at higher latitudes in summer. The greater variation of incoming solar energy over New York explains the larger amplitude in the annual cycle of temperature (see Figure 3.4).

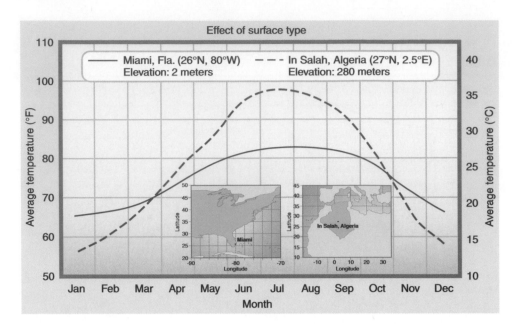

Figure 3.6

Comparison of the annual temperature cycle of In Salah, Algeria in the Sahara Desert and the annual cycle at Miami, Florida, which is at approximately the same latitude. Note the large annual temperature range in the desert.

Surface Type

As discussed in Chapter 2, the surface of the Earth absorbs approximately 50% of the solar energy incident at the top of the atmosphere. So, the surface contains heat that can be transferred to the atmosphere. Because the atmosphere is heated by the Earth's surface, the surface type plays an important role in determining the surface air temperature.

Deserts such as the Sahara in northern Africa have a large annual range in temperature compared to coastal locations such as Miami (Figure 3.6). Why is this? Deserts gain large amounts of solar energy because of the persistent clear-sky conditions over the desert. But dry sand is a poor conductor of heat and has a low specific heat. Therefore, the dry sand at the surface rapidly heats up during the day and cools down during the evening. Since the surface and the atmosphere transfer energy to each other, the surface air temperature also has a large annual range.

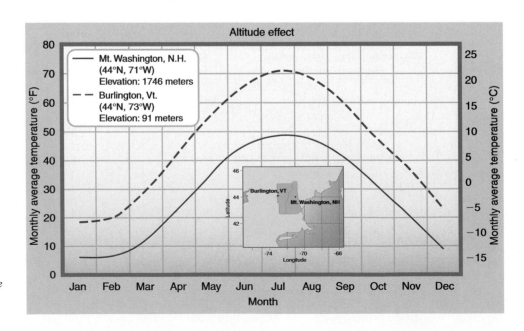

Figure 3.7

Comparison of the annual temperature cycle of Burlington, Vermont, and Mt. Washington, New Hampshire, which is at approximately the same latitude but at a much higher altitude than Burlington.

On land surfaces, vegetation also modifies the annual range in temperature. Vegetation reduces the temperature range in several ways. Plants transpire and use some of the solar energy that reaches the surface. Evaporation in the vicinity of plants takes in energy that would otherwise go into the raising of temperature. This prevents vegetated surfaces from having the high surface temperatures observed over the bare, dry soil of deserts.

Elevation and Aspect

When we consider energy exchanges over mountains and hills, the altitude and the direction a slope faces are important. The effect of altitude is demonstrated by comparing the annual temperature cycle of Burlington, Vermont (elevation 92 meters [300 feet]), and Mount Washington, New Hampshire (elevation 1748 meters [5727 feet]) (Figure 3.7). The two locations have similar latitudes but very different altitudes. The higher elevation station, Mount Washington, is on average colder than Burlington in every month of the year. At the higher elevation of Mount Washington, the air is less dense, and there are fewer molecules to absorb incoming solar radiation. In addition, terrestrial radiation emitted by the surface can more easily escape to space and therefore does not heat the atmosphere. Mount Washington is also much windier than Burlington (see Box 12.2 in Chapter 12). The high winds and associated turbulence rapidly carry energy away from the surface and mix the energy throughout the lower troposphere.

The **aspect** is also an important influence on the energy budget of a region, particularly the solar-energy side of the ledger. Aspect is the direction that a mountain slope faces. In the northern hemisphere, under cloudless skies, a north-facing slope receives less solar energy than a south-facing slope (Figure 3.8). Because south-facing slopes receive more solar energy, they are warmer. South-facing slopes are also generally drier. More solar energy results in increased evaporation and reduced moisture in the soil.

The effect of aspect can be seen by comparing the vegetation type of south-facing slopes versus the type of vegetation growing on north-facing slopes (Figure 3.9). In many regions of the western United States where the amount of vegetation depends heavily on available moisture, only sparse vegetation grows on south-facing

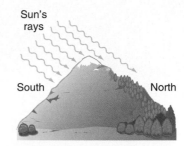

Figure 3.8

Schematic illustrating that in the Northern Hemisphere the Sun's rays are more intense on a south-facing slope than one facing north. As a result, temperature, moisture levels, and vegetation types can vary widely from one side of a mountain to another.

Figure 3.9

Photograph confirming the reality of the effect shown schematically in Figure 3.8. More trees are found on the north-facing slope of this mountain than on the south-facing slope.

slopes while plants grow densely on the moister north-facing slopes. These differences in vegetation further enhance temperature differences between the north- and south-facing slopes.

Since the elevation and aspect of the land affect temperatures and moisture of the ground, they also influence the economic planning of a region. Ski resorts are built on cooler north-facing slopes where melting and evaporation are less and snow cover persists longer. Vineyards and apple growers in New York State plant their fruit on south-facing slopes where the growing season is longer.

Effects of Large Bodies of Water

The seasonal temperature cycle of a city is related to its proximity to a body of water. The annual temperature cycles of Dallas and Los Angeles demonstrate this, as shown in Figure 3.10. These two large cities are at approximately the same latitude, and therefore have about the same amount of solar energy entering at the top of the atmosphere. Los Angeles is located on the shore of the Pacific Ocean whereas Dallas is inland, far from a large body of water. The two cities' seasonal temperature cycles are distinctly different. Dallas' annual temperature range is 22° C (39° F) while Los Angeles' is only 8° C (15° F). Los Angeles' monthly mean temperatures are modified because of the nearby Pacific Ocean. The summertime maximum temperatures are cooler and the winters are warmer than in Dallas.

Energy exchanges with the Earth's surface strongly influence the surface air temperature. Large water bodies act to thermally stabilize the temperature of the surrounding air so that the differences between months are reduced. This causes the lags in temperature discussed earlier. The factors that contribute to temperature differences between continental and maritime regions are:

1. The specific heat of water is almost three times greater than land. More heat is therefore required to raise the temperature of water. Water also cools down more slowly than land.
2. Evaporation of water reduces temperature extremes over and near lakes and oceans.
3. Solar radiation absorbed by water is distributed throughout a large depth of the water body due to mixing and the transparency of water to solar radiation. Over land, the solar radiation is absorbed by the surface, and heat can quickly be transferred to the atmosphere above it.

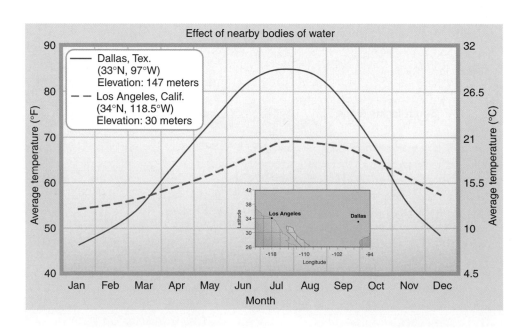

Figure 3.10

A comparison of the annual temperature cycles of Dallas, Texas, and Los Angeles, California. Despite the similarity in latitude, the two cities' temperature cycles are quite different, owing to the effect the Pacific Ocean has on Los Angeles weather.

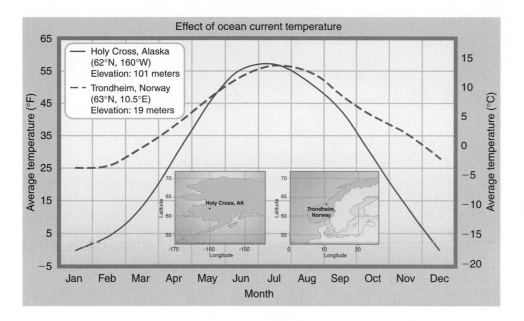

Figure 3.11

The annual temperature cycles in Trondheim, Norway, and Holy Cross, Alaska. The warmer winter in Trondheim is caused by its proximity to a nearby warm ocean current.

The temperature of a nearby body of water also plays an important role in modifying a region's temperature. Consider the cities of Holy Cross, Alaska, and Trondheim, Norway (Figure 3.11). Both cities are at similar latitudes and are near large bodies of water. Trondheim is located near the Norway coast and is influenced by the warm Gulf Stream ocean current. As air moves over the Gulf Stream on its trek to Norway, sensible heating warms it and water evaporates into it, causing further warming when the vapor condenses in clouds.

Holy Cross is not so lucky. Its nearby body of water is the frigid, sometimes frozen Bering Sea. Air advecting over ice gains very little heat, and water cannot evaporate into the air. The two locations have similar summertime temperatures, as Figure 3.11 reveals. But Trondheim's winters, unlike Holy Cross's, are made milder by the influence of the relatively warm ocean waters nearby. Because of this crucial difference, Holy Cross's annual temperature range is approximately 15° C (25° F) more than Trondheim's.

Cloud Cover

Clouds have a large impact on the solar and terrestrial energy gains near the surface. Clouds reflect and absorb solar energy. They reduce the amount of solar radiation reaching the surface, and cause daytime cooling. The thicker the cloud, the more pronounced the cooling. But clouds also have a warming effect, because they emit longwave radiation toward the surface (Figure 3.12). This warming effect can be very pronounced at night, keeping minimum temperatures higher than in clear-sky regions.

Consider the annual temperature cycles of Grand Rapids, Michigan, and Madison, Wisconsin (Figure 3.13). Winds typically blow from west to east across this region in winter. Air blowing eastward over Madison arrives from the dry Great Plains. In contrast, air blowing eastward over Grand Rapids has passed across Lake Michigan and becomes moister as a result. This leads to chronic "lake-effect" clouds in Grand Rapids, but not Madison, from late fall into early spring. The strong warming effect of clouds at night causes the mean temperature in Grand Rapids to be as much as 5° F warmer than in Madison in January. However, in the summertime the winds are more often from the south and cloudiness at the two locations is similar. Consequently, Madison and Grand Rapids temperatures in July are nearly identical.

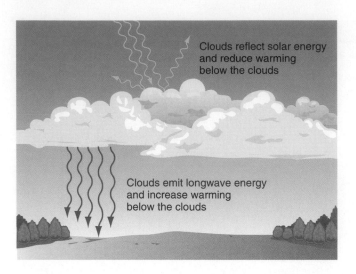

Figure 3.12

The effect of clouds on the daytime energy budget of the surface. During the day, clouds reflect solar radiation back to space and reduce energy gains of the air below and the surface. Clouds emit longwave radiation and inhibit cooling of the air below.

■ Interannual Temperature Variations

The previous discussion on annual temperature variations was based on temperatures averaged over several decades. These climatological temperatures are computed by averaging all temperatures over a 30-year period. These averages are called **normal temperatures.**

We know from personal experience, however, that the temperature can be abnormal. Some winters are colder than others, and some summers are hotter than others. *Interannual temperature variations* are temperature changes from one year to the next. Sometimes these changes are subtle and the causes unknown, while at other times the changes are dramatic and the causes are known.

Meteorologists study interannual temperature variations by plotting the departure from the climatological normal temperatures by year. Departures for a given year are found by subtracting the normal value from that year's mean value. These departures are called **anomalies.**

Figure 3.13

The annual temperature cycles of Grand Rapids, Michigan, and Madison, Wisconsin. Wintertime cloudiness in Grand Rapids (indicated by low percentage of possible sunshine in bar graph at bottom) keeps it warmer there than in Madison.

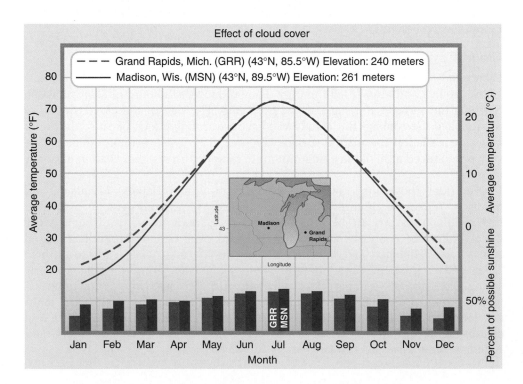

Figure 3.14 plots the surface temperature anomalies over the world's land-masses (where most observations are made) for the years 1880 to 1999. The 30-year "normals" from 1951 to 1980 are used as the reference period in Figure 3.14, which explains why the anomalies are negative before that period and positive after it. The thin line in Figure 3.14 represents the anomalies of individual years. It wiggles up and down, indicating year-to-year changes. The thick line reveals longer-term trends by averaging these yearly anomalies over five years, which smooths out the sharp bumps but leaves untouched any trends longer than a few years.

What does Figure 3.14 tell us? In general, temperatures across the globe increased from 1880 until about 1940, after which the temperature trend reveals a slight decrease over the next two decades. The global temperature has steadily increased since 1970, with the 1990s being the warmest decade on record. Over the last 120 years the average global surface temperature has increased approximately 0.6° C (1° F).

Global temperature changes of 0.6° C near the surface do not sound very significant. To help put this change in perspective, consider that temperatures during the latest ice age (20,000 years ago) were about 5° C colder than today's temperatures. So on a global scale, the change of a degree or so is big news—it is a sizable fraction of the difference between today's climate and a harsher climate for which we are not adapted.

It is important to note that changes at particular geographic locations may be greater or smaller than the global mean, and year-to-year temperatures may vary

The Mount St. Helens Eruption: A Satellite View

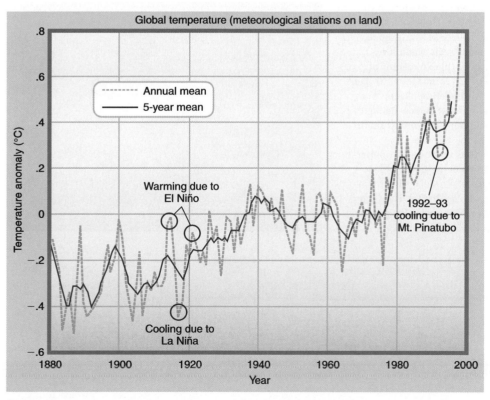

■ Figure 3.14

Temperature departures from the global mean temperature over land during the last 120 years. The upward trend is evidence that the average global temperature is increasing. Whether this trend is natural or a result of human activity is under debate. Note the shorter-term effects of the ocean temperature phenomena El Niño and La Niña, which have periodically affected global temperatures throughout this period. The episodic cooling effect of the Mt. Pinatubo volcanic explosion is also indicated.

considerably. For example, notice in Figure 3.14 that the temperatures of 1992 and 1993 are much cooler than the years before or after. These cool temperatures are the direct result of the eruption of a single volcano, Mt. Pinatubo in the Philippines, on June 15, 1991. This eruption provided the first direct observational evidence that strong volcanic eruptions tend to cool the Earth (Box 3.1). During a violent eruption, huge quantities of ash, dust, and sulfur dioxide are ejected into the stratosphere (Figure 3.15, A). Volcanic ash falls back to Earth quickly, but the other debris stays in the stratosphere for a couple of years. There, its presence modifies the energy balance of the planet. The sulfur injected into the stratosphere by the volcanic eruption is chemically transformed into small droplets of sulfuric acid (Figure 3.15, B). These tiny drops absorb solar radiation and, more importantly, increase the amount of solar energy reflected back to space. Less solar energy is transmitted to the surface, resulting in a cooling of the air near the surface.

The oceanic temperature phenomena known as *El Niño* and *La Niña* (to be covered in detail in Chapter 8) can also alter the world's temperatures, even over

BOX 3.1

Volcanoes and Temperature

In this chapter we discuss the possible linkages between volcanoes and temperature. The most famous United States volcano of the 20th century, Mount St. Helens in Washington, provided proof of this linkage on a small scale when it exploded on the morning of May 18, 1980 (blob in center of satellite image at right). The cloud of ash from Mount St. Helens spread rapidly eastward and enshrouded Spokane, Washington for days. However, the ash cloud was narrow enough that Boise, Idaho, less than 450 kilometers (280 miles) south of Spokane, escaped the influence of the volcano.

The effect of the Mount St. Helens eruption on temperature at Spokane versus Boise is shown in the figure below. The volcanic ash flattened out the usual diurnal cycle of temperature at Spokane as compared to Boise. In fact, right after the arrival of the ash cloud at Spokane (arrow on the graph) the temperature actually *decreased* during the afternoon!

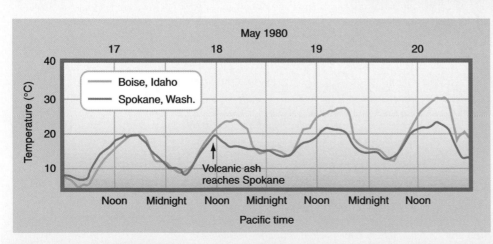

Adapted from A. Robock and C. Mass, "The Mount St. Helens Volcanic Eruption of 18 May 1980: Large Short-Term Surface Temperature Effects," *Science*, 216, pp. 628–630. Copyright © 1982 American Association for the Advancement of Science.

land. El Niño causes unusually warm waters over the eastern Pacific Ocean and, because the ocean affects the atmosphere, leads to warming temperatures across a sizable portion of the globe. La Niña cools the Pacific and other parts of the globe. During the period from 1916 to 1921, a La Niña pattern was sandwiched between two potent El Niño events, leading to the sharp zig-zag in global temperatures during that period. The rapid climb in temperatures during the 1990s may also be related to the persistent El Niños during that decade.

In summary, the interannual temperature pattern of the Earth appears to be a persistent upward trend, punctuated by interludes of less warm periods. What can explain this upward trend? Many scientists attribute it to increased atmospheric concentrations of greenhouse gases, such as carbon dioxide, CFCs, and methane. In Chapter 2, we briefly discussed how increasing the concentrations of these gases could lead to a warming trend. Before revisiting this problem, we must discuss the relationship between air temperature and water vapor content, something we will do in the next chapter.

BOX 3.1 (continued)

This strange turn of events can be explained in terms of an energy budget. The amount of solar energy reaching the surface at Spokane was greatly reduced on the day of the explosion and significantly reduced for several days afterward. This kept solar energy gains low for several days. Meanwhile, outgoing terrestrial radiation remained relatively high. The net result of gains minus losses was a small surplus on the days after the explosion, and a deficit on the day of the explosion. As a result, the temperature at Spokane rose slowly or not at all, even during the daytime.

The debris from Mount St. Helens remained mostly in the troposphere and had little global effect. However, much larger volcanoes can eject material into the stratosphere, where it can reflect solar radiation for years. As a result, the Earth's temperature can experience a pronounced, if temporary, cooling—just like Spokane, but on a global scale and for a few years, not just a few days.

Historical evidence supports the premise that large volcanic eruptions cool the Earth. Between 1812 and 1817 there were three major volcanic eruptions. Soufriere on St. Vincent Island in the eastern Caribbean Sea erupted in 1812; Mayon in the Philippines in 1812; and Tambora (the largest) in Sumbawa, Indonesia in April 1815.

Debris from the eruption of Mt. Tambora (8° south latitude, 118° east longitude) took one year to spread globally. The following year, 1816, is known as the "year without a summer" in eastern North America. While extensive meteorological observations did not exist at this time, people's diaries and weather journals documented the cold weather of the summer of 1816.

In New England snow fell in June and frost occurred in July and August. Even though late frost killed a large number of crops, the entire summer was not below freezing. Indeed, on June 5, the day before the snowfall, the temperatures in Vermont were in the low 30's Celsius (upper 80s Fahrenheit)!

After the early June cold spell in New England, farmers, hoping for a good crop, replanted their crops as temperatures returned to normal. Another cold spell hit in early July, bringing freezing temperatures to the area. Harvests were poor that year and resulted in severe food shortages in parts of New England. The poor harvest had an economic impact throughout the United States. In Philadelphia in May 1817 a bushel of corn cost twice as much as it had in April 1816.

Weather in Europe and other regions of the globe also went haywire in 1816. In Europe, the cold and wet weather contributed to a disastrous harvest as crops rotted in the field. Famine, food riots, grain hoarding, and government embargoes followed. The cold, moist weather patterns may have contributed to the typhus epidemic of 1816 to 1819 in Europe that killed approximately 200,000 people. At about the same time, a cholera outbreak originated in Bengal, India and spread throughout the world.

Bad "volcano weather" also had an unexpected and lasting impact on the world of literature and the movies. In the summer of 1816 a small group of friends gathered in Switzerland, where the cold wet summer prevented outdoor activities. One of the friends proposed a ghost story contest, a fitting match to the dreary weather. Storytelling came to mind naturally, because two of the friends—Lord Byron and Percy Bysshe Shelley—were already among the greats of British Romantic poetry. However, their ghost stories are lost to history. Shelley's nineteen-year-old wife, Mary, bested them all with a tale that came to her in a dream during that year without a summer. In 1818, Mary Shelley's ghost story was published, giving birth to a monster as chilling as the volcano that helped create it: *Frankenstein.*

■ Figure 3.15

A, *A schematic illustrating how volcanic explosions affect solar radiation.* **B,** *A photograph showing the volcanic aerosol layer due to Mt. Pinatubo. Astronauts on the space shuttle Atlantis took this photograph during mission STS-43, about three weeks after the eruption of Mt. Pinatubo. The volcanic aerosol is seen in the stratosphere (the thunderstorm clouds over South America are below the aerosol layer). The volcanic debris was a 20-megaton cloud of sulfur dioxide that reduced the amount of solar energy gains of the planet and resulted in a global cooling in 1992 and 1993 (see Figure 3.14).*

■ Diurnal Temperature Cycle

Closer to home than the question of global warming is the daily cycle of local temperature. If you made temperature measurements every hour of a given day for many years, and averaged the measurements, a temperature cycle would emerge that shows a regular pattern of change according to the time of day. This is the diurnal temperature cycle that is driven by the daily changes in the energy budget near the ground. This variation is driven by the Earth's rotation.

The daily variation of air temperature near the ground for a cloud-free day is demonstrated in Figure 3.16. We have learned that temperature changes are driven by the difference in incoming solar radiation gains versus outgoing terrestrial energy losses. Therefore, these quantities are also plotted in Figure 3.16.

Figure 3.16 summarizes the following story. After sunrise, the ground warms as a result of absorption of solar energy. As the ground warms it transfers heat to the atmosphere. Cool air emits less energy than warm air, as we learned in Chapter 2. The rising Sun adds more energy to the air than the air is emitting, and therefore the air temperature increases all morning long. Incoming solar energy peaks at noon, as does the Sun's trajectory in the sky. After noon, the solar energy gains are reduced, but the energy gains are still greater than the energy losses and so the surface temperature continues to increase. The energy losses usually exceed the energy gains by 4:00 PM, when the maximum daily temperature is reached. This is when the maximum amount of outgoing energy is emitted. The energy losses exceed gains all night long, because the source of gains is the Sun and it is below the horizon. Therefore, the temperature decreases all night long. Temperatures reach a minimum around sunrise, and then the cycle starts all over again.

The factors affecting the diurnal temperature changes during a cycle are the same as those that determine the annual temperature cycle: latitude, surface type, elevation and aspect, relationship to large bodies of water, and cloud cover. To review their effect and to see how they impact daily as opposed to monthly, temperatures, let's go over them one by one:

1. *Latitude:* How much solar energy a region gains plays an important role in establishing the daily variation in the region's energy budget. Latitude, in turn, de-

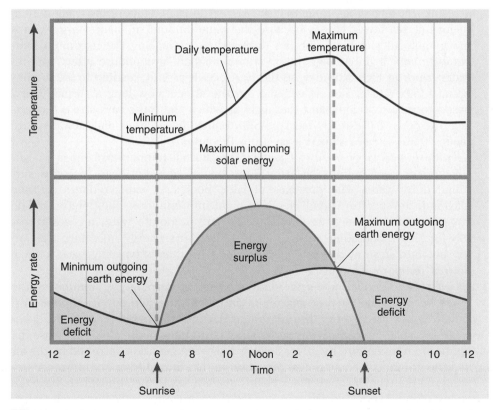

■ **Figure 3.16**

The diurnal variation in temperature is primarily controlled by energy gains from the Sun and energy losses caused by emission of infrared radiation. Incoming solar energy is confined to the daytime. The amount of outgoing infrared radiation is related directly to the temperature (see Chapter 2), as indicated by the green dashed lines. Temperature increases when the energy gains exceed the energy losses (orange shading), and the temperature decreases when the net energy is negative (purple shading).

termines the intensity of the Sun's rays and number of daylight hours. In equatorial regions the position of the Sun in the sky changes dramatically throughout the course of a day, from below the horizon to nearly overhead. Solar energy gains therefore vary greatly, resulting in large variations in air temperature during the day. Equatorial regions, in general, have a greater daytime variation of temperature than polar regions because of the larger variations in solar zenith angle in equatorial regions.

2. *Surface type:* How much the ground, and therefore the air near the ground, warms will depend on the condition and composition of the surface. Solar energy falling on bare, dry, sandy soil is absorbed within a thin layer near the surface. Since this type of soil has a low specific heat and because the solar energy is all absorbed in a thin layer, the surface quickly warms. The heated surface then transfers its energy to the air via convection and conduction, warming the air. During the night, the soil surface cools down quickly as it loses energy in the form of terrestrial radiation. For all these reasons, desert regions experience a large diurnal temperature range. When vegetation is present, some of the solar radiation is converted to chemical energy by photosynthesis, some heats the plant tissue and never reaches the surface, and some is used to evaporate water from the plant (transpiration). The result is that the maximum temperature is less over the vegetative field than over bare soil.

3. *Surface elevation and aspect:* Air temperatures are normally warmer at lower elevations. At lower elevations, greater amounts of water vapor absorb the terrestrial radiation emitted by the ground, thus warming the air. By comparison, high elevation regions have less water vapor, and less atmosphere. As a result, these regions tend to have larger diurnal temperature ranges than regions at sea level. This is because the same amount of solar energy warms fewer molecules at high altitudes during the day, making the maximum temperature very high. At night there is only a small "greenhouse effect" due to water vapor at high altitudes, so the Earth cools rapidly, leading to a low minimum temperature. The aspect of a surface affects how long a surface is exposed to direct sunlight and the angle at which the Sun's rays strike the surface. Regions that directly face the Sun tend to have larger daily temperature ranges than surfaces sloped away from the Sun.

4. *Relationship to large bodies of water:* The diurnal temperature range is usually greater over regions far from large bodies of water than locations surrounded by water. The presence of large bodies of water reduces daytime maximum temperatures and increases nighttime minimum temperatures. The large amounts of energy required to heat and evaporate water are a key reason for this moderating effect; much of the Sun's energy goes into the water, not into heating the air. This explains why islands typically have a small diurnal temperature range.

5. *Cloud cover:* Clouds reduce the diurnal temperature range by minimizing the range between the daytime maximum and nighttime minimum temperatures. When clouds are present, the maximum daytime temperature is reduced because less solar energy reaches the ground. During the night, clouds increase the minimum temperature by increasing the longwave radiation absorbed by the air and the ground. The amount of reduction in the diurnal temperature range depends on the type of cloud. Thick clouds that are low in the atmosphere have the largest effect. Low clouds are warmer and therefore emit more terrestrial energy than high clouds. Thick clouds also let less solar energy pass through them. High, thin clouds have the smallest impact on an otherwise clear-sky diurnal temperature range.

Now that we understand the controls of daily temperature, we can appreciate the rare combination of circumstances that lead to record high and low temperatures, which are explored in Box 3.2.

BOX 3.2

Record Cold and Record Heat Across the United States

Records fascinate meteorologists as much as sports fans. And when temperatures approach all-time records, even the least weather-conscious person becomes a "weather junkie." Let's examine the patterns of record high and low temperatures across the fifty United States to see what we can learn about temperature from its extremes.

The figures that follow show the location and thermometer reading for the state record minimum and maximum temperatures. The first thing to notice is that record lows tend to occur in mountainous regions. Record highs are more likely in low-lying areas. Thirty of the record lows were observed at or above 1000 feet of elevation, while 37 record highs occurred at or below the 1000-foot mark. This is a natural consequence of the adiabatic cooling of air that is above sea level. Mountains thus have a built-in advantage when it comes to cold temperatures. In addition, the relative lack of "greenhouse warming" in the thin dry atmosphere at high altitudes promotes the rapid loss of energy that must characterize a record-cold night.

Other interesting differences arise in the numerical and geographical spread of the state records. Every state has a record high of 100° F or higher, but the highest all-time U.S. temperature is 134° F, yielding a range of just 34° F. There is also no obvious dependence of the record highs on latitude. In contrast, the difference between the "warmest" and coldest state record low temperatures is a whopping 92° F. Latitude strongly affects record lows, with the values steadily dropping from south to north (e.g., from Texas to North Dakota).

Still other oddities arise from a closer look at the records. Utah has the largest spread between its record high and its record low, an amazing 186° F. Remarkably, both of Utah's state temperature records were set in the same year, 1985. East of the Mississippi River, the state with the largest range in record temperatures is, surprisingly, Wisconsin at 169° F. The effect of "continentality"—hot summers and bitter winters far away from the moderating effects of an ocean—gives Wisconsin a more extreme climate than states with high mountains, such as New Hampshire or North Carolina.

When do records occur in a given year? Almost half of all state-record lows across the United States cluster in two time periods: January 17 to 23 and February 8 to 14.

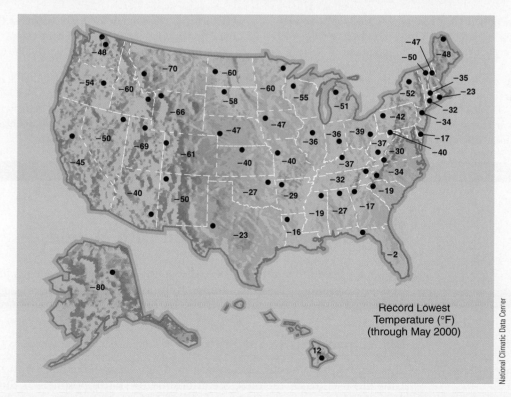

Record Lowest
Temperature (°F)
(through May 2000)

National Climatic Data Center

(continued)

BOX 3.2
(continued)

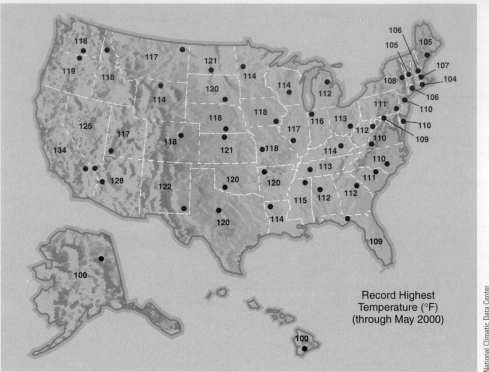

Record Highest
Temperature (°F)
(through May 2000)

National Climatic Data Center

Both of these periods "lag" the winter solstice by several weeks. This is because the Earth-atmosphere system does not respond instantaneously to the minimum of solar heating. Instead, it takes weeks of longwave cooling over heavy snow cover to create the kind of bitter-cold air that is required for a record low.

What this explanation does not account for is the strange lull in records in late January. This gap may be a manifestation of the "January thaw," a period of warmer-than-usual weather that is observed in some winters across the eastern United States at around the same time. However, the small number of state record lows is not enough information upon which to base a conclusion.

Record highs are much more spread out across the summer months, but the period of July 10 to 15 is home to fifteen state records. As with record lows, a lag of a few weeks after the solstice is required for the Earth–atmosphere system to "catch up" with the cycle in solar heating.

Temperature Variation with Height

So far we have limited ourselves to discussing temperature changes just 1.5 meters above the ground. However, temperature also varies with height, and these variations can be closely related to surface conditions.

Lapse Rates and Stability

Introduction to Stability

In Chapter 1 we briefly discussed how temperature variations with height or altitude are used to characterize the atmosphere. The troposphere is the atmospheric layer where, on average, the temperature decreases with increasing altitude at a rate of 6.5° C per kilometer (3.6° F per 1000 feet). This is the average **lapse rate;** on any given day the lapse rate will vary. The specific change of temperature with alti-

BOX 3.2
(continued)

Hawaii is an anomaly. Its record high is in April, but its record low was recorded later in the year, in May. We can explain this by recalling what we have learned about solar zenith angles. Hawaii is located just south of the Tropic of Cancer. As a result, it receives the most direct sunlight in spring and in fall, not in summer. A spring record high is therefore logical. Even in winter the Hawaiian sun is high and the Pacific Ocean keeps temperatures moderate, making a May record low possible.

A fear concerning global warming is that it will lead to extreme record high temperatures unlike anything our planet has experienced. Does this trend show up in state record temperatures? Below is a graphic showing all state record temperatures by decade:

This graphic tells a surprising story: the decade of extreme heat was not the 1990s, it was the 1930s! Seventy percent of all U.S. state-record highs were set in the 1930s or earlier, despite the overall trend since then toward warmer *mean* temperatures worldwide (Figure 3.14). The exceptional weather of the 1930s led to the catastrophic "Dust Bowl" of the Texas and Oklahoma Panhandles, which we discuss in later chapters.

This graphic also reveals that a decade with many record high temperatures usually has a lot of record low temperatures, too. For example, there were more state record lows than highs set during the 1980s. There does seem to be a trend toward more record highs in recent decades. But U.S. state-record temperatures do not yet substantiate the fears of extreme heat due to global warming.

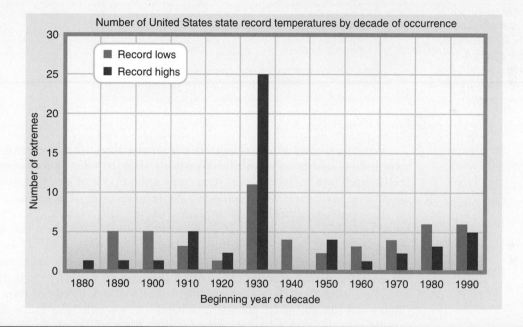

tude at any particular time and location is referred to as the **environmental lapse rate.** The environmental lapse rate changes from day to day and from hour to hour. It can be measured by attaching a thermometer to a helium-filled balloon. The temperature measurements are radioed down to the surface and recorded. Vertical temperature measurements of the environmental lapse rate occur twice a day at many locations throughout the world.

In Chapter 2 we discussed how the temperature of an air parcel changes as it is lifted. When lifted, a dry parcel cools at the dry adiabatic lapse rate, 10° C per kilometer. If the environmental lapse rate is greater than 10° C per kilometer, a parcel that is forced to ascend will cool more slowly than its surroundings are cooling. In other words, the parcel will become warmer than its surroundings. A parcel that is warmer than its surroundings is also less dense than the air around it (recall Box 1.2). Because it is less dense, it will keep rising, like a hot air balloon.

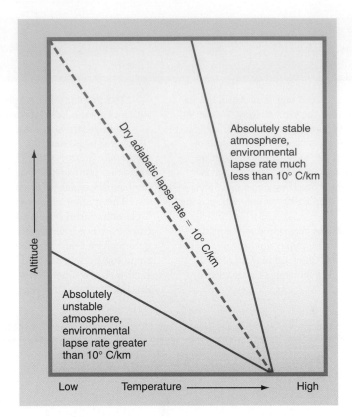

Figure 3.17

The vertical temperature profile in an absolutely unstable atmosphere tilts more to the left than the dry adiabatic lapse rate profile (dashed line). This means that in an absolutely unstable atmosphere, the temperature decreases with altitude at a higher rate than 10° C per kilometer. In a stable atmosphere, the temperature profile tilts less to the left than does the dry adiabatic lapse rate.

When the environmental lapse rate is greater than 10° C per kilometer, the atmosphere is said to be **absolutely unstable** (Figure 3.17). An absolutely unstable atmosphere is very favorable for strong upward motions of air, because if a parcel of air is lifted upwards it will accelerate away from its original position.

By contrast, a **stable** atmosphere inhibits the vertical movements of air parcels. Stable atmospheres occur in completely dry air when the environmental lapse rate is less than the dry adiabatic lapse rate of 10° C per kilometer. The environmental lapse rate tells us the rate at which the atmosphere is cooling with increasing altitude. Air parcels that rise in a stable atmosphere rapidly become cooler than their surroundings, and sink back to where they came from. (When we discuss moist parcels in Chapter 4, we will have to modify this definition somewhat to take into account the effect of latent heating on the parcel's lapse rate.)

In the stratosphere the temperature does not decrease, but instead increases with altitude. In these regions the environmental lapse rate is a negative value. This is the reverse, or the *inverse,* of what we citizens of the troposphere normally expect. Accordingly, regions of the atmosphere in which the temperature increases with altitude are called **temperature inversions.**

A temperature inversion is an extreme case of a stable atmosphere. When a temperature inversion is present, parcels that are displaced from their position will soon become much cooler than their surroundings and will quickly sink back to their original positions.

Temperature Inversions Near the Ground

The range of diurnal and annual cycles in temperature depends on altitude, the distance from the ground. Figure 3.18 shows the temperature measured at various times on a cloud-free day in summer. We start at 3:00 PM (Figure 3.18, *A*). At this time the temperature is highest close to the ground, where solar heating has raised

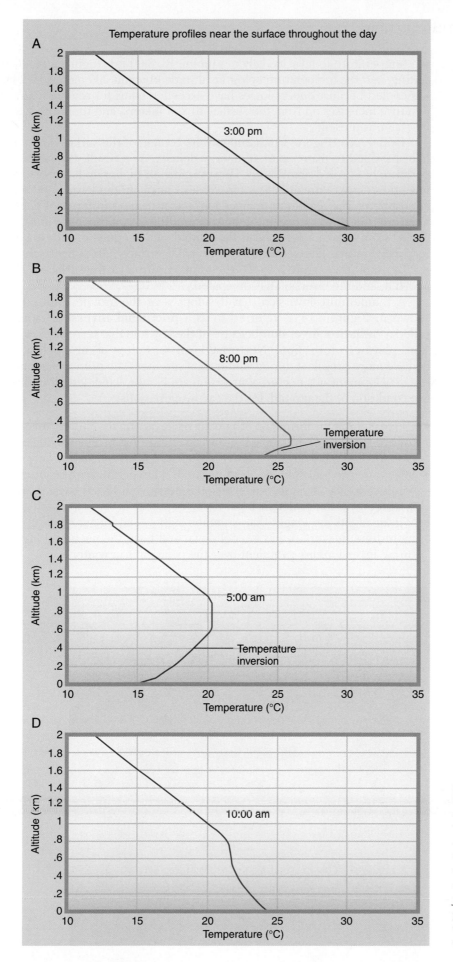

Figure 3.18

Life cycle of a temperature inversion. **A,** *Mid-afternoon.* **B,** *Evening.* **C,** *Sunrise.* **D,** *Mid-morning. Overnight cooling at the surface due to longwave radiation emissions causes a surface temperature inversion to be present in* **B** *and* **C.** *In an inversion, temperature increases with altitude.*

Blue Skies

**Atmospheric Basics/Layers
of the Atmosphere**

the temperature to near its daily maximum. By 8:00 PM (Figure 3.18, *B*) the Sun is setting, and the temperature below an altitude of about 200 meters (0.2 kilometers) has cooled because energy losses have exceeded gains for several hours. The temperature increases up from the surface to 200 meters, and therefore a temperature inversion is present.

By 5:00 AM (Figure 3.18, *C*) nighttime cooling of the surface and a lack of sunlight have led to many hours of energy deficits. As a result, the near-surface temperature has cooled significantly. However, the air higher up has cooled less. Therefore, the temperature at 5:00 AM increases sharply in the lowest 600 meters of the atmosphere. The temperature inversion is near its peak in terms of depth and temperature difference between the bottom and the top of the inversion.

At sunrise, solar energy heats the ground, and conduction and convection transfer heat upward. This warms the air near the surface more effectively than higher up. By 10:00 AM (Figure 3.18, *D*) the temperature near the surface is warmer than above it, and the inversion has dissipated.

A temperature inversion that develops near the ground during the night, as shown in Figure 3.18, is referred to as a **nocturnal inversion** (sometimes also called a **radiation inversion** because of the key role that terrestrial radiation emissions play in its formation). Nocturnal inversions often occur on clear, calm nights and are more prevalent during winter than summer. The primary factors controlling the development of a nocturnal inversion are clouds, wind, length of the night, and the condition of the ground, for the following reasons:

- Clouds inhibit the formation of a nocturnal inversion by emitting terrestrial energy toward the surface, reducing the surface energy losses. The cooling of the ground is suppressed, and development of the surface nocturnal inversion is inhibited.
- Winds are the mixers of the atmosphere. If a temperature inversion exists and the winds suddenly increase, the inversion is destroyed. This is because warm air aloft is mixed with the cooler air below.
- Winter nights are longer than summer nights. The longer the Sun is down, the more time the surface has to cool down and form the nocturnal inversion. Nocturnal inversions are prevalent over the polar caps in winter.

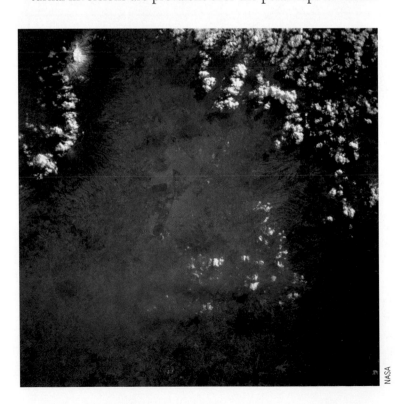

■ Figure 3.19

Photograph taken from the space shuttle of pollution (fuzzy gray areas in center) over Mexico City. This pollution is often trapped by a temperature inversion over the city, which is located at a high altitude (2239 meters, or 7347 feet) valley ringed by towering mountains as high as 5,452 meters (17,887 feet). The mountains are seen at left and right in the photograph.

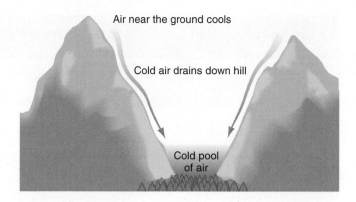

Air near the ground cools

Cold air drains down hill

Cold pool
of air

■ **Figure 3.20**

Cold air drains down the slope into the valley, where pollution can get trapped near the ground due to a temperature inversion near the surface.

- The condition of the ground is another variable to consider when forecasting nocturnal inversions. For example, snow is a very good emitter of terrestrial energy, but, because of the air trapped within it, is a poor conductor. The top of the snow surface cools rapidly but does not warm from below. As a result, the air just above a deep snow cover cools rapidly, a condition favorable for a radiation inversion and for record cold temperatures.

Why are temperature inversions important? As noted above, inversions suppress the upward movement of air. The stratosphere keeps weather below it, in the troposphere. A nocturnal inversion keeps parcels of air from rising very high before they sink back down. Severe-weather experts sometimes refer to a temperature inversion as a "lid," since it is difficult for air to move vertically through it. Episodes of dense air pollution are often associated with temperature inversions. The pollutants are released near the surface but cannot escape because vertical mixing is reduced by the existence of the inversion (Figure 3.19). We explore the implications of inversions for severe weather in Chapter 11 and for pollution in Chapter 10.

Temperature inversions also often develop in valleys. Consider the conditions of a valley on a clear calm night in October. During the evening the air begins to cool by radiation losses. Air that is colder than its environment sinks. The coldest air drains down the hill, settling at the bottom of the valley (Figure 3.20). By late evening, the air at the bottom of the valley is often much colder than the air above it.

The record low temperature for the state of Utah (see Box 3.2) was set in 1985 in a bowl-shaped 2.4 kilometer (8000 foot) high valley descriptively named Peters Sink. Bitterly cold air over snow cover at high altitude sank into the valley and caused the temperature to plummet to an astounding −56° C (−69° F), only one degree shy of the all-time record low for the lower 48 United States.

▌ Wind-Chill Temperature

Wind-Chill Temperature

On a cold, windy day, you try to keep yourself warm by wearing appropriate clothing or seeking shelter from the wind. It feels colder in the wind because the wind sweeps away heated air in contact with your body and replaces it with colder air. While still air is a poor conductor, moving air is not! The cooling power of the wind is measured by the **wind-chill** factor. *The wind-chill describes the increased loss of heat by the movement of the air.* The wind-chill is relevant to humans and other animals that need to maintain a constant temperature that is higher than their surroundings.

The wind-chill factor cannot be measured with a thermometer; it must be computed. The **wind-chill equivalent temperature,** expressed in degrees, translates your body's heat losses under the current temperature and wind conditions into the air temperature with a 3-knot wind that would produce equivalent heat losses. This is not an easy conversion! The original wind-chill formula was devised by Antarctic explorer Paul Siple in 1945 and has been used ever since.

TABLE 3.1   The Improved Wind-Chill Equivalent Temperature Chart Adopted by the National Weather Service in November, 2001

Surface Temperature (°F)	Wind Speed (mph)							
	5	10	15	20	25	30	35	40
40	36	34	32	30	29	28	28	27
35	31	27	25	24	23	22	21	20
30	25	21	19	17	16	15	14	13
25	19	15	13	11	9	8	7	6
20	13	9	6	4	3	1	0	−1
15	7	3	0	−2	−4	−5	−7	−8
10	1	−4	−7	−9	−11	−12	−14	−15
5	−5	−10	−13	−15	−17	−19	−21	−22
0	−11	−16	−19	−22	−24	−26	−27	−29
−5	−16	−22	−26	−29	−31	−33	−34	−36
−10	−22	−28	−32	−35	−37	−39	−41	−43
−15	−28	−35	−39	−42	−44	−46	−48	−50
−20	−34	−41	−45	−48	−51	−53	−55	−57
−25	−40	−47	−51	−55	−58	−60	−62	−64
−30	−46	−53	−58	−61	−64	−67	−69	−71
−35	−52	−59	−64	−68	−71	−73	−76	−78
−40	−57	−66	−71	−74	−78	−80	−82	−84

The darker area indicates temperatures at which frostbite can occur within thirty minutes.

However, recent research has revealed some flaws in Siple's work, such as assuming that the wind at face level is equal to the wind at 10 meters (33 feet) above the surface. For this and other reasons, the wind-chill temperatures commonly reported on television and in newspapers have overstated the severity of the conditions considerably.

An updated wind-chill temperature chart, shown in Table 3.1, gives the wind-chill equivalent temperature as a function of current temperature and wind speed. From this table we see that a temperature of 20° F with a 30-mph wind has an equivalent heat loss to a temperature of 10° F and a 5-mph wind. The wind-chill temperature for both of these situations is 1° F, as opposed to −20° F for the first case using Siple's approach. But the lesson is the same using either formula: *cold temperatures plus wind cause danger to exposed flesh.* Box 3.3 discusses some of the hazards of extreme temperatures and actions you can take to avoid harm.

Temperature and Agriculture

For both large-scale farming and home gardeners, a cold air outbreak at the wrong time can be costly. A freeze in Florida or California can reduce harvests and increase the price of citrus or vegetable products. Cold air outbreaks accompanied by nocturnal inversions are the farmers' worst enemies.

Because of nocturnal inversions in winter, the air temperature measured at the standard height of 1.5 meters may be above freezing (0° C, 32° F) while temperatures at the ground near crop plants may fall below freezing, injuring them. To pro-

BOX 3.3

Temperature and Your Health

Extreme temperatures threaten your well-being in a surprising number of ways. Knowledge of the risks can help keep you from becoming a victim of extreme weather.

Frostbite occurs when your skin and other tissues cool down to the point that ice crystals form in your bodily fluids. Cold temperatures combined with strong winds can cause frostbite in a matter of minutes (see Table 3.1). Frostbite can also occur as a result of long exposure to less severe, but subfreezing, conditions. Frostbite usually strikes your body's extremities first: fingers, toes, and ears. Your face is also at risk because it is often exposed to the cold. Numbness in the skin is followed by the death of cells as the water in your body freezes. The simple way to prevent frostbite is to avoid direct exposure of your skin to cold and wind in winter.

Hypothermia occurs when your body temperature drops dangerously low. It occurs when the energy budget of your body runs at a deficit, which leads to a temperature decrease.

Hypothermia is possible when your body is immersed in cold for long periods, for example by floating in a cold lake or ocean. Body temperature can drop to a point where metabolic activities and organs, such as the brain, cannot operate normally. The first signs of hypothermia are confusion and a loss of judgment, followed by stupor and possibly death, as depicted graphically in the movie *Titanic*. Again, insulation from the cold is essential for survival.

Warm temperatures can also threaten your health. During summer there are many regions in the United States and the world where the temperature reaches 38° C (100° F) or higher. The environmental temperature in this case is greater, not less, than your body temperature. In this case, your body's energy gains exceed energy losses, and your temperature can rise to dangerous levels. We focus on what to do in these situations when we discuss heat waves in Chapter 9.

tect the plants, their energy budget must be modified to keep their temperature from falling below freezing.

One method of frost protection is putting certain plastic coverings over the plants. The plastics used do not transmit longwave radiation well, and so they trap heat. This approach is similar to adding cloud cover, in the sense that certain plastics decrease the overall longwave radiative energy losses. This slows the plants' temperature decrease.

Another method of frost protection is to supply heat to the area using orchard heaters, large heaters that heat the air directly. In addition to supplying heat to the air around the plants, the rising hot air mixes the air throughout the inversion. This mixing inhibits the development of a near-surface temperature inversion in the same way as a strong wind.

Another frost protection strategy is direct mechanical mixing of the air. Large fans are used to stir up the air surrounding the crops, and the turbulence generated by the fans mixes warm air from above down into cold surface air. Some desperate farmers even resort to using a helicopter, flying at very low altitudes, to mix warmer air from the nocturnal inversion down to the surface.

A popular strategy to protect plants from freezing is, ironically, to cover them with ice! Many plants are not damaged if the temperature is at 0° C (32° F). Moderate to severe damage by frost generally occurs at approximately −2° C (28° F). If the temperature of the plants is at freezing, spraying them with water will cause the water to freeze. As we learned in Chapter 2, a phase change of one gram of water from liquid to ice releases 80 calories of heat to its environment. This small amount of heat can be just enough to keep plants from falling to −2° C, saving them from frost damage.

The never-ending battle between agriculture and weather is a testament to the importance of temperature to society. The strategies for frost protection, all of which involve principles we have covered in the first three chapters, illustrate how an understanding of meteorology can have very practical and useful applications.

PUTTING IT ALL TOGETHER

SUMMARY

Temperature near the surface is governed by the energy gains and losses at ground level. The ground gains energy during the day from the Sun and emits longwave radiation both day and night. An imbalance between gains and losses causes a change in temperature.

The surface exchanges energy with the air in a variety of ways. Conduction between the ground and the atmosphere is slow and only affects the air temperature close to the ground. Convection, turbulent mixing by the winds, and radiative and latent heat fluxes exchange most of the energy between the ground and the rest of the air.

If energy gains and losses ebb and flow in a regular pattern over time, then the surface temperature will change in a cyclical pattern. The surface temperature has cycles on daily, annual, and interannual time scales. The annual and diurnal temperature cycles are primarily driven by the periodic nature of solar energy gains at the ground. The diurnal and annual temperature cycles are related to:

1. Solar energy input, and therefore latitude, the time of day, and the time of year;
2. The surface type, including proximity to bodies of water, which determines albedo and specific heat, and influences evaporation;
3. Cloud cover, which suppresses temperature changes;
4. Altitude and aspect, which cause generally cooler temperatures at higher altitudes and on north-facing slopes in the Northern Hemisphere.

Temperature changes with altitude above the Earth's surface. The environmental lapse rate describes how the temperature varies with altitude at any given time and location. When warmer air overlies colder air, a temperature inversion exists. Vertical motions of air parcels are inhibited by these "stable" temperature inversions. If the environmental lapse rate is greater than the dry adiabatic lapse rate of 10° C per kilometer, then the atmosphere is "absolutely unstable."

Conditions favorable for the formation of a near-surface temperature inversion are clear skies, long nights, calm or light winds, and little vegetation. These conditions promote radiative energy losses that are greatest near the surface. Drainage of cold air into valleys can also form inversions near the ground. Temperature inversions play a role in air pollution episodes and agriculture.

During winter in northern United States and Canada, nightly weather reports often include the wind-chill equivalent temperature. The wind-chill temperature indicates how cold it feels outside due to the combination of surface temperature and wind.

KEY TERMS

You should understand all of the following terms. Use the glossary and this chapter to improve your understanding of these terms.

Absolutely unstable atmosphere
Annual average temperature
Annual temperature cycle
Annual temperature range
Anomalies
Aspect
Daily mean temperature
Diurnal temperature cycle
Diurnal temperature range
Environmental lapse rate
Insolation
Lag
Lapse rate
Monthly mean temperature
Nocturnal inversion
Normal temperatures
Radiation inversion
Seasonal temperature cycle
Stable atmosphere
Surface temperature
Temperature inversion
Temperature range
Turbulence
Wind-chill
Wind-chill equivalent temperature

REVIEW QUESTIONS

1. Why is the energy budget at the ground a critical factor in determining the air temperature *above* the ground?
2. The temperature on a cloudy, snowy winter day is 0° C from midnight to 11:00 PM. Then, during the last hour of the day, the clouds break up and the temperature drops to a reading of −6° C. What is the daily mean temperature as calculated by the average of the daily maximum and minimum temperature? What is the daily mean temperature if you average the sum of each hourly temperature?
3. Why does the maximum daily temperature lag several hours behind the peak in incoming solar energy from the Sun?
4. Take three pans of water and fill one with cold water, one with hot water, and the final one with warm water. Place one hand in the cold water while the second hand is in the hot water. Hold your hands in the water for 2 minutes. Remove your hands and place them both in the third pan of water. Does the water feel warm or cold? Explain your observations.
5. *Leucochroa candidissima* is a species of desert snail. During the day they climb up the branches of plants. Why do you suppose the snails do this?
6. When I was a child, I was envious of my friend during the winter. We lived in houses that were very similar, though he lived across the street from me on the north side. After a winter storm passed and the sky cleared, he often had much less work to do in shoveling his sidewalk than me! Why do you think this was the case?
7. If you were building a house in the Northern Hemisphere, why would you want to put your large windows on the south-facing side? Would your answer be the same if you lived in Sydney, Australia?
8. Why do cities located on islands or near the coast have a smaller annual temperature range than cities located in the interior of continents?
9. Observations show that the diurnal temperature range is wider over dry sand than over wet sand. Why would this be so?
10. Place a thermometer on a windowsill in direct sunlight and another thermometer on the floor below the window next to the wall. Come back in one hour and write down the

temperatures measured in the two locations. Explain your temperature observations in terms of energy gains and losses.

11. Name two factors that appear to control interannual changes in global temperature.

12. Explain why the diurnal cycle of temperature of a city in Maine is related in part to the time of year.

13. Why are nocturnal inversions in the middle latitudes more common during winter than summer?

14. Let's say that the temperature at the surface in winter is 0° C and the temperature at 1 kilometer above the surface is −5° C. What is the environmental lapse rate for this situation? Is this situation absolutely unstable or stable? Is there a temperature inversion present?

15. Using the *Blue Skies* CD-ROM exercise "Atmospheric Basics/Layers of the Atmosphere," locate several different weather observation sites that have inversions on June 14, 1999 at the following different altitudes: surface, 3 kilometers, and 13 kilometers. Compare the wind speed and the air temperature for the inversions at the surface and at 13 kilometers. What wind speeds are typical for inversions near the surface? What wind speeds are typical for inversions near the tropopause at 13 kilometers? What do you think causes the inversion at 13 kilometers, and why doesn't the wind at that altitude destroy the inversion?

16. The best way to prevent frost damage is to plan the location of your crop. Based on our discussions of the factors controlling temperature, can you explain why it is better to plant frost-sensitive crops on a hillside rather than at the bottom of a valley?

17. Calculate what the temperature of a dry air parcel from near the tropopause (temperature: 200 K; altitude: 10 kilometers) would be if it were brought to the surface and it warmed adiabatically all the way down. Compare this temperature to the "freezing point" of water. Why don't farmers simply pump air from the tropopause down onto their crops when frost threatens?

■ WEB ACTIVITIES

Choose Chapter 3 on the textbook's Web site:
http://info.brookscole.com/ackerman
and select from the following resources:

- Interactive Modules that illustrate and extend your understanding of key topics in this chapter
- Tutorial Quizzes to test your mastery of terms and concepts

 For additional readings, go to the InfoTrac College Edition, your online library, at:
http://www.infotrac-college.com

Water in the Atmosphere

After completing this chapter, you should be able to:

- Define *saturation* and its importance in the atmosphere
- Explain why there can be more water vapor in warm air than cold air, and how this affects the atmosphere
- Describe how clouds and precipitation form
- Identify the major cloud and precipitation types and explain the significant differences among them

Introduction

In early January of 1998, the Canadian province of Quebec was literally crushed by water falling from the sky. For a solid week, water vapor slid upward over the northeastern United States and central and eastern Canada. The vapor turned to clouds (yellow and orange areas in satellite picture below), and precipitation dropped out of the clouds. Rain fell on Quebec, but the temperatures at the ground were cold enough for the rain to freeze on contact with roofs, roads, trees, and towers.

CP Picture Archive (Jacques Boissinot)

GOES-8 4KM ENHANCED IR 9 JAN 98 @ 00:15 UTC

CIMSS, University of Wisconsin-Madison

The freezing rain coated Canada with as much as 10 centimeters (4 inches) of ice. The ice clung to everything it touched. Its heavy weight broke 30,000 wooden power poles. Incredibly, the ice crumpled 1,000 steel electrical transmission towers into pretzel shapes and destroyed them (above). About 1.5 million people in Canada lost electricity, and half of those still were powerless a month later. A fifth of the nation's workers could not get to work easily in the wake of the storm. Not even the ageless Rolling Stones were immune. The ice crushed the roof at Olympic Stadium and caused them to cancel their Montreal concert.

In this chapter, we explore water in the atmosphere in all its phases: water vapor, clouds, and precipitation. We will explain how fog, clouds, and precipitation form. We will also learn how slightly different weather conditions can turn a cold rain into pellets of ice or a destructive freezing rain that can paralyze half a nation.

 ## Evaporation: The Source of Atmospheric Water

Water is, near the surface, the atmosphere's most abundant trace gas. How does water enter the atmosphere? **Evaporation** puts it there. As we learned in Chapter 1, evaporation is the process by which water is converted from liquid form into its gaseous state, water vapor. Evaporation occurs constantly over the surface of the Earth. When water molecules at the surface of liquid water gain enough energy to escape as vapor into the air above, evaporation results. As we learned in Chapter 2, it takes a lot of latent heat energy to change liquid water to vapor. Evaporation therefore occurs more rapidly over warmer surfaces, which supply water molecules with enough energy to escape into the atmosphere. Evaporation is also greater when the atmospheric pressure is low, the wind speed is high, and there is relatively little water vapor already in the air.

To better understand evaporation, let's consider the following example. Put some liquid water in a closed container. Keep the container at a constant temperature and pressure. Initially the container has only liquid water in it (Figure 4.1, *A*). Some individual molecules in the liquid water will have more (and some will have less) kinetic energy than the average. For instance, a water molecule in the liquid phase might gain kinetic energy considerably above the average because of several rapid collisions with neighboring molecules. Now imagine such a molecule at the liquid's surface, the boundary between the water and the air. If it has enough kinetic energy to overcome the attractive force of nearby molecules and is moving toward the air, it may escape from the liquid (Figure 4.1, *B*).

As time goes on, some water molecules at the liquid surface will be escaping and evaporating. Others will condense (Figure 4.1, *C*). Eventually, because the container is sealed, the number of molecules leaving the surface of the liquid will be the same as the number entering. This means there will be no net change in the number of molecules in the liquid or the vapor phase (Figure 4.1, *D*).

A situation in which there is no net change is described as being in *equilibrium.* When the number of molecules leaving the liquid is in equilibrium with the number condensing, the air above the surface is **saturated.** That is, the rate of return of water molecules is exactly equal to the rate of escape of molecules from the water. As we will see, the concept of saturation is central to understanding the formation of clouds and precipitation.

Counting the number of molecules in the container above the water is one way to measure the amount of water. There are several methods of measuring the amount of water vapor in the atmosphere that are easier to carry out and do not require counting molecules. We look at these different methods below.

 ## Measuring Water Vapor in the Air

Why do we need to measure the amount of water vapor in the atmosphere? There are several reasons:

1. The change of phase of water is an important energy source for storms, atmospheric circulation patterns, and cloud and precipitation formation.
2. Water vapor is the source of all clouds and precipitation. The potential for cloud formation and dissipation depends on the amount of water vapor in the atmosphere.

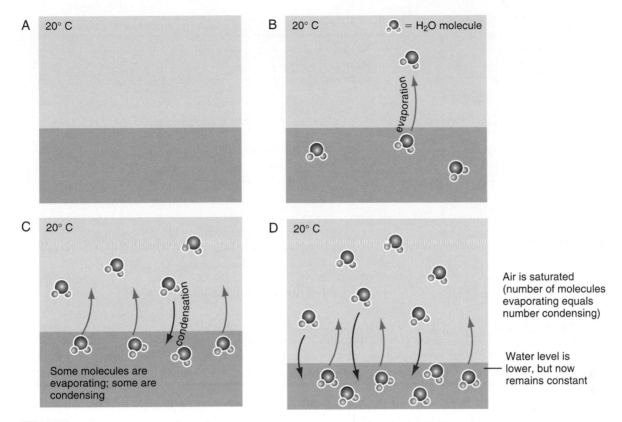

A 20° C

B 20° C   💧 = H₂O molecule

evaporation

C 20° C

condensation

Some molecules are
evaporating; some are
condensing

D 20° C

Air is saturated
(number of molecules
evaporating equals
number condensing)

Water level is
lower, but now
remains constant

Figure 4.1

*The sequence of events that leads to saturation of air. In **A**, initially dry air lies above water at
20° C. In **B**, evaporation into the air begins, and in **C**, condensation back to the water occurs.
However, evaporation exceeds condensation and the number of water molecules in the air increases. The air is saturated in **D**, where the number of evaporating and condensing water molecules is equal.*

3. The amount of water in the atmosphere determines the rate of evaporation.
   Rates of evaporation are important to weather and to many forms of plant and
   animal life.
4. Water vapor is a principal absorber of longwave radiant energy. It is the most important greenhouse gas.

   News reports of current weather conditions often include the dew point temperature and the relative humidity. These are just two of several ways to express the
amount of water vapor in the atmosphere. Each way has advantages and disadvantages. In this section, we will discuss four different methods of representing the
amount of water vapor in the atmosphere: mixing ratio, vapor pressure, relative humidity and dew point/frost point.

▪ Mixing Ratio

One way of expressing the amount of water vapor in the atmosphere is the ratio of
the weight of water vapor to the weight of the other molecules in a given volume of
air. This is the **mixing ratio.** The unit of mixing ratio is grams of water vapor per
kilogram of dry air. Typical values of the mixing ratio near the surface of the Earth
range between less than 1 gram per kilogram in polar regions to over 15 grams per
kilogram in the tropical regions. Since the surface of the Earth is the source of water vapor for the atmosphere, the mixing ratio generally decreases as you get farther
above the surface (Figure 4.2).

http://info.brookscole.com/ackerman

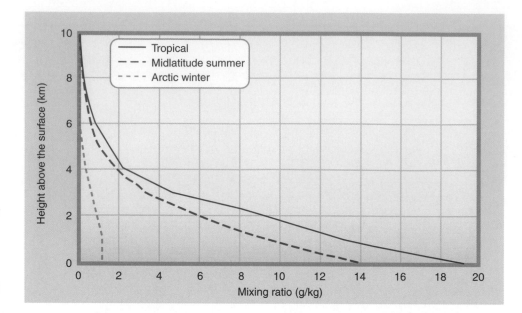

Figure 4.2

Atmospheric water vapor comes from the surface of the Earth. This explains why values of mixing ratio typically decrease with altitude even at different regions of the world and during different seasons.

Mixing ratio is an absolute measure of water vapor. This means that it is proportional to the actual number of water molecules in the air. Adding or removing water vapor molecules from a fixed volume of air changes its mixing ratio. Evaporating water into the volume increases the mixing ratio. Since the mixing ratio has to do only with the weight of water vapor relative to the total weight of an air mass, cooling the air or expanding the air has no effect on the value of the mixing ratio, because the total mass and total number of molecules remain unchanged.

Vapor Pressure

Gas molecules exert a pressure when they collide with objects. The atmosphere is a mixture of gas molecules, and each type of gas contributes its part of the total atmospheric pressure. The pressure the water molecules exert is another useful method of representing the amount of water vapor in the atmosphere. The pressure caused by these water vapor molecules is called the **vapor pressure.** Atmospheric vapor pressure is expressed in millibars (mb).

As we learned in Chapter 1, water vapor is at most only 4% of the total atmosphere. The average surface pressure due to all atmospheric gases is a little over 1000 mb. Therefore the vapor pressure attributable to water vapor alone is never more than about 4% of 1000 mb, or 40 mb.

A variety of factors can change the vapor pressure. The higher the temperature, the greater the average kinetic energy of the molecules. Both the total pressure and the vapor pressure increase with temperature. Increasing the number of water vapor molecules in a specific volume of air will also raise the vapor pressure. When water evaporates into a volume of air, both the vapor pressure and the mixing ratio increase. However, if the air cools, the vapor pressure decreases along with the total air pressure, but the mixing ratio remains constant. Atmospheric scientists use vapor pressure to express the amount of water in the atmosphere when they discuss the formation of clouds.

When air is saturated (as in Figure 4.1, D), the pressure exerted by the water vapor molecules is called the **saturation vapor pressure.** Saturation vapor pressure in the atmosphere is reached whenever the atmospheric water vapor exerts a pressure equal to what the saturation vapor pressure would be at that particular temperature in a closed container.

Figure 4.3 reveals several facts about vapor pressure and saturation vapor pressure. The data in Figure 4.3 represent over 6 years of hourly observations of vapor

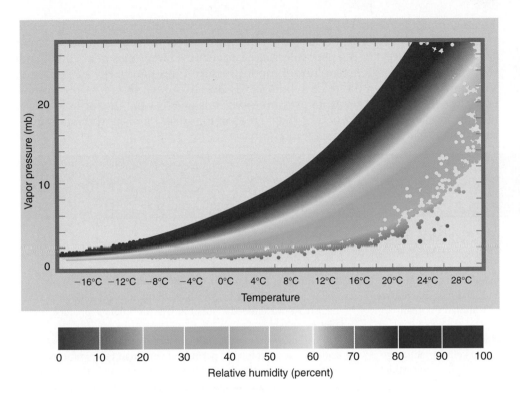

■ **Figure 4.3**
*Observations of vapor pressure
as a function of temperature on
a ridgetop at Black Rock Forest
along the lower Hudson River
in New York. The observations
were made hourly from December 1994 through mid-April
2001. Relative humidity is indicated by the color-coding. Notice how the highest observed
values of vapor pressure at each
temperature form an arc that
curves upward from left to
right. This upper limit on vapor
pressure is the saturation vapor
pressure. (Based on data from
the Open Lowland environmental monitoring station at Black
Rock Forest, Cornwall, NY.
Used by permission of the Black
Rock Forest Consortium.)*

pressure and temperature on a ridgetop 80 kilometers (50 miles) north of New York
City. What does this graph tell us? The lack of observations in the top left part of the
graph implies that vapor pressure cannot exceed a certain value for a given temperature. The maximum value at each temperature is the saturation vapor pressure for
that temperature. Following the maximum (pink) values from left to right, you see
a curve that swoops upward rapidly as the temperature increases.

Blue Skies

**Moisture and Stability/
Moisture Graph**

This last point is the most important fact about saturation vapor pressure: *it increases rapidly as the temperature increases.* Why? As the temperature of water increases, the number of molecules with enough kinetic energy to evaporate from the
water surface increases. Increasing the temperature also increases the number and
speed of the water molecules in the vapor phase. As a result, more molecules move
at greater speed and exert a higher pressure.

It is often said that "warm air holds more water vapor than cold air," but this is
a misleading simplification. This saying implies that warm air expands and has more
room for water vapor, which is incorrect. Instead, the amount of water vapor in the
air is, as we have discussed here, an equilibrium between the air and the surface beneath it. Moreover, the saying is sometimes false, for example when comparing a hot
dry desert to a cold moist coastline. It is more accurate to say that a saturated parcel of warm air will contain many more water vapor molecules than a saturated parcel of cooler air.

■ Relative Humidity

Neither the vapor pressure nor the mixing ratio tells us how close the air is to being
saturated. The ratio of the actual vapor pressure exerted by molecules of water vapor versus the saturation vapor pressure at the same temperature indicates just how
close the air is to saturation. This ratio is called the *saturation ratio.* Multiplying the
saturation ratio by 100% yields the **relative humidity.**

$$\text{Relative Humidity} = 100\% \times \frac{\text{Vapor Pressure}}{\text{Saturation Vapor Pressure}}$$

Relative humidity describes how far the air is from saturation. Saturated air has a relative humidity of 100%, because the vapor pressure equals the saturation vapor pressure. In Figure 4.3, the pink colors indicate relative humidities near 100%, and they are also the maximum values of vapor pressure for a particular temperature. This implies that when the relative humidity is close to 100%, the vapor pressure is very close to the saturation vapor pressure—which is exactly what the formula above tells us. A relative humidity of 50% (light green in Figure 4.3) tells us the vapor pressure is half that required for saturation.

BOX 4.1

Atmospheric Moisture and Your Health

Water in the atmosphere can be a killer in all of its phases: vapor (humidity), liquid (fog and rain), and ice (freezing rain and ice storms). The victim can be a National Football League star—or you.

On August 1, 2001, 27-year-old Minnesota Vikings football player Korey Stringer died just hours after practicing with his team. How could such a physically fit young athlete, who was selected by his peers to play in the Pro Bowl the previous year, die so suddenly?

When our bodies get hot we cool down by sweating. It is not the sweating that cools our bodies, but rather the evaporation of the sweat. If the air has a high vapor pressure then the rate of evaporation is reduced. This hampers our bodies' ability to maintain a nearly constant internal body temperature. This is why we are uncomfortable on hot, muggy days. Like an engine without proper ventilation, we are overheating.

When athletes practice for hours in summertime heat and humidity, their bodies can overheat to the point that organs can fail and death can occur. This is called heatstroke. For example, Stringer's temperature when he reached the hospital was above 108.8° F, more than ten degrees above normal.

The Apparent Temperature Index or Heat Index (see the table on the facing page) indicates how hot it *feels*. It is expressed as a function of air temperature and the relative humidity. R.G. Steadman developed this index in 1979. When the temperature is high but the relative humidity is low, the Heat Index is less than the actual temperature. This is because cooling by evaporation of sweat is very efficient in these situations. However, high relative humidities prevent evaporation and make it seem hotter than it really is. In these cases, the Heat Index is greater than the actual temperature.

A combination of high temperature and high humidity leads to extreme heat indices as much as 40° F above the actual air temperature. In these situations, exercising outside can be fatal. For example, the Vikings practiced in Mankato, Minnesota in the morning to avoid daytime maximum temperatures. The air temperature at noon in Mankato on the day of Stringer's last practice was 89° F, which does not sound particularly unusual for summertime. However, the dew point was a sultry 80° F, giving a relative humidity of 75%. The high humidity in Mankato, combined with the temperature, yielded a Heat Index of 106° F. As the following table shows, heatstroke was indeed a possibility at the Vikings' practice that morning.

Athletes are not the only ones at risk due to high Heat Index values. Prolonged periods of very high temperatures in association with high humidities can be extremely dangerous even to those who are not exercising at the time, as we will discuss in Chapter 9.

Fog can be a killer, too. Each year 680 fatalities occur in the United States as a result of traffic accidents during which fog is present. A combination of high speed and low visibility is often to blame. On December 27, 1996, fog caused a 54-vehicle series of pileups on the Sunshine Skyway Bridge over Tampa Bay, Florida. The wrecks killed one person and snarled traffic for seven hours. Fog also played a key role in the United States' worst modern passenger train wreck. A towboat lost in fog bumped a railroad bridge near Mobile, Alabama early on September 22, 1993. The bridge was pushed out of alignment, causing an Amtrak train to derail as it crossed the bridge a few minutes later. The train plunged into the water, killing 47 people.

Rain also kills. Up to 20% of all fatal highway accidents occur on wet pavement. A little rain after a dry spell, combined with the built-up residue of oil on roads, turns concrete and asphalt into a slick and oily mess. Cars can "hydroplane" on the wet surface and skid out of control because of reduced friction between the tires and road surface.

Ice storms consisting of freezing rain and sleet are perhaps the most dangerous of all. Driving in an ice storm is life-threatening, much more uncontrollable than in rain. Even off the roadways, you can easily fall and break bones on slippery steps or a sidewalk.

In 1999, automobile accidents attributed to bad weather caused approximately 5,000 deaths and 500,000 injuries in the United States, according to the National Highway Traffic Safety Administration. Whether exercising in hot humid weather or driving in fog, you should give careful consideration to possible threats to your safety related to the water in your environment.

The amount of relative humidity also affects the rate of evaporation. At the same pressure and temperature, water evaporates more slowly in air that has a high relative humidity and more quickly in air that has a low relative humidity. This fact is of prime importance to the public, because high humidity makes perspiration an inefficient way of removing heat by evaporation. This can lead to uncomfortable and even life-threatening conditions (Box 4.1). Relative humidity is more generally an important indicator of the rate of moisture and heat loss by plants and animals.

BOX 4.1

Heat Index Table

Relative humidity (%)	Air temperature (°F)										
	70	**75**	**80**	**85**	**90**	**95**	**100**	**105**	**110**	**115**	**120**
0	64	69	73	78	83	87	91	95	99	103	107
10	65	70	75	80	85	90	95	100	105	111	116
20	66	72	77	82	87	93	99	105	112	120	130
30	67	73	78	84	90	96	104	113	123	135	148
40	68	74	79	86	93	101	110	123	137	151	
50	69	75	81	88	96	107	120	135	150		
60	70	76	82	90	100	114	132	149			
70	70	77	85	93	106	124	144				
80	71	78	86	97	113	136					
90	71	79	88	102	122						
100	72	80	91	108							

■ Great risk to health, heatstroke imminent.
■ Risk of heatstroke.
□ Prolonged exposure and physical activity could lead to heatstroke.
■ Prolonged exposure and physical activity may lead to fatigue.

Relative humidity can change in response to a wide range of circumstances. For example, in the case of a constant volume of air at a constant temperature, changing the vapor pressure changes the relative humidity. Why does the relative humidity change in this case? Adding water molecules to a fixed volume of air increases the vapor pressure but has no effect on the saturation vapor pressure, because the temperature has not changed. The saturation ratio is changed, and so is the relative humidity.

Changing the saturation vapor pressure also changes the relative humidity. As shown in Figure 4.3, the saturation vapor pressure decreases rapidly when the temperature of the air decreases. Therefore, a decrease in temperature results in an increase in the relative humidity, and increasing the temperature decreases the relative humidity. This explains why regions with cold winters may have very low indoor relative humidities. As cold outside air finds its way into a building it is eventually heated, and this greatly decreases the air's relative humidity.

To summarize: *Adding water vapor and/or cooling the air increases relative humidity; removing water vapor and/or warming the air decreases relative humidity.*

Dew Point/Frost Point

When air is saturated, the vapor pressure is equal to the saturation vapor pressure. The air cannot contain any more moisture. If the vapor pressure is greater than the saturation vapor pressure, the relative humidity exceeds 100%. Ordinarily, this is not possible in the atmosphere. The excess moisture must condense out of the air until the relative humidity is once again reduced to 100%. This condensed water is called **dew** (Figure 4.4).

Dew forms when moisture is added to the air, when the air is cooled, or when a combination of both moistening and cooling occurs. The most commonplace occurrences of dew are caused by cooling. An everyday example of dew is the moisture that forms on the outside of an ice-cold drink glass. Overnight cooling of the air near the ground causes morning dew on grass, car windshields, and spider webs.

The temperature to which air must be cooled to become saturated without changing the pressure is called the **dew point.** The dew point temperature is determined by keeping the pressure fixed, because changing the pressure affects the vapor pressure and therefore the temperature at which saturation occurs.

To know how close the air is to saturation, we need to know the dew point and the air temperature. *The closer the dew point is to the air temperature, the closer the air is to saturation.* When the dew point equals the air temperature, the air is

Figure 4.4

When the temperature of the air around this web cooled to the dew point temperature, dew formed, making the web more visible.

Steve Ackerman

TABLE 4.1　Various Humidity Quantities for Two Air Temperatures and Two Relative Humidities, for an Atmospheric Pressure of 1000 mb.

Temperature	−10° C (14° F)		20° C (68° F)	
Relative humidity	25%	75%	25%	75%
Mixing ratio (g/kg)	0.45	1.35	3.67	11.15
Vapor pressure (mb)	0.72	2.16	5.87	17.60
Saturation vapor pressure (mb)	2.88	2.88	23.47	23.47
Dew point temperature	−26.2° C	−13.5° C	−0.5° C	15.6° C
	(−15.2° F)	(7.8° F)	(31.3° F)	(60.1° F)
Dew point depression	16.2° C	3.5° C	20.5° C	4.4° C

saturated, so the dew point temperature cannot be greater than the air temperature. The temperature difference between the air and the dew point is called the **dew point depression.** Table 4.1 compares all the different measures of water vapor in the atmosphere for two different temperatures and two different relative humidities.

If temperatures are below freezing, saturation or moistening of cooled air will lead to deposition (Chapter 2) rather than condensation. The ice crystals that form are called **frost.** The temperature to which air must be cooled at a constant pressure to cause frost to form (normally below 0° C [32° F]) is called the **frost point.** Dew may form and then freeze if the temperature falls below freezing, forming *frozen dew.* Frozen dew is different than frost. Frozen dew first condenses as liquid water before freezing, instead of becoming ice via deposition, as in the case of frost.

We can use the energy budget concepts from Chapter 3 to explain dew and frost. Dew and frost form on objects in air close to the ground, such as blades of grass. Whether or not a blade of grass cools below the dew or frost point is a function of its energy gains and losses. At night, a blade of grass loses energy by emission of longwave radiation while gaining energy by absorbing the longwave radiation emitted from surrounding objects. Under clear nighttime skies, objects near the ground emit more radiation than they receive from the sky, and so a blade of grass cools. If the temperature of a grass blade falls below the dew or frost point, dew or frost will form on the grass.

This explains why frost forms in an open field but not under a tree (Figure 4.5). Trees emit more radiation toward the ground than does the clear sky. Energy gains

Figure 4.5

Frost, ice crystals formed by deposition of water vapor on subfreezing surfaces, will form in open fields before forming under a tree. In the background, steam fog is forming over the water.

Steve Ackerman

of the grass in the open field are less than those of the grass under the tree. The grass in the open field cools faster and reaches the frost point before the grass blades under the tree.

The dew point is useful in forecasting minimum temperatures. On a clear calm night, the temperature will often drop to near the dew point. This is because condensation releases energy, and this energy release counteracts cooling below the dew point.

Formation of frost and dew are examples of phase transitions between the gas phase of water and its solid and liquid states. These transitions are vital to understanding how clouds form, which we now examine in detail.

Condensation and Deposition: Cloud Formation

Clouds are the atmosphere's equivalent of movie stars, instantly recognizable worldwide and the subject of endless curiosity. Children and poets look for shapes in the clouds. Parents struggle to explain how something so large can keep from falling out of the sky, but then can disappear in a matter of minutes.

The key to understanding clouds is water. Clouds, from the fair-weather wisps to the mightiest thunderstorms, are composed of nothing more than tiny 20-micron–sized particles of liquid water called **cloud droplets** and particles of ice called **ice crystals.** However, the formation and growth of these particles is one of the most complicated aspects of weather and climate. Before we can look at clouds, we must first examine in some detail how they are made.

Solute and Curvature Effects

As a volume of unsaturated air cools, its relative humidity increases. If the air is sufficiently cooled, the temperature equals the dew point and the relative humidity equals 100%. Based on what we have learned so far, condensation should occur at this point, forming a cloud. But cloud droplets can actually form at relative humidities other than 100%. Why?

Over the oceans and seas, waves and winds add salt to the mix of atmospheric components. When salt dissolves in water, the salt particles become dispersed among the water molecules. Salt particles dissolved in water attract water molecules even more strongly than neighboring water molecules. The greater the concentration of salt, the more the rate of evaporation is reduced, all other things being equal.

The ability of dissolved salt to hold onto water molecules is called the **solute effect.** By suppressing evaporation, the solute effect enhances the growth of droplets by condensation, thereby allowing cloud formation at relative humidities well below 100%. But as the droplet grows, the solution becomes more dilute and the solute effect decreases.

In our previous discussion of evaporation, we discussed a single molecule near the edge of a flat surface of still water. But cloud droplets aren't flat surfaces. A molecule on any surface feels attracted to its neighbors, which attempt to keep it part of the water. A molecule on a curved surface such as a cloud droplet has fewer neighbors to attract it (Figure 4.6) and can therefore escape the fluid more easily.

As a result, even if air is saturated with respect to a flat surface of water, it may be unsaturated with respect to a curved surface. This is called the **curvature effect.** This effect opposes the formation of small droplets by condensation. As a result, the relative humidity must be above 100%—a condition known as **supersaturation**—for cloud formation to occur.

It is surprisingly difficult to form a water droplet out of air that contains only water vapor. It takes a relative humidity of more than 200% for water vapor molecules to form a tiny cloud droplet of pure water. This is because a tiny droplet has a strongly curved surface. But relative humidities this high are never observed in the

$\bullet$ = H_2O

This molecule has more neighbors

This molecule has fewer neighbors

Figure 4.6

The smaller the drop, the more curved the surface, reducing the number of neighbors for each water molecule at the surface. This curvature effect makes it easier for small drops to evaporate.

atmosphere. So how do liquid droplets form from pure vapor? This question leads us to the subject of nucleation.

◼ Nucleation

Droplets form around particles. The initial formation of a cloud droplet around any type of particle is called **nucleation.** There are two types of nucleation: homogeneous and heterogeneous nucleation. In **homogeneous nucleation,** the droplet is formed only by water molecules. Homogeneous nucleation requires that enough water molecules bond together to form a cluster, or particle, that then acts as a nucleus for further condensation. Water-only bonding only works if the water molecules have low kinetic energy. If the kinetic energy of the molecules is too high, the cluster cannot form. For this reason, homogeneous nucleation only occurs at temperatures colder than –40° C (–40° F).

You can see homogeneous nucleation for yourself when you open a chilled bottled beverage that has very clean air in the bottle's neck. By removing the cap, you allow the air inside the neck to expand adiabatically and cool rapidly, but temporarily, to –40° C. The smoky cloud in the neck of the bottle is the result of homogeneous nucleation!

We learned in Chapter 1 that temperatures are as low as –40° C only in the upper troposphere, close to the stratosphere. Clouds usually form in much warmer air. Therefore, most clouds must develop through a different process. **Heterogeneous nucleation** occurs when small, nonwater particles serve as sites for cloud droplet formation. The particles are usually aerosols such as those we studied in Chapter 1. The aerosols that assist in forming liquid droplets are called **condensation nuclei.**

In the next sections, we will first consider the formation of liquid droplets around condensation nuclei and then address how ice crystals form around ice nuclei.

Condensation Nuclei

There are two types of condensation nuclei: hygroscopic and hydrophobic. **Hygroscopic nuclei** dissolve in water, and **hydrophobic nuclei** do not. Nucleation is more favorable on hygroscopic ("water-seeking") nuclei. Droplet formation can occur on hygroscopic nuclei even when the relative humidity is below 100% because the solute effect reduces the rate of evaporation. Hydrophobic ("water-repelling") nuclei resist condensation but can form droplets when relative humidities are near 100%.

There are plenty of condensation nuclei in the atmosphere in the form of dust, salt, pollen, and other small particles. The surface of the Earth is the major source of aerosols. The concentration of condensation nuclei is therefore usually greatest near the surface and decreases with altitude. In general, there is no lack of condensation nuclei for forming water droplets. Polluted cities have more condensation nuclei than wilderness environments. Over the oceans the air has fewer condensation nuclei than over land. Many of the nuclei over the oceans also contain salt thrown from waves, making them hygroscopic nuclei.

Ice Nuclei

When ice crystals form, water molecules cannot deposit onto the crystal haphazardly, as they can when condensing onto an existing water droplet. The molecule must fit into the shape of the crystal. **Ice nuclei,** the particles around which the ice crystals form, are important in the beginning stages of ice crystal formation. The ice nuclei make it easier for deposition to occur. Ice particles can form in four ways: deposition nucleation, freezing nucleation, immersion nucleation, and contact nucleation.

In **deposition nucleation,** ice forms from vapor by deposition onto the ice nucleus when the air is supersaturated with respect to ice. This happens most often on particles, such as clay, that have a molecular geometry resembling the molecular structure of ice. This geometry helps water molecules in the surrounding air to align in the proper molecular structure for forming ice when they deposit on the surface of the nuclei.

Liquid water at a temperature below 0° C is referred to as **supercooled water.** **Freezing nucleation** is the process by which a supercooled drop freezes without the aid of a non-water particle.

The existence of supercooled water requires some explanation. There is a big difference between freezing a small water droplet and freezing a larger body of water. The freezing point of a large body of water (such as the water in an ice tray) is 0° C at standard pressure. However, a 1-millimeter diameter droplet will generally not freeze until the temperature falls below −11° C (−24° F). A tiny droplet, but not a larger body of water, can be supercooled.

How can this happen? For ice to form all the water molecules must align in the proper crystal structure. First a few molecules align, and then the rest quickly follow, turning the liquid into a block of ice. The larger a volume of water is as its temperature falls below freezing, the higher the chances that a few of the molecules will line up in the proper manner. In a small volume of water the chance that some of the molecules aligning in the correct structure is reduced, simply because there are fewer molecules. For this reason, 0° C is more accurately called the *melting* point, not the freezing point, of water.

In **immersion nucleation** the nucleus is submerged in a liquid drop. Once the drop reaches a given temperature, the immersed nucleus allows the supercooled liquid to rapidly align in the crystalline structure of ice, causing the drop to freeze.

Ice nuclei may also collide with supercooled drops. The drop freezes immediately upon contact with the ice. This is referred to as **contact nucleation.**

■ Cloud Particle Growth by Condensation and Deposition

Once a cloud particle forms, it can grow if the air around it is saturated. If the particle is a liquid droplet, it will continue to grow by condensation, if the vapor pressure of the air is greater than the vapor pressure just above the surface of the particle. If the particle is ice, it grows by deposition.

Growth by condensation and deposition produces droplets, but they are small. It takes a long time to create droplets large enough to fall as precipitation. We look at precipitation particle growth a little later in this chapter. First, we examine how growth by condensation produces fog, a cloud at the ground.

■ Fog Formation

The air in contact with the ground can become saturated if it cools or when water from the surface evaporates into it. Water vapor then condenses on cloud condensation nuclei to form a suspension of tiny water drops. This is a cloud at the ground, which we call **fog.**

The formation of heavy fog often reduces visibility to the point where certain modes of transportation become hazardous (Figure 4.7). The appearance of fog on highways can trigger chain-reaction accidents involving scores of vehicles. Fog also contributed to the famous collision between the *Titanic*

■ **Figure 4.7**

Fog consists of tiny water droplets that can reduce visibility to below one kilometer (0.6 miles). Heavy fog can be a travel hazard.

and an iceberg. In early December 1952, a fog in London became so thick (partly because of pollution) that people walked into canals and rivers because they could not see the ground!

The distribution of heavy fog over the continental United States is shown in Figure 4.8. Heavy fog in Hawaii and Puerto Rico occurs on less than 10 days per year. Fog is most common in the Appalachian Mountains and near bodies of water, especially along the northwest and northeast coasts.

Fogs are named for the ways in which they form. We explore four different types of fog below: radiation fog, advection fog, evaporation fog, and upslope fog.

Radiation Fog

Radiation fogs form in the same way that dew does. On clear, long nights the ground rapidly cools by radiation and the air just above the ground cools by conduction and radiation (Figure 4.9). As the temperature of the air drops, the relative humidity increases. Radiation fogs tend to develop on clear nights, when radiative cooling near the ground is more rapid. Light winds are also required because they can gently mix moist air near the ground. Winds that are too strong mix the air near the ground with the drier, warmer air above. This keeps the air near the surface from saturating.

Radiation fogs are common in autumn in river valleys and small depressions. The cold air sinks to the bottom of the valley, providing the cool air. Rivers and streams provide the water vapor needed to increase the relative humidity via evaporation. These fogs are often called *valley fogs*.

There are some rules of thumb for forecasting a radiation fog. If the dew point temperature is approximately 8° C (14° F) below the air temperature at sunset, and if the winds are predicted to be less than 9 km/hr (5 knots), there is a good chance that a radiation fog will form during the night.

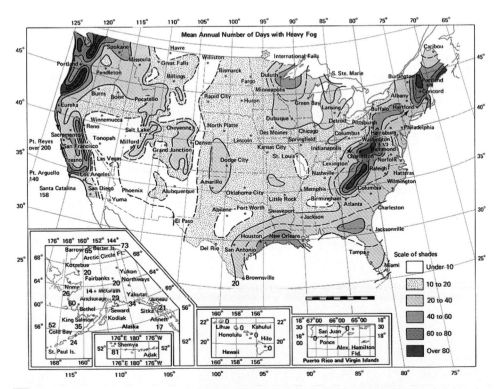

Figure 4.8

The mean annual number of dense fog days with visibility less than 300 meters across the United States, Puerto Rico, and the Virgin Islands. (Source: Gedzelman, S.D., The Science and Wonders of the Atmosphere, John Wiley and Sons, 1980, Figure 11.3.)

Lake Ontario

Lake Erie

RIVER FOG - GOES-8 IMAGER VISIBLE - 8:45 AM EDT 20 SEP 1994

SSEC and CIMSS, University of Wisconsin-Madison

■ **Figure 4.9**

Satellite photograph on the morning of September 20, 1994, showing radiation fog (narrow white areas) in the Ohio River valley and its tributaries.

Advection Fog

When warm air is advected (blown horizontally) over a cold surface, the air near the ground cools because of energy exchanges with the surface. The relative humidity increases, and an **advection fog** may form (Figure 4.10).

Advection fog is common off the coast of California as warm moist air over the Pacific is advected over the cold coastal waters. Off the East Coast, warm air over the warm Gulf Stream current may be advected over the colder coastal waters. Another foggy region is off the coast of Japan, where the cold water of the Oyoshio current meets the warm Kuroshio current. These fogs form at all times of the year and can last for more than a week.

Advection fogs can also occur when warm air flows from over the water to cooler land. Fog is common along the coast of the Gulf of Mexico during fall and winter. During these times saturation of the air occurs when warm moist air flows from the Gulf of Mexico over the cooler land. These types of fog are also common in New England and give London its reputation for fog.

Evaporation Fog

If you take a long hot shower, you may "fog up" the bathroom. Some of the warm water from the shower evaporates into the cooler bathroom air, moistening it to saturation and forming a fog. **Evaporation fogs** also occur in the vicinity of warm

■ **Figure 4.10**

Advection fog is common off the coast of California near San Francisco as warm moist air over the Pacific is advected over the cold coastal waters.

NOAA Photo Library

fronts and are sometimes called *frontal fogs*. These fogs form when water evaporates from rain that falls from warmer air above the ground into cold air near the surface. Frontal fogs only form after it has been raining for hours, because it takes time for the evaporating drops to saturate the air. Similarly, it is difficult to fog up the bathroom by taking a short shower.

Evaporation fogs also form over lakes when much colder air moves over warmer water. The vapor pressure of the cold air is less than that of the air over the water. As a result, evaporation is rapid. This rapid evaporation saturates the air above the surface. The condensation further warms the air. This warmed air rises and mixes with the cold air above it, reaching saturation and causing more fog to form.

Evaporation fog over a lake gives the appearance of steam rising out of the water and is sometimes referred to as a **steam fog** (Figure 4.11). It is common over lakes during late autumn or early winter in the more northern midlatitude regions of the globe. Steam fog is common when very cold air rushes over unfrozen waters.

Upslope Fog

Consider air rising over a mountain barrier. As the air rises, it expands and cools and the relative humidity rises. If the air becomes saturated, an **upslope fog** forms. Upslope fog is common in moist mountainous regions such as the Appalachian Highlands. This type of fog forms best when the air near the ground, before flowing upslope, is cool and moist. Therefore, it does not require much lifting before saturation occurs.

■ Lifting Mechanisms That Form Clouds

Upslope fog occurs when air moves along a rising ground surface. In general, most clouds form when air cools to the dew point as a parcel of air rises vertically.

As a parcel of air rises, it expands because atmospheric pressure decreases with altitude. Expanding the parcel requires work, which requires energy. This energy comes from decreasing the kinetic energy of the molecules inside the parcel. As it rises the parcel's temperature decreases at the dry adiabatic lapse rate, 10° C per kilometer. As the parcel of air continues to rise, it cools and the relative humidity increases to 100%. At this point, vapor condenses on cloud condensation nuclei and a cloud forms. The altitude at which a cloud begins to form is called the **lifting condensation level** (LCL).

Cloud Base Altitude

■ **Figure 4.11**

An aerial view of steam fog rising and moving downwind from a small lake in Australia. Using your knowledge of relative humidity plus the shadows in this photograph, can you determine what time of day this picture was taken?

Figure 4.12 depicts four mechanisms that cause air to ascend. Air is lifted as it moves against a mountain range (Figure 4.12, *A*). The air cannot go through the mountain and so it flows over the mountain. This is **orographic lifting.**

Other mechanisms can also cause clouds to form. For example, at the same pressure cold air is more dense than warm air. Fronts represent the boundaries between these air masses of different densities. As fronts move, **frontal lifting** occurs when less dense warm air is forced to rise over the cooler, denser air (Figure 4.12, *B*). Frontal lifting is common in winter. We study fronts in detail in Chapter 9.

During summer, **convection** is an important lifting mechanism. In summertime convection, solar energy passes through the atmosphere and heats the surface. The air near the surface warms, becomes less dense than the air around it, and rises (Figure 4.12, *C*).

The final mechanism, **convergence,** occurs when air near the surface flows together from different directions. When the air near the ground converges, or is squeezed together, it causes upward motion (Figure 4.12, *D*).

In each of these examples of lifting, the rising air creates an **updraft.** The updraft keeps the cloud particles suspended in mid-air despite the force of gravity that acts to bring them to the ground.

Certain atmospheric conditions are less favorable for cloud development than others. For example, temperature inversions and stable atmospheres suppress atmospheric vertical motions. As the parcel rises into an inversion it soon becomes colder and denser than its environment and sinks. For this reason, inversions can be cloud-free. On the other hand, unstable atmospheres are very favorable for cloud and precipitation formation, because unstable atmospheres promote rising motions.

After this thorough introduction to how clouds are made, we are now ready to examine and classify the clouds we routinely see in the skies above us.

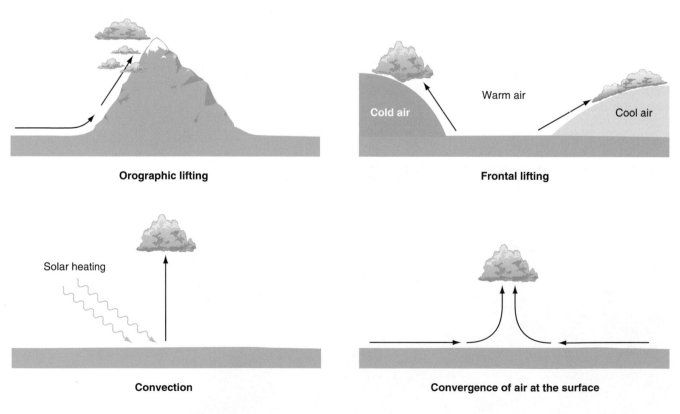

Orographic lifting

Frontal lifting

Convection

Convergence of air at the surface

Figure 4.12

Four mechanisms that cause air to ascend and form a cloud.

Cloud Classification

Fogs are named for the process that caused the air to become saturated. In 1803, British pharmacist and chemist Luke Howard devised a different kind of classification system for clouds above the ground. It has proven so successful that meteorologists have used Howard's system ever since, with minor modifications. According to his system, clouds are given Latin names corresponding to their appearance—layered or convective—and their altitude. Clouds are also categorized based on whether or not they are precipitating.

Layered clouds are much wider than they are tall. They generally have flat bases and tops and can extend from horizon to horizon. The Latin word *stratus* describes the layered cloud category, just as "stratosphere" describes a layered region of the atmosphere. Stratus-type clouds form in relatively stable air.

Convective clouds are as tall, or taller, than they are wide. These clouds look lumpy and piled up, like a cauliflower. Convective cloud types are indicated by the root word *cumulo,* which means "heap" in Latin. Convective clouds may become very tall and are rounded on top. They generally form in unstable air.

Clouds can also be classified by their altitude and their ability to create precipitation. The root word *cirro* (meaning "curl") describes a high cloud that is usually composed of wispy ice crystals. The Latin word *alto* ("high") is used to indicate a cloud in the middle of the troposphere that is below the high cirro-type clouds (just as altos in a choir sing lower notes than sopranos, but higher notes than basses). The prefix or suffix *nimbus* ("rain") denotes a cloud that is causing precipitation.

Using the combination of appearance, altitude, and ability to make precipitation, a wide range of cloud types can be identified. Table 4.2 classifies ten common cloud types and their typical heights above the ground, and Figure 4.13 depicts them pictorially. Now we will examine each cloud type, from the ground up.

 Low Clouds

Stratus (St)

Stratus clouds, abbreviated *St,* are fog that hovers just above (rather than on) the ground. They are fuzzy and featureless in appearance (Figure 4.14). From the ground these clouds appear light to dark gray in color and cover the sky. They are common along coastlines and in valleys. Early morning fogs may lift and form a stratus cloud. Stratus clouds may also originate when moist, cold air is advected at low altitudes over a region. No precipitation normally occurs with stratus clouds, although a fine mist or drizzle sometimes is visible on car windshields during thick stratus.

TABLE 4.2   Common Cloud Types

Cloud Type	Layered cloud	Convective cloud	Mixed/Neither	Typical Altitudes Feet	Typical Altitudes Kilometers
High	Cirrostratus	Cirrocumulus	Cirrus	20,000–40,000	6–12
Middle	Altostratus	Altocumulus	—	6500–20,000	2–6
Precipitating		Cumulonimbus	—	Surface–50,000	0–15
	Nimbostratus			Surface–10,000	0–3
Low	Stratus	Cumulus	Stratocumulus	Surface–6500	0–2

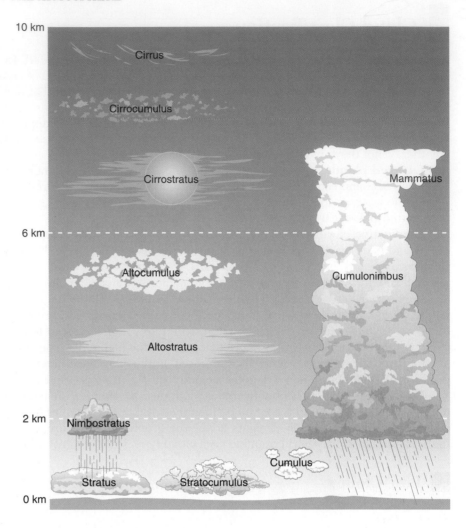

Figure 4.13

The major cloud types arranged by their typical altitude.

Figure 4.14

Stratus (St) are low-altitude layered clouds.

Lil Ackerman

■ **Figure 4.15**

Stratocumulus (Sc) are low-lying clouds with both layered and convective aspects. Stratocumulus are distinguished from stratus by variations in color across the sky.

Stratocumulus (Sc)

Stratocumulus (Sc) clouds are low-lying clouds that cover the sky and appear white to gray in color (Figure 4.15). They are a combination of layered and convective cloud types. This is because stratocumulus often occur in a shallow layer of unstable air near the surface that is overlain by stable air.

Unlike featureless stratus clouds, stratocumulus clouds often appear in rows or patches. You can distinguish stratocumulus from stratus by looking for more variations in color and a lumpier appearance.

Stratocumulus are common in certain regions, such as coastlines and in valleys. Marine stratocumulus layers are very persistent off the California and South American coastlines. In those regions, moist air flows over cooler waters and becomes saturated. Stratocumulus clouds are also associated with fronts. When accompanying a large weather system, stratocumulus are often the last clouds to appear before the skies clear completely.

Precipitation does not occur with stratocumulus. However, if the unstable surface air grows deeper (for example, as a result of daytime heating), the stratocumulus can grow taller and develop into convective clouds that produce rain or snow.

Cumulus (Cu)

Cumulus (Cu) clouds generally have well-defined, flat bases and intricately contoured domed tops resembling cauliflower. The edges of the cloud are distinct. The bases are generally dark gray and the sunlit sides are bright white.

These clouds form whenever fairly humid air rises, usually by convection. The height of the bottom of the cloud (the *cloud base*) is related to the temperature and the dew point of the rising air. Cumulus clouds in the dry southwest United States generally have much higher bases than those in the Southeast. Cumulus clouds may also form over mountains or large hills if the air is unstable. These orographically forced clouds appear stationary though they continually form and dissipate.

The two basic forms of cumulus clouds are *fair-weather cumulus* and *cumulus congestus.* Fair-weather cumulus clouds symbolize pleasant weather conditions all over the world (Figure 4.16). They have a height similar to their width. These clouds are common in summer when solar heating of the surface triggers convection. During autumn and winter, cumulus often form in cold air over large open lakes that are

Figure 4.16

Cumulus (Cu) clouds are often observed on summer days.

Steve Ackerman

still warm. Fair-weather cumulus are not deep enough to cause rain, although some may grow into large storms.

Cumulus congestus or towering cumulus are tall relative to their width. For these clouds to form the atmosphere must have a deep unstable layer, deeper than is required for the formation of the fair-weather cumulus. These towering clouds are common in summer and may have light rain falling from them. In regions like Florida, cumulus congestus may produce heavy rains for a few minutes. When cumulus congestus form in the morning it is a good indicator that storms may form later in the day. If the cloud tops appear fuzzy, ice is forming and the cloud may be developing into a cumulonimbus.

Figure 4.17

Nimbostratus (Ns) are deep layered clouds that bring precipitation and appear dark gray to ~e blue in color.

Elise Collet

■ Precipitating Clouds

Nimbostratus (Ns)

Nimbostratus (Ns) are deep clouds that bring precipitation and appear dark gray to pale blue in color (Figure 4.17). The cloud base is difficult to see because precipitation is falling from the cloud. For this reason, nimbostratus sometimes look similar to stratus, stratocumulus, or altostratus. Nimbostratus often precede warm fronts.

The precipitation that falls from nimbostratus is usually continuous and light to moderate in intensity. More episodic and intense precipitation is associated with cumulonimbus clouds.

Cumulonimbus (Cb)

Cumulonimbus (Cb) are thunderstorm clouds. They extend upward to high altitudes, often to the tropopause and sometimes into the lower stratosphere. Cumulonimbus produce large amounts of precipitation, severe weather, and even tornadoes (Figure 4.18).

A distinguishing feature of cumulonimbus is the flattened **anvil** shape of the top of the cloud. The anvil develops when the updraft slows down and spreads outward horizontally as it encounters the very stable air in the stratosphere. Underneath the anvil, sinking air may create pouches called **mammatus** (Figure 4.19). Although mammatus clouds are not severe weather, they can form under the anvils of strong thunderstorms.

Cumulonimbus clouds develop in unstable, moist atmospheres and are fairly common in the United States in spring and summer. They often occur ahead of cold

David W. Martin, SSEC, University of Wisconsin-Madison

■ **Figure 4.18**

Cumulonimbus (Cb) over Lake Wingra, Madison, Wisconsin.

http://info.brookscole.com/ackerman

■ **Figure 4.19**

Pouchy mammatus clouds (top) sometimes form on the underside of cumulonimbus anvils.

fronts. In summer they can form over mountains because of orographic lifting in combination with solar heating. Cumulonimbus clouds can be isolated or organized in groups. When cumulonimbus clouds develop into an organized system the chance of severe weather often increases, as we will see in Chapter 11.

■ Middle Clouds

Altostratus (As)

Altostratus (As) are layered clouds made up mostly of liquid water droplets and are gray to pale blue in appearance (Figure 4.20). Altostratus form when the middle layers of the atmosphere are moist and slowly lifted. If the Sun appears through these clouds, it has a "watery" appearance, whereas stratus clouds normally obscure the Sun. Altostratus clouds are common ahead of a warm front, before the nimbostratus.

Altocumulus (Ac)

The appearance of **altocumulus** (Ac) clouds varies considerably. They can be thin or thick, white or gray, organized in lines or randomly distributed. They occur in the

Figure 4.20

Altostratus clouds (As) are layered clouds that exist in the middle layers of the troposphere and give the Sun or Moon a "watery" appearance.

middle levels of the atmosphere when the air is moist, not too stable, and is being lifted. They are similar in appearance to stratocumulus, though with a higher cloud base (Figure 4.21).

You can distinguish between various types of cumulus clouds using the "fist-thumb-pinkytip" rule. Because of distance, clouds that are higher up appear smaller to your eye than those closer to the ground. If you extend your arm on a line from your eye to the cloud, cumulus clouds are generally about as big as your fist. Altocumulus clouds, in contrast, are only as big as your thumb. If the lumps of cumulus are even smaller, as small as the tip of your little finger, then the cloud is probably cirrocumulus (described in the next section).

Altocumulus clouds often appear ahead of a warm front, prior to altostratus. If other cloud types accompany altocumulus, a storm is probably approaching. Also,

Figure 4.21

Altocumulus (Ac) occur in the middle levels of the atmosphere when the air is moist.

http://info.brookscole.com/ackerman

Figure 4.22

Altocumulus castellanus clouds grow into turret-like tufts and indicate unstable air in the middle troposphere.

if the altocumulus rises up in little castle-like turrets called *castellanus* (Figure 4.22), it signals that the air aloft is unstable.

High Clouds

Cirrocumulus (Cc)

Cirrocumulus (Cc) clouds are thin, white clouds that often appear in ripples arranged in a regular formation (Figure 4.23). The smaller size of the individual cumulus lumps in cirrocumulus distinguish this cloud type from altocumulus. Cirrocumulus are composed of ice crystals, and occur high in the atmosphere in regions that are relatively moist and unstable.

Figure 4.23

Cirrocumulus (Cc) are thin, white clouds that appear high in the troposphere.

Although these clouds occur year-round, they are not very common and are usually present with other cloud types. Their tiny, delicately shaped features make cirrocumulus among the most beautiful of clouds. A "mackerel sky" is one that contains cirrocumulus clouds (or small altocumulus clouds) in a pattern that resembles fish scales.

Cirrocumulus appear in association with large precipitation-causing weather systems, especially warm fronts. This cloud type usually follows cirrus and precedes altocumulus as a precipitation-causing warm front approaches. For this reason, the saying "Mackerel sky, not three days dry" became a popular piece of weather folklore in the days before modern weather forecasting.

Cirrostratus (Cs)

Cirrostratus (Cs) clouds can cover part or all of the sky. They are uniform in appearance and can be thin or thick, white or light gray in color (Figure 4.24). Sometimes cirrostratus are almost invisible and the Sun shines through easily, unlike the "watery sky" of altostratus. They occur high in the atmosphere and are composed of ice crystals.

J. Deguara/Storm Chaser

■ **Figure 4.24**

Cirrostratus (Cs) are layered clouds that are sometimes observed in connection with optical effects, such as halos and sundogs.

Figure 4.25

Cirrus (Ci) are wispy, fibrous, white clouds that are composed of ice crystals.

Cirrostratus are common during winter in association with large-scale weather systems. If the cirrostratus cloud thickens into altostratus, an approaching storm is indicated. They may also appear far out in advance of a tropical or subtropical weather disturbance.

Cirrostratus clouds are most famous for the optical effects that occur when the Sun or Moon shine through them. Halos, bright arcs, and brilliant spots form when light passes through the ice crystals composing the cirrostratus. We examine these optical effects in the next chapter.

Cirrus (Ci)

Cirrus (Ci) are wispy, fibrous, white clouds that are made of ice crystals. They often occur as wisps here and there across the sky that are aligned in the same direction as the upper-level winds (Figure 4.25). They are a very common cloud type associated with all weather systems, including fair-weather high pressure areas. Cirrus precede warm fronts and accompany jet streams. Mountains can also generate cirrus when air is forced over high peaks. When isolated cirrus occur, they do not indicate approaching bad weather. *Mares' tails* are cirrus clouds that are long and flowing, like a horse's tail.

There are many other types of clouds, including some that are associated with specific weather and climate patterns such as the ozone hole and mountain wind circulations. We will examine those clouds in later chapters when we discuss the phenomena associated with them.

Clouds and the Greenhouse Effect

Before we delve into the details of cloud composition, let's discuss their crucial role in the global warming debate. As we learned in Chapter 2, greenhouse gases such as water vapor and carbon dioxide warm the atmosphere by absorbing the longwave radiation emitted from the surface. Water vapor is an important greenhouse gas because it absorbs longwave energy effectively. Absorption of that energy warms the atmosphere. Increases in greenhouse gases over time can potentially result in a climate change, because the atmosphere becomes more effective at absorbing longwave energy emitted by the surface.

If the atmosphere warms, initially the relative humidity should decrease. Evaporation depends on relative humidity. With a lower relative humidity, more evaporation occurs, which adds more water molecules to the atmosphere and enhances the greenhouse warming. Increases in the temperature of the atmosphere would affect the dynamics of weather and climate.

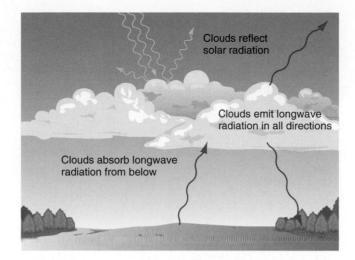

Figure 4.26

In the solar spectrum clouds tend to cool Earth; in the long-wave spectrum they tend to warm the planet.

But greenhouse gases are not the whole story. Clouds have a large impact on the energy gains of the atmosphere. Clouds reflect solar energy into space, away from the air beneath them, and clouds reduce the amount of solar radiation reaching the surface. Because of this, clouds tend to cool the Earth (Figure 4.26). The thicker the cloud, the more energy reflected back to space, and the less solar energy available to warm the surface and atmosphere below the cloud. By reflecting solar energy back to space, clouds tend to cool the planet.

Clouds also have a warming effect on atmosphere below them, because they are very good emitters and absorbers of terrestrial radiation. Clouds block the emission of longwave radiation to space and inhibit the ability of the planet to emit its absorbed solar energy to space in the form of longwave radiation (Figure 4.26). Thus, in the longwave, clouds act to warm the planet, much like the greenhouse gases do.

To complicate matters, the altitude of a given cloud is important in determining how much it warms the planet. Cirrus are cold clouds. Thick cirrus emit very little energy because of their cold temperature, according to the Stefan-Boltzmann Law from Chapter 2. At the same time, cirrus are effective at absorbing the surface emitted heat. Thus, with respect to longwave radiation losses to space, cirrus tend to warm the planet.

Stratus also warm the planet, but not as much as cirrus. This is because stratus are low in the atmosphere and have temperatures that are closer to surface temperatures than cirrus clouds. Stratus absorb radiation emitted by the surface but, because these low clouds have a similar temperature as the surface, they emit similar amounts of terrestrial radiation to space as the surface.

To complicate matters still further, a cloud's effectiveness at reflecting sunlight is related to how large the cloud droplets or cloud ice crystals are. We will investigate the reasoning behind this in Chapter 5, but it is easily demonstrated with ice in a familiar form. If you look at a glass filled with crushed ice next to a glass filled with ice cubes, you can see that the crushed ice is brighter (whiter) than the glass of ice cubes. The crushed ice particles are smaller than the cubes. Similarly, clouds consisting of small droplets are brighter than clouds consisting of large particles. Clouds composed of small particles therefore have a higher albedo and reflect more solar radiation back into space, causing more cooling than clouds with large particles.

In summary, clouds can act to cool or warm the planet, depending on how much of the Earth they cover, how thick they are, how high they are, and how big the cloud particles are. Recent measurements by NASA indicate that on average, the reflection of sunlight by clouds more than compensates for the clouds' greenhouse warming. Thus, today's distribution of clouds tends to cool the planet.

But this may not always be the case. As the atmosphere warms, the distribution of cloud amount, cloud altitude, and cloud thickness all may change. We do not

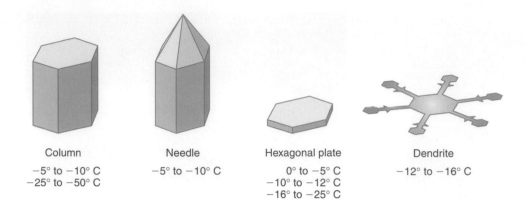

Column	Needle	Hexagonal plate	Dendrite
−5° to −10° C	−5° to −10° C	0° to −5° C	−12° to −16° C
−25° to −50° C		−10° to −12° C	
		−16° to −25° C	

 Figure 4.27

The four basic ice crystal habits are column, needle, hexagonal plate, and dendrite. The shape in which an ice particle grows depends on the temperature of its environment.

Growing a Snowflake

know what the effect of clouds will be on the surface temperatures if the global climate changes. Clouds could dampen any greenhouse warming by increasing cloud cover or decreasing cloud altitude. On the other hand, clouds could increase a warming if the cloud cover decreases or cloud altitude increases. Climate is so sensitive to how clouds might change that an accurate prediction of climate change hinges on correctly predicting the fine details of cloud formation and composition.

Cloud Composition

Every cloud has a unique composition. The composition of a cloud includes the phase of the water in it, the number and size of particles, and the shape of any ice particles, if they are present.

Water-bearing clouds differ in composition over land versus the oceans. There are more cloud condensation nuclei over land than over the oceans. For this reason, continental clouds tend to have a greater number of water droplets than maritime clouds. Clouds over land have approximately 500 million to 1 billion cloud droplets per cubic meter of air; maritime clouds have about one tenth as many. But because the water content of maritime and continental clouds are similar, the drops in continental clouds are usually smaller and more numerous than the maritime counterparts. Maritime clouds have large, soluble, heterogeneous nuclei that favor the formation of large droplets.

Ice-containing clouds vary greatly in terms of the number, size, and shape of the ice crystals in them. The size and shape of a crystal is called its **crystal habit.** Temperature determines the particular crystal habit of ice. Figure 4.27 shows the four basic shapes of ice crystals, each of which occur preferentially in the following temperature ranges: the **hexagonal plate** (0° to –5° C; –10° to –12° C; –16° to –25° C), the **needle** (–5° to –10° C), the **column** (–5° to –10° C; –25° to –50° C), and the **dendrite** (–12° C to –16° C). The dendrites are hexagonal with elongated branches, or fingers, of ice. They most closely resemble what we think of as snowflakes. We will soon learn that this is because ice crystals grow fastest around –15° C, the range in which dendrite formation is preferred.

Precipitation

Precipitation is any liquid or solid water particle that falls from the atmosphere and reaches the ground. Precipitation can be long-lasting and steady, or may fall as a brief and intense **shower.** Because precipitation is formed from water vapor, it removes water vapor from the atmosphere. Rain, snow, sleet, freezing rain, and hail are all forms of precipitation.

Dew and frost also remove water vapor through condensation or deposition onto surfaces on the ground. Dew and frost are not precipitation, however, because they do not fall from a cloud under the force of gravity. In precipitation, water vapor condenses onto a particle that eventually grows large enough to fall out of a cloud and to the surface. After discussing how particles grow into precipitation, we examine the main precipitation types.

■ Precipitation Growth in Warm Clouds

Rainmaking, natural or artificial (Box 4.2), is not easy. Cloud particles are typically 10 microns (μm) in size. (For comparison, the period at the end of this sentence is about 500 μm in diameter.) Small raindrops are typically 1000 μm in diameter. About 1 million cloud droplets have to combine to form a single raindrop. How does this happen? Just as with cloud particle growth, the formation of precipitation is complicated by the different properties of water at different temperatures.

The simple, but incorrect, explanation for precipitation is growth by condensation. Why can't a cloud droplet grow into a raindrop by condensing water onto its surface? Because this process does not work fast enough, except at the smallest drop sizes. It would take several *days* for a full-sized raindrop to form by condensational

BOX 4.2

Controlling the Weather

It's an age-old question: Can humans control the weather? In the past, people rang bells or fired cannons to prevent lightning or cause rain—producing sound and fury, but nothing in the way of success.

The scientific era of weather modification began in the 1940s and 1950s with the advent of cloud-seeding experiments. In cloud seeding, airplanes drop particles of dry ice or silver iodide into cold clouds. These particles act as extremely effective ice nuclei, potentially increasing the amount of rainfall or snowfall. However, progress in cloud seeding has been slow. Today, scientists agree that the seeding of orographic clouds can enhance precipitation by a modest 10% on a seasonal basis. But how this actually happens is still poorly understood.

Farmers have long sought for a way to suppress hail. One hailstorm can destroy a year's crops in a few minutes. Cloud seeding during the early stages of cumulonimbus development can reduce hail damage by keeping hailstone sizes small, but the results so far have been mixed. Research on hail suppression continues.

Another active area of weather modification is fog dispersal. Fog can shut down an airport for hours, causing delays and cost overruns. Like clouds, fog can be seeded with materials that cause the water droplets in fog to turn into ice, which cleanses the air of fog. This is now a routine practice at many airports worldwide. However, it is impractical over large regions.

At one time or another you may have wondered, "why can't meteorologists make tornadoes and hurricanes go away?" Perhaps a nuclear explosion or, less violently, cloud seeding could disrupt the circulation of a tornado or a hurricane. Alternatively, oceans could be covered with water-impermeable chemicals, cutting off a hurricane's fuel source.

Severe weather is here to stay, however. Hurricane seeding in the 1950s and 1960s produced few firm results. Since then, research on this subject has been curtailed. Tornadoes are even less well understood than hurricanes, and there is currently no anti-tornado research in progress.

Legal and ethical dilemmas arise whenever severe weather modification is proposed. The atmosphere cannot be put into a test tube, and so experiments must be conducted on actual storms. But what if a cloud-seeded hurricane hits New York City instead of moving out to sea harmlessly? Who is to blame? How would the environment respond if the Gulf of Mexico were covered with a film of chemicals? Obviously, the consequences of exploding nuclear bombs in storms would be horrific. The difficulty of weather modification, combined with the risks and dilemmas associated with it, have kept work on this subject to a minimum.

As we will see throughout this text, humans have changed the atmosphere in many ways. These changes usually come about as unintended by-products of modern civilization, however. So far, humans have been more effective at modifying weather and climate by mistake than by design.

growth! Precipitation forms much more quickly than that. There must be another mechanism for the relatively rapid growth of precipitation-sized particles.

Collision–Coalescence

One process that could produce a larger drop quickly would be to combine several smaller particles. To do this the cloud particles have to bump into each other and merge together, or *coalesce*. This is called the **collision–coalescence** process.

To explain how this process works, let's consider water droplets in an updraft. Water droplets in clouds with different sizes move at different speeds, as gravity and vertical motions act upon them. The difference in speed increases the chance of collisions, just as the combination of fast trucks and slow cars increases the chance of collisions on a highway.

But just because two droplets approach one another does not guarantee they will collide. A large droplet in motion creates an air current around it that can be strong enough to force tiny droplets to flow around it. (For the same reason, it's rare for cars on a highway to hit flying birds because the air current around the car carries the bird around, not into, the car.) This reduces the probability that droplets will collide, even though initially they may be headed straight for one another (Figure 4.28). Once a drop grows to a size where the downward force of gravity exceeds the updraft force of air currents, it falls downward through the cloud. As the drop falls through the cloud, it sweeps up smaller droplets in its path, collecting them and growing by collision–coalescence.

■ Precipitation Growth in Cold Clouds

The process of combining particles through collision-coalescence is an important mechanism for forming precipitation in clouds composed solely of liquid water droplets. Therefore, it is most effective in "warm" clouds in the tropics. Outside of the tropics, clouds contain ice particles, even in the summertime. How does precipitation form in these cold clouds?

Accretion and Aggregation

Collision also helps create precipitation in cold clouds. When an ice crystal falls through a cloud it may collide with and collect supercooled water drops. This process of ice crystal growth by sweeping up supercooled water drops is called **accretion.** When ice crystals collide with supercooled drops, the drops freeze almost instantly. Accretion thus provides a mechanism for the particle to grow quickly.

An ice particle produced by the accretion process that has a size between 1 millimeter and 5 millimeters (0.04 to 0.2 inches) and no discernible crystal habit is called **graupel** (pl. *graupeln*). Upon collision and freezing, the supercooled water often traps air bubbles. Because of this trapped air, the density of a graupel is low and it can easily be crushed, unlike a hailstone.

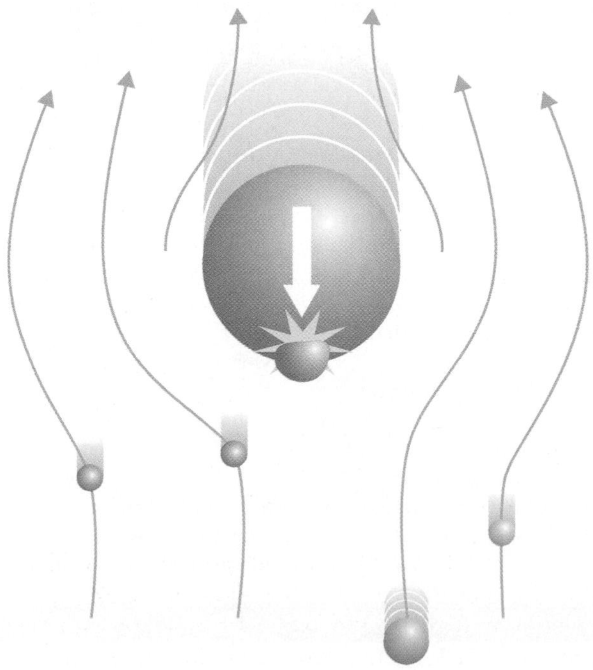

■ **Figure 4.28**

Large water drops fall faster than smaller ones. Because of the different fall speeds, water drops sometimes collide and coalesce. Very small droplets may flow around the larger drops and avoid colliding with them.

Aggregation is the process by which ice crystals collide and form a single larger ice particle (Figure 4.29). The probability that two crystals will stick together depends on the shape of the crystals. If two dendrites collide it is likely that their branches will become entangled and the two crystals will stick together. When two plates collide there is a good chance that they will simply bounce off one another.

Temperature also plays a role in aggregation. If the temperature of one crystal is slightly above freezing it may be encased in a thin film of liquid water. If this particle collides with another crystal, the thin film of water may freeze at the point of contact and bond the two particles into one. (This is why you should never lick a cold metal flag pole!)

A **snowflake** is an individual ice crystal or an aggregate of ice crystals. Snowfalls do not consist of single crystals. More commonly, they are composed of flakes that are aggregates of ice crystals. Snowflakes composed of aggregates can sometimes reach three or four inches in size.

In both warm and cold clouds, how big a droplet or crystal grows by collision processes depends on how long it stays in the cloud. The longer a particle is in the cloud, the more particles it can collect and the larger it grows. The strength of the updrafts and the thickness of the cloud determine how long it stays in the cloud. This is why only tall clouds with strong updrafts, such as cumulonimbus, produce large precipitation particles. In contrast, stratus clouds are shallow in depth and have much weaker updrafts than cumulonimbus. Particles typically do not stay in a stratus cloud long, and large particles rarely form.

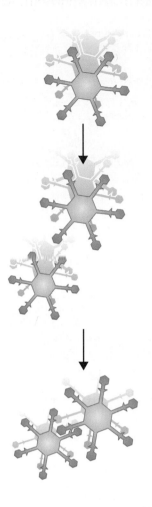

■ **Figure 4.29**

Ice crystals of different sizes or different shapes may collide and stick together, or aggregate.

The Bergeron–Wegener Process

The presence of ice in a cloud gives it a unique precipitation-making ability. To understand it, let's consider two sealed containers connected by a tube with a valve that can be opened (Figure 4.30). One container holds supercooled water at a temperature of –5° C (23° F) and the second contains ice at the same temperature.

The bonding forces in ice are much stronger than water, so fewer molecules have the energy to escape the ice than the number leaving the water. This means there will be fewer molecules of water in the vapor phase in the container with the ice than the container that has the water. That is, the vapor pressure over water is greater than the vapor pressure over the ice. Because vapor pressure is proportional to the number of water molecules in the air, this also means that *the saturation vapor pressure is greater over water than over ice.* If the valve were opened between the containers, water molecules in the vapor would flow toward the region of lower vapor pressure over the ice.

Now consider an ice crystal surrounded by supercooled droplets (Figure 4.31). If the air is saturated (100% relative humidity) with respect to the water droplets, it is *super*saturated with respect to the ice crystals. Water vapor molecules will deposit onto the crystal, lowering the relative humidity of the air. In response, water molecules evaporate from the water droplet, supplying more water molecules to the air that then deposit onto the crystal.

Put simply, *ice crystals grow at the expense of water droplets* in a cloud that has both. This ice crystal growth process is called the **Bergeron–Wegener process.** It was first proposed by meteorologist Alfred Wegener (who also famously proposed the theory of continental drift) in 1911 and explained more extensively by

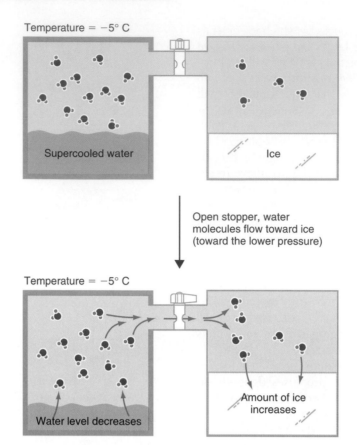

Figure 4.30

At a given temperature, the saturation vapor pressure over ice is less than the saturation vapor pressure over water. As a result, water vapor is preferentially attracted to ice versus water.

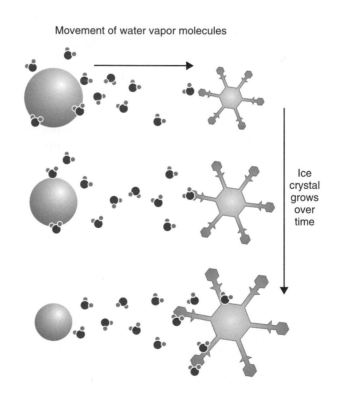

Figure 4.31

The larger saturation vapor pressure over a liquid water surface than over an ice surface causes the ice crystal to grow and the supercooled drops to evaporate.

Tor Bergeron, a member of the renowned Bergen School of Norwegian meteorology. This process contributes to the rapid growth of ice crystals and, therefore, to the ability of a cloud to form precipitation.

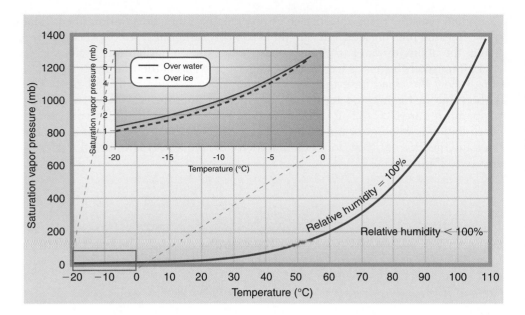

Figure 4.32

The differences between the saturation vapor pressure over ice and over water. The larger graph shows the saturation vapor pressure over water, which is the upper boundary of the observations in Figure 4.3. The inset compares the saturation vapor pressure over ice and over water. Since the line for water lies above the line representing ice, when the air is saturated (100% relative humidity) with respect to liquid water, it is supersaturated with respect to ice.

This ice crystal growth process is fastest when the difference in the saturation vapor pressure between water and ice is large. As shown in Figure 4.32, these differences maximize when the air temperature is between –12° and –17° C (10.4° and 1.4° F). Based on what we have already learned about crystal shapes, this means that dendrites grow the fastest by the Bergeron–Wegener growth process. This explains why snowflakes look more like asterisks than plates, columns or needles!

Precipitation Types

What happens when a particle falls out the base of a cloud? It is not officially precipitation until it reaches the ground. In some cases this never happens. For instance, if the atmosphere below the cloud is very dry, the particle may evaporate inbetween the cloud base and the ground. **Virga** is rain that evaporates before reaching the surface.

Similarly, falling ice crystals may sublimate in dry air while being carried horizontally by the strong winds aloft. These **fallstreaks** (Figure 4.33) often produce visually striking patterns.

Figure 4.33

Fallstreaks are wisps of ice particles that fall out of a cloud but evaporate before reaching the surface. Fallstreaks typically exhibit a hooked form produced by changes in wind speed with height.

Andy Kilgore

When particles reach the surface as precipitation, they do so primarily as rain, snow, freezing rain, or sleet. Why are there so many different types of precipitation? Again, the ability of water to change phase is the key. In midlatitude regions, precipitation usually begins as ice particles. Figure 4.34 shows the temperature conditions below the cloud base that lead to rain, snow, freezing rain, and sleet. The dashed line in Figure 4.34 represents the melting line, the altitude at which the temperatures are 0° C (32° F). The main difference between the different types of precipitation is whether or not an inversion exists near the ground—and if so, how thick the inversion is. Now, let's examine each precipitation type in detail, with reference to Figure 4.34.

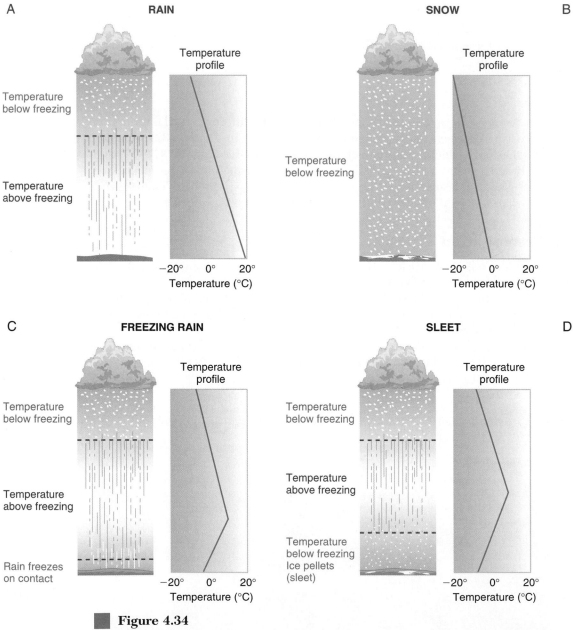

■ **Figure 4.34**

*The vertical variation of temperature will determine if precipitation falls as rain (**A**), snow (**B**), freezing rain (**C**), or sleet (**D**). In this example, all the precipitation particles are initially ice crystals. Particles that fall into the melting layer become liquid drops. To form freezing rain or sleet a surface temperature inversion is required.*

Rain

If the temperature remains above 0° C (32° F) from the cloud base to the surface, precipitation particles melt into liquid droplets called **rain** (Figure 4.34, *A*). Raindrops are not teardrop-shaped, but instead are spheres flattened by the pressure of the wind as they fall. Raindrops are at least 500 μm (0.5 millimeters or 0.02 inches) in diameter. Precipitation drops smaller than this are collectively called **drizzle,** which is often associated with stratus clouds.

Average annual rainfall as estimated by satellite measurements is shown in Figure 4.35. The data in this figure mesh well with what we have learned about water in the atmosphere. For example, the warm tropical regions of the globe receive the most rainfall. This is partly a consequence of the fact that there is more water vapor in warm, moist air than in cooler air (see Figure 4.3). But why is there a narrow band of high rainfall near the equator? Also, why are the regions offshore of California and western South America so dry? To explain both of these enigmas we will need to understand global-scale wind patterns, the topic of Chapter 7.

Rain intensity is classified by the volume of rain that falls in an hour, according to the following scale:

Intensity	Hourly Rainfall
Light	0.25–2.5 millimeters (0.01–0.10 inches)
Moderate	2.5–7.6 millimeters (0.11–0.30 inches)
Heavy	More than 7.6 millimeters (0.30 inches)

However, extreme rainfall rates are possible. When this happens, severe flooding usually follows. On the Fourth of July in 1956, Unionville, Maryland was drenched with a world-record 31.24 millimeters (1.23 inches) of rain in only one minute! More recently, on July 28, 1997, the campus of Colorado State University (CSU) at Fort Collins, Colorado received 134.6 millimeters (5.3 inches) of rain in less than 6 hours. The resulting flood inundated the CSU student center and library. All of the Fall 1997 semester textbooks were lost in the flooded campus bookstore, and 425,000 library books were soaked.

Snow

If the temperature underneath a cloud stays below freezing all the way to the ground, the snowflakes never melt and **snow** falls (Figure 4.34, *B*). As with rain, snow intensity is recorded in three categories. This is done according to volume of

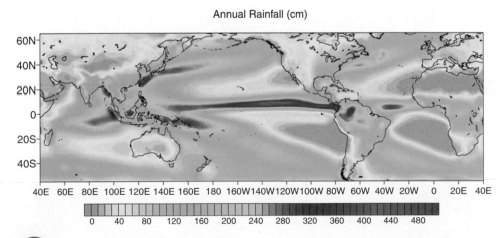

Annual Rainfall (cm)

Figure 4.35

Annual average rainfall across the world based on satellite observations. (Source: http://www. jisao.washington.edu/legates_msu/index.new.html.)

snowfall, or, because snow reflects and scatters light effectively, according to the reduction in visibility that results from the snowfall. The categories are as follows:

Intensity	Hourly Snowfall	Visibility
Light	Less than 0.5 centimeters (0.2 inches)	0.8 kilometers (0.5 miles) or more
Moderate	0.5–4 centimeters (0.2–1.5 inches)	0.4–0.8 kilometers (0.25–0.5 miles)
Heavy	Greater than 4 centimeters (1.5 inches)	Less than 0.4 kilometers (0.25 miles)

Water expands when it freezes, and snowflakes trap air between them when they clump together on the ground. For these reasons, an inch of snow and an inch of rain are not the same thing. A general rule-of-thumb is that 10 inches (25 centimeters) of new snow has the same water content as just 1 inch (2.5 centimeters) of rain. Cold, dry snows may be closer to a 20-to-1 ratio. However, wet snows with relatively warm temperatures can contain much more water, as much as 4 inches of water per 10 inches of snow.

These statistics can help us interpret the annual average snowfall across the United States, as shown in Figure 4.36. The northern Great Plains receive up to five feet of snow annually. When this thick snow cover melts, however, it is equivalent to just a few inches of rainfall. This is one reason why the Great Plains are dry. Elsewhere in Figure 4.36, it is apparent that the snowiest regions are in the Rocky Mountains, where cold temperatures and orographic lifting enhance snowfall.

Like rainfall, snowfall rates can be extreme. In Silver Lake, Colorado on April 14-15, 1921, a whopping 195.6 centimeters (76 inches, or over 6 feet) of snow fell in just 24 hours, a North American record.

Sleet and Freezing Rain

Ice storms occur when precipitation particles melt and then fall through a layer of cold air near the ground. The two precipitation types most common during ice storms are freezing rain and sleet.

Freezing rain forms when a thin layer of cold air near the surface causes melted precipitation to become supercooled (Figure 4.34, *C*). It then freezes on

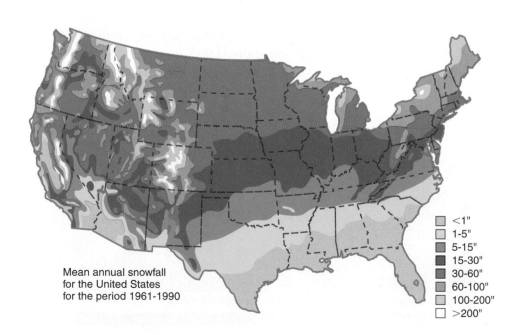

Figure 4.36

Average annual snowfall across the lower 48 United States. Latitude and proximity to water strongly affect snowfall amounts over the eastern states, while elevation dominates in the western states. (Source: http://ulysses. atmos.colostate.edu/~odie/ snowtxt.html.)

Mean annual snowfall
for the United States
for the period 1961-1990

- <1"
- 1-5"
- 5-15"
- 15-30"
- 30-60"
- 60-100"
- 100-200"
- >200"

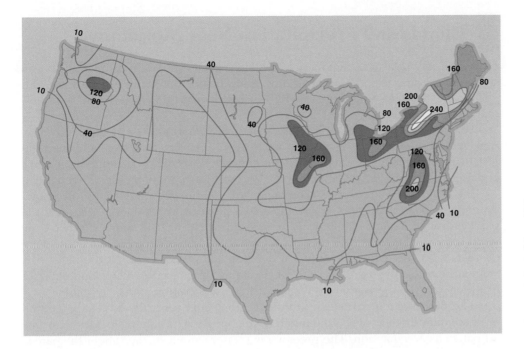

■ **Figure 4.37**

*Total number of times that
freezing rain was reported in
hourly observations across the
lower 48 United States during
the winters from 1982 to 1990.
(Source: http://www.nssl.noaa.
gov/~cortinas/preprints/waf15/
climo.html.)*

contact with exposed objects. Freezing rain covers everything in a sheet of ice, creating shimmering landscapes. However, even a little freezing rain causes treacherous road conditions and tree and power line damage. It is also responsible for aircraft icing, which is a cause of fatal plane accidents.

Figure 4.37 shows the number of hourly observations of freezing rain across the United States for the winter months of 1982 through 1990. Freezing rain is most common in the Northeast, the upper Mississippi and Columbia River valleys, and along the mid-Atlantic Coast. These are regions where cold air sinks or can be trapped at the surface (to be discussed in greater detail in Chapter 12). Meanwhile, warmer precipitating air moves above the cold air, creating an inversion. This thin layer of below-freezing air at the surface is a key ingredient for freezing rain.

Sleet consists of translucent balls of ice that are frozen raindrops. It occurs when the layer of subfreezing air at the surface is deep enough for the raindrop to freeze (Figure 4.34, *D*). Therefore, the difference between sleet formation and freezing rain formation is quite small, even though the two precipitation types do not look at all alike. When sleet hits the surface, it bounces and does not coat objects with a sheet of ice, as freezing rain does. Instead it covers flat surfaces such as roads and driveways like millions of icy ball bearings.

Keep in mind that the different scenarios in Figure 4.34 are idealized. Snow and sleet can occur at temperatures above 0° C when the air underneath the cloud is dry. In these cases, the precipitation particle partly evaporates as it falls. The evaporation cools the particle enough to keep it frozen all the way to the ground. Sleet seems particularly resistant to melting. One of your authors has seen sleet three separate times with surface temperatures at or above 10° C (50° F)!

In addition to rain, snow, freezing rain, and sleet, other precipitation types exist. The most notable is **hail,** which is precipitation in the form of large balls or lumps of ice that look like sleet on steroids. The formation of hail is quite different than sleet formation, however. Hail develops in the complex air motions inside a towering cumulonimbus cloud. For that reason, we will discuss hail in connection with thunderstorms in Chapter 11. The "ladder" in Figure 4.38 summarizes the steps in the formation of the precipitation types we have discussed in this chapter.

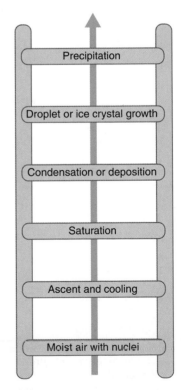

■ **Figure 4.38**

*The steps by which clouds and
precipitation are made, starting
at the bottom and climbing to
the top.*

Clouds and Precipitation Near Mountains

In closing, let's investigate how concepts in this chapter integrate with concepts presented in previous chapters by analyzing airflow over a mountain (Figure 4.39). Our goal: to explain why semi-arid **rain shadow** regions exist downwind (the "lee side") of a large mountain range.

When air rises up a mountain, it expands and cools at the dry adiabatic lapse rate. As the temperature decreases, the relative humidity increases, and the temperature approaches the dew point temperature. Eventually, the temperature will equal the dew point temperature and a cloud will form, marking the cloud base.

If the air continues to rise, water vapor will continually condense to form cloud droplets. A phase change of water vapor to a liquid releases energy, warming the parcel through latent heating. This causes the parcel to cool more slowly, at the moist adiabatic lapse rate. As the parcel continues to be orographically lifted, cloud droplets grow by condensation and by collision and coalescence. As the moist parcel continues to rise and cool, some of the liquid drops freeze. The cloud particles can then grow by the Bergeron–Wegener process, accretion, and aggregation.

The precipitation particles continue to grow and eventually fall to the ground on the windward side of the mountain as precipitation. Water molecules leave the air parcel as precipitation. The precipitating water cannot further affect the parcel, because the water molecules are now on the ground.

At the top of the mountain, the relative humidity of the parcel is 100%. As the air sinks, it warms, so the relative humidity decreases. The cloud particles evaporate as the air sinks and warms. The cloud disappears. With no cloud in the parcel, the parcel now warms all the way down the mountain at the dry adiabatic lapse rate. This is a crucial difference, because on the way up the mountain the parcel cools for part of its journey at the (lesser) moist adiabatic rate. As a result, the parcel ends up on the lee side warmer than it was at the very beginning.

In addition, there are fewer water molecules in the parcel because of the precipitation on the windward side. Increasing the temperature and removing water vapor from the air both act to lower the relative humidity. This is why regions located downwind of a mountain range are both warmer and drier than their windward counterparts. These are the rain shadow regions. (We will revisit this problem in more quantitative detail in Chapter 7.)

The rain-shadow effect is strongest when the wind is nearly perpendicular to the mountain range. In these cases, the windward side will have more cloud cover

Figure 4.39

As moist air flows over a mountain, the temperature decreases, the air becomes saturated, clouds form, and precipitation falls on the upwind side. Sinking motions on the leeward side of the mountain cause the air to become warmer and drier than before. This generates a rain shadow downwind of the mountain.

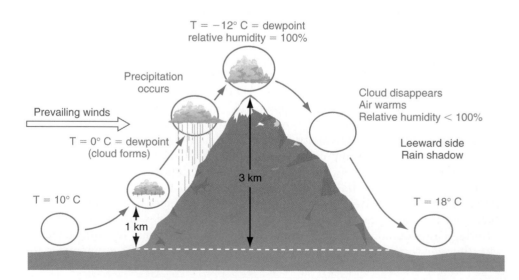

TABLE 4.3	Difference in Temperature, Cloud Cover, and Precipitation on the Windward and Leeward Sides of the Cascade Mountains in Washington State	
	Windward Side (Tacoma)	*Leeward Side (Yakima)*
Mean winter temperature	41° F	31° F
Mean summer temperature	62° F	69° F
Mean annual range	25° F	45° F
Number of cloudy days per year	292	252
Average annual precipitation	38.7 inches	7.2 inches

than the leeward side. The increased cloud cover will in turn reduce the annual temperature range (see Chapter 3) on the windward side compared to the leeward side. This is supported by weather observations on each side of the Cascade Mountains of Washington, which run north to south in a region of west-to-east winds (Table 4.3).

A similar rain shadow exists in the southern Appalachian Mountains. There, south and west winds ride up and over the mile-high mountains running southwest to northeast in western North Carolina. As Figure 4.40 shows, Asheville, North Carolina—at 700 meters (2300 feet) above sea level, more than 1 kilometer lower than nearby peaks—receives half the annual rainfall of locations just one county to its southwest.

These observations demonstrate the influence of the topography on water in the atmosphere. They also illustrate the far-reaching consequences of the principles we have studied in the first four chapters of this text. Observations are central to the study of weather and climate—so important, in fact, that we devote the next chapter to how we sense the atmosphere.

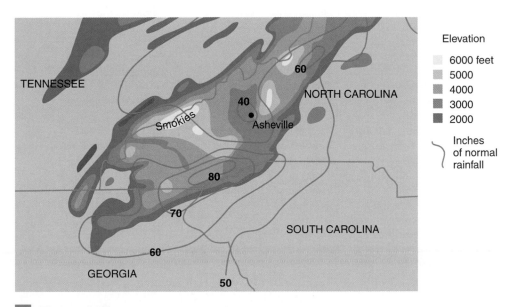

Figure 4.40

Annual rainfall observations (contours) and elevation (shading) in the southern Appalachian Mountains. The rainfall minimum at Asheville, North Carolina, is due to a rain-shadow effect caused by winds riding up and over mile-high mountains near Asheville. (Source: Gaffin, D.M. and Hotz, D.G., National Weather Digest, September 2000, pp. 4–5.)

PUTTING IT ALL TOGETHER

SUMMARY

Water in the atmosphere exists as water vapor, clouds, and precipitation. We can measure water vapor in a variety of ways. Mixing ratio, vapor pressure, relative humidity, and dew point temperature are the most common "yardsticks" of water vapor. The vapor pressure of saturated (100% relative humidity) air increases rapidly as the air is warmed. Also, changing the amount of water vapor and/or the air temperature changes the relative humidity.

A cloud is a suspension of water droplets, ice crystals, or both. Updrafts keep clouds aloft. The formation of a cloud requires water vapor, saturated air, and nuclei onto which the vapor can condense or deposit. These nuclei can be water droplets or aerosol particles. Clouds over land have more, and smaller, cloud droplets than clouds over the oceans.

Fog is a cloud at ground level. Fogs develop through two processes that lead to saturation. Air can cool to the dew point and become saturated, producing radiation fog, advection fog, or upslope fog. Air can also saturate via evaporation of water into the air. This produces steam fog or evaporation fog.

Most other clouds are formed when air cools to saturation as it is lifted. The four primary mechanisms for lifting air are orographic lifting, frontal lifting, convection, and convergence near the surface.

Clouds can be classified as layered (*strato-*) or convective (*cumulo-*), and also as low, middle (*alto-*), high (*cirro-*), or precipitating (*nimbo-*). The ten basic cloud types are cirrus, cirrostratus, cirrocumulus, altostratus, altocumulus, cumulus, stratus, stratocumulus, nimbostratus, and cumulonimbus, the thunderstorm cloud.

Clouds play a major role in the greenhouse effect. They warm the surface but also reflect sunlight, cooling the surface. Today, clouds tend to cool the planet, but this could change in the future.

Precipitation processes differ in warm and cold clouds. In warm water-only clouds, precipitation-sized particles grow by collision and coalescence of large and small water droplets. In cold clouds with ice crystals, accretion, aggregation, and the Bergeron–Wegener process cause precipitation. In the latter process, ice crystals attract water vapor more strongly than do liquid water drops in a temperature range commonly found in midlatitude clouds.

The most common forms of precipitation are rain, snow, freezing rain and sleet. Freezing rain and sleet form when there is a temperature inversion near the surface. The location of mountains and valleys affect the type and amount of both clouds and precipitation. In particular, regions downwind of a mountain range are drier and sunnier than the upwind slopes.

KEY TERMS

You should understand all of the following terms. Use the glossary and this chapter to improve your understanding of these terms.

Accretion	Hail
Advection fog	Heterogeneous nucleation
Aggregation	Hexagonal plates
Altocumulus	Homogeneous nucleation
Altostratus	Hydrophobic nuclei
Anvil	Hygroscopic nuclei
Bergeron-Wegener process	Ice crystals
Cirrocumulus	Ice nuclei
Cirrostratus	Immersion nucleation
Cirrus	Layered clouds
Cloud droplet	Lifting condensation level
Collision-coalescence	Mammatus
Columns	Needles
Condensation nuclei	Nimbostratus
Contact nucleation	Nucleation
Convection	Mixing ratio
Convective clouds	Orographic lifting
Convergence	Precipitation
Crystal habit	Radiation fog
Cumulonimbus	Rain
Cumulus	Rain shadow
Curvature effect	Relative humidity
Dendrites	Saturation
Deposition nucleation	Saturation vapor pressure
Dew	Shower
Dew point depression	Sleet
Dew point temperature	Snow
Drizzle	Snowflake
Evaporation	Solute effect
Evaporation fog	Steam fog
Fallstreaks	Stratocumulus
Fog	Stratus
Freezing nucleation	Supercooled water
Freezing rain	Supersaturation
Frontal lifting	Updraft
Frost	Upslope fog
Frost point	Vapor pressure
Graupel	Virga

REVIEW QUESTIONS

1. Name two ways that you can cause a parcel of air to become saturated.

2. One day, the dew point is 20° C. The next day at the same location, the dew point is 10° C. On which day are there more water vapor molecules in the air? On which day is the relative humidity higher? (Hint: you may not have enough information to answer both questions.)

3. Using the *Blue Skies* exercise "Moisture and Stability/ Moisture Graph," experiment with the relationships between temperature, dew point, vapor pressure, saturation vapor pressure, and relative humidity. What is the relative humidity when the temperature is 10° C and the dew point is 5° C? Does the relative humidity change if you double the temperature and the dew point to 20° C and 10° C, respectively? If so, why does it change? Finally, compare the saturation vapor pressure for a temperature of 30° C versus 10° C. Discuss how this can explain the greater frequency of thunderstorms and hurricanes in the tropics than in the polar regions.

4. If the vapor pressure at a location is 45 mb and the saturation vapor pressure there is 15 mb, what is the relative humidity? Using the *Blue Skies* exercise "Moisture and Stability/ Moisture Graph," determine what the temperature and dew point are for this example.

5. On November 28, 2001, the University of Virginia played Michigan State University in a basketball game held at the Richmond Coliseum. It was an unusually warm, muggy evening. Underneath the basketball court was a sheet of ice used for hockey games. The game was halted because of slippery floor conditions. Using your understanding of water in the atmosphere, can you explain why the floor became slippery?

6. On a cold night when frost is predicted, you park your car underneath a tree instead of out in the open. Will frost form on your windshield? Explain why or why not.

7. Sometimes a fog will appear over a roadway after a summer rain shower. What type of fog is this?

8. Why is there no such thing as a cirronimbus cloud?

9. Name the cloud type associated with each of the following: Optical effects; Thunderstorms; Fair weather.

10. Walking outside, you hold up your hand to the sky. You see a lumpy cloud that has features as big as your thumb. What is the name of this cloud?

11. Tor Bergeron observed that if a fog formed in a forest and the temperature was above 0° C (32° F) the fog extended down to the ground. If the temperature was below –5° C (23° F) the fog would not reach the forest floor. Explain his observation.

12. Explain how clouds help warm the ground. How do they help cool the ground? How does the altitude of the cloud affect its ability to warm or cool the ground?

13. An ice crystal grows for 5 minutes in a supersaturated environment with a temperature of –1° C. The crystal is carried to a different part of the cloud where the temperature is –14° C and the environment is still supersaturated. The crystal stays in this region of the cloud for another 5 minutes. Draw a picture of what the ice crystal might look like.

14. Many public restrooms have automatic hand-dryers. Why do they use hot air instead of cold air? The instructions say to place your hands in the airflow and gently rub them together. Explain how this dries your hands more rapidly than just holding them motionless in the air.

15. What temperature pattern must be present to cause freezing rain or sleet?

16. Can you easily dry your wet laundry outside when the temperature is below freezing? Explain your conclusion.

17. You are driving when very large raindrops suddenly splash onto your windshield. Is the updraft in the cloud above you strong or weak? Which cloud type is probably above you: nimbostratus or cumulonimbus?

18. What factors determine the growth of an ice crystal? How are these factors different than the factors that affect cloud droplet growth?

19. It starts snowing in very dry air that is above 0° C. Do you think the air temperature will rise, fall or remain the same? (Hint: what will happen to the snow, and how will that cause an energy gain or loss by the atmosphere?)

WEB ACTIVITIES

Choose Chapter 4 on the textbook's Web site:
http://info.brookscole.com/ackerman
and select from the following resources:

- Interactive Modules that illustrate and extend your understanding of key topics in this chapter
- Tutorial Quizzes to test your mastery of terms and concepts

For additional readings, go to the InfoTrac College Edition, your online library, at:
http://www.infotrac-college.com

CHAPTER 5

Observing the Atmosphere

After completing this chapter, you should be able to:

■ Explain how surface and upper-air weather observations are made

■ Compare and contrast how and where weather satellites, radar, and wind profilers sense the atmosphere

■ Interpret different kinds of weather satellite imagery

■ Shed light on why there are blue skies and red sunrises and sunsets

■ Understand why, when, and where rainbows, halos, and mirages may be seen

Introduction

You have probably heard the saying that "no two snowflakes are alike." But how do we know this is true? Because a farmer from Vermont with a passion for the weather saw it with his own two eyes, proved it with photographs, and told the world.

Wilson "Snowflake" Bentley grew up in Jericho, Vermont. When he was just 19 years old, he began photographing snow crystals in a barn on his farm using a microscope attached to a camera. Over the next 46 years Bentley photographed over 5000 snow crystals with the same equipment. No two were exactly alike.

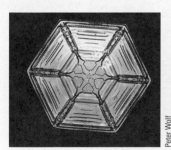

The complex steps in ice-crystal formation we studied in Chapter 4 virtually guarantee the uniqueness of a snowflake by the time it reaches the ground. But seeing is believing in meteorology. Bentley's photographs (above left and right) are still renowned worldwide, more than 70 years after his death. While capturing nature's artwork, Bentley became a scientific pioneer in photomicrography and in the study of precipitation.

So far in this text we have focused on the essential facts of weather and climate. In this chapter, we will also emphasize *how* we know these facts. As in the case of Bentley's snowflakes, we learn about the atmosphere by observing it.

In many cases scientists observe the atmosphere using advanced equipment and techniques, such as automated instruments, orbiting satellites, and Doppler radar. We will examine these approaches in some detail. But your eyes are as finely crafted as any scientific instrument, and they too can reveal the atmosphere's secrets. So we will also discuss the awe-inspiring world of rainbows, halos, and mirages. Following Snowflake Bentley's example, enjoy the beauty of the photographs in this chapter—and learn from them, too.

Meteorological Observations

Meteorology, like every other science, relies on careful and precise measurement of its subject. Meteorologists observe the atmosphere using two basic approaches. **Direct methods,** also called *in situ* for "in place," measure the properties of the air that are in contact with the instrument being used. **Indirect methods,** also referred to as *remote sensing*, obtain information without coming into physical contact with the region of the atmosphere being measured.

Our skin directly senses the temperature of objects we touch. A *thermometer* directly measures the temperature of the air it touches. Our eyes measure temperature indirectly when they observe steam above a cup of coffee or the different colors in a flame. Satellites measure temperature indirectly by sensing radiation coming from the surfaces below.

Direct Measurements of Surface Conditions

How do you think local weather observations are made? Some people imagine that a meteorologist goes outside

Figure 5.1

The Automated Surface Observing System (ASOS) makes continuous observations of the atmosphere using both direct and indirect methods. ASOS measures cloud height, visibility, precipitation, pressure, temperature, dew point, wind direction and speed, and rainfall accumulation. Observations can be made automatically in relatively remote locations, such as at this site in Haines, Alaska.

Tom Burgdorf

BOX 5.1

The Meteogram

As we learned in Chapter 1, meteorologists use the station model to condense many atmospheric observations made at the same time. The *meteogram* allows a meteorologist to see how weather variables vary in time at a single location.

The accompanying figure is an example of a meteogram for Madison, Wisconsin, on August 5 and 6, 2000. Time, in UTC (5 hours ahead of local time), runs along the x-axis. There are different weather parameters plotted as functions of time. The top portion of the figure lists visibility in miles. Below visibility are precipitation values in inches and current weather conditions, using the symbols discussed in Chapter 1. The meteogram also includes observations of wind speed and direction, and peak wind gusts. The second panel from the top includes cloud base altitude and cloud coverage. The third panel is a graph of station pressure. The bottom panel plots temperature and dew point temperature.

What does this meteogram tell us? Clouds were present over Madison during most of this time period. Precipitation was occurring through most of the evening on August 5, with a thunderstorm at 2100 UTC. By 0000 UTC on August 6, it rained 1.5 inches. The rain gave way to drizzle at 0300 UTC on August 6, and fog set in by 0500 UTC.

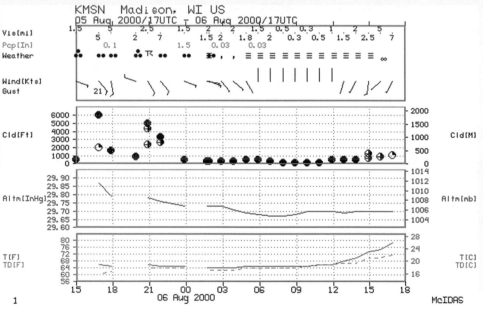

Many of these meteogram observations can be explained using what we have learned so far in this text. Notice that the winds were very light during periods of fog. The fog reduced visibility by scattering light. The dew point and temperature were equal during rainy and foggy periods. Near sunrise on August 6 at 1200 UTC, the temperature increased as the Sun warmed the Earth and lower atmosphere. The dew point temperature also increased, probably because dew was evaporating into the air. The fog lifted by late morning as surface heating reduced the relative humidity and promoted mixing of the air at and above the surface.

every hour and takes the temperature and "pulse" of the atmosphere. This was indeed the routine worldwide many decades ago. Today in the United States, however, most weather observations are made automatically by a combination of electronic sensors called the Automated Surface Observing System (ASOS).

ASOS (pronounced "A-sauce") is the United States' primary surface weather observing network. This system, installed during the 1990s, measures cloud height, visibility, precipitation, pressure, temperature, dew point, wind direction and speed, and rainfall accumulation (Figure 5.1). ASOS provides weather information from more locations, more consistently, and more frequently (as often as once a minute) than human observers can. In many other parts of the world, meteorologists still use a combination of automatic and manual methods.

Meteorologists often combine and graphically depict observations of the atmosphere to make them more understandable. The station model (Chapter 1) is one way to do this with surface data. The meteogram, a chart of one or more weather variables at a given location over a given period of time, is another (Box 5.1). Observations from these and other sources are then analyzed to determine weather patterns and create forecasts, as we will learn in Chapter 13.

Below, we examine how ASOS and human observers measure the primary weather variables. Automated observing is still in its infancy, and we note a few flaws in the current generation of ASOS instruments as well as more traditional approaches. Observing the atmosphere isn't as simple as it initially seems!

■ Temperature

Liquid-filled and metallic thermometers measure temperature by measuring how much the liquid and metal expand or contract when they are heated or cooled by their surroundings. The mercury thermometer is the best-known example. But mercury freezes at $-40°$ C ($-40°$ F). In bitter-cold situations, observers use alcohol thermometers. Today, however, most thermometers are electronic.

ASOS uses an electronic **resistance thermometer.** Resistance thermometers measure the electrical resistance of a wire made of a metal such as nickel or platinum. As the temperature of the wire increases, its resistance decreases. The amount of resistance can be translated to specific temperatures. The temperature of a wire is the same as that of the surrounding air, so the air temperature can be determined by measuring the wire's resistance to the flow of electricity. The data are then automatically reported to the National Weather Service.

To measure the air temperature accurately, thermometers are shielded from sources of energy other than the air. A thermometer in direct sunlight reports a higher temperature than one in the shade, even when the air temperature is the same around both thermometers. Figure 5.2 shows the typical "Stevenson screen" shelter used to shield thermometers at non-ASOS facilities. These standard instrument shelters are painted white so that they reflect nearly all sunlight, thereby minimizing the effect of direct sunlight on temperature inside the shelter. Because thermometers measure air temperature directly, shelters are well ventilated. The louvered sides of the shelter permit air to flow in and out, so that the air inside is essentially the same as the air outside.

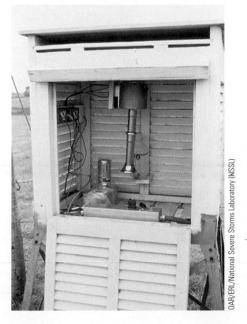

■ **Figure 5.2**

A standard instrument shelter, known as the "Stevenson Screen," houses and protects temperature, pressure, and relative humidity instruments. The front door is open to show the interior where the instruments are located. The shelter is a white, wooden box with louvered sides; it stands 4 feet above the ground.

OAR/ERL/National Severe Storms Laboratory (NSSL)

http://info.brookscole.com/ackerman

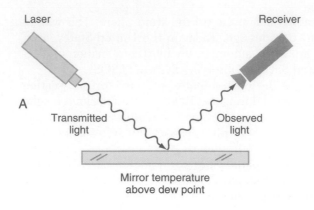

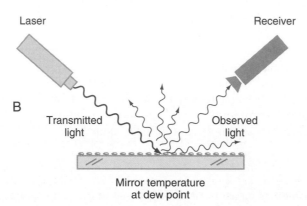

Wet-Bulb Temperatures and Humidity

■ **Figure 5.3**

*A, A laser beam and a mirror is used by ASOS to measure dew point temperatures. **B,** When dew forms, less light reaches the receiver because the dew drops scatter light into all directions, as do clouds.*

■ Humidity

ASOS measures dew point using a **dew point hygrometer.** It is based on the simple observation that a bathroom mirror will "fog up" when the temperature equals the dew point during a bath or shower.

In a dew point hygrometer, a laser light shines on a mirror. The light reflects off the mirror onto an instrument that measures the intensity of the reflected light (Figure 5.3, *A*). The mirror surface is then chilled. When the mirror surface cools to the dew point temperature, dew forms on the mirror. The water droplets (or ice crystals for the frost point) block light from reaching the detector (Figure 5.3, *B*). This is interpreted to mean that the mirror is at the dew point temperature. The mirror is then warmed above the dew point to evaporate the dew, and then cooled again to make a new measurement.

A weakness of this approach is that substances other than dew can cover the mirror and interfere with the laser beam. This fools the hygrometer into "thinking" that the dew point has been reached. ASOS dew point temperatures can sometimes "freeze" at or near 0° C (32° F).

A **psychrometer** is an alternative instrument for measuring relative humidity. A psychrometer consists of two ventilated mercury thermometers, one of which has a wet wick around its bulb and is called the *wet-bulb thermometer*. Evaporation of water off the wick removes heat from the thermometer. The temperature of the wet-bulb thermometer, but not the other thermometer, drops according to the rate of evaporation. To operate correctly, the thermometers have to be ventilated by either whirling the instrument around (*sling psychrometer*) or by using a fan (*aspirated psychrometer*). After a few minutes, the temperature of the wet bulb will stabilize at a particular temperature, referred to as the *wet-bulb temperature*. A table is then used to convert the temperature difference between the two thermometers into relative humidity. (This table can be found on the text's Web site).

■ Pressure

Galileo's student Evangelista Torricelli invented the mercury barometer in 1643. The **mercury barometer** consists of a long tube open at one end. Air is removed from the tube, and the open end is immersed in a dish of mercury. As explained in Chapter 1, the weight of the air above will then balance a column of mercury in the tube. The height of the mercury is therefore a measure of the atmospheric pressure.

The **aneroid barometer** (Figure 5.4) is a flexible metal box, called a *cell,* which is tightly sealed after air is partially removed. Changes in external air pressure cause the cell to contract or expand. The size of the measured cell is thus related to the atmospheric pressure. A pointer and a dial convert the cell's size to atmospheric pressure.

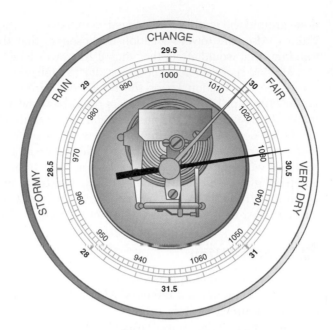

■ **Figure 5.4**
An aneroid barometer converts the size of a partial-vacuum container to atmospheric pressure. The black pointer indicates the pressure, and the gold pointer can be set by the user so that the change in pressure can be determined at a glance in inches of mercury (outer numbers) and millibars (inner numbers). Because weather is closely related to air pressure, many barometers include short weather forecasts. However, these forecasts apply only to pressure that has been adjusted to sea level. (Source: Lutgens, F. and Tarbuck, E., The Atmosphere, 8th ed., Prentice-Hall, 2001, p.159.)

The aneroid barometer is smaller and more durable than the mercury barometer, and unlike mercury is not poisonous. For these reasons, airplanes use aneroid barometers called *altimeters* that convert pressure to altitude. However, aneroid barometers are also less accurate than mercury barometers. ASOS uses electronic barometers that combine durability with accuracy and quick response to changing pressure. They are the most reliable ASOS sensors.

■ Wind

Wind has both speed and direction. **Anemometers** measure wind speed and **wind vanes** measure wind direction (Figure 5.5). A typical wind vane has a pointer in front and fins in back. When the wind direction changes, the force on the upwind side of the fin is greater than the downwind side and so the vane rotates until the forces are balanced. When the wind is blowing, the wind vane points *into* the wind. For example, in a north wind the wind vane points northward.

A **cup anemometer** measures wind speed. The cups catch the wind and produce pressure difference inside and outside the cup. The pressure difference, along with the force of the wind, causes the cups to rotate. Electric switches measure the speed of the rotation, which is proportional to the wind speed.

The ASOS wind vane and anemometer are mounted on a tower 10 meters high. This is

Lewis Kozlosky, National Weather Service

■ **Figure 5.5**
The wind vane (left) and cup anemometer (right) are used to measure wind direction and wind speed.

done around the world to minimize the influence of the ground on the wind observations. Also, the tower's location is chosen so that there are no nearby trees or buildings that would affect the wind at the tower level.

At wind speeds below about 5 km/hr (3 mph) the cup anemometer is prone to error, because friction keeps the cups from turning. At wind speeds above 160 km/hr (100 mph), cup anemometers often blow away or give unreliable measurements. In freezing rain, the anemometer can literally freeze up and stop turning.

Other devices are also employed in wind measurement. **Propellers** are used to measure wind speed. The propeller blades rotate at a rate proportional to the wind speed. A **windsock** is often used at airports. A windsock is a cone-shaped bag with an opening at both ends. When it is limp, winds are light; when it is stretched out, winds are strong. Pilots can quickly determine the wind direction and speed along a runway just by observing the shape and direction of a windsock.

Precipitation

The key to precipitation measurement is to catch the falling precipitation and then to record its amount and intensity. Although it sounds simple, precipitation is difficult to measure accurately.

The precipitation sensor used by ASOS is the **rain gauge.** It consists of a funnel-like receptacle above a bucket. Precipitation falls into the upper portion of the rain gauge, which is called the collector. The collector is heated to melt any frozen precipitation, such as snow or hail. The collected water is funneled into a tipping bucket. The tipping bucket measures water depth in increments of 0.01 inch. It is called a tipping bucket because as water is collected, the tipping bucket fills to the point where it tips over and empties out, indicating 0.01 inches of water was measured.

Although the rain gauge is commonly used, it is subject to many errors. For example, the gauge can leak. During heavy rainstorms, rain can splash out of the gauge. Winds can blow precipitation across, rather than into, the gauge. A wind shield is placed around the collector to reduce this error.

Direct Measurements of Upper-Air Weather Observations

Meteorologists monitor the atmosphere above the surface by using a radio-equipped meteorological instrument package carried aloft by a helium-filled "weather balloon." These instrument packages are called **radiosondes.**

Radiosondes measure vertical profiles of air temperature, relative humidity, and pressure from the ground all the way up to about 30 kilometers (19 miles or 10 mb). Temperature and relative humidity are measured electronically; a small aneroid barometer measures pressure. At low air pressures in the stratosphere, the balloon expands so much that it explodes and the radiosonde drifts back to the ground underneath a small parachute.

Wind speed and direction can also be determined by tracking the position of the balloon. When winds are also measured, the observation is called a **rawinsonde** (Figure 5.6). Rawinsonde measurements are made worldwide at several hundred locations twice each day at 0000 UTC and 1200 UTC.

Rawinsondes are the workhorses of the weather data network above the ground. However, they are usually launched only from land-based weather stations, which leaves out the 70% of the atmosphere that lies above the oceans. Also, strong upper-level winds can carry the balloons far downstream and data can be lost. Since these situations are of prime interest to weather forecasters, this is a drawback of using balloon-borne instruments.

Figure 5.6
The vertical distribution of temperature, humidity, pressure, and winds are obtained with an instrument called a rawinsonde. *A lighter-than-air balloon carries a small instrument box beneath it. A parachute is in the center of the string between the balloon and the instrument package, and opens when the balloon bursts in the stratosphere. Measurements are transmitted back to the surface for analysis.*

Thermodynamic Diagrams

The vertical distribution of rawinsonde observations above a location is known as a **sounding.** Meteorologists plot individual soundings on special *thermodynamic diagrams.* Although these diagrams are beyond the scope of our discussion here, they rely on concepts covered in this text and are covered in detail on the text's Web site.

Indirect Methods of Observing Weather

There are two basic types of indirect methods of sensing the atmosphere: active sensors and passive sensors. **Active sensors** emit energy, such as a radio wave or beam of light, into the atmosphere and then measure the energy that returns. **Passive sensors** measure radiation emitted by the atmosphere, the Earth's surface, or the Sun.

For example, if you yell in an auditorium and hear an echo you are doing active sensing. You emit sound waves from your mouth and use your ears to capture the energy that returns. But if you make no sound and simply listen to the sounds around you, you are doing passive sensing. The eye is also an excellent example of a passive sensor.

Much of what we observe about the atmosphere using indirect measurement techniques deals with how light interacts with molecules or objects, such as water drops, suspended in the atmosphere. To understand the ways in which indirect methods work, we need to discuss some basic laws that govern how light interacts with objects. This is important for explaining satellite and radar observations as well as our visual observations.

Laws of Reflection and Refraction

Suppose light traveling through air encounters a pool of water. When the light rays strike the boundary between the air and water, several things can happen. Some rays are turned back in the direction from which they came, or *reflected* (Figure 5.7, *A*). Other rays are transmitted into the water. Some of the transmitted rays change direction, or are *refracted* (Figure 5.7, *B*).

 http://info.brookscole.com/ackerman

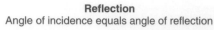

Reflection
Angle of incidence equals angle of reflection

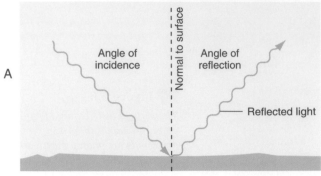

A

Refraction
Light ray bends toward the normal when entering water

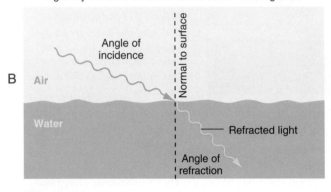

B

■ **Figure 5.7**

*Reflection (**A**) and refraction (**B**). The angles of the reflected and refracted rays depend on the angle of incidence, and are measured with respect to the* normal, *an imaginary line drawn perpendicular to the air–water interface.*

Reflection

Stand directly in front of a mirror and shine a flashlight at your mirror image. Where does the reflected beam of light go? It comes right back at you. This is called **reflection.** Now stand a little to the side of the mirror and shine the flashlight at the mirror at an angle. Where does the flashlight beam go now? The angle at which the light strikes the surface always equals the angle of reflection (see Figure 5.7, *A*). This simple law of reflection describes how a single ray reflects off a surface, and the law holds whenever light is reflected. As we saw in Figure 5.3, for example, ASOS hygrometers employ the reflection of laser light to measure dew point.

Refraction

If you shine a beam of light on a container of water or a block of glass and make careful observations, you will note the following (as illustrated in Figure 5.7, *B*):

1. When a ray of light enters water at an angle other than 90°, it bends toward the line that is perpendicular to the surface of the water or glass, which is called the *normal.*
2. When a ray of light leaves water and enters air at an angle other than 90°, it bends away from the normal.
3. When a ray of light enters or leaves the water at a right angle (parallel to the normal), the ray does not change direction.

These three observations summarize the laws of **refraction.**

Refraction explains many everyday phenomena. For example, refraction causes objects partly immersed in water (such as a person standing in a shallow pool or a straw in a glass) to look bent or broken into two.

Refraction depends on changes in optical properties along a light ray's path to your eye. The ratio of the speed of light in a vacuum to the speed of light in a substance is defined as the **index of refraction** of that substance. The index of refraction is a measure of the optical density of the substance. The higher the index of refraction, the more optically dense the substance. The index of refraction of water is approximately 1.33 while that of glass is 1.5. Air has an index of refraction of slightly greater than one, which is a function of temperature. Because the indices of refraction for air and water are different, this means that light moving from air into water, or from water into air, will be bent.

When light is traveling from water into air at a slanted angle, it bends away from the normal. At the **critical angle** the ray exiting the water travels along the air-water surface (Figure 5.8). When the ray makes an angle with the normal that exceeds the critical angle, the ray cannot pass through the interface and reflects back into the water. This condition, called **total internal reflection,** occurs only when light encounters a medium with a lower index of refraction. This is why light can reflect off of the back of the raindrop as the light moves from water to air. As we will see, this is important in the formation of optical effects such as the rainbow. Refraction also explains why stars twinkle (Box 5.2).

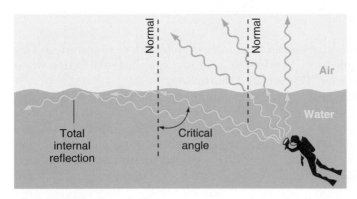

■ Figure 5.8

The critical angle occurs when light travels from one medium into another that has a lower index of refraction, such as from water into air. At the critical angle the refracted ray travels along the surface of the water. An angle of incidence that is greater than the critical angle will be totally reflected and will not enter the air.

BOX 5.2

Twinkle, Twinkle, Little Star…

On a still cloudless night, it's easy to assume that the atmosphere is uniform in structure. In reality, it is always changing. This explains why stars twinkle.

As starlight traverses the atmosphere it continually undergoes small deviations in its direction of travel because of refraction. The refraction is caused by differences in the temperature, density, and moisture content of the air through which it travels. The small refraction the light undergoes causes small, rapid changes in the apparent position or brightness of the star. We call these changes of position and brightness "twinkling."

We can explain twinkling using the diagram at right. Small differences in atmospheric structure are often caused by turbulence, which usually exists between the surface and the top of the atmosphere. Now consider a parcel of turbulent air, whose index of refraction is different from that of its environment, that moves into our view of a star. Because of refraction, when the ray encounters the air parcel the star light appears to come from position *B* instead of its true position, *A*. As the parcel moves out of our line of sight, the next ray of starlight may encounter another parcel with a different index of refraction, changing the apparent position of the star. If the refraction is strong, the star may fade or even disappear for an instant.

The twinkling of stars is referred to as *astronomical scintillation*. It is most apparent on clear, cold, and windy nights. The effect is greatest for stars near the horizon, because the

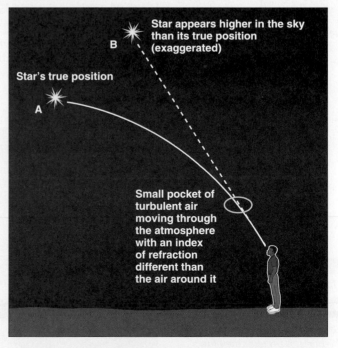

light from them passes through more atmosphere and is therefore more likely to encounter a larger number of air parcels with different indices of refraction.

Scattering

In Chapter 2, we said that radiation can be absorbed, reflected, or transmitted. This is true, but it isn't the whole truth. In addition, light rays can change direction when they encounter small particles. This is called **scattering.**

The type and direction of scattering depends critically on the size of the particle. Scattering by particles that are small with respect to the wavelength of the incident radiation (such as molecules) is called *Rayleigh scattering*. In these cases, the particle scatters equal amounts of energy in the original, or forward, direction as in the opposite, or backward, direction. *Geometric scattering* occurs with a large particle, such as a cloud drop. The particle scatters more energy forward than backward in these situations.

Scattering answers some of the most obvious questions in all of meteorology. For example, why is the sky blue? Until scientific genius Lord Rayleigh described scattering by small particles in 1869, this was a mystery. While all colors are scattered in all directions by air molecules, as demonstrated by Lord Rayleigh, violet and blue are scattered the most, up to sixteen times more than red light. This scattered light reaches our eyes from all skyward directions on a clear day. As a result, the sky looks blue. We do not perceive it as violet because our eyes are more sensitive to blue than violet.

Another weather enigma explained by scattering is: Why are sunrises and sunsets red? Sunlight passes through more air at sunset and sunrise than during the day when the Sun is higher in the sky. More atmosphere means more molecules to scatter the violet and blue light away from your eyes. If the path is long enough, all of the blue and violet light scatters out of your line of sight. The other colors continue on their way to your eyes. This is why sunsets are often yellow, orange, and red (Figure 5.9). Volcanic particles such as those present in the sky in this figure accentuate the scattering and the colors.

Have you ever seen a morning or evening sky that looked like brilliant beams of light? This phenomenon, known as **crepuscular rays,** is also caused by scattering (Figure 5.10). Irregular objects between you and the Sun, such as a cloud or mountain near the horizon, block some but not all of the light. Where the light peeks through, scattering illuminates its path from the Sun to your eyes. This creates beams in the sky. These beams appear to converge at the horizon. However, this is an illusion, similar to the impression that the rails on train tracks appear to come together in the distance.

Figure 5.9

The sky in this photograph (taken in the summer of 1993 in Madison, Wisconsin) is a mixture of yellows and reds near the horizon because the blue end of the spectrum has been scattered away from the beams of the setting Sun by the atmosphere. The colors were accentuated by the presence of volcanic particles from the Mount Pinatubo eruption in 1991.

Pam Knox

J. Cook/Weatherwise Magazine/Heldref Publications

Figure 5.10

Crepuscular rays shine between objects such as clouds or mountains. Scattering causes the beams to be visible by diverting light toward your eyes.

The repeated scattering of light, called *multiple scattering*, causes whitish light because enough light of all colors is scattered to your eye. This explains why piles of salt, sugar, and snow crystals appear white, although the individual crystals are clear. Multiple scattering also explains the blue-white appearance of distant mountains, and why haze near the horizon causes the sky to appear whitish. Both high and low clouds in sunshine look white due to scattering, either by water droplets or ice crystals. Scattered light by clouds also is important in climate change studies (Box 5.3).

BOX 5.3

Multiple Scattering and Climate Change

Climate change ultimately depends on changes in global energy gains and losses. Clouds modify the energy budget of the atmosphere and the Earth's surface. The effect of clouds on climate depends not only on how much cloud exists but also on the size of the particles composing the cloud. For this reason, the scattering of light by clouds can be a critical aspect of climate change.

If you smash a clear ice cube, the pile of small pieces appears brighter and whiter than the whole cube. This is because the shards scatter light more extensively than one whole cube. This kind of multiple scattering also happens in clouds. If the amount of water in two clouds is the same, but one cloud contains very large drops and the other very small drops, the cloud containing the small particles will appear brighter.

Multiple scattering has implications for climate and climate change. If the average particle size of a cloud were to become smaller, the cloud would become brighter and more solar energy would leave the top of the cloud and be lost from the atmosphere to outer space. That energy would not be available to warm the ground or ocean.

How can the particle sizes in clouds decrease? The average size of cloud particles is reduced if the number of cloud condensation nuclei increases. An increase of nuclei, with a fixed amount of water in the cloud, leads to more, but smaller, particles.

Where would these additional cloud condensation nuclei come from? Human activities are one likely source. If everything else remained the same, changing the number of particles existing in the atmosphere would modify the global energy budget and create the possibility of climate change. This possibility is discussed in Chapter 15.

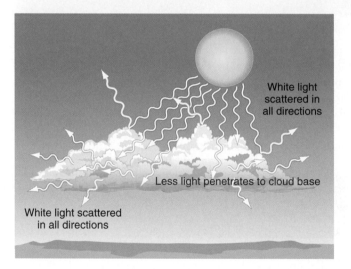

■ **Figure 5.11**

The bottom of thick clouds appears gray (top) because very little light exits the base. Most of the sunlight is scattered out the sides and top of the clouds (bottom).

The bottoms of even the whitest clouds appear grayish, sometimes ominously so. Is a terrible storm coming? Not usually. The cloud base is dark not because of absorption of light by a tall severe storm. Instead, it is because of multiple scattering that occurs above the cloud base. Multiple scattering redirects the light out the tops and the sides of the cloud. This allows very little light to be transmitted out of the cloud base. The base therefore appears dark (Figure 5.11).

Now that we understand scattering, we can grasp how ASOS measures two more important weather parameters: cloud ceiling and visibility.

■ ASOS Indirect Sensors

Cloud Ceiling Sensor

Cloud ceiling is the height of the lowest widespread cloud base. Pilots determine cloud height as they fly through a cloud layer. Trained observers on the ground can

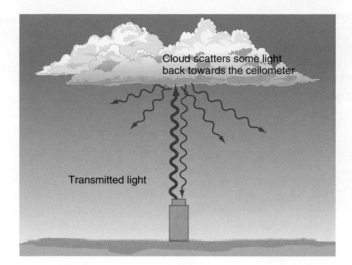

Cloud scatters some light back towards the ceilometer

Transmitted light

Figure 5.12

A ceilometer uses light scattered by clouds to determine the height of the cloud base.

estimate cloud ceilings visually. ASOS uses a **ceilometer** (pronounced "seal-OM-it-er") to measure cloud ceiling.

The ceilometer is an active remote sensing instrument. It uses a beam of radiation—in the case of ASOS, a laser beam—to detect cloud heights. The ceilometer sends pulses of radiation upward (Figure 5.12). If a cloud is present, part of this beam is scattered by the cloud and sent back to the ceilometer. The time interval between when the pulse is transmitted and when it is received back at the instrument is a measure of the cloud height.

The ASOS ceilometer was designed with the needs of aviation takeoffs and landings in mind. Thin clouds, and all clouds above 3.66 kilometers (12,000 feet) are invisible to ASOS because too little energy returns from the laser beam to be detected. This means that ASOS cannot "see" clouds such as cirrus and altocumulus even when they are plainly visible to human observers—a distinct disadvantage. Clouds at the horizon are also not "seen" by ASOS.

Visibility Sensor

Visibility is simply the horizontal distance at which a person with normal vision can see and identify specified objects. We can estimate visibility if we know the distances between objects and ourselves. If an object a quarter of a mile away cannot be seen, then the visibility must be less than a quarter of a mile. Visibility is reduced when particles between the object and us scatter or absorb light.

ASOS uses an active remote sensing method to measure visibility. The visibility instrument projects a flash of light over a very short distance. Particles suspended in the atmosphere, such as fog particles, scatter the light. The greater the number of particles, the more light is scattered, and the lower the visibility. The light flash that is scattered by the atmosphere is measured by a receiver and converted into a visibility value. This is tricky, since human eyes do not work exactly the same as light sensors. Also, human visibility reports contain information in all directions along the horizon, unlike the ASOS sensor.

Meteorological Satellite Observations

Satellite instruments provide extensive observations of use to meteorologists. Not only do they fill in the gaps of surface and weather balloon observations, but they also provide information that cannot be obtained by any other method.

Weather satellites fly around the Earth in two basic orbits: a geostationary Earth orbit (abbreviated *GEO*) and a low Earth orbit (also called *LEO*) (Figure 5.13). Geostationary satellites orbit the Earth as fast as the Earth spins. Therefore, they

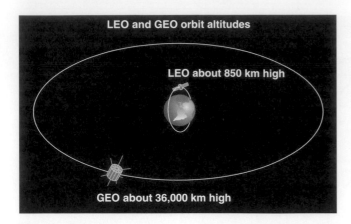

■ **Figure 5.13**

Weather satellites typically fly in geostationary orbits (GEO) or low-Earth orbits (LEO).

hover over a single point above the Earth at an altitude of about 36,000 kilometers (22,300 miles). LEO satellites orbit the Earth at an altitude of 850 kilometers (530 miles), a little higher than the altitude of the orbiting Space Shuttle. To maintain its position, a **GEO satellite** must be located over the equator. In contrast, most **LEO satellites** go around and around from pole to pole. Each low-Earth orbit is slightly to the west of the previous one, as the Earth rotates beneath the satellite.

GEO and LEO satellites have definite advantages and disadvantages relative to each other. A GEO satellite has a continuous high-quality view of the tropics and midlatitudes. However, a GEO satellite has a poor view of the polar regions, which the LEO satellite excels at periodically as it passes over the poles. A LEO satellite flies over midlatitude and tropical regions only twice a day. For this reason, its coverage of weather is hit-and-miss. When the timing is just right, however, the low altitude of a LEO satellite allows it to obtain excellent detailed snapshots of weather events, as we will see in images of a tornadic thunderstorm in Chapter 11.

Therefore, meteorologists use both GEO and LEO satellites; the choice depends on the specific need. The satellite pictures on your favorite television weather broadcast are from a GEO satellite. This is because a GEO satellite tracks storm systems continuously. For those studying global weather, a LEO satellite in a Sun-following orbit views all regions of the Earth in a single day and is a valuable tool.

The United States typically operates two geostationary satellites called GOES (Geostationary Operational Environment Satellite). One has a good view of the East Coast; the other is focused on the West Coast. Another satellite in geostationary orbit is the European METEOSAT (METEOrological SATellite), which views the eastern Atlantic Ocean, Africa, and Europe. The Japanese GMS (Geostationary Meteorological Satellite) has a good view of Asia, Australia, and the western Pacific Ocean.

The National Oceanic and Atmospheric Administration typically maintains two polar-orbiting satellites. One views the United States at approximately 2:00 PM and 2:00 AM local time and the second views regions of the United States around 10:00 AM and 10:00 PM local time. The U.S. Defense Meteorological Satellite Program has used LEO satellites for decades, as we will see in Chapter 10. Most research satellites in use today fly in low-Earth orbits, including those that investigate the ozone hole and land-use patterns.

Satellites can observe or infer an amazing range of meteorological variables. Winds and chemicals in the stratosphere, rainfall, ocean temperatures, severe weather potential—satellites can do it all, and do it on a global basis. Below, we will focus on the most common satellite images: the ones you see on TV weather programs.

■ Interpreting Satellite Images

Satellite instruments measure electromagnetic energy that the Earth and the atmosphere reflect, scatter, transmit, and emit. These passive remote sensing instruments are called **radiometers.** Two common types of radiometers are used in satellite meteorology. One type measures the amount of visible light from the Sun that is reflected back to space by the Earth's surface or by clouds. The second measures the amount of radiation emitted by the surface or clouds.

The radiometers on satellites are not cameras, although they do produce images. Radiometers use moving mirrors to view different regions of Earth. They scan the Earth much like the way we read lines of text in a book. The instrument begins at a starting point in one direction and then scans across the Earth below line-by-line, making observations as it proceeds. The images created by combining these scans are normally updated every 15 to 30 minutes.

Satellite radiometers can "see" in a wide range of electromagnetic radiation channels, as we saw in Chapter 2. The most common channels are **visible** light (0.6 microns), **infrared** (10 to 12 microns), and a special channel near one of the infrared absorption bands of H_2O that is called "**water vapor.**" Better detail or *resolution* is obtained at smaller wavelengths. The resolution of visible satellite images on TV weather broadcasts is often 1 kilometer, whereas infrared and water vapor images usually have resolutions between 4 and 8 kilometers. This means that weather features on the scale of 1 kilometer, such as individual cumulus clouds, can be seen in visible images, but not in infrared or water vapor images.

Visible Imagery

A visible satellite image represents sunlight scattered or reflected by objects on Earth. Differences in the albedo of clouds, water, land, and vegetation allow us to distinguish these features in the imagery. Dark areas in a visible satellite image represent geographic regions where only small amounts of visible light from the Sun are reflected back to space. The oceans are usually dark, while snow and thick clouds are bright (Figure 5.14).

The brightness of a cloud in a visible satellite image depends on the number of water drops or ice crystals in it. Stratus have lots of particles that scatter solar radiation, so stratus appear white in a visible image. For the same reason, fog is also very

Interpreting Satellite Visible Images

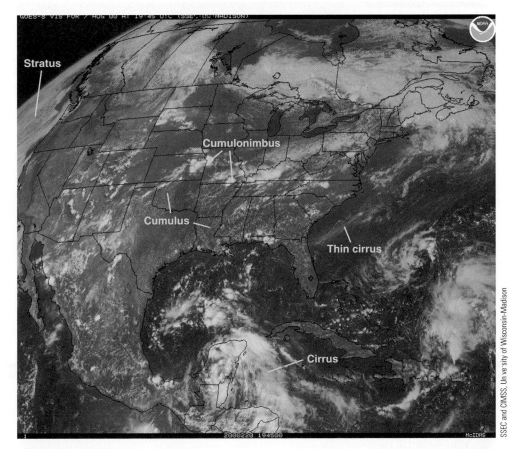

Figure 5.14

A visible image of North America from the GOES satellite on August 7, 2000 at 1945 UTC. Notice the bright white cirrus and cumulonimbus clouds, as well as the stratus cloud deck at far left off the Pacific Northwest coast.

http://info.brookscole.com/ackerman

easy to see on visible satellite images. On the other hand, thin cirrus are difficult to see because there are fewer ice particles available to scatter solar radiation.

Analysis of visible satellite images allows meteorologists to locate thunderstorms, hurricanes, fronts, and fog. Weather events can be tracked using time sequences of satellite images. This allows weather forecasters to predict their movement over short time periods of 30 minutes or less. Horizontal wind speed and direction can also be determined by tracking individual cloud features in a time sequence of satellite images.

Interpreting Satellite IR Images: Cloud Altitude

Interpreting Satellite IR Images: Surface Temperature

Infrared Imagery

The infrared (IR) radiometers on satellites measure radiation with wavelengths between 10 and 12 microns. The information in this "channel" is very different than in the visible image. Instead of light that has been reflected or scattered, the IR radiometer measures heat. In IR radiometric images, cold objects are white and hot surfaces appear black (Figure 5.15). An advantage of the IR satellite image over the visible image is that it gives useful information both day and night. (As we saw in Chapter 2, visible images at night are not very informative!) Television weathercasters frequently show animations of time sequences of IR images.

As we learned in Chapter 2, all objects emit radiation proportional to their temperature and also the degree to which they can emit radiation. An IR instrument provides information on the temperature of land, water, and clouds by measuring the IR radiation emitted from surfaces below the satellite. The radiant energy measured by

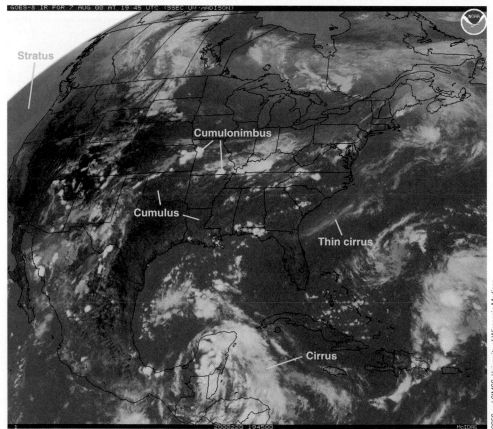

SSEC and CIMSS, University of Wisconsin-Madison

Figure 5.15

An IR image of North America from the GOES satellite on August 7, 2000 at 1945 UTC. Compare this image with Figure 5.14, which is for exactly the same time and region. High clouds show up more prominently in this image than in the visible image, but low clouds such as cumulus and stratus are hard to distinguish from the surface. This is because IR images depict heat output, and low clouds have essentially the same temperature as the Earth's surface.

IR radiometers is converted to a temperature. IR imagery can be used to distinguish low clouds from high clouds. Low clouds are relatively warm and appear gray in satellite IR images. Thick, cold clouds, like the tops of thunderstorms, appear bright white.

Differences in the solar and IR properties of different clouds allow us to use both visible and IR imagery to distinguish among cloud types (Figure 5.16). For example, stratus clouds can be easily distinguished from thin cirrus. Stratus are gray in the IR image and bright white in the visible image, while thin cirrus are white in the IR image and gray in the visible image.

A drawback of both visible and IR satellite images is that they rely on the presence of clouds. What if we want to know what the atmosphere is doing in clear regions? Water vapor imagery can provide this information, and much more.

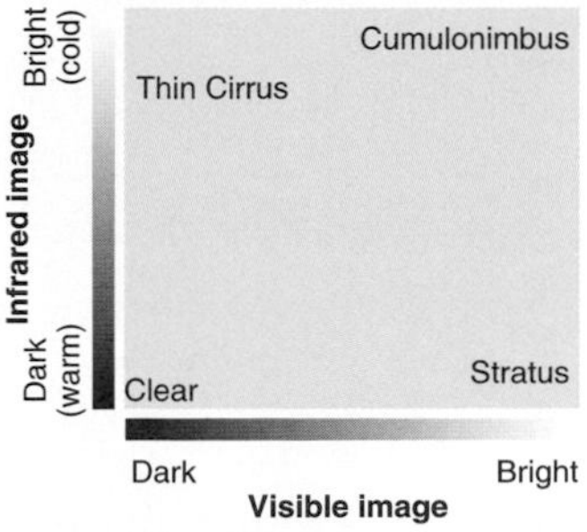

Water Vapor Imagery

Water vapor in the atmosphere is transparent to radiation at visible and 10-to-12-micron wavelengths. This is why we use visible and IR satellite imagery to observe surface features and clouds. However, water vapor absorbs and emits radiation very efficiently in wavelengths between 6.5 and 6.9 microns. Satellite radiometers that measure the amount of radiation emitted by the atmosphere at these wavelengths can be used to detect the amount of water vapor in the atmosphere.

As we mentioned in Chapter 1, the amount of water vapor peaks at the surface and decreases higher up in the atmosphere. Satellites look down on the atmosphere, and so a water vapor image "sees" the atmosphere where water vapor increases in concentration. This level is usually between 300 and 600 mb. As a result, water vapor images tell us about the upper and middle troposphere, which is a key region for storm development and growth. And water vapor, unlike clouds, is always present in the atmosphere. A water vapor image gives us detailed information everywhere, not just in cloudy regions.

In a water vapor image (Figure 5.17), black indicates low amounts of water vapor and milky white indicates high concentrations. Bright white regions correspond

Figure 5.16

A matrix for how different types of clouds appear in visible versus IR satellite images. For example, cumulonimbus clouds are bright in both images, but stratus clouds are bright only in the visible image.

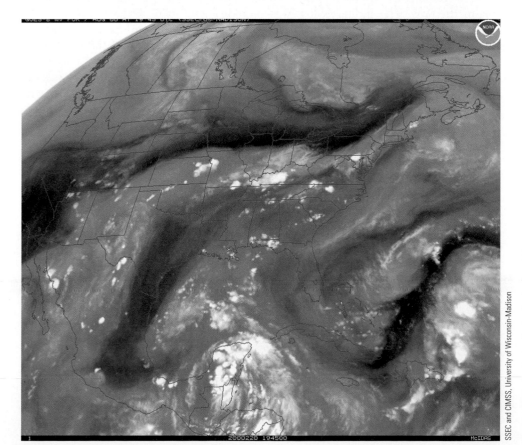

Figure 5.17

A water vapor image of North America made from a GOES satellite on August 7, 2000 at 1945 UTC. Light areas are more humid than dark regions. Bright white regions indicate high clouds. Compare with Figures 5.14 and 5.15, which were made simultaneously. Notice that water vapor images give information in all regions, even those that lack clouds.

SSEC and CIMSS, University of Wisconsin-Madison

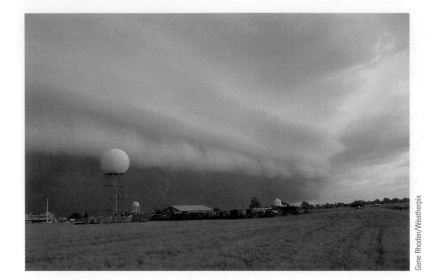

■ **Figure 5.18**

The round, white dome of a National Weather Service Doppler radar (at left in photo) is silhouetted against an approaching thunderstorm gust front.

Interpreting Satellite Water Vapor Images

to thunderstorms. Dark regions are dry air with low relative humidity, such as stratospheric air. In the midlatitude regions, zones with strong contrast in the amount of water vapor can indicate the presence of a jet stream.

Satellite observations view the world from the top down. In the next section, we explore a type of indirect observation that sees the atmosphere from the bottom up: radar.

■ Radar Observations

The acronym *radar* stands for *RAdio Detection And Ranging*, a World War II invention. Radar was supposed to detect aircraft in flight, but precipitation frequently got in the way. The military's loss was meteorology's gain. Today, weather radar tracks the development, direction, and movement of storms and estimates precipitation amounts. Radar is becoming the modern, indirect version of the rain gauge, as we will see in Chapter 8.

A radar transmitter housed in a golf-ball–shaped dome (Figure 5.18) sends out pulses of radio waves over 1000 times each second on a path inclined several degrees above the horizon. Precipitation-sized particles scatter these radio waves

■ **Figure 5.19**

Weather radars send out a narrow-beam radio wave that is scattered off precipitation. Some of the scattered energy returns to the weather radar and forecasters interpret it to locate regions of precipitation and the intensity of precipitation. The purple lines represent the transmitted pulse while the red lines represent the reflected signal.

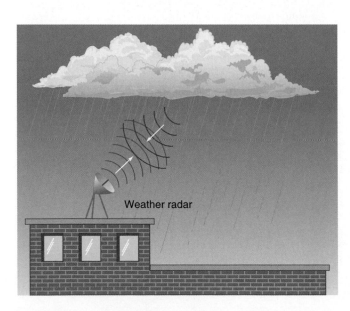

Weather radar

(Figure 5.19). Some of the radar waves are scattered back to the transmitting point, where they are detected. The received signal is called the **radar echo.**

The radar echo indicates the location and intensity of precipitation. The time the radar signal takes to go out to the precipitation and return (on the order of milliseconds) depends on how far away the precipitation is. Precipitation more than about 240 kilometers (150 miles) away cannot be detected, because at such long ranges the radar beam is too high and weak to "see" the precipitating portions of clouds. The direction of the precipitation is simply determined from the direction the radar is pointing when it receives the echo.

The intensity of the radar echo indicates the intensity of the precipitation (e.g., millimeters or inches of rain per hour). Relatively high amounts of returned energy indicate high rainfall rates. The amount of energy scattered is proportional to the size and the concentration of particles. Intense radar echoes imply large particles in high concentrations. In these situations, rainfall is extremely heavy and hail is a possibility.

The radar echo is displayed in a color image (Figure 5.20). The colors represent the amount of energy scattered back to the radar site, or the **radar reflectivity.**

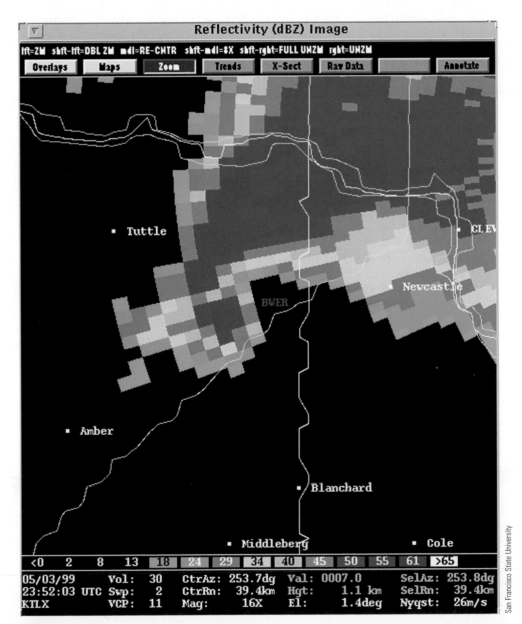

Figure 5.20

An example of a weather radar image from the May 3, 1999 Oklahoma tornado outbreak. Colors represent the amount of energy scattered back to the radar site, or the reflectivity, in dBZ (see colored legend at bottom). High reflectivities are colored red and indicate high precipitation rates.

San Francisco State University

http://info.brookscole.com/ackerman

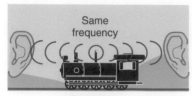

Same
frequency

Stationary

A

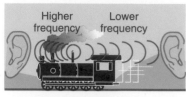

Higher
frequency

Lower
frequency

← **In motion**

B

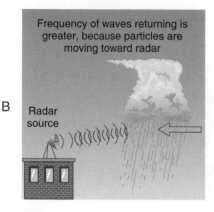

Frequency of waves returning is
greater, because particles are
moving toward radar

Radar
source

Figure 5.21

A, *The frequency of sound
waves change when the source
or the observer is moving. If the
source and observer are converging, the frequency is
higher; if they are moving
apart, the frequency is lower.*
B, *This same effect is used by
Doppler radar to detect wind
speeds in precipitating storms.*

Radar reflectivity is measured in decibels (dBZ). A reflectivity of 20 dBZ usually indicates rain. High reflectivities are colored red and indicate high precipitation rates and, above 55 dBZ, hail. The image is updated roughly every 5
minutes, even more frequently than satellite images. Composites of individual
radar images can be used to give regional and national views of precipitation,
as we will see in Chapter 11.

Doppler Radar

Today's **Doppler radar** can measure not only the location and intensity of precipitation, but also the speed of the wind in precipitating regions. To explain
how it works, we can use an everyday example of city noise.

Sound travels as a wave, just like radio waves. As a car with a loud stereo, or
a siren, or a train approaches you and then passes by, the sound you hear
changes. The pitch is higher (i.e., the frequency increases) as the sound-maker
approaches you, and the pitch goes down as the source of the sound moves away
(Figure 5.21). This phenomenon is the **Doppler effect,** named after the
19th-century Austrian physicist Johann Christian Doppler who first explained it.

To prove the Doppler effect to skeptics, a Dutch meteorologist named
Buys Ballot devised the following experiment. A group of trumpeters stood on
a moving train and played a single note together. The observers on the ground
heard the note go up, and then go down, as the train approached and then
passed by (Figure 5.21, *A*).

The Doppler effect works the same with radio waves. Just like the observer
on the left in Figure 5.21, *A,* the radar receiver "hears" waves of a higher frequency than the waves that were sent out, if precipitation particles are moving
toward the source (Figure 5.21, *B*).

The relative motion of particles measured by a Doppler radar are color-
coded and displayed in an image for quick analysis by a forecaster. Typically, the
cool colors (usually greens) represent motion towards the radar. The warm colors (reds and purples) indicate motion away from the radar. Warm colors next
to cool colors indicate spinning motion within the storm, which may lead to a
funnel cloud or a tornado.

Figure 5.22 is a radar display depicting Doppler velocity observations of a
thunderstorm that spawned a devastating tornado in and near Oklahoma City
on May 3, 1999. The region likely to have a tornado is located where the winds
are traveling toward the radar and then away from the radar over a small distance. We will discuss the radar identification of tornadoes in more detail in
Chapter 11.

Wind Profiler

A recent application of Doppler technology is the **wind profiler.** Wind profilers are
Doppler radars pointed skyward. By measuring the change in the frequency of the
returning radar beam, a wind profiler can determine the wind speed even when precipitation is not present. By alternately pointing the transmitted radio beam in different directions, the three-dimensional wind field can be determined, providing a
continuous measurement of winds at a given location.

Figure 5.23 shows a day's worth of hourly wind observations from the ground to
the stratosphere at a location in Nebraska, with time elapsing from right to left. The
strongest (red) winds are located between 200 and 300 mb, where the jet stream is
located. This type of detailed hourly wind information greatly augments radiosonde
observations, which are made only twice a day. However, wind profilers are more expensive than radiosondes and their coverage is mostly limited to the central United
States.

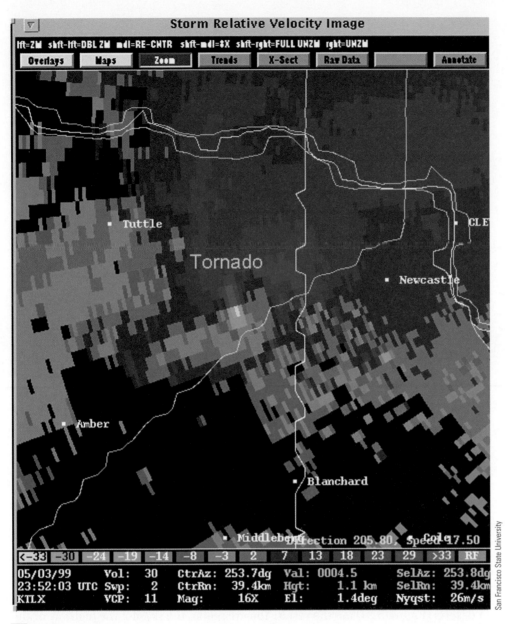

Figure 5.22

A radar display in Doppler mode (also known as velocity mode) of a thunderstorm that spawned a tornado and caused destruction in Oklahoma City on May 3, 1999. Greens represent motion toward the radar and the reds indicate motion away from the radar (see color legend on bottom of figure). The tornado is located in the vicinity of the red-green couplet in the center. Compare this image to Figure 5.20, which is reflectivity from the same radar at the same time.

Atmospheric Optics

As we have seen throughout this chapter, scientific observations of the atmosphere cannot be divorced from the human senses. The best example is vision. Our eyes are passive remote sensing instruments sensitive to light with wavelengths between 0.39 and 0.78 microns. Our brains then process the collected data. This section focuses on some of the beautiful and fascinating atmospheric phenomena that you can observe with your own two eyes.

http://info.brookscole.com/ackerman

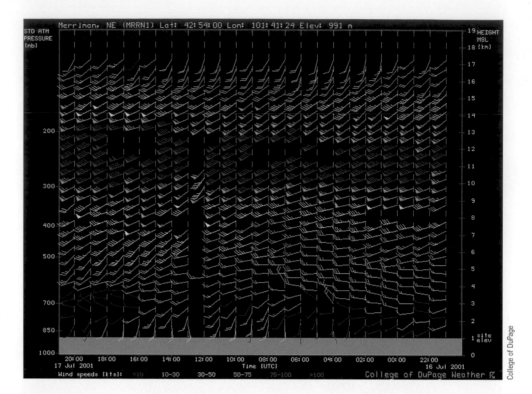

Figure 5.23

Hourly wind profiler data for Merriman, Nebraska during July 16–17, 2001. Time elapses from right to left in this image. Winds are color coded (see key at bottom, in knots) and are also represented using the "flagpole" approach in Chapter 1. The strongest winds are from 200 to 300 mb and are from the west-southwest.

Mirages

Have you ever seen the surface of a highway look watery on a hot dry summer day? This is a classic example of a **mirage.** Mirages are not illusions, but are instead images formed by the refraction of light.

As we learned earlier, refraction occurs when light moves through substances with different indices of refraction. Since the index of refraction for air depends on temperature, light can refract if it passes through layers of air that have different temperatures.

On a summer day, the air just above a highway is much hotter than the air a few meters higher up. The strong temperature gradient causes light from above to be bent downward and then upward toward your eye. "Puddles" on dry highways and in deserts (Figure 5.24) are the result. You are seeing the sky, not the highway or the sand. The refraction is even more obvious when a recognizable object, such as a palm tree (Figure 5.25), is seen in a mirage. It's upside down! For this reason, we call this an **inferior** (meaning "below") **mirage.**

Mirages also develop when the surface is much colder than the air above it, such as over snow, ice, or a cold lake or ocean. In these temperature inversions, light rays passing upward through the warm air in the inversion refract downward toward the surface. A mirage image thus appears above, not below, the true position of the object (Figure 5.26, *A*). This is a **superior mirage.** Superior mirages can bring objects into view that are below the horizon (Figure 5.26, *B*). A special type of superior mirage, the **Fata**

Figure 5.24

An inferior mirage in the desert. The giraffe is not walking on water; instead, refraction causes light from the sky to bend and appear watery.

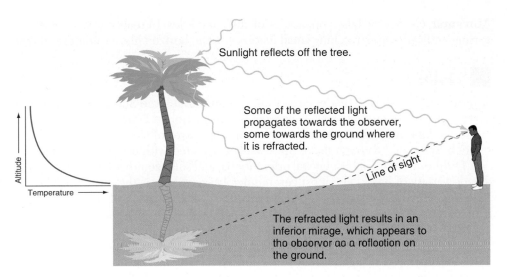

Sunlight reflects off the tree.

Some of the reflected light propagates towards the observer, some towards the ground where it is refracted.

Line of sight

Altitude

Temperature

The refracted light results in an inferior mirage, which appears to the observer as a reflection on the ground.

■ **Figure 5.25**

An inferior mirage, where the refracted image appears below the object's true position, occurs because light is refracted downward and then upward from the ground toward the observer. This requires a strong temperature gradient near the ground.

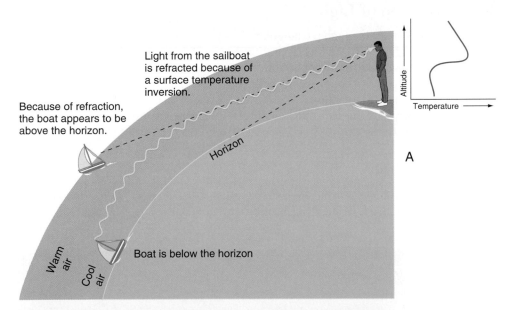

Light from the sailboat is refracted because of a surface temperature inversion.

Because of refraction, the boat appears to be above the horizon.

Horizon

Altitude

Temperature

Warm air

Cool air

Boat is below the horizon

A

B

S. Miyauchi/Weatherwise Magazine/Heldref Publications

■ **Figure 5.26**

A, A superior mirage occurs when air is refracted through a temperature inversion. B, A superior mirage causes objects to appear to hover in mid-air above the horizon, such as these icebergs.

Morgana, creates the false appearance of cliffs or castles. In reality, refraction is distorting and vertically stretching small features on the horizon like a funhouse mirror.

Halos

A **halo** is a whitish ring that encircles but does not touch the Sun or Moon. It is an optical phenomenon that owes its existence to refraction of light by ice crystals. Because the light must shine through a fairly uniform layer of ice crystals that are thin enough to let light through, halos are usually associated with cirrostratus clouds.

Different crystal habits, crystal orientations, and solar zenith angles can produce a wide variety of halos. The most commonly observed halo is the 22° halo. The 22° halo encircles the Sun at about a hand's width from the center of the Sun, if your arm is fully extended. On rare occasions, pencil-shaped crystals can cause halos inside (Figure 5.27) or around the 22° halo, a remarkable sight!

Small column-like ice crystals form the 22° halo. Light rays enter a crystal, refract, and then refract again as they exit the crystal (Figure 5.28). Because the crystals are randomly oriented in space, there are many different directions from which light rays can enter the crystals. More light rays are refracted at this 22° angle than at any other, producing the concentration of light known as the halo.

Dispersion of Light

When sunlight passes through a triangular glass prism, it separates into the colors of the rainbow (Figure 5.29). This separation happens because different colors, defined by their wavelength, refract by different amounts. The shortest (blue and violet) wavelengths refract the most; red light refracts the least. The separation of colors is referred to as **dispersion.** Not only prisms but also the atmosphere, water drops, and ice crystals can cause dispersion. This causes a wide range of breathtaking visual effects in the atmosphere, as we will see.

Green Flash

When the Sun is finally setting or just rising over the horizon, some people have reported seeing a brief, brilliant flash of green from the top part of the Sun's disk.

Figure 5.27

A spectacular series of halos, as seen in the Netherlands. The outermost ring is the 22° halo. The lightpost in the middle is used to shield the camera from the direct light of the Sun; otherwise, the image would be overexposed.

Harald Edens

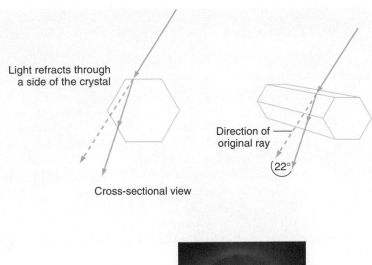

Light refracts through
a side of the crystal

Direction of
original ray

22°

Cross-sectional view

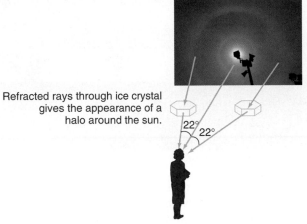

Refracted rays through ice crystal
gives the appearance of a
halo around the sun.

22° 22°

Figure 5.28

Diagram showing how light is refracted through a crystal to form a 22° halo.

Often discounted as myth, this **green flash** is real and is a product of refraction and dispersion.

As the Sun sets, refraction bends and disperses its light. Red bends least and disappears first. Blue and violet should be the last seen, but multiple scattering keeps this light from reaching your eyes. The color with the next-shortest wavelength is green. Green light is refracted more than red, but it is not scattered as effectively as blue. Therefore, a sliver of green appears for a second or two as the Sun disappears over the horizon (Figure 5.30). The best conditions for seeing a green flash are a nearly flat horizon and a lengthy sunset or sunrise.

Sundogs

On sunny mornings and afternoons with high clouds, you are likely to see shiny, colored regions at either side of the Sun. These are **sundogs,** another optical effect

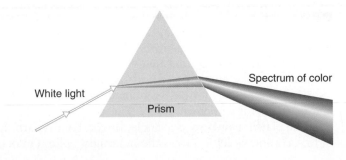

Spectrum of color

White light

Prism

Figure 5.29

Dispersion of light via refraction.

Figure 5.30

The green flash, as captured in a montage of photographs. The green flash is the sliver of green light refracted above the top rim of the Sun at far right.

caused by refraction and dispersion. Sundogs appear because hexagonal ice crystals in the high clouds tend to drift downward with their flat bases parallel to the ground. The sunlight passing through the crystal refracts sideways. If the Sun is low enough in the sky, you see spots of bright light instead of a complete halo. Therefore, sundogs are usually 22° away from the Sun. They can be on one or both sides of the Sun, depending on where the clouds are (Figure 5.31). Refraction causes blue light to be bent more than red light, and so sundogs show the spectrum of colors from red nearest the Sun to blue on the outside of the sundog. Both cirrus and cirrostratus clouds produce sundogs.

Sun Pillar

A setting Sun behind high clouds will sometimes project a narrow column of reddish light straight above it. This is a **Sun pillar.** Although it sounds very similar to sundogs, the Sun pillar is caused by reflection, not refraction, of sunlight. Flat-plate ice crystals floating in the cloud reflect the Sun's reddened light to your eyes in a line above, or sometimes below, the Sun. Pillars can also appear over streetlights when ice crystals are in the air.

Now that we are familiar with both reflection and refraction with dispersion, we can investigate the most famous optical phenomenon of all: the rainbow.

Rainbows

When did you last see a rainbow? What was the time of day? What were the weather conditions? The chances are that you saw it in the late afternoon or early evening, in the eastern sky, and it was or had just been raining. Our ability to make these good guesses is based on the science behind the rainbow. Let's review what it looks like and learn how it works.

The classic **rainbow** is a single, bright, colored arc. Red is the outermost color of this arc and violet is always the innermost color. On occasion, you may have seen

Figure 5.31

A sundog appears in high clouds to the right of the Sun. Its colors are caused by dispersion.

two rainbows at once. The lower rainbow is the **primary rainbow** and the higher, fainter, colored arc is the **secondary rainbow** (Figure 5.32). The color sequence of the secondary rainbow is opposite to the primary; red is on the inside of the arc and violet on the outside. The rainbows usually stand out against a fairly dark sky. To explain these observations we need to trace rays of light as they enter and leave large drops of water.

Raindrops act as prisms, refracting and reflecting light just like ice crystals. Figure 5.33 shows how light rays are refracted at the surface of the drop. This happens because the indices of refraction for water and air are different. Then a total internal reflection occurs at the back of the drop, and the light refracts again as it exits the drop. Because of this, sunlight that shines through water drops tends to brighten the region of the sky opposite from the Sun and especially at an angle of about 40°. The shape of the rainbow is a circle, but we only see the arc above the ground.

According to Figure 5.33, the rainbow is located opposite to the Sun; this explains why rainbows aren't seen at noon with the Sun overhead. The concentration of light rays in Figure 5.33 reveals why the rainbow seems so bright compared to the sky around it. But we haven't yet explained its *color.*

Raindrops refract sunlight, dispersing white light into its spectrum of colors (Figure 5.34). Blue light is bent at a 40° angle, red at 42°, and the other colors are bent at angles intermediate between these two values. These are the colors of the primary rainbow. If the light is reflected twice instead of once, the bending is at different angles: blue is bent at a 54.5° angle, and red is bent at 52°. This second region of color is the secondary rainbow.

Figure 5.34 also explains why the blue colors are on the inside of the primary rainbow, but on the outside in the secondary rainbow. The rainbow is an assembly of different rays of light leaving millions of falling raindrops. If you looked at only one drop you could change the color you see by changing the height of your line of sight. If you stood straight up, you would see the drop as red. If you squatted down, you would come to a height where the drop looked violet. When you see a rainbow you are looking at drops at different heights. The top drops look red, just as if you

Figure 5.32

A double rainbow over a church in Longmont, Colorado. The primary rainbow is the brighter arc on the inside of the fainter secondary rainbow. Notice that the order of colors is reversed in the two rainbows. Why do we look in the opposite direction from the Sun, as in this picture, in order to see rainbows?

were standing straight up while looking at the single drop. And the bottom drops look purple, as though you were squatting. This explains why the outside of the primary rainbow arc is red and the inner portions are violet. Red light appears below violet in the secondary rainbows because of the second reflection. This explains why the color sequence of the secondary rainbow is a mirror image of the colors in the primary rainbow. Using the concepts of reflection, refraction, and dispersion we have explained most aspects of the rainbow!

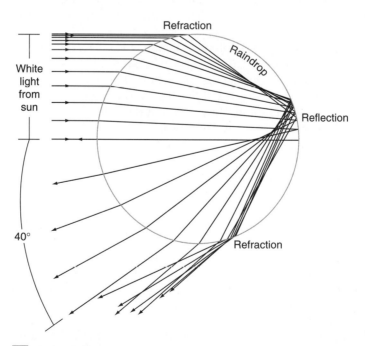

Figure 5.33

How light rays are refracted (surface of raindrop, left) and reflected (back of raindrop, right) to form a rainbow. The cluster of light rays at an angle of about 40° explains the location, shape, and brightness of the rainbow. (Source: Adapted from Greenler, R., Rainbows, Halos, and Glories, Cambridge University Press, 1980, Figure 1-1.)

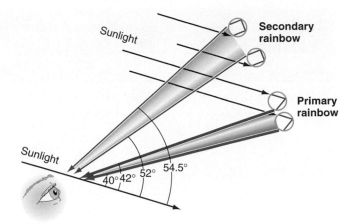

Figure 5.34

Colors of the rainbow arise from drops at different altitudes in the sky. Distinct sections of a rainbow are produced by different drops suspended in the atmosphere. When viewing a rainbow, the Sun must always be behind you.

Coronas, Glories, and the Brocken Bow

Hold the underside of a compact disc up to a light at an angle. Tilt it slightly back and forth. Do you see streaks of color? They resemble a rainbow, but the color stems from a different optical effect: **diffraction.** Diffraction occurs when light is bent around small objects, such as the tiny lines of data in your compact disc.

Cloud droplets also diffract light. When the droplets are of different sizes, the colors blend together and the light is a diffuse white. A **corona** forms when thin water clouds obscure the Sun or Moon and cause a whitish region to form around the Sun or Moon. The key visual difference between the corona and a halo is that the corona is *not* a ring around the Sun or Moon, but is instead a broader, bright region that appears to touch the Sun or Moon.

The color produced by diffraction is called **iridescence.** Clouds become iridescent frequently in many parts of the world. The only ingredients needed are light and water clouds.

We close our tour of atmospheric optics with a spectacular example of iridescent diffraction: the **Brocken bow,** named for a mountain peak in Germany. The photographer in Figure 5.35 was hiking in the mountains above the clouds, which

Steve West

Figure 5.35

A glorious Brocken bow around the shadow of photographer Steve West in the Arizona mountains. The rings of color are caused by diffraction of reflected light by the water droplets in the fog in the picture.

Blue Skies

**Sky Identification/
Atmospheric Optics**

TABLE 5.1	Atmospheric Optical Phenomena Categorized by What Causes Them		
By:	*Air*	*Ice Crystals*	*Liquid Water*
Refraction	Mirages	Halos	R
	Green flash	Sundogs	a
	Twinkling stars		i
			n
Reflection	—	Sun pillar	b
			o
			w
			s
Scattering	Blue skies	White clouds	White clouds with dark bases
	Blue mountains		
	Red sunrises/sunsets		
	Crepuscular rays		
Diffraction	—	—	Corona
			Iridescence
			Brocken bow

happened to be composed of water droplets of about the same size. His body blocked the sunlight, forming a shadow on the fog below. The light *around* his shadow was reflected, refracted, and then diffracted by the water droplets. This produced rings of color similar to a rainbow, hence the name "Brocken bow." It is yet another of the atmosphere's gorgeous optical tricks.

When the Brocken bow forms around the shadow of an airplane, it is usually called a **glory.** Look for it the next time you take an airplane flight over water clouds!

Table 5.1 classifies all of the optical phenomena we have examined in this chapter by the mechanism that forms them.

PUTTING IT ALL TOGETHER

■ SUMMARY

We can observe the atmosphere in several ways. Direct methods measure the properties of the air in contact with the instrument that is sensing it. Indirect methods obtain data from afar. Indirect methods can be either active or passive. Active sensors emit pulses of energy and observe what comes back to the sensor. A passive sensor "sees" whatever energy naturally comes to it. Our eyes passively sense visible light, whereas meteorological instruments can observe in many different regions of the electromagnetic spectrum.

Observations of the atmosphere, both direct and indirect, contribute greatly to our understanding of weather and climate.

The Automated Surface Observing System (ASOS) in the United States observes and records surface temperature, humidity, pressure, wind, precipitation, cloud ceiling, and visibility data as often as once a minute. Rawinsondes provide twice-a-day observations of temperature, humidity, and wind from the surface into the stratosphere.

Satellites observe the global atmosphere by indirect methods. The three most common weather satellite images are the visible, infrared, and water vapor images from geostationary satellites. Visible images provide detailed views of clouds, but only during daytime. Infrared satellite images reveal storm systems via their cloud patterns 24 hours a day. Water vapor images provide meteorologists with information in both cloudy and cloud-free regions of the upper and middle troposphere.

Radar actively senses radio waves scattered by large water and ice particles and converts them into information about precipitation. Doppler radar reveals wind patterns within a storm by tracking the relative motions of precipitation particles. Wind profilers provide detailed wind information from the surface to the lower stratosphere.

Light can be reflected, bent (or refracted), scattered, and diffracted, and its colors can be dispersed. One or more of these processes can cause beautiful optical effects in the sky. The most famous, the rainbow, is caused by reflection, refraction, and dispersion of light in raindrops. Refraction of light by ice crystals causes halos and sundogs. Refraction of light by layers of air of different temperatures produces mirages. Scattering of light by air molecules causes blue skies, red sunsets, and white clouds with dark bases. Diffraction of light creates white fuzzy coronas and brilliantly colored Brocken bows.

"The deeper one enters into the study of Nature, the further one ventures into and along the by-paths that, like a mystic maze, thread Nature's realm in every direction, the broader and grander becomes the vista opened up to the view."

—Wilson "Snowflake" Bentley

KEY TERMS

You should understand all the following terms. Use the glossary and this chapter to improve your understanding of these terms.

Active sensors
Anemometer
Aneroid barometer
ASOS
Brocken bow
Ceilometer
Cloud ceiling
Corona
Crepuscular rays
Critical angle
Cup anemometer
Dew point hygrometer
Diffraction
Direct methods
Dispersion
Doppler effect
Doppler radar
Fata Morgana
GEO satellite
Glory
Green flash
Halo
Index of refraction
Indirect methods
Inferior mirage
Infrared satellite image
Iridescence
LEO satellite
Mercury barometer

Mirage
Passive sensors
Primary rainbow
Propellers
Psychrometer
Radar echo
Radar reflectivity
Radiometers
Radiosonde
Rainbow
Rain gauge
Rawinsonde
Reflection
Refraction
Resistance thermometer
Scattering
Secondary rainbow
Sounding
Sundogs
Sun pillar
Superior mirage
Total internal reflection
Visibility
Visible satellite image
Water vapor satellite image
Wind profiler
Windsock
Wind vane

REVIEW QUESTIONS

1. Classify each of the following as a direct or indirect method of observation: hearing; seeing; touching. Which of these methods are active sensing? Which are passive sensing?
2. Why should temperature measurement be made in a shaded and ventilated location?
3. A bird manages to build a nest over an ASOS dew point sensor. How could this affect the accuracy of the observations?
4. Why can't radiosondes observe the mesosphere?
5. Describe reflection and refraction.
6. Explain how refraction changes the length of daylight.
7. Which of the following phenomena is *not* caused by scattering: red sunrises; blue skies; a hot torch with a blue flame; the "blue" in the Blue Ridge Mountains?
8. Explain why clouds are white even though they are composed of water drops, which are transparent.
9. Why are there both LEO and GEO weather satellites? Why not use just one type of orbit?
10. You want to look at a satellite image of a hurricane over the Gulf of Mexico at 0600 UTC. Which type of image do you choose: visible, IR, or water vapor? If you want to see the wind patterns in the upper troposphere that guide the hurricane's direction, which type of image do you choose?
11. If you wanted to launch a satellite to observe the ozone hole, which orbit would you want to place the satellite in: LEO or GEO?
12. Why are there weather radars located every few hundred miles across the United States, whereas one GEO satellite can see the entire country?
13. Police radars that measure the speed of cars on a highway are based on what scientific principle?
14. Compare and contrast mirages with images in a mirror.
15. You are in Minnesota during a snowy winter. On a bitterly cold, clear morning, a lone cow stands in a snowy field on the horizon. What kind of optical effect do you expect to see? Will you see an upside-down cow, or a cow looming over the horizon?
16. Using the *Blue Skies* exercise "Sky Identification/Atmospheric Optics," study the appearance of various optical effects.
17. Visit an art gallery to determine if rainbows, halos, or other optical phenomena are correctly drawn. For example, are the colors in the correct order?
18. Explain how you can see colored "bows" near waterfalls or in the spray of a lawn sprinkler. Can you also expect to see halos in these situations? Explain why or why not.
19. Can a totally dry atmosphere have a corona? Can it produce a green flash?

WEB ACTIVITIES

Choose Chapter 5 on the textbook's Web site:
http://info.brookscole.com/ackerman
and select from the following resources:
- Interactive Modules that illustrate and extend your understanding of key topics in this chapter
- Tutorial Quizzes to test your mastery of terms and concepts

 For additional readings, go to the InfoTrac College Edition, your online library, at:
http://www.infotrac-college.com

Atmospheric Forces and Wind

After completing this chapter, you should be able to:

- List and describe the forces that act on the atmosphere
- Define various force-balances and relate them to the real-life horizontal wind patterns that they help us understand
- Use force-balance concepts to explain why air rises vertically in a low-pressure system but sinks in a high-pressure system, and how this creates different types of weather in highs versus lows

Introduction

Why does the wind blow? That question was asked across the nation of France on Christmas night, 1999. A large storm racing toward Europe blasted France with extremely high winds on that terrifying night. Along the English Channel, the wind gusted well over 200 km/hr (125 mph). As shown below, most of France experienced winds of more than 80 km/hr (50 mph).

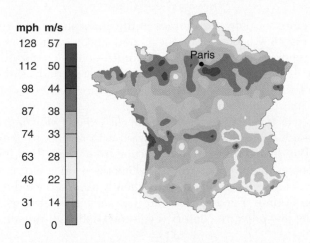

mph	m/s
128	57
112	50
98	44
87	38
74	33
63	28
49	22
31	14
0	0

In the city of Paris the high winds damaged many cultural monuments, including the cathedral of Notre Dame. The winds destroyed 80% of the rare species of trees at the Palace of Versailles, uprooting approximately 10,000 trees on the palace grounds (right). Of these, 90% were more than 200 years old, including trees planted by Napoleon and Marie Antoinette. The devastation took only two hours.

Versailles did not suffer alone. Nearly 60% of France's forests were destroyed by the storm. There was

little or no warning. And less than 2 days later, another windstorm streaked across southern France and caused even more damage throughout Europe!

AP/Wide World Photos

Wind is air in motion that arises from a combination of forces. Violent destructive winds, as well as gentle summer breezes, result from a complex interplay of different forces. In this chapter we explain the forces of nature that affect the atmosphere, how they combine to create wind, and how to use weather maps to interpret wind direction and speed.

Magnitude and Direction of Forces

Wind and weather result from the interaction of several forces in the atmosphere. Before we discuss the specifics of these forces, we need to understand how to deal with forces in general.

Forces are characterized by direction and magnitude (or strength). When discussing the interactions of forces on an object or a region of the atmosphere, we

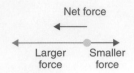

Figure 6.1

A parcel of air (blue circle) will move in the direction of the net force, which is the sum of the forces acting on the parcel.

must analyze not only how strong they are but also the directions in which they push. To illustrate forces and understand their interactions, we represent their strength and direction by arrows. The arrow points in the direction that the force is acting. The length of the shaft represents the magnitude. (On weather maps, where space is tight, the wind magnitude is represented with "flags," as we saw in Chapter 1.)

Two or more forces can act on the same point. The overall force that results from interacting forces can be expressed as a single force, called the *net force* or the *resultant.* If the two forces act in opposite directions and with different magnitudes, the object will move in the direction of the stronger force (as in a tug-of-war). A single resultant force can represent these two forces. Its direction will be the same as the stronger force, and its magnitude will be the difference between the two forces (Figure 6.1).

In the atmosphere, forces often act at an angle to each other. In this situation, we need a method of determining the magnitude and direction of the net force. One way is to construct a parallelogram graphically (Figure 6.2) using two forces to represent two of the sides. The diagonal of the parallelogram represents the net force of the two given forces. The length of the diagonal represents the magnitude of the combined forces. The direction of the net force is along the diagonal and away from where the two forces are applied. Now that we know how to visualize the interaction of forces, we can examine the laws that explain how forces create motion.

Laws of Motion

A force applied to an object often results in movement. When an object is changing its position, we say it is *in motion.* As with force, motion has a magnitude and a direction. An object's **velocity** is the magnitude and direction of its motion. The **speed** of the object, the distance traveled in a given amount of time, is the magnitude of the motion. A change in an object's velocity—either magnitude, or direction, or both—is its **acceleration.**

Wind is air in motion. Before we delve into forces in the atmosphere, let's review wind terminology. As we saw in Chapter 1, weather reports include wind speed and direction. Wind speed is reported on weather maps in **knots**—one nautical mile per hour (equivalent to about 1.15 miles per hour or 0.5 meters per second). If the wind speed is strong (greater than 15 knots) and highly variable, the weather report will include **wind gusts,** the maximum observed wind speed.

Wind direction is reported as *the direction from which the wind is blowing.* A north wind blows from the north to the south. It is reported with respect to compass directions (Figure 6.3). **Windward** refers to the direction from which the wind is coming, while **leeward** denotes the direction toward which the wind is blowing. The *prevailing wind direction* of a region is the most frequently observed wind direction during a given period of time. Finally, winds are normally associated with pressure systems, for example, low-pressure centers **(cyclones)** and high-pressure systems **(anticyclones).**

Newton's First Law: Law of Inertia

Fundamental laws of physics explain atmospheric motions. In the 17th century, Sir Isaac Newton—who also first explained the colors of the rainbow—defined the laws of inertia and of momentum. These "Newton's Laws" explain the movement of galaxies, planets, cars, baseballs, and parcels of air.

Newton's First Law of Motion states that *a body at rest tends to stay at rest while a body in motion tends to stay in motion, traveling at a constant speed and in a straight line, until acted upon by an outside force.* The resistance to a change in motion is called **inertia,** and so this law is often called the **Law of Inertia.**

Figure 6.2

The graphic method of determining the resultant of two forces acting at an angle to one another on a parcel of air.

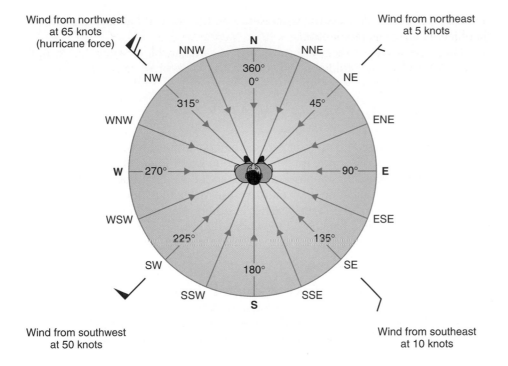

Figure 6.3

Wind direction is expressed as degrees around a circle (clockwise from the top, or due north), or as compass points. Outside the circle, wind direction and speed are depicted using "flagpoles" with flags attached, as explained in Chapter 1.

The Law of Inertia can be easily related to everyday experience. For example, if a car suddenly starts moving, the passengers lurch backward. The passengers are initially at rest and tend to remain at rest when the car first starts. The moving car is exerting a force that eventually gets the passengers moving at the same speed. If the car then suddenly stops, the passengers surge forward. That is, they tend to remain in motion. If the car makes a quick, sudden right turn, the passengers, whose inertia makes them "want" to continue traveling in a straight line, are crammed toward the left side of the car.

Similarly, if forces begin to act on a region of the atmosphere, it will begin to move. But it takes a while to get it moving. We will usually describe this with reference to small regions or "parcels" of air. However, an air mass will also tend to remain in the same place, if atmospheric forces acting upon it are very weak or absent. A heat wave is often the result of an unmoving or slowly moving mass of air remaining over a location, because the forces acting on it are relatively weak.

Newton's Second Law: Law of Momentum

Newton's Second Law of Motion states that a force exerted on an object (or a parcel of air) of a given mass causes the object to accelerate in the direction of the applied force. The relationship between acceleration, the force, and the mass is expressed mathematically as

$$\text{Force} = \text{Mass} \times \text{Acceleration}$$

or, as it is often written in meteorology,

$$\text{Acceleration} = \text{Force} \div \text{Mass}$$

This law, and most equations in meteorology that involve acceleration, go by the name of the **Law of Momentum.** The **momentum** of an object is its mass multiplied by its velocity. Two objects can have the same velocity (or mass) but the one with greater mass (or velocity) has greater momentum. For an object whose mass does not change, mass multiplied by acceleration equals the change in momentum.

Since force also equals mass times acceleration, this means that applying a force to an object changes its momentum.

The momentum of the wind can propel a sailboat, blow over a truck, uproot large trees, and topple buildings. We now discuss what forces cause the air to move.

Forces That Move the Air

Five different forces combine to move air: the gravitational force, the pressure gradient force, the Coriolis force, friction, and centrifugal force. Some of these forces are *real*, meaning that they are observable no matter what your perspective is. Other forces are *apparent*, meaning that you may observe them in one frame of reference (for example, on a rotating Earth) but not in others. We must examine these forces to determine where and how hard the wind will blow.

Gravitational Force

The familiar **gravitational force** is directed downward perpendicular to the ground and is approximately equal to the mass times the gravitational acceleration, 9.8 meters per second per second. The "per second per second" is not a typographical error. Remember that acceleration is the change of velocity over time, and the units of velocity are meters per second.

BOX 6.1

Planes and Pressure Differences

Pressure differences and wind explain how planes fly. If you chopped an airplane wing in half from front to back and looked at it from the side, you would see that it has a distinctively curved shape. The wings are designed this way so they can lift the plane off the ground. The forces responsible for lifting the plane off the ground are produced by pressure differences between the wings' top and bottom resulting from the wind flow over the wing.

You can demonstrate the basic concept of airplane flight for yourself by shaping a piece of cardboard like a wing, securing it to a table, and blowing air over it with a hair dryer (see below). The cardboard will lift off the table!

Bernoulli's Principle (named after the Swiss scientist Daniel Bernoulli) states that when a fluid speeds up the pressure lowers, and where the fluid slows down the pressure is high. If an airplane's wing is cutting through the air at a constant speed, as the air flows over the top of the wing it has to speed up to get to the same point at the same time as the air flowing under the wing. This lowers the pressure above the wing, creating a pressure difference between the bottom and top of the wing. The result is an upward force called *lift*. There is also a *drag* force acting to slow down the moving wing. The net effect of lift and drag, if the wing is designed correctly and drag is smaller than lift, forces the wing and the plane upward off the ground—Takeoff!

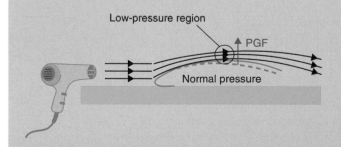

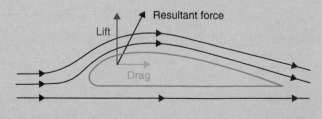

If the gravitational force were the only force acting on the atmosphere, Chicken Little would have been right: the sky would indeed be falling! However, the gravitational force is almost equally opposed everywhere by the vertical portion of the force we will examine next: the pressure gradient force.

Pressure Gradient Force

Spray-paint cans have labels warning that the contents are under pressure. When the nozzle is squeezed, or the sides are punctured, the large pressure difference over the small distance between the air and the inside of the can forces the contents out of the can. The force that results from pressure differences over distances in a fluid, such as our atmosphere, is called the **pressure gradient force (PGF).**

A **pressure gradient** is simply a change in pressure over a distance. Atmospheric pressure differs from place to place for many reasons, all of which can usually be traced back to uneven heating of the Earth's surface by the Sun. Generating pressure differences is also important in flying and sailing (Box 6.1)

The PGF always pushes from higher pressure toward lower pressure (Figure 6.4). Its magnitude is equal to the change in pressure over some distance, divided by the air density. Mathematically, it is written per unit mass as

$$PGF = -\frac{1}{\text{Air density}} \times \frac{\text{Change in pressure}}{\text{Distance}}$$

The minus sign is a reminder that the force is pointed "downslope" toward lower pressure. When pressure changes rapidly over a small distance, the PGF is large. Strong winds almost always result from large pressure gradients. However, the reverse isn't always true: large pressure gradients don't always cause strong winds, as we will now see for ourselves.

The largest pressure gradients in the atmosphere are nearly always right over our heads. As we learned in Chapter 1, typical atmospheric pressure decreases from over 1000 mb at the surface to only half that value only 5 kilometers overhead. This amazingly large pressure gradient is directed upward, because the highest pressure is at the surface. But we don't feel sucked or pushed upward off the ground! When we discuss hydrostatic balance below, we will learn why this vertical pressure gradient does not cause an enormous upward rush of air.

The pressure gradients that most often cause wind are in the horizontal direction. On the surface weather map, atmospheric pressure measured at the surface is converted to sea-level pressure, as we discussed in Chapter 1, to remove pressure differences caused by altitude variations. This allows meteorologists to isolate the wind-causing horizontal pressure gradients on the surface weather map. Then the sea-level pressure patterns are analyzed. Isobars of constant sea-level pressure are drawn at 4-mb intervals on either side of 1000 mb (Figure 6.5). Since the pressure difference between any two isobars is fixed, the closer the isobars, the larger the pressure gradient, the stronger the PGF, and in general the greater the wind speed.

The surface map plots atmospheric pressure adjusted to the altitude of sea level. Another type of weather map reverses this approach and plots the *altitude* of a given *pressure* surface. This map is commonly used when analyzing the weather above the surface, and is called a *constant-pressure chart* or **isobaric chart.** Constant-pressure charts are commonly drawn for 850, 700, 500, 300, 250, and 200 mb. Figure 6.6 is an example of a 500-mb isobaric chart. The units of altitude are called *geopotential meters* and are nearly equivalent to geometric meters measured with a meterstick.[1]

Isobaric maps are useful for portraying horizontal pressure gradients above the ground. The spacing between the lines of constant height (isoheight) is proportional

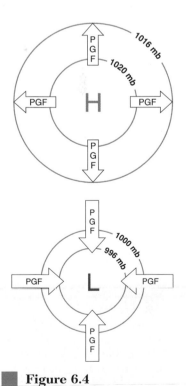

■ **Figure 6.4**

The PGF always pushes directly from higher toward lower pressure.

---

[1]The altitude of the pressure surface is measured in units that are proportional to the potential energy of a unit mass of air at this altitude relative to that at sea level.

Sea-level pressure (mb)/surface wind speed (knots)   1200 UTC September 6, 2001

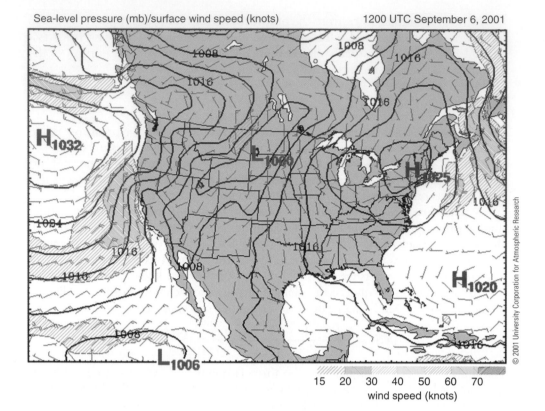

Figure 6.5

A typical surface weather map showing the isobars and wind direction and speed (higher speeds are shaded).

Heights (tens of meters) and winds (knots) at 500 mb   0000 UTC April 22, 2001

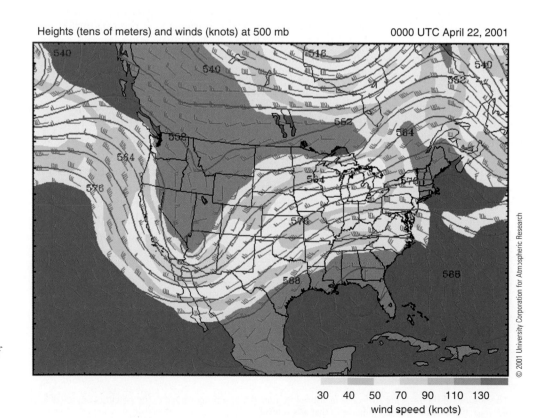

Figure 6.6

A 500-mb isobaric chart for a severe-weather outbreak over Kansas. Note that the isobars are replaced with gray lines of constant height, or isoheights. Notice that the highest winds (shaded) occur where the isoheights are closest together.

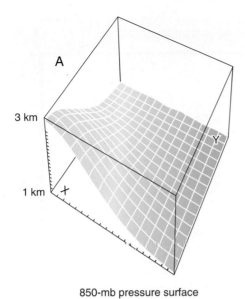

850-mb pressure surface

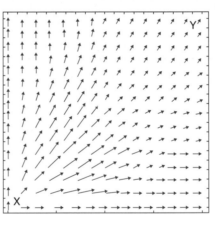

Contours of the altitude of 850-mb pressure
(drawn every 0.2 km)

C

PGF along 850-mb surface

Figure 6.7

Isolines of constant height are proportional to the PGF. In **A,** *the 850-mb pressure surface is represented by the colored diagram. The 850-mb pressure is higher over location X than over location Y.* **B,** *Lines of constant height of the 850-mb pressure show this relationship.* **C,** *The PGF, represented by arrows, occurs where the lines of constant altitude are closer together. The PGF acts perpendicular to the lines of constant altitude of the pressure.*

to the PGF. Figure 6.7 illustrates this relationship between the spacing of isoheights and isobars. The colored surface (Figure 6.7, *A*) is a constant pressure surface of 850 mb. Anywhere on this surface the pressure is 850 mb. The altitude of this 850-mb surface varies; for example, its altitude is higher at point X than at point Y. Figure 6.7, *B*, illustrates height lines of the 850-mb pressure surface. The magnitude of the PGF is proportional to the spacing of the contour lines. This is shown in Figure 6.7, *C*. The steeper the slope of the pressure surfaces, the greater the PGF, because the pressure gradient is the change in pressure over distance. The magnitude of the PGF is therefore weak near location Y and strong near location X. The direction of the PGF is perpendicular to the lines of height, and points toward lower heights.

To summarize, on an isobaric map the lines of constant height of the pressure surface allow us to infer the direction and magnitude of the PGF. Referring again to Figure 6.6, the stronger winds are located in regions where the spacing of the height lines is at a minimum, as we would expect, because this is where the PGF is the largest. But notice that the winds are *not* in the direction of the PGF! The winds, in general, are blowing to the right of the PGF, and are nearly parallel to the height lines. This indicates that there must be other forces acting on the winds.

■ Coriolis Force

The rotation of the Earth produces a force that is felt in the frame of reference of the spinning planet. This force is the **Coriolis force,** named for the 19th-century French engineer who explained it mathematically. The Coriolis force changes the direction—but not the overall speed—of anything moving with respect to the ground, including air.

Why does the Coriolis force arise? Because the Earth is a rotating sphere, locations on the Earth's surface move at different speeds depending on their latitude (Figure 6.8). A point on the equator and a point at the latitude of, for example, New

Blue Skies

Atmospheric Forces/Coriolis Force

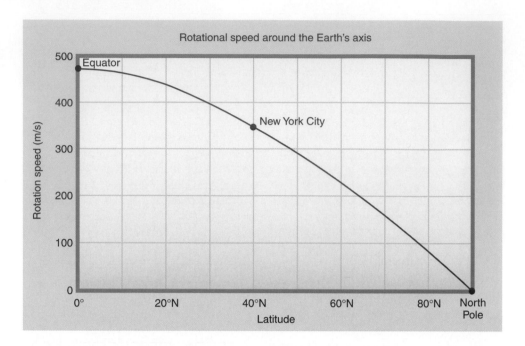

Figure 6.8

The rotational speed of the Earth is greatest at the equator and is zero at the poles.

York City, both make one complete revolution in 24 hours. But a trip around the Earth at the equator is longer than a trip around the Earth at the latitude of New York (40°N). So, the point on the equator has to move faster than the point at New York City in order to complete its journey. This difference causes the Coriolis force.

Here's how the Coriolis force works. Consider a parcel of air moving on its own above two cities at different latitudes, say, from above City X to above City Y (Figure 6.9). Newton's First Law says the parcel will continue to travel at the same speed and in a straight line. But, because of the Earth's rotation, its path does *not* appear to be a straight line to observers who are rotating along with the Earth.

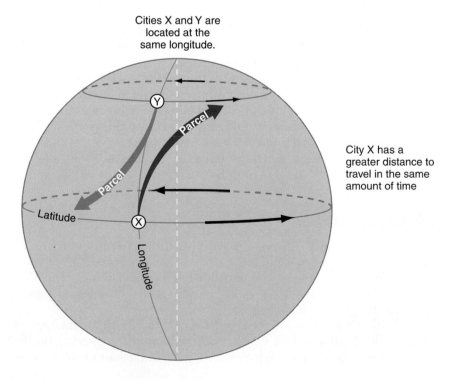

Figure 6.9

The red arrow represents the path traveled by a parcel of air originating at City X and heading toward City Y. The blue arrow represents a parcel of air traveling from City Y to City X. In both cases, the Coriolis force makes the parcels appear to deflect to the right, from the point of view of an observer facing in the direction of travel. This force is due to the differences in speed illustrated here and in Figure 6.8.

Here's why the Coriolis force causes an apparent change in the parcel's direction in the eyes of the Earth-bound observers. As the parcel heads north toward Y, it carries with it the eastward momentum that the Earth's rotation imparts to it at X. City X is moving eastward faster than Y, because it has more distance to travel in the same amount of time. Because of this difference in latitude, therefore, the parcel of air appears to earthbound (rotating) observers to gain speed toward the east as it moves more slowly northward. The combination of its northward and eastward velocities (recall Figure 6.2) results in a bending path that deflects to the right of its destination. If the journey were reversed and the parcel were headed from Y to X, it would still appear to veer to the right of the direction of travel. This gives us a fundamental rule: *the Coriolis force deflects moving air to the right in the Northern Hemisphere.* (This rule works for motions in all horizontal directions, not just those to the north and south. It can be explained by the Conservation of Angular Momentum, which we discuss in Chapters 7 and 10.)

In the Southern Hemisphere, this rule is reversed, because the Earth's rotation is clockwise as viewed at the South Pole. (Spin a globe and look at it from below to prove this to yourself!) This gives us another fundamental rule: *The Coriolis force deflects moving air to the left in the Southern Hemisphere.* More generally, the Coriolis force is nearly always responsible for any wind pattern that changes significantly with latitude. The Coriolis force is also the reason why certain wind patterns show a preference for spinning in only one direction in the Northern or Southern Hemisphere.

The magnitude of the Coriolis force *(CF)* per unit mass is proportional to the distance from the equator and the speed of the wind. This can be written mathematically as

$$CF = \pm fV$$

in which V is the wind speed, and *f* is the "Coriolis parameter," which is defined as follows:

$$f = 2 \times \text{Earth's rotation rate} \times \text{the sine of the latitude}$$

In the equation for determining the magnitude of the Coriolis force, the plus sign is for winds in the Northern Hemisphere from the south or east directions, indicating a Coriolis force toward the east or north, respectively. The minus sign is for Northern Hemisphere winds from the north or west directions, representing a Coriolis force toward the west or south, respectively.

Some important properties of the Coriolis force are shown in Figure 6.10. The Coriolis parameter *f* is zero at the equator (0°) because the sine of the angle zero is

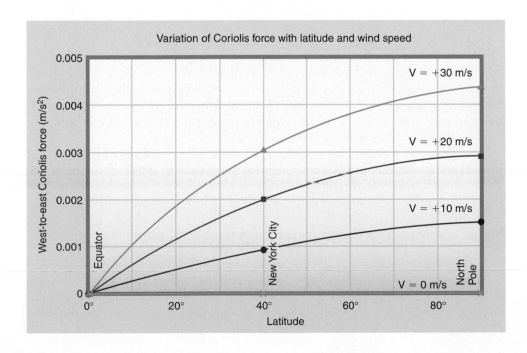

Figure 6.10

The strength of the Coriolis force increases with both latitude and wind speed, but is zero at the equator and for calm winds.

Going Down the Drain with the Rossby Number— Clockwise or Counterclockwise?

Most people, including famous cartoonists (see below), seem to think that water draining in a sink or bathtub spins in opposite directions in the Northern and Southern Hemispheres.

For this to be the case, the Coriolis force would have to be a major influence on the draining water. However, in this chapter we state that for the Coriolis force to be important, the motion (of air or water, it makes no difference) must be on a fairly large scale and persist for a length of time comparable to that of a day—24 hours. Does this sound like your sink or bathtub? No, not unless your bathtub is the size of the Atlantic Ocean and takes half a day to empty.

Therefore, the short answer to the age-old question is: the water draining from a tub or sink spins in either direction in either hemisphere, because it is influenced more strongly

by other forces, such as the centrifugal force, than by the Coriolis force. And it does the same at the equator, too! *Water going down the drain is too small and quick to "feel" the Earth's rotation via the Coriolis force, no matter where it is.*

We can make this point even more dramatically by using some simple mathematics. Meteorologists quantify which forces are important in a weather situation by calculating their ratios. The ratio of the centrifugal force to the Coriolis force is a special quantity called the *Rossby number*, abbreviated *Ro* in the equation below and named after the famous discoverer of the jet stream, Carl-Gustav Rossby (in his shorts at right)

$$Ro = \frac{CENTF}{CF} = \frac{\frac{V^2}{R}}{fV}$$

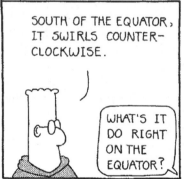

equal to 0. This means that the Coriolis force $\pm fV$ is zero at the equator. This also implies that the Coriolis force increases toward the poles. Therefore, winds at the equator are unaffected by the Coriolis force, but winds at higher latitudes are strongly affected by the Coriolis force. This equation also reveals that if the wind speed is zero, the Coriolis force is zero. The magnitude of the Coriolis force increases with increasing wind speed.

We have to consider the Coriolis force *only* when dealing with objects or winds that travel over long distances and do so over periods of time comparable to the length of a day. In Box 6.2 we address a famous question: Does the Coriolis force affect water spiraling down the drain of a sink or bathtub? The answer may surprise you.

■ Centrifugal Force/Centripetal Acceleration

When an object changes its direction of motion it is accelerating even if its speed does not change. This acceleration is called **centripetal acceleration** (*centripetal* means "pushed toward a center"). You experience this acceleration when you are riding in a car that makes a sudden turn and you are forced away from the direction

BOX 6.2
(continued)

Walter Munk

When the Rossby number is very small, the Coriolis force is much bigger than the centrifugal force. In these cases, wind and water motions change direction in the Northern vs. Southern Hemispheres. In the case of the drain, the spiral is an area of low pressure and—*if* the Coriolis force mattered—would spin counterclockwise in the Northern Hemisphere. (So the "Dilbert" cartoon is wrong, regardless.)

When the Rossby number is very large, the Coriolis force is much smaller than the centrifugal force. In these cases, wind and water motions do not "feel" the Coriolis force and spin in whichever direction other forces impart on them.

So, to summarize: low *Ro* means flip-flopping flow directions across the Equator; high *Ro* means that water down the drain spins in any old direction, heedless of the Coriolis force.

Now let's do the key calculation: what is the Rossby number for your sink or bathtub? The size of this number will tell us if water spiraling down the sink spins in different directions across the Equator (large = no; near zero = yes).

To make this calculation, all we have to do is come up with some ballpark estimates for the size and speed of the water in your sink. Let's assume you are at the latitude of Madison, Wisconsin, where the Coriolis parameter *f* is equal to .0001 per second. We'll be generous and say that the size of the spiral down the drain is large, around 1 meter. We'll be equally generous and say that the speed of the spiraling flow is fast, about 1 meter per second. Using these numbers, we find that:

$$Ro_{\text{your sink}} = \frac{\dfrac{(1 \text{ m/s})^2}{(1 \text{ m})}}{} = 10,000$$

The Rossby number for your sink is about 10,000! Clearly this is a large value, much, much larger than zero. This estimate tells us with mathematical certainty that the Coriolis force is nowhere close to being a factor in the water spinning in your sink. And so the water can spin in either direction in either the Northern or Southern Hemisphere.

of the turn. From a nonmoving observer's point of view, an acceleration has been applied to cause a turn and your inertia has carried you straight ahead. From your point of view, however, a force has been exerted on you—and we call this apparent force the *centrifugal force*. (*Centrifugal* means "pushed or fleeing outward from a center"; the ending *fugal* has the same basic meaning as the word *fugitive*). Depending on the context and point of view, we refer to this one effect by either name.

The direction of the centripetal acceleration is always toward the center of the curve, perpendicular to the direction of motion. The centripetal acceleration (CENTF) per unit mass can be written mathematically as

$$\text{CENTF} = V^2 \div R$$

where *V* is the wind speed and *R* is the radius of curvature of the curved path.

The faster the speed and the tighter the curve of the path traveled (i.e., smaller *R*), the larger the centripetal acceleration. In the atmosphere, we define *R* to be positive when the direction of the curving wind is in the same sense as the Earth's rotation, or counterclockwise in the Northern Hemisphere. This occurs with cyclones.

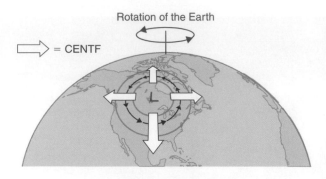

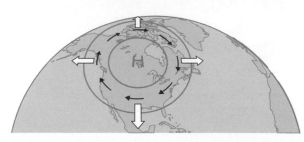

Figure 6.11

The centrifugal force CENTF always pushes outward from the center of curved winds, and is greater for stronger winds and more tightly curved wind patterns. Notice that CENTF does not reverse direction, as does the PGF, for low- vs. high-pressure centers.

In anticyclones, R is negative. Since V^2 is never negative, the sign of centripetal acceleration is positive for cyclones and negative for anticyclones. This implies that the centrifugal force pushes in the opposite direction of the PGF in lows, but in the same direction as the PGF in highs. All of these aspects of the centrifugal force are illustrated in Figure 6.11.

From the equation, we see that the centrifugal force will give air parcels a strong push only during high winds that curve sharply (i.e., those that have a small "turning radius" R). As a result, this force is most important in hurricanes and tornadoes, although this force plays a significant role in curving jet streams as well.

Frictional Force

The **frictional force** in the atmosphere is, for our current purposes, caused by the flow of wind over the roughness of the Earth's surface. In these cases friction opposes, or decelerates, the wind in the same way that a rough road surface slows down a car (Figure 6.12).

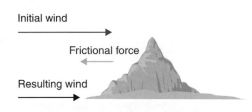

Figure 6.12

The frictional force opposes the direction of the wind but usually is not strong enough to stop the wind altogether.

Mathematically, the frictional force (*FF*) per unit mass can be written very simplistically as

$$FF = -kV$$

in which k is a parameter describing the roughness of the Earth's surface and V is the wind speed. The minus sign emphasizes that friction opposes the the wind regardless of its direction.

The roughness of the surface and the speed of the wind determine the magnitude of frictional force. The force of friction over smooth, still water is smaller than it is over trees in a forest or a rough, craggy mountain peak, and it is nearly zero above the lowest kilometer or two of the atmosphere. The magnitude of the frictional force increases with increasing wind speed.

Through turbulence, friction plays a crucial role in determining the winds near the Earth's surface—so crucial, in fact, that we return to this subject in more detail at the beginning of Chapter 12. The text's Web site explores the role of friction in the flight of a home run in baseball stadiums at different altitudes.

Friction and Fly Balls

Putting Forces Together: Atmospheric Force-Balances

Before we proceed, let's look back at the forces we have discussed. Gravity is the strong, silent type that constantly affects air, but as we'll see, it does not usually lead to wind. The PGF arises from pressure gradients, which ultimately trace their origins to solar heating. This gets the atmosphere moving horizontally and causes wind.

Only when the PGF has caused wind can the other forces play roles. This is easily demonstrated—just look at the equations for the Coriolis, centrifugal, and frictional forces. All of them contain V, the wind speed. Therefore, zero wind speed means no Coriolis, no centrifugal, and no frictional forces. For this reason it's accurate to say that the PGF *acts* to cause the wind, whereas the Coriolis, centrifugal, and frictional forces *react* to the wind to change its speed and/or direction.

To understand the wind, however, it is not enough to study forces. It is the *balance* of these forces that usually determines the strength and direction of the wind. Newton's Second Law is more precisely written as

$$\text{Sum of Forces} = \text{Mass} \times \text{Acceleration}$$

Therefore, we need to put together all the forces that act on an air parcel or a region of the atmosphere. When they add up to zero, in terms of both magnitude and direction, a balance has been achieved.

According to Newton's Second Law, a balance of forces equals zero acceleration. This does not necessarily mean zero wind, because acceleration is the change of wind, not the wind itself.

Next we examine a few different ways that these five forces balance each other and create wind, as outlined in Table 6.1. As we study force-balances, remember what we have learned about the forces: A strong PGF means a strong wind; the Coriolis force acts to change the direction of the wind, but has no effect on the wind speed and is zero for calm conditions or at the equator; the centrifugal force acts only on curving winds; and friction slows down the wind regardless of its direction.

Balance in a Tornado

Force-Balance	Explains Why	Gravitational Force	Vertical PGF	Horizontal PGF	Coriolis Force	Centrifugal Force	Frictional Force
				Forces That Balance Each Other (Checkmarks)			
Hydrostatic	• The sky isn't falling (or rising)	✓	✓				
Geostrophic	• Wind blows parallel to, not across, isoheights on isobaric charts • Wind is stronger when isobars or isoheights are closer together • Wind is clockwise (counterclockwise) around high (low) in the Northern Hemisphere			✓	✓		
Gradient	• High-pressure and low-pressure areas differ in both wind direction and wind strength			✓	✓	✓	
Guldberg–Mohn	• Wind blows across isobars toward lower pressure at the surface • Lows are cloudy and wet, but highs are sunny and dry		✓	✓			✓

TABLE 6.1 Some Atmospheric Force-Balances

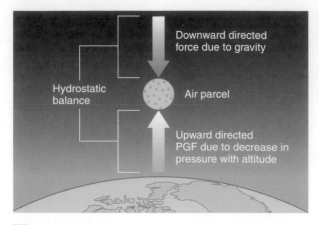

Figure 6.13

An air parcel in hydrostatic balance does not accelerate vertically because the vertical pressure gradient and gravitational forces acting on the parcel are balanced. The parcel is able to move vertically, although in reality these motions are tiny. The parcel can still be accelerated horizontally. (Source: Lester, P., Aviation Weather, 2nd ed., Jeppesen, 2001, p.3-3.)

Hydrostatic Balance

Near the Earth's surface the pressure decreases by about 10 mb for every 100-meter increase in altitude. This is a much stronger pressure gradient than in a hurricane! Why don't we observe strong vertical winds as a result of this strong vertical pressure gradient?

The explanation is that motion results from an imbalance of forces, and in the atmosphere the vertical pressure gradient is usually in balance with gravity. When these two forces are equal and push in opposite directions, a **hydrostatic balance** exists (Figure 6.13).

In the case of hydrostatic balance, not only are accelerations zero but vertical motions themselves are small—usually only about 1 centimeter per second. Much stronger vertical winds usually occur only in and near thunderstorms, which disrupt the balance between the gravitational and the vertical PGFs.

We can express hydrostatic balance per unit mass as an equation by adding the expressions for the vertical pressure gradient and the gravitational acceleration and setting them equal to zero (i.e., they are balanced, equal, and opposite to each other):

$$\text{PGF} + \text{GF} = 0$$

or

$$\frac{-1}{\text{Air Density}} \times \frac{\text{Change in Pressure in Vertical}}{\text{Vertical Distance}} - 9.8 \text{ m/s per second} = 0$$

In a balanced equation if one term decreases, another must increase to keep their sum equal to zero. For example, if air density in the previous equation decreases, the vertical change in pressure can increase to compensate. Because the density of hot air is less than the density of cold air, this equation implies that the pressure must decrease with altitude more rapidly in cold air than in hot air. We will use this fact later, when we discuss the thermal wind and the sea breeze.

Geostrophic Balance, the Geostrophic Wind, and Buys Ballot's Law

The most fundamental *horizontal* force-balance arises when the PGF is counterbalanced by the Coriolis force. This is called **geostrophic balance** (the Greek *geo* means "Earth" and *strophic* means "turning"). In other words, this balance exists because the Earth turns and causes the Coriolis force. It can be written as

$$\text{PGF} + \text{CF} = 0$$

or

$$\frac{-1}{\text{Air Density}} \times \frac{\text{Change in Pressure in Horizontal}}{\text{Horizontal Distance}} \pm \text{fV} = 0$$

Why is geostrophic balance so important? To a good first approximation, it allows us to explain both the direction and the strength of the wind on most weather maps, particularly on isobaric charts representing weather above the Earth's surface.

Here's how geostrophic balance explains horizontal winds: In Figure 6.14, a typical pressure pattern aloft is depicted. The PGF always pushes from higher toward lower pressure. The Coriolis force opposes it. However, for the Coriolis force to do

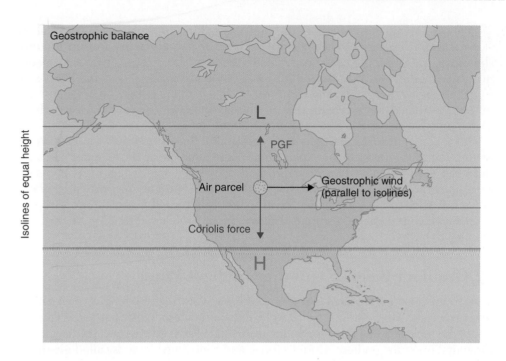

■ **Figure 6.14**

The geostrophic wind represents a balance between the horizontal pressure gradient and Coriolis forces. The straight lines represent heights of constant altitude of the 500-mb pressure surface. The geostrophic wind always is directed parallel to these lines, with lower pressure to its left in the Northern Hemisphere.

this, the wind direction must be perpendicular, not parallel, to the PGF. Because the Coriolis force always pushes to the right of the wind in the Northern Hemisphere, this requires a wind in geostrophic balance—which we call the **geostrophic wind**—to blow with lower pressure to its left.

This *"low pressure on the left of the wind"* rule is called **Buys Ballot's Law** after the Dutch meteorologist who devised it. Pilots use this rule during flight in order to avoid bad weather associated with low-pressure systems.

Buys Ballot's Law also explains a fundamental feature of wind in the atmosphere. Because the wind must blow with low pressure on its left in the Northern Hemisphere, it must blow clockwise around highs and counterclockwise around lows (Figure 6.15). In the Southern Hemisphere this pattern is reversed, with counterclockwise highs and clockwise lows. (Strictly speaking, we should include the centrifugal force, because highs and lows are curved, but the result is the same with respect to wind direction.)

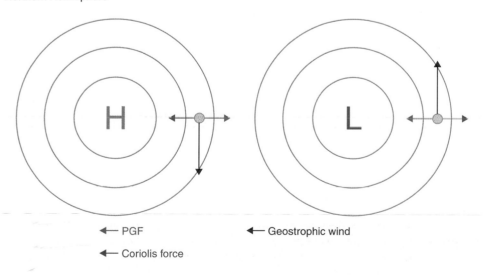

■ **Figure 6.15**

A force diagram representing the forces acting on an air parcel moving around high pressure (left) and low pressure (right). Because of geostrophic balance, in the Northern Hemisphere the air rotates clockwise around high pressure and counterclockwise around low pressure. This is called the geostrophic wind. (For the moment we ignore the centrifugal force, which does not change the direction of the wind.)

We can explain the strength of the wind using the geostrophic balance equation shown earlier. Remember that balance requires that both terms add up to zero. Therefore, the larger the pressure gradient, then the larger that V, the wind speed, must be. In other words, *winds on a weather map are strong where the isobars or isoheights are close together, and winds are weak where the isobars or isoheights are far apart.* This is an excellent rule of thumb that works even in cases where geostrophic balance is not a good approximation.

Like any approximation to reality, geostrophic balance is only as good as its assumptions. It's often a poor explanation of the winds near the equator, because there the Coriolis force is nearly zero and does not balance the PGF by itself. More generally, any situation in which other forces besides the pressure gradient and Coriolis forces are acting on the atmosphere will not be in geostrophic balance. Nevertheless, geostrophic balance is an excellent first step toward explaining the wind.

◼ Gradient Balance and the Gradient Wind

Geostrophic balance requires that the wind blow in a straight line. However, this is rarely the case. More often, upper-level winds undulate in curvy patterns. This requires the inclusion of a third force, the centrifugal force. The three-way balance of horizontal pressure gradient, Coriolis force, and centrifugal force is called **gradient balance** (Figure 6.16), and the wind that results from this balance is called the **gradient wind.** The gradient wind is an excellent approximation to the actual wind observed above the Earth's surface, especially at middle latitudes and in the jet stream.

Gradient balance illustrates that high and low pressure areas differ not only in the direction but also the strength of the wind. Remember that the centrifugal force always pushes toward the outside of curves and is proportional to the square of the wind speed. Therefore, in the case of a curved low-pressure area, the centrifugal force pushes in the same direction as the Coriolis force. Since both of these forces are related to wind speed, the wind does not need to be as strong in this case as in the case of

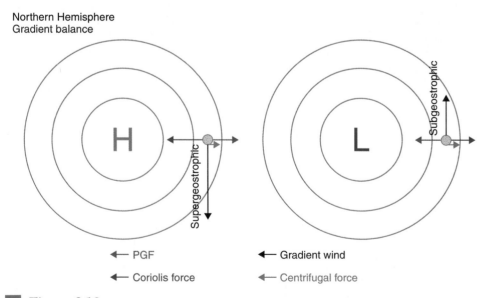

◼ Figure 6.16

As in Figure 6.15, except now we include the centrifugal force. The wind caused by the three-way balance of pressure-gradient, Coriolis, and centrifugal forces is called the gradient wind. *Notice that the gradient wind arrows are longer (*supergeostrophic*) or shorter (*subgeostrophic*) than the geostrophic wind arrows in Figure 6.15, depending on whether the pressure pattern is a high or a low.*

uncurved flow in order to achieve balance with the PGF. The wind is therefore slower in this situation than in a purely geostrophic case, and is called **subgeostrophic flow.**

Conversely, in the case of a curved high-pressure area the centrifugal force teams up with the PGF, both pushing outward from the high. The Coriolis force attempts to balance this double-team of forces. To accomplish this, the wind speed— to which the Coriolis force is proportional—must be considerably higher than in the purely geostrophic case. This is called **supergeostrophic flow.**

Therefore, *for high- and low-pressure areas that have the same spacing of isobars or isoheights, winds around a high will be stronger than winds around a low* because of gradient balance. This is summarized in Figure 6.16. In reality, however, the pressure gradients around lows are stronger than those around highs. We explore some mathematical details of gradient balance on the text's Web site.

Computing the Gradient Wind

Adjustment to Balance

You may wonder, does the wind magically develop in perfect geostrophic balance? The answer is no; as with shoes, cars, and roommates, a continual process of adjustment leads toward a "perfect fit."

This adjustment to balance by the wind is illustrated in Figure 6.17. When a pressure gradient arises, initially there is an *imbalance* of forces. An air parcel in the region is therefore pushed toward lower pressure. The Coriolis force responds as soon as there is motion, turning the wind toward the right (in the Northern Hemisphere). The air parcel wiggles a little as it oscillates toward a balance between the pressure gradient and Coriolis forces. Although the net force acting on the parcel of air tends to be zero at this stage, the parcel continues to move because of inertia (Newton's First Law). Geostrophic balance is reached, or nearly reached.

However, in real life the atmosphere is always being jostled around. This means that its adjustment to geostrophic—or another—balance is imperfect and temporary at best. In this context, it's remarkable how closely the real atmosphere resembles our force-balance approximations. The reason is that the atmosphere adjusts to imbalances rapidly, during a time span of minutes to a few hours. Twice-per-day radiosonde observations rarely catch it red-handed in the act of adjusting.

Like a cat rebounding from an unexpected fall, the atmosphere quickly adjusts to imbalance (licking its wounds, as it were), purrs "I meant to do that," and returns toward the state of balance that we normally see on weather maps.

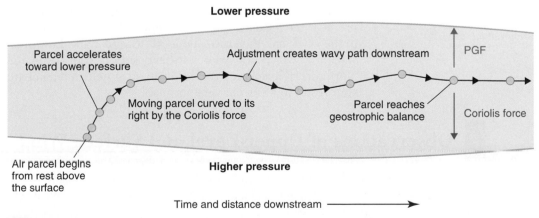

Figure 6.17

The adjustment to balance of a parcel of air that is suddenly placed in a horizontal pressure gradient high above the ground. (Source: Gedzelman, S.D., The Science and Wonders of the Atmosphere, *John Wiley and Sons, 1980, p.247.)*

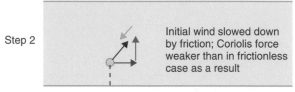

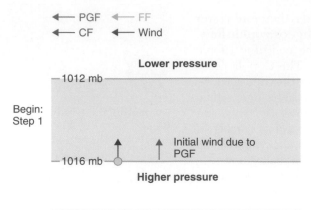

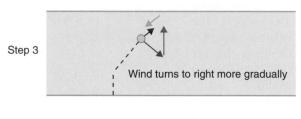

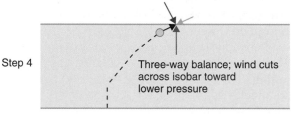

Figure 6.18

The adjustment to Guldberg–Mohn balance of a parcel of air that is near the Earth's surface. The parcel crosses the isobar toward lower pressure under the combined influence of the horizontal pressure-gradient, Coriolis, and frictional forces.

Balance of Forces

Guldberg–Mohn Balance and Buys Ballot's Law Revisited

Near the Earth's surface, the frictional force plays a pivotal role in wind. The adjustment to balance in these situations is profoundly affected by the ability of friction to slow the wind and therefore weaken the Coriolis force. This three-way balance of horizontal PGF, Coriolis force, and frictional force is sometimes called **Guldberg–Mohn balance** in honor of its 19th-century discoverers.

Figure 6.18 depicts the idealized, stepwise evolution of this balance. As in Figure 6.17, the creation of a PGF creates a wind from higher toward lower pressure. However, near the surface this wind triggers not only the Coriolis force, but also the frictional force. Friction opposes the wind, so the wind slows down which in turn makes the Coriolis force weaker.

As the situation evolves, the PGF wins the tug-of-war. This is because of the balance of the three forces. Without the influence of the Coriolis force, the wind would blow straight into lower pressure. With the Coriolis force at full strength opposing the PGF, the wind would blow parallel to the isobars as in geostrophic balance. But in this case, friction weakens the wind and the Coriolis force, and the air parcel crosses the isobars toward lower pressure. However, it does so at an *angle,* not straight across from higher to lower pressure.

Therefore, Guldberg–Mohn balance is characterized by near-surface winds that blow across the isobars on a slant toward lower pressure. The angle at which the winds cross the isobars depends on the type of surface and the latitude. Over open water, where friction is low, the winds typically cross the isobars at an angle between 15° and 30°. Over land the angle is usually between 25° and 50°. Friction also damps out the wiggles seen in Figure 6.17 (Figure 6.19). For a variety of reasons, the effect of friction is important at the boundary between air and water (Box 6.3).

Because the direction of the surface wind is not the same as the geostrophic wind, we must revise Buys Ballot's Law for use near the ground. The fix is simple: Stand with your back to the wind, and then turn about 30° to your right. Low pressure now will be on your left-hand side if you are in the Northern Hemisphere. Try this in an open field and compare your result to a current weather map!

Observations of Upper-Level and Surface Wind

So far we have discussed force-balances individually. However, the atmosphere is a three-ring circus of air motions, and a single weather chart or map can reveal several force-balances simultaneously.

In Figure 6.20, an upper-level isobaric chart depicts fairly typical conditions. Where the isoheights are nearly straight, the wind "flagpoles" point parallel to the lines of equal height. The "flags" on the flagpoles indicate that the strongest winds are observed by the rawinsondes where the isoheights are closest together. Both of these observations are consequences of geostrophic balance.

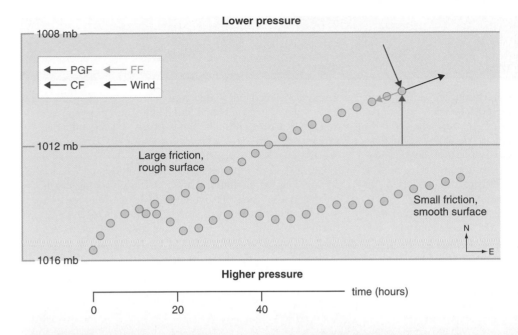

Figure 6.19

A numerical simulation of how varying amounts of friction affect the adjustment to Guldberg–Mohn balance. At the surface, friction causes the wind to cross the isobars and converge toward regions of low pressure. The amount of surface friction determines the angle at which surface winds cross the isobars. Over smoother surfaces the angle is smaller than over rougher surfaces. When friction is weak, the paths of parcels are wavy, as in Figure 6.17. (Source: Knox, J. and Borenstein, S., "Unphysical descriptions of the approach to geostrophic equilibrium," Journal of Geoscience Education, vol. 46, March 1998, pp. 190–192.)

BOX 6.3

Wind and Waves

In this chapter we have emphasized how friction at the ground modifies the wind. The wind also modifies the condition of the surface. We see this when wind blows through trees or across a wheat field. Solid objects remain in place even in a stiff wind, but fluids such as water can be moved by the wind. As a result, winds can have a profound effect on the water conditions at the surface of oceans and lakes.

A good example of this is wind-generated waves. Waves form as the wind's energy is transferred to the surface of water.

The size of a wind-generated wave depends on:

1. The *wind speed:* The stronger the winds, the larger the force, and thus the bigger the wave. The wind must also be constant, not just a wind gust here or there.
2. The *duration* of the winds: The longer the wind blows over the open water, the larger the waves.
3. The *fetch:* This is the distance of open water over which the wind blows. The longer the fetch, the larger the waves.

The curve in the figure below shows how wind speed (horizontal axis) affects the maximum possible height of

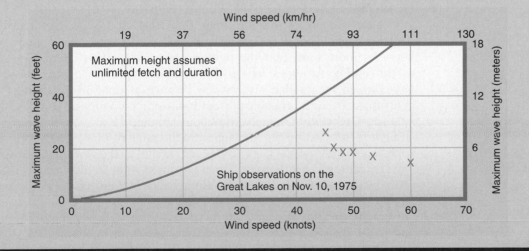

BOX 6.3
(continued)

waves on an infinitely wide ocean (vertical axis). This curve is not an "ivory tower" result, but instead is based on real-world observations.

The real-world observations plotted with crosses on this figure are for a tragic event in American history. On November 10, 1975, the Great Lakes iron ore freighter *Edmund Fitzgerald* sank in a windstorm on eastern Lake Superior. We will study this shipwreck in great detail in Chapter 10. For now, note how the high winds relate to the wave heights. In this particular storm, not only were the wind speeds high, but the fetch was also long for a lake. Both of these factors increased the size of the waves on Lake Superior during the

Fitzgerald's final hours. The figure below shows the path of the *Edmund Fitzgerald* (thin dashed line) and the wind direction (black arrows) at the time the great ship sank. Why the winds were so strong—the central question of Chapter 10—is still a mystery!

High waves are not only hazardous to shipping, but are also perilous for coastlines. Waves can cause flooding, crush buildings, scour the soil from under structures, and wash away beaches. The rapid rise in sea level associated with a storm is called the *storm surge*, which we investigate in our study of tropical cyclones in Chapter 8.

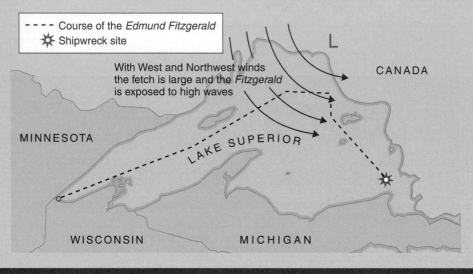

Now imagine that you are standing with your back to the wind on this chart. Anywhere on this chart, you will find that lower heights, which are analogous to lower pressures, are toward your left. This is Buys Ballot's Law, another offshoot of geostrophic balance.

In the curved regions on Figure 6.20, the wind is making a turn just as you would in a car. Therefore, the centrifugal force is significant in these regions, and becomes more important the tighter the curve is. This chart depicts conditions in the middle latitudes where the horizontal pressure gradients are generally large, which means that a three-way balance of pressure gradient, Coriolis, and centrifugal forces is probably present in the curved regions. We called this *gradient balance*. The wind in the curved regions around low pressure is slower than in the straight regions where the isoheights are the same distance apart. As in the straight-wind regions, the wind "flagpoles" point parallel to the isoheights.

Near the surface, however, friction causes wind to blow across the isobars toward lower pressure via Guldberg–Mohn balance. Figure 6.21 shows this effect very clearly, as does every surface weather map.

Also notice that the midlatitude winds at the surface in Figure 6.21 are slower and more likely to be from the east than the winds in Figure 6.20, which by and large blow from west to east. This is partially explained by the simple fact that friction slows down the wind. Depending on the surface, friction usually slows the wind

250-mb heights (tens of meters)/winds (knots)   0000 UTC September 5, 2001

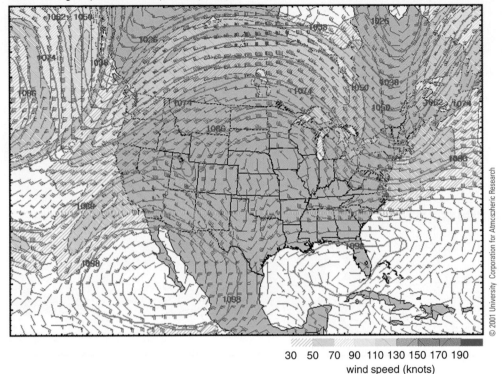

© 2001 University Corporation for Atmospheric Research

30   50   70   90 110 130 150 170 190
wind speed (knots)

Figure 6.20

A typical isobaric chart. Notice that the winds are directed parallel to the isoheights, with lower heights to their left. Also, the highest winds (shaded) occur where the isoheights are closest together and also at the crests of the high-pressure ridges (central Canada, top of figure). All of these features are explained by geostrophic and gradient balances.

by 25% to 75% of what the speed would be in the absence of friction. However, this increase of westerly wind speed with increasing altitude is also seen all the way up in the middle and upper troposphere, far away from the friction near the Earth's surface. Why is this true? To explain this, we need to put together not forces, but two force-*balances*.

Putting Force-Balances Together: The Thermal Wind

We can explain our observation that winds increase throughout the troposphere by using geostrophic and hydrostatic force-balances. Let's start from the very beginning, with energy. An energy imbalance can lead to a force that generates a wind. Energy imbalances are, in fact, the trigger for atmosphere and ocean circulations.

Consider a layer of air between two pressure surfaces, say, 850 mb and 300 mb (Figure 6.22). Initially the isobars are parallel to one another. If we heat a column of air (make its energy gains greater than its energy losses), the density decreases. To reduce the air density within the column, the vertical height difference between the two pressure levels must increase. By heating just a portion of the layer of air, we generate a horizontal pressure gradient. In Figure 6.22 an outward PGF is generated near 300 mb and an inward PGF near 850 mb. This leads initially to a wind pattern with air flowing toward the region at 850 mb and outward at 300 mb.

 http://info.brookscole.com/ackerman

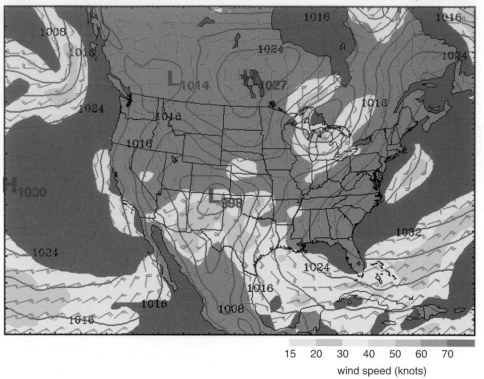

Figure 6.21

The surface weather map corresponding to Figure 6.6. Notice that the winds blow inward at an angle away from higher pressure and toward lower pressure. In this case, the convergence of surface air helped cause severe weather in Kansas (see the Introduction to Chapter 11).

Another way of saying this is that hydrostatic balance tells us that pressure must decrease more rapidly with increasing altitude in cold air than in hot air. In other words, cold air is more compressed than hot air. Therefore, the 300-mb pressure surface is at a higher altitude at 30°N than near the pole, and a PGF acts from the south to the north.

Moving air at middle latitudes feels the Earth's rotation, and so we must also discuss the role of geostrophic balance in this situation. The geostrophic wind is proportional to the slope of the pressure surfaces. The greater the slope, the stronger the geostrophic wind. Figure 6.23 shows that the thickness between pressure surfaces in warm, tropical air is greater than in cold, polar air. And so the slope of the pressure surfaces keeps increasing all the way up in the troposphere. For this reason, the geostrophic wind increases with altitude when the temperature changes horizontally. This explains why winds at 300 mb are usually stronger than at 500 mb, and also why winds at 500 mb are usually stronger than at 850 mb.

Finally, Buys Ballot's Law tells us the direction in which the wind blows. At

Figure 6.22

When the air between two pressure surfaces is warmed, the distance between the two pressure levels, or thickness, increases. Cooling the air would reduce the thickness. In either case, horizontal PGFs are created in the troposphere.

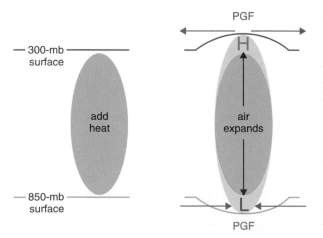

high altitudes in the middle latitudes, lower pressure is toward the pole. Buys Ballot's law tells us that the winds in this region must blow from west to east. In the tropics, this pattern is reversed and upper-level winds usually blow from east to west.

This relationship between the vertical changes in the geostrophic wind and the horizontal temperature changes is called the **thermal wind** because it relates temperature and winds to each other. Its main result can be summarized for the middle latitudes of both hemispheres as follows: *The winds are more westerly as you go up where it's colder toward the poles.* It is also an example of how force-balances can be combined just like forces to explain how and why the wind blows. Keep this thermal wind relationship in mind when we study large-scale circulations, jet streams, and fronts in the atmosphere in Chapters 7, 9, and 10.

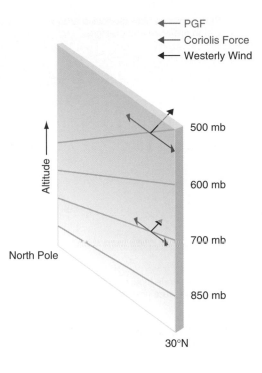

← PGF
← Coriolis Force
← Westerly Wind

500 mb

Altitude

600 mb

700 mb

North Pole

850 mb

30°N

Figure 6.23

A simplified view of why the winds in the middle latitudes tend to blow from west to east. On average, the tropospheric temperature decreases poleward. This leads via hydrostatic balance to sloping pressure surfaces and poleward PGFs. Winds are directed eastward because of the geostrophic balance of the pressure gradient and Coriolis forces. This wind is called the thermal wind.

Make a Jet Stream

Putting Horizontal and Vertical Winds Together

We've learned that at the surface, the wind blows across the isobars away from high-pressure areas and into low-pressure areas. But what happens to the air once it reaches the center of the low? The atmosphere, like a balloon that is being inflated, prefers to expand in other directions rather than increasing its density in one spot. However, the air at the low's center cannot go into the solid ground. Neither can it retrace its steps and move away from the low, which would violate the principle of balance. Therefore, its only option is to go upward. This partly explains why air in a low-pressure center rises. Similarly, air in a high-pressure center sinks, because if it didn't, the diverging air at the surface would create a vacuum at the center! Figure 6.24 illustrates this connection between horizontal and vertical wind for both highs and lows.

Figure 6.24 demonstrates that the vertical motion of air in pressure systems can be explained as a consequence of Guldberg–Mohn balance. This balance, along with our study of adiabatic processes and clouds in previous chapters, explains a fundamental fact of weather: *low-pressure areas are usually cloudy and wet, while high-pressure areas are usually clear and dry.* This is because rising air cools adiabatically, causing the water vapor in it to condense and form clouds and precipitation. Sinking air warms adiabatically, lowering its relative humidity and evaporating any clouds and precipitation. Voilà—using just three forces combined in one force-balance, plus a few details from earlier chapters, we have explained one of the most basic and universal observations of weather!

Sea Breezes

We conclude our study of wind and forces by looking at a small-scale wind pattern that connects our knowledge of the atmosphere from previous chapters with the

http://info.brookscole.com/ackerman

Surface winds blow clockwise around
a high pressure and diverge.

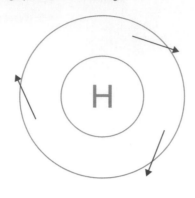

View from above

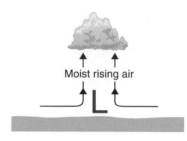

View from side

Surface winds blow counterclockwise around
a low pressure and converge.

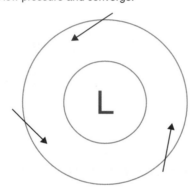

View from above   View from side

■ Figure 6.24

*How surface wind patterns in-
duce vertical wind motions. The
divergence of air out of a surface high induces sinking air
above, causing dry and clear
conditions (top). The convergence of air into a surface low
causes rising air above it, leading to moist, cloudy conditions
(bottom).*

wind. If you have spent much time at the beach during the summer you've probably noticed that at around 3:00 PM there often is a strong, steady breeze blowing in from the water. This steady wind, the **sea breeze,** is a result of the uneven heating during the daytime between the land and the adjacent water. At night the wind often reverses direction and blows from the land to the water (a **land breeze**). Land and sea breezes are too small and quick for the Coriolis force to play a dominant role.

In the morning the land and sea can be the same temperature (Figure 6.25, *A*). Later in the day the land surface, which has a low specific heat and is a poor conductor of heat (Chapter 2), heats much more quickly than water. As the land warms up, the air next to it heats by conduction and rises. This air then heats the air above it by convection (Chapter 2), as shown schematically in Figure 6.25, *B*. Just as in our discussion of the thermal wind, this leads to a bulging upward of pressure surfaces over the land (Figure 6.25, *C*) creating a PGF pushing toward the ocean (Figure 6.25, *D*).

Figure 6.25, *E*, completes the picture of the sea-breeze circulation. At the surface over land, low pressure is developing. This is suggested in the figure by the lowering of isobars over the land. Meanwhile, over the ocean, high pressure is developing. This generates a horizontal PGF at the surface acting from over the ocean toward the land. Air moves toward the land, creating a sea breeze. To replace the surface air that is moving from over the water toward land, air sinks from above, completing the circulation.

Notice the slope of the pressure surfaces in Figure 6.25, *E*. As the temperature difference between the land and water increases throughout the afternoon (Chapter

3), the circulation increases in strength and winds increase, reaching a maximum in the middle to late afternoon. Over land, the distance between two isobars (e.g., 980 and 960 mb), is greater than over the ocean. This difference is what keeps the circulation moving and is due to the air over land being warmer than the air over the ocean.

If you are not at the beach to feel the sea breeze, you can observe it by analyzing satellite imagery (Figure 6.26). A rising parcel of air expands and cools, and the relative humidity increases—conditions favorable for the formation of clouds (Chapter 4). For this reason, the upward branch of the sea breeze is often visible in satellite pictures (Chapter 5) in the form of cumulus clouds. During the day, the upward branch moves inland and is an indication of the strength of the sea breeze. If the atmospheric conditions are favorable for the formation of thunderstorms, the sea breeze may provide just enough lifting to cause thunderstorms to develop.

Whenever large land and water bodies are adjacent to one another, sea breezes may develop and may cause thunderstorms. Florida's abundant summertime rainfall is a result of sea breezes. One sea breeze advances from the Atlantic Ocean on the east and another from the Gulf of Mexico on the west coast of Florida.

The important concept is that heating (or cooling) of a column of air leads to horizontal differences in pressure, generating a PGF, which in turn causes the air to move and a circulation to develop. During the evening, the land cools more quickly than the water and the process is reversed (Figure 6.27). The net result is a land breeze, surface winds that blow from the land out to sea.

 ## Scales of Motion

Atmospheric motions span an enormous range of space and time. The size of an atmospheric weather system is related to how long it exists. Small swirls of wind may last for a few seconds while a hurricane may last several days. In general, as the size of the phenomenon increases, so does its life span. The size and life spans of different atmospheric phenomena are shown in Figure 6.28. Meteorologists divide weather systems into four primary size categories:

- **Microscale** is generally applied to circulations that are less than 1 kilometer in size, generally smaller than a puffy cumulus cloud. At these scales, the pressure gradient and centrifugal and frictional forces are usually important, and the Coriolis force is negligible. We look at these scales in Chapter 12.
- **Mesoscale** systems, such as thunderstorms, fronts, and sea breezes, range in size between about 1 kilometer and a few hundred kilometers. The Coriolis force becomes more important at the large end of this category (e.g., in hurricanes). We study mesoscale systems in Chapters 8, 9, 11, and 12.

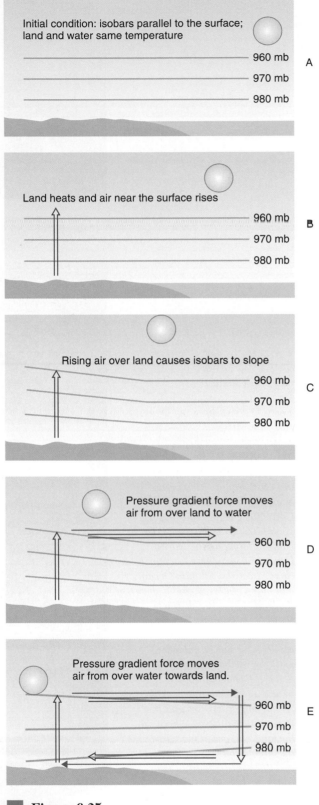

Figure 6.25

Sequence of schematic images depicting the formation of the sea breeze.

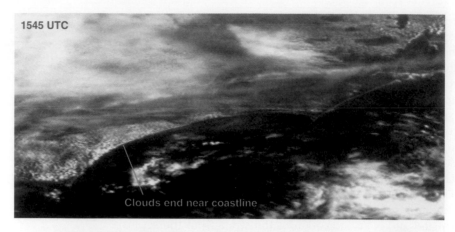

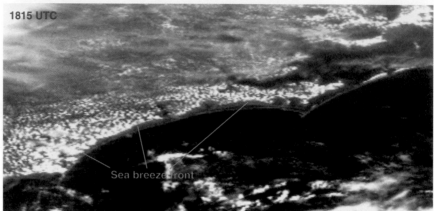

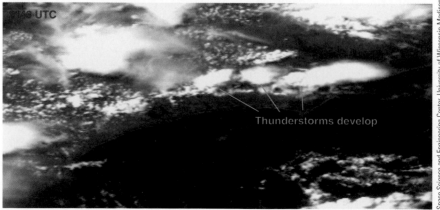

■ Figure 6.26

A sequence of satellite images demonstrating the sea breeze along the east coast of North Carolina. The first image is at 1545 UTC, or 11:45 AM EDT. Notice that along the North Carolina coast, there are few clouds over the Atlantic Ocean, and many scattered clouds over the land. There is a distinct cloud-free boundary at the coastline. By 1815 UTC the clouds have moved inland, marking the boundary of the upward branch of the sea breeze, or the sea-breeze front. As the day progresses and the temperature difference between the land and ocean increases, the circulation gets stronger and the sea-breeze front penetrates farther inland (2015 UTC). On this day the atmosphere rises easily, and the lifting associated with the sea breeze is enough to trigger thunderstorms by 2145 UTC, or 5:45 PM local time.

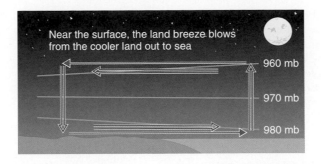

Figure 6.27

A land breeze forms during the night as the land cools faster than the sea.

- **Synoptic-scale** systems, like midlatitude low-pressure systems, are close to 1000 kilometers in size. In this category, the geostrophic balance between the pressure gradient and Coriolis forces is dominant. Chapter 10 examines midlatitude lows and highs in detail.
- **Planetary-scale** systems are much greater than 1000 kilometers in size and again geostrophic balance helps explain them. We examine these systems in depth in the next chapter.

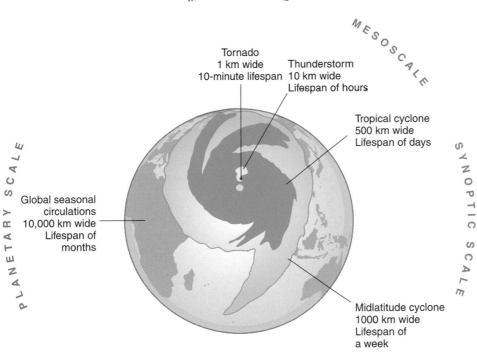

Figure 6.28

Sizes and lifespans of various atmospheric wind circulations. Their sizes are suggested graphically by the size of the radar or satellite signature of each pattern, and their lifespans are suggested by the locations of their names on the "clock." Note that the smallest circulations are also the briefest. (Source: http://k12.ocs. ou.edu/teachers/graphics/ Wxscales.gif.)

PUTTING IT ALL TOGETHER

SUMMARY

The wind is air in motion. Newton's laws of motion describe the physical laws that govern the movement of objects, including air. The important forces at work, besides gravity, in the atmosphere are:

Pressure Gradient Force (PGF): directed from higher pressure toward lower pressure at right angles to lines of constant pressure, or constant height on an isobaric chart. The greater the change in pressure over a given distance, the stronger the PGF.

Coriolis Force: an apparent force caused by the rotation of the Earth. The Coriolis force acts to the right in the Northern Hemisphere and to the left in the Southern Hemisphere. The Coriolis force is zero at the equator and increases in magnitude toward the poles. This force is zero if the velocity of the air parcel is zero, and increases as the air parcel's velocity increases.

Centrifugal Force: always directed outward from the center of curving motion. It increases as the speed of the object increases and/or the radius of curvature decreases.

Frictional Force: always acts in the direction opposite to movement. The magnitude of this force depends on the type of surface and the wind speed. Friction is weak over a frozen lake and strong over a forest or mountain.

Winds arise as a result of different balanced combinations of these forces. These balances are:

Hydrostatic Balance: a balance of the gravitational force and the vertical PGF. This balance explains the lack of strong vertical winds in most of the atmosphere.

Geostrophic Balance: a balance between the horizontal pressure gradient and Coriolis forces. Geostrophic winds blow parallel to the isoheights. Buys Ballot's Law states that in the Northern Hemisphere, lower pressure is to the left when your back is to the wind. In addition, the wind in geostrophic balance is stronger when the isobars or isoheights are close together, and weaker when they are far apart. Because of this balance, winds blow counterclockwise around lows and clockwise around highs in the Northern Hemisphere. The reverse is true in the Southern Hemisphere.

Gradient Balance: a three-way balance of horizontal pressure gradient, Coriolis, and centrifugal forces. This balance explains why the winds in a high and a low are different not only in the sense of rotation, but also in terms of their speeds.

Guldberg–Mohn Balance: a three-way balance of horizontal pressure gradient, Coriolis, and frictional forces. This balance explains why winds blow at an angle toward lower pressure, rise upward at the center, and cause cloudy, wet weather in lows. It also explains why highs are sunny regions of sinking, dry air.

The thermal wind, a combination of hydrostatic and geostrophic balances, explains why winds in the middle latitudes generally become more westerly with increasing altitude. This is because the surface temperature typically decreases toward the poles. The sea breeze is a similar, though smaller scale, circulation that is driven by pressure gradients created by unequal heating of land and ocean.

Winds and wind patterns occur at all scales in the atmosphere. In the next chapter we look at wind systems that span the globe.

KEY TERMS

You should understand all of the following terms. Use the glossary and this chapter to improve your understanding of these terms.

Acceleration	Geostrophic balance
Anticyclones	Geostrophic wind
Buys Ballot's Law	Gradient balance
Centrifugal force	Gradient wind
Centripetal acceleration	Gravitational force
Coriolis force	Guldberg–Mohn balance
Cyclones	Hydrostatic balance
Force	Inertia
Frictional force	Isobaric charts
Land breeze	Pressure gradient
Law of Inertia	Pressure gradient force (PGF)
Law of Momentum	
Leeward	Sea breeze
Mesoscale	Speed
Microscale	Subgeostrophic flow
Momentum	Supergeostrophic flow
Newton's First Law of Motion	Synoptic scale
	Thermal wind
Newton's Second Law of Motion	Velocity
	Wind gust
Planetary scale	Windward

REVIEW QUESTIONS

1. You hear on a weather report that the wind is "15 miles per hour." To fully describe the velocity of the wind, what other information must be reported?

2. If air has zero velocity, is the wind blowing? If air has zero acceleration, does this mean that the wind is calm?

3. According to Newton's Second Law, what causes air (or anything else) to accelerate?

4. Which forces that act on the atmosphere are dependent on wind speed? Which forces are not? Which forces act in the vertical direction? Which forces act in the horizontal direction?

5. Wind is triggered in the beginning by which force?

6. If the pressure difference over a fixed distance doubles, how does this change the PGF? If the pressure difference is fixed, but the distance across which the pressure changes doubles, how does this change the PGF?

7. Using a calculator, compute the magnitude of the Coriolis parameter at the latitude of the Madison, Wisconsin, area (latitude approximately 43.3°N). Use 0.0000729 per second for the rotation rate of the Earth. Then multiply this number by 3600. Your result is the change in eastward wind speed caused by the Coriolis force acting on a 1 meter per second southerly wind for an hour. What would your result be at the equator?

8. If you drive a car around a sharp curve at 40 mph and a friend of yours in another car drives around the same curve at only 20 mph, which one of you will have experienced a stronger centrifugal force? What is the ratio of the force in your situation versus your friend's situation?

9. Relate the discussion of cars and centrifugal forces in Question 8 to the atmosphere. Where is the centrifugal force important? Where is it zero? In what direction does it push air parcels curving around a low-pressure system?

10. If the surface wind blows toward the north, in what direction does the frictional force act on it? If the wind reverses direction and blows toward the south, in what direction does the frictional force act now?

11. The Empire State Building in New York City is approximately 306 meters tall. Assume that the air density from the bottom to the top of the building is a constant all the way up and is equal to 1 kilogram per cubic meter. Then, using hydrostatic balance, compute what the pressure change is from the bottom to the top of the skyscraper. (Hints: insert *1* for the air density in the hydrostatic balance equation; then in-

sert *300* for the distance; solve for the pressure difference and divide by 100 to get your answer in millibars.) Your answer is an approximation, because density actually decreases in the vertical. Even so, it explains why your ears 'pop' (Chapter 1) when riding the high-speed elevators up to the observation deck of the Empire State Building.

12. A hurricane has very strong winds. On a weather map, would the isobars around a hurricane be very close together, or far apart? Explain using the concept of geostrophic balance.

13. If an airplane is flying over the United States with a strong "tailwind" pushing it forward, on which side of the plane are you more likely to see bad weather off in the distance: the left-hand side of the plane, or the right-hand side?

14. Explain in words how the effect of the frictional force on surface winds causes lows to be regions of bad weather and highs to be regions of good weather.

15. What forces are most important in causing the sea and land breezes? Do you think the Coriolis force is an important factor in causing these winds? Why or why not?

16. Does Buys Ballot's Law have to be modified to apply to winds in the Southern Hemisphere? Why or why not?

17. Using the *Blue Skies* CD-ROM exercise "Atmospheric Forces/Coriolis Force," try your hand at flying a plane from the North Pole to New York City or Tokyo. Then click on "South Pole" and fly a plane from the South Pole to Sydney. Why are these tasks not as easy as they sound at first?

 # WEB ACTIVITIES

Choose Chapter 6 on the textbook's Web site:
http://info.brookscole.com/ackerman
and select from the following resources:

- Interactive Modules that illustrate and extend your understanding of key topics in this chapter
- Tutorial Quizzes to test your mastery of terms and concepts

 For additional readings, go to the InfoTrac College Edition, your online library, at:
http://www.infotrac-college.com

CHAPTER

7

Global-Scale Winds

After completing this chapter, you should be able to:

■ Use a conceptual model of the atmospheric circulation to explain the existence and location of global-scale high- and low-pressure areas

■ Explain why and where jet streams form

■ Describe the season-to-season changes in wind patterns such as the Indian monsoon

Introduction

What do the discovery of the New World and the theory of evolution have in common? An understanding of global-scale winds helped bring both about.

Mariners have known for centuries that the **trade winds** off the Atlantic coast of North Africa blow steadily from the northeast direction, out to sea, while farther north along the coast of Europe the winds typically blow from west to east. In 1492 Christopher Columbus took advantage of this knowledge at the start of his famous voyage by sailing with the wind from Spain to the Canary Islands and then west across the Atlantic (Figure 7.1). When he returned to Spain the next year, Columbus sailed north toward the Azores where westerly winds quickly guided his ships homeward.

Nearly three and a half centuries after Columbus's first voyage, the *HMS Beagle* sailed around the world. On board were Charles Darwin and others on a global scientific expedition. Throughout its five-year sea trek, the *Beagle* used the prevailing winds of both the Northern and Southern Hemispheres to complete its journey (Figure 7.2). In particular, the crossings of the Pacific and Indian Oceans were accomplished at around 20°S latitude, where the winds are steady from the southeast.

Both ships steered clear of the subtropical regions around 30°N and 30°S latitudes. Columbus and the captain of the *Beagle*, Robert FitzRoy, knew the winds in these **horse latitudes** were typically light or calm,

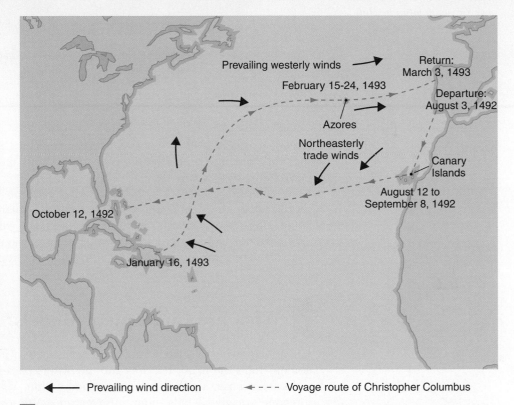

Figure 7.1

Christopher Columbus's route to and from the New World shows an understanding of prevailing wind directions.

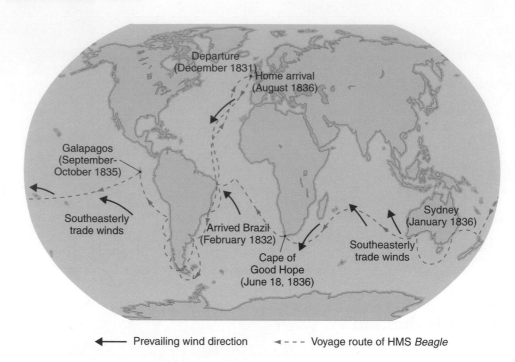

Figure 7.2

The journey of the HMS Beagle. Note the effort made to stay in the region of the trade winds near 20° south latitude while crossing both the Pacific (left) and Indian (center right) Oceans.

forcing mariners to throw horses overboard when their ships were stranded and short of drinking water. Columbus's and FitzRoy's understanding of winds helped make their voyages famous.

Using observations of global-scale wind patterns, this chapter develops a conceptual model of global-scale winds that helps explain the location of the world's largest deserts, tropical rain patterns, and even the existence of the jet stream. We also learn how atmospheric motions help correct the energy imbalances caused by the Earth's tilt, warming the poles and cooling the tropics.

What are Conceptual Models?

A simple drawing of a face is not very sophisticated (Figure 7.3). Yet it successfully demonstrates the major features and the overall pattern of a human face: two eyes and ears, a nose, a mouth, eyebrows, and a chin. When you describe a friend to someone, you give details about those features. A description such as "He has brown eyes, bushy eyebrows, a big nose, and small ears" may not give a very complete picture of someone, but it might be enough to recognize that person in a crowd. It is the same with describing the atmosphere. A general description of a few current atmospheric features might allow you to recognize the overall weather pattern.

We can draw simple faces based on observations. For example, we all know from experience that the nose lies above the mouth and between the eyes. To make a simple, conceptual model of global wind patterns requires similar observations. We will rely on our own observations, historical records, and data from satellites to identify the major features of a simple model that represents the atmospheric circulation.

While all faces have the same basic features, few people look identical. It is the same with the atmosphere. You can identify many of the basic features of the atmosphere on most days, but you will not find two days with identical patterns. An exercise on the text's Web site is designed to help you identify these general atmospheric patterns on satellite images. Go investigate!

Figure 7.3

A drawing like this one represents the major features of a human face in a simple fashion. In this chapter we will create a similar simplified picture of the global wind patterns.

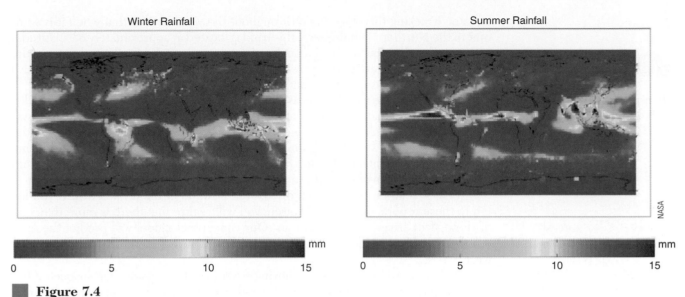

Winter Rainfall   Summer Rainfall

Figure 7.4

Global precipitation patterns in winter (left) and summer (right) in millimeters per day.

Observations Our Model Should Explain

Recognizing the Face of Weather

A simple conceptual model of the global wind pattern must explain the steady winds that have long been observed by mariners. A model of global atmospheric circulation must also account for regions that consistently lack winds. Such a model must also be consistent with other observed patterns, such as the position of deserts and regions of high precipitation (Figure 7.4).

Another observation our conceptual model should explain is the global pattern of cloudiness that we first noted in Chapter 1. Figure 7.5 is a general cloud pattern as observed from a series of weather satellites. We can see several distinct patterns in these images. Note the lack of clouds near the horse latitudes, consistent with the precipitation maps. A band of thunderstorms extends around the world near the

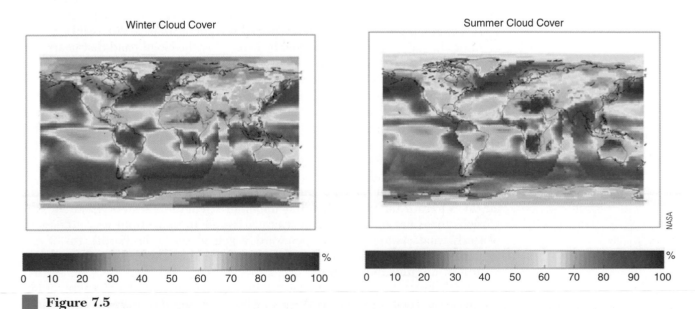

Winter Cloud Cover   Summer Cloud Cover

Figure 7.5

Cloud cover in winter (left) and summer (right) as a percentage. Orange and red areas are cloudy most of the time.

http://info.brookscole.com/ackerman

Blue Skies

**Atmospheric Circulation/
Global Atmosphere**

equator. Tracking this cloud band throughout the year, we find that when it is summer in the Northern Hemisphere, this band is located at approximately 8°N latitude. The average position of this cloud band is approximately 4°S latitude when the Southern Hemisphere experiences summer. The maps of average precipitation also depict these patterns in January and July.

The midlatitude cloud patterns in Figure 7.5 have characteristics distinctly different from the cloud systems we observe in the tropics. The midlatitude cloud systems move quickly, change from one day to the next, and are further poleward in the hemisphere that is experiencing summer than in the hemisphere that is experiencing winter. In both summer and winter, the midlatitude cloud systems tend to move from west to east. You can observe these patterns for yourself on TV weather reports.

Today's weather reports often discuss the position of the **jet stream.** If you listen to these reports carefully and often, you will realize that there is more than one of these swift "rivers of air" above us. Our conceptual model will predict the existence of two of these jet streams. Both are fast-moving currents of air that flow from west to east.

The existence of a jet stream moving from west to east was long suspected because of the movement of storm and cloud systems. But the suspicions remained unconfirmed until World War II, when the United States prepared for major air raids against Japan. The B-29 airplanes flew from east to west at altitudes of 10 kilometers, where they encountered a strong stream of westerly winds that slowed or even stopped the planes in mid-air! Observations of storm patterns in the winter reveal that winter storms in the middle latitudes move from west to east. These storms are usually accompanied by the jet stream. Our model must describe this observation as well. Let's begin building our model.

A Simple Conceptual Model of Global Circulation Patterns

A conceptual model makes sense of nature by simplifying wherever possible. Initially we will assume that the Earth is entirely covered with water. With no land present we can assume that what happens in one hemisphere happens in the other, and we can ignore the variations in circulation patterns that land masses cause. While this water world is unrealistic, it will explain the basic observations we have discussed.

Let's begin our conceptual model by explaining the cloud band that nearly encircles the tropics. These are convective clouds that require upward vertical motions. This cloud band is nearly always present, so there must be steady upward motions near the equator. Accordingly, our conceptual model has upward vertical motions near the equator (Figure 7.6). This rising air produces the cloud and precipitation patterns of the tropics. Eventually, the rising air encounters the stable stratosphere, stops rising, and spreads northward and southward along the tropopause. Let's follow the northward branch of air.

As the air flows northward, the Coriolis force turns it to the right (eastward). As you remember from Chapter 6, the magnitude of the Coriolis force increases as the air flows toward the pole. When this upper-air stream reaches 30°N latitude, the Coriolis force has turned the flow, causing the air to move from west to east—a westerly wind. As the air moves poleward, it gets closer to the Earth's axis of spin (Figure 7.7). Before going on we need to understand a concept known as **conservation of angular momentum.**

In Chapter 6 we saw that the momentum of an object moving in a straight line is equal to its mass × its velocity. A rotating body has **angular momentum,** defined as: mass × the rotation velocity × the perpendicular distance from the axis of

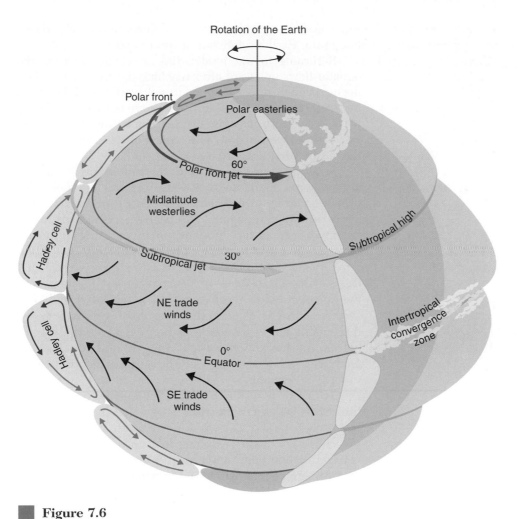

■ **Figure 7.6**

A conceptual model capable of explaining weather patterns spanning the globe.

rotation. An undisturbed rotating object will conserve its angular momentum. In other words, the product of mass, velocity, and distance from the rotation axis is a constant value. If the distance from the axis of rotation decreases, the velocity of rotation must increase to conserve angular momentum. For example, figure skaters and divers spin faster as their arms and legs are moved closer to the axis of rotation (Figure 7.8). As a parcel of air moves north or south from the equator, its distance from the Earth's axis of rotation decreases. The parcel's angular velocity must increase to conserve angular momentum.

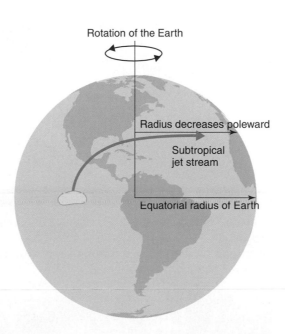

■ **Figure 7.7**

As a parcel of air moves poleward, its distance from the Earth's spin axis decreases. To conserve angular momentum, the parcel must increase its west-to-east speed, creating a subtropical jet stream.

Figure 7.8

Keeping his or her limbs close to the axis of rotation (the body) allows a diver or skater to conserve angular momentum and thereby spin faster. The conservation of angular momentum is important in explaining the existence of the subtropical jet stream.

This means that the air must travel faster in the same direction as the Earth's rotation as it moves poleward.

Returning to our model, the ascending air near the equator flows poleward after reaching the tropopause. Deflected by the Coriolis force on its poleward trek, it becomes a westerly wind in the upper troposphere between 20°N and 30°N latitude. To conserve angular momentum the air blows rapidly from west to east. This stream of air is referred to as the **subtropical jet stream** (Figure 7.9).

As air continues to flow into the subtropical jet stream region, some of the air must leave the jet—air molecules cannot continue to pile into this region forever. Since the air is near the tropopause, the stable stratosphere above it suppresses upward motions, so in our model the air must sink toward the surface. Sinking air suppresses cloud development and precipitation. It is therefore logical that this sinking air is above our current desert locations and the horse latitudes (see Figure 7.6), about 30°N and 30°S latitudes. Stratus clouds often appear over oceans below this sinking air, as discussed in Box 7.1. Once the subsiding air reaches the surface, it spreads north and south. Now let's follow the air near the surface that is flowing toward the equator.

As the air flows south toward the equator, the Coriolis force pulls it to the right (west) (to the left in the Southern Hemisphere). The Coriolis force weakens as the air approaches the equator, resulting in a northeast wind in the Northern Hemisphere and a southeast wind in the Southern Hemisphere. These are the trade winds sought by traders traveling from Europe to the Americas. Notice how the trade winds from the two hemispheres converge, supplying moist air for cloud development. The region where the two

Figure 7.9

The subtropical jet stream can be recognized in infrared satellite images by the cirrus clouds that often accompany jet streams. In this example, a strong subtropical jet extends northeastward from the equatorial Pacific Ocean across Mexico and the Gulf of Mexico and over Florida. Compare this real-life example to the path of the air parcel in our conceptual model in Figure 7.7.

BOX 7.1

Marine Stratocumulus Cloud Regions

High-pressure systems are supposed to be cloudless, according to Chapter 6. So why aren't the world's subtropical highs always sunny? It is common to find stratocumulus clouds in the vicinity of the descending branch of the Hadley Cell that lies over cold ocean regions. This is evident in the satellite image of the Western Hemisphere at right. The marine air near the surface is cool and humid. As the air in the upper troposphere descends, it warms adiabatically. When a deep layer of the atmosphere sinks, a temperature inversion can develop as a result of adiabatic compression. This is illustrated in the chart below. The upper region of the layer descends over a greater distance than the lower region and thus warms more. After subsiding, the top of the layer is therefore warmer than the bottom. This is called a *subsidence inversion* because it results from descending, or subsiding, air. In meteorology, subsidence denotes sinking motions. This particular subsidence inversion is also called the *trade-wind inversion* because it occurs in the region

Stratus

Stratocumulus

NOAA Photo Library

of the trade winds. The temperature increases sharply with altitude in the trade-wind inversion. Stratus clouds form when the air approaches saturation as a result of mixing of the dry subsiding air with the moist air near the ocean surface. The stratocumulus clouds observed in the satellite image above lie just below the trade-wind inversion where there is enough instability for the stratus to become partly convective, forming stratocumulus clouds. Because of the trade-wind inversion, which inhibits vertical motion, these clouds cannot grow taller.

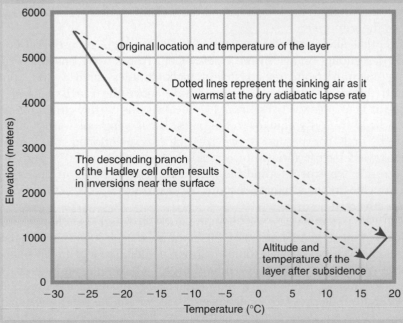

CIMSS, University of Wisconsin-Madison

Figure 7.10

An image of the Earth on October 14, 1999, from a series of satellite photographs. The colored regions represent the surface temperature, and the white and gray regions represent cloud cover. The ITCZ is observed in infrared satellite images as a band of convection near the equator.

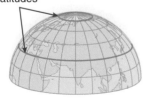

Two different latitudes

Figure 7.11

The circumference of a latitude line close to a pole is smaller than the circumference of a latitude line closer to the equator.

trade winds come together is called the **Intertropical Convergence Zone** (ITCZ) (Figure 7.10).

In our conceptual model a circulation cycle extends from the equatorial to subtropical regions. Air flows upward at the ITCZ and spreads poleward at the tropopause to form a jet stream, in the vicinity of which the air sinks. When the air reaches the surface, some of the air flows toward the equator and converges into the ITCZ (see Figure 7.6). This circulation cell is called the **Hadley Cell.** Eighteenth-century scientist George Hadley was the first person to propose a reasonable explanation of the trade winds, though he assumed the sinking air descended over each pole instead of the subtropics.

Our conceptual model predicts regions of relatively high atmospheric pressure near the descending branch of the Hadley Cell. Here's why: Imagine a ring of air moving poleward from the equator at high altitudes, such as the upper troposphere. If this ring stays at the same altitude above the Earth's surface, it gets smaller as it goes poleward (Figure 7.11) because the Earth's circumference shrinks as you head from the equator to the poles. Air moving poleward is therefore "squeezed," or is converging. As we have already mentioned, the Coriolis force turns poleward motion into westerly winds in the vicinity of 30°N and 30°S latitude. As a result, we expect to find the maximum converging of air near these latitudes. Since converging air implies more air overhead at 30°, we expect to find high pressure at the surface.

Figure 7.12 shows the average sea-level pressures observed over the globe for January and July. Notice the high pressures at the surface near 30°N and 30°S latitudes, particularly over the oceans, confirming our model. These semipermanent pressure systems are referred to as **subtropical highs.** These subtropical highs, or *anticyclones*, have a major influence on the weather and climate of the subtropics and middle latitudes. The pressure gradients are weak within the subtropical highs and the surface winds are therefore light or calm over large regions of the subtropical oceans—the horse latitudes. This is consistent with the observations of mariners.

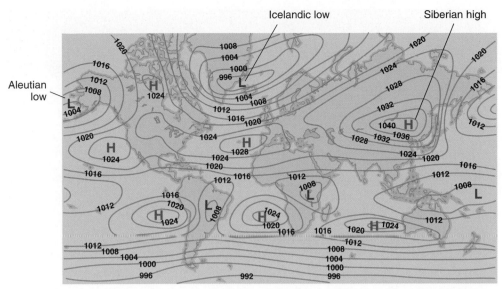

Sea-level pressure—January

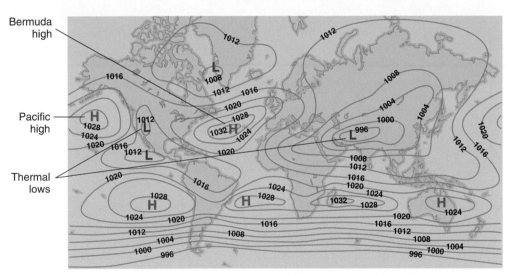

Sea-level pressure—July

Figure 7.12

Sea-level pressure and wind maps for a typical January and July. Notice how our conceptual model of Figure 7.6 is consistent with the observed surface winds. In the wind maps on the next page, the thicker the wind arrow, the more constant the wind. (Source: Gedzelman, S.D., The Science and Wonders of the Atmosphere, *John Wiley and Sons, 1980, p. 262.)*

Air near the surface flows outward from the subtropical highs under the influence of friction (see Chapter 6). Equatorward of these highs, these surface winds are known as the *trade winds*—the same winds used to advantage by Columbus and the *Beagle.* Trade winds exist in both hemispheres and converge into the ITCZ. In equatorial regions where neither trade wind dominates, the wind is calm. These wind regions are known as the **doldrums**; crews of sailing ships dreaded the doldrums because of the light winds, hot temperatures, and high relative humidity.

Let's turn our attention to the polar regions, where (as we learned in Chapter 2) more energy is emitted to space than is gained from solar radiation. As a consequence, the air above the poles is cold and sinks. Sinking air warms adiabatically. Warmer air over the cold polar surface is an inversion and inhibits precipitation. This explains the small amount of precipitation observed near the poles (see Figure 7.4).

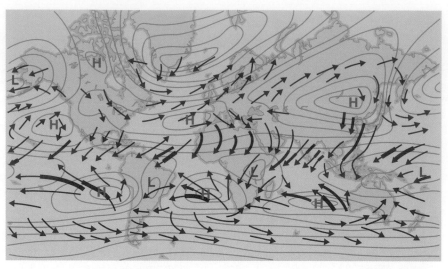

Wind—January

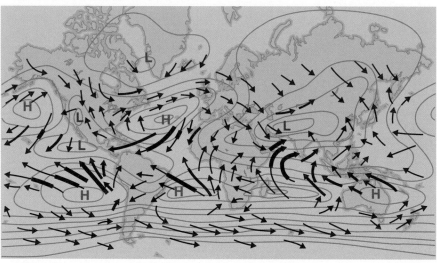

Wind—July

■ **Figure 7.12—cont'd**

Make a Jet Stream

When the sinking air over the poles reaches the ground, it must flow equatorward. Because this air is flowing southward from northern polar regions, the Coriolis force acts to turn these winds westward, forming **polar easterlies** at the surface poleward of 60° latitude (see Figure 7.6). Eventually, air must rise and flow poleward to replace the sinking air over the poles, completing the circulation of air through the polar cell (see Figure 7.6). The sea-level pressure maps show that the polar easterlies are well developed in the winter and over continents. Eventually, as these winds flow equatorward, they are met by subtropical air flowing poleward.

On the poleward sides of the descending branch of the Hadley Cell, surface air moves poleward (see Figure 7.6). The Coriolis force, which increases in magnitude as the poles are approached, moves the surface wind in a westerly direction in the midlatitudes of both hemispheres. Our conceptual model therefore produces **midlatitude westerlies** at the surface. The midlatitude westerlies are consistent with the observations in Figure 7.12, validating our conceptual model.

The midlatitude westerlies encounter the polar easterlies around 60° latitude. This clash of winds separates warm tropical air from cold polar air and is referred to as the **polar front.** As we saw in Chapter 6, strong winds should exist above regions

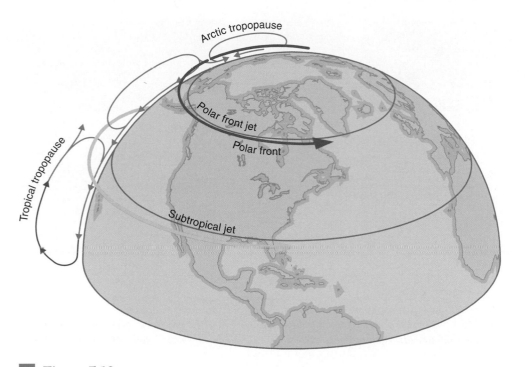
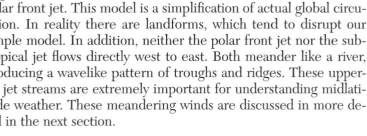

Figure 7.13

The subtropical and polar jet streams in relation to our conceptual model of global wind circulations.

of strong temperature gradients. Therefore, we expect to find another westerly jet stream above the polar front, referred to as the **polar front jet** (Figure 7.13). The polar front jet stream is displaced further poleward than the subtropical jet stream (Figure 7.14).

Our conceptual model explains important circulation features of the atmosphere: the ITCZ, the subtropical jet stream, the subtropical deserts, the trade winds, midlatitude westerlies, and the polar front jet. This model is a simplification of actual global circulation. In reality there are landforms, which tend to disrupt our simple model. In addition, neither the polar front jet nor the subtropical jet flows directly west to east. Both meander like a river, producing a wavelike pattern of troughs and ridges. These upper-air jet streams are extremely important for understanding midlatitude weather. These meandering winds are discussed in more detail in the next section.

Upper-Air Midlatitude Westerlies

The winds in the upper troposphere, above approximately 500 mb, flow in wavelike patterns with troughs and ridges (Figure 7.15). The air flow through these upper-level waves results in storms that move warm air poleward and cold air toward the equator. The Northern Hemisphere is typically encircled by several (Figure 7.15 shows three) of these waves at any particular time. These long-waves are called **Rossby waves,** named after Carl-Gustav Rossby, the famous 20th-century meteorologist who discovered them. Rossby waves typically drift slowly eastward. As discussed in Chapter 6, rising motions tend to occur downstream of the wave troughs, with sinking motions occurring upstream of the troughs. The movement of these

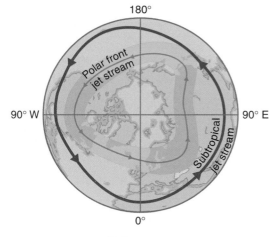

Figure 7.14

The approximate positions of the polar front jet stream and the subtropical jet stream over the Northern Hemisphere during winter.

 http://info.brookscole.com/ackerman

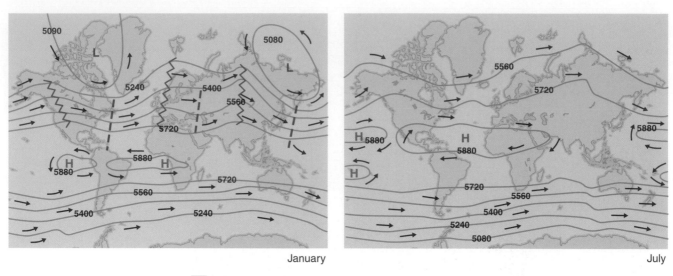

January   July

■ **Figure 7.15**

These 500-mb weather maps show that the winds of the upper troposphere meander in wavelike patterns with troughs and ridges. Troughs (dashed red lines) occur where the height contours dip equatorward, and ridges (jagged lines) occur where the height contours bulge poleward.

waves is very important in determining the development of surface weather systems, so understanding these waves is central to weather prediction.

Waves are described by their wavelength (distance between troughs or ridges) and amplitude (north–south extent). Amplitude and wavelength determine the type of weather associated with these waves. In the case of Rossby waves, a small-amplitude pattern results in mostly west-to-east winds known as a **zonal flow pattern,** nearly parallel to the lines of constant latitude (Figure 7.16, *A*). In zonal flow, cold air masses tend to stay toward the polar regions and the warm air remains equatorward. A **meridional flow pattern** occurs when the waves have large amplitude with deep troughs and peaked ridges (Figure 7.16, *B*). In a meridional flow pattern, cold air flows equatorward toward the subtropics and warm air flows poleward. The usual westerly jet may be absent, replaced by strong northerly and southerly winds.

A simple way to determine the type of flow pattern is the **zonal index.** The zonal index is the average speed of the west-to-east 500-mb wind in the latitude belt between 35° and 55° latitude. High zonal index situations indicate zonal flow patterns; low or negative zonal index values indicate meridional flow patterns.

A 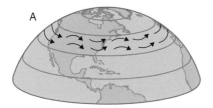   B 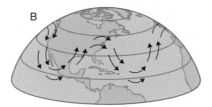   C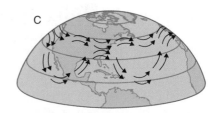

■ **Figure 7.16**

A, *In a zonal flow pattern, the air flows nearly parallel to latitudes and moves from west to east.* **B,** *In a meridional flow pattern, the air moves north and south as it flows eastward.* **C,** *The flow pattern can exhibit a combination of zonal and meridional flow patterns.*

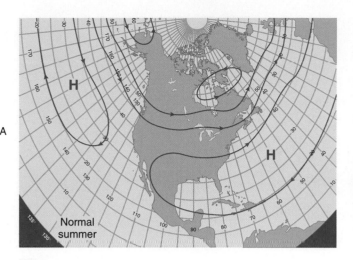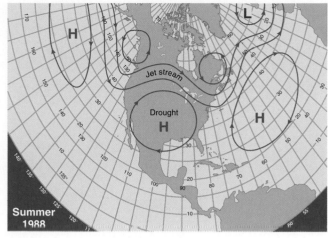

Figure 7.17

*Normal (**A**) and "blocking" (**B**) wind patterns above North America in summer. The upper-level "blocking" high over the United States in 1988 diverted the jet stream into Canada and led to drought conditions over the central United States. (Source: Namias, J., "Written in the Winds: The Great Drought of '88," Weatherwise, vol. 42, pp .85–87.)*

Different configurations are also possible. A **split flow pattern** can occur where zonal flow exists near the pole with a meridional flow pattern further to the south. Sometimes in these cases the meridional pattern becomes so strong that masses of air separate and become cut off from the main westerly air flow. Cut-off lows or cut-off highs (Figure 7.16, *C*) represent large pools of cold or warm air that block the eastward progression of weather systems. These **blocking patterns** can persist for extended periods of time and can result in extreme weather events such as floods and droughts.

A blocking pattern contributed to the summer flooding of the Midwest and drought in the southeast United States in 1993. This blocking pattern had a cold pool of air stalled over the Northern Rocky Mountains and the Pacific Northwest. In another year, a different blocking pattern created a warm high-pressure system over the central United States with troughs over the east and west coasts. This particular blocking pattern contributed to the spring and summer drought in the Midwest, the Northeast, and the Great Plains in 1988 (Figure 7.17).

If you look at upper-level weather charts every day for several weeks, you will notice that high zonal index periods are punctuated by low-index spells, and sometimes split flow patterns develop. This irregular oscillation between patterns is well-known, but poorly understood, and is called the **index cycle.** Current research is examining the possibility that the index cycle is related to complex atmosphere–ocean interactions (see Chapter 8).

Superimposed on the Rossby long-waves are ripples referred to as **short-waves** (Figure 7.18). Short-waves travel rapidly through the longer Rossby waves. The short-waves travel along with the Rossby waves and are sometimes difficult to observe. It is therefore difficult to predict their position and the associated weather. Both short-waves and long-waves are needed for the development of storms. Rossby waves, short-waves, and long-waves are examined in more detail in Chapter 10.

The Poleward Transport of Energy

In Chapter 2 we learned that radiative energy gains exceed the radiative losses at the top of the atmosphere in the tropical regions of the globe. In the polar regions the radiative losses exceed the gains. Circulations in the atmosphere help transfer heat

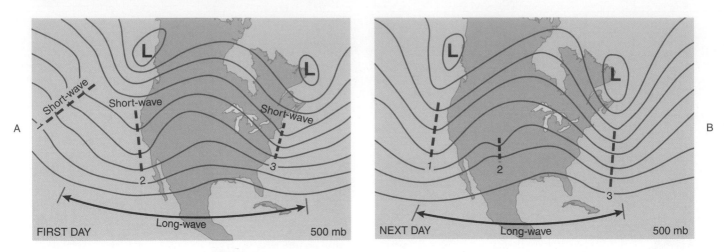

Figure 7.18

Short-waves are weather disturbances that move through the atmosphere faster than the long-wave Rossby Waves. (Source: From Ahrens, C., Meteorology Today, *6th ed., Brooks/Cole, 2000, p. 331.)*

poleward to compensate for these regional differences in energy budgets. The movement of warm air poleward and of cold polar air equatorward is one way this heat transfer occurs. The air flow of our conceptual model includes these patterns. Thus, in addition to explaining the main weather features of the globe, our model of atmospheric circulation implies a heat transport from the tropical regions to the poles.

If there were no poleward heat transfer, the poles would be much colder and the tropics much warmer. Figure 7.19 shows the annual average poleward energy transport accomplished by the atmosphere and the oceans. The region of maximum energy transport by the atmosphere falls between 30° and 60° latitude in both hemispheres. This poleward heat transport is accomplished by the midlatitude cyclones. Chapter 10 discusses these storms in detail.

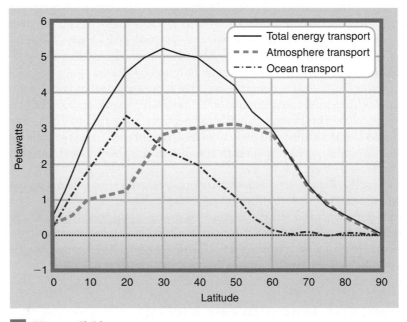

Figure 7.19

Atmospheric winds transport heat from the tropics to the North Pole, helping to maintain an energy balance around the globe.

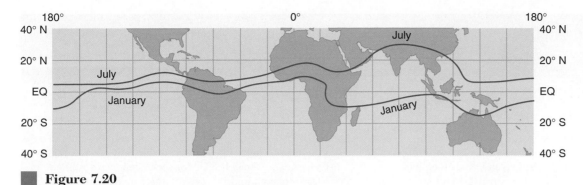

Figure 7.20

The approximate position of the ITCZ in January and July.

Seasonal Variations

There are seasonal variations in the amount of solar energy received by the Northern and Southern Hemispheres, and these variations account for seasonal changes in weather. How does our simple model handle these seasonal differences? The global circulation pattern—made up of the ITCZ, the polar front, the subtropical highs, and the jet streams—shifts with the Sun, moving poleward in spring and equatorward in autumn. This is reflected in the mean sea-level pressure maps for January and July (see Figure 7.12). The subtropical highs are shifted further poleward in each of the summer hemispheres.

The position of the ITCZ has a definite seasonal variation. The average position of the ITCZ is approximately 4°N. Seasonal averages show that the ITCZ is shifted toward the summer pole (Figure 7.20). North–south variations in the position of the ITCZ are larger over land than over the oceans because the thermal properties of continents differ from those of oceans. Because oceans heat up and cool down less drastically than continents, over water the ITCZ is generally anchored to within a few degrees of the equator.

In each hemisphere, the polar jet stream is displaced further poleward in summer than in winter. During the summer, the positions of the subtropical highs also shift poleward. As a result, the boundary between the surface air flowing poleward from the subtropical highs and the polar easterlies is displaced poleward. This boundary indicates the position of the polar jet stream. The poles are warmer in summer than winter. Because the polar air is warmer, the temperature difference between the polar air and the subtropical air is smaller. According to the thermal wind relationship (Chapter 6), this means that the polar jet stream can be expected to be weaker during the summer. The transport of energy from the middle latitudes to poles should be smaller during the summer than during the winter, as the summer poles are illuminated and thus gain solar energy. Observations generally confirm these expectations, again validating our conceptual model.

The semipermanent low-pressure systems also vary seasonally in their position (see Figure 7.12). In January there is a string of low-pressure areas around 8° south latitude over South America, Africa, and Indonesia. This corresponds to the average position of the ITCZ. During July there is a low-pressure region west of Central America that is associated with the mean position of the ITCZ in this region. During January there are also low-pressure regions near Iceland and near the Aleutian Islands. These are the subpolar lows and are regions that often spawn storms. During July there is one low near the North Pole and it is weaker than during winter. The low-pressure region around the South Pole is associated with the cold Antarctic continent and its topography.

BOX 7.2

Precipitation Patterns and Topography

As discussed in Chapter 4, orographic lifting (lifting of air as it travels over a mountain) is one mechanism of lifting air to reach saturation, cloud formation, and thus precipitation. Careful examination of global precipitation patterns indicates a correlation between total precipitation and topography. Precipitation, for example, is increased on the windward side of a mountain. However, the correlation is not perfect. Precipitation is also related to the temperature and humidity of the rising air, the speed and the direction of the winds relative to the mountain topography, and the stability of the air.

The coastal mountain ranges of the Pacific Northwest are a good example of the relationship between topography and precipitation. Because of their location in the midlatitude westerlies, the upwind side (the side the wind strikes first) of these mountains receives a lot of rain while the downwind side can resemble a desert. To explain this observation, let's consider a simplified model of the weather situation.

cooled even further to reach the point of saturation. In other words, the dew point temperature decreases. The dew point decreases approximately 2° C for each kilometer it is lifted.

After being lifted 1 kilometer, the temperature and the dew point are the same, so the air is saturated. The mixing ratio then decreases, because water vapor molecules are removed from the air to form cloud droplets. Further lifting causes the air to cool at the moist adiabatic lapse rate, approximately 6° C per kilometer. Collision and coalescence result in large droplets and precipitation as the air ascends the mountain. At the summit, assumed here to be 3 kilometers, the air has cooled to approximately −2° C. As long as the air remains cloudy, the dew point and the temperature are equal. As the cool air begins its descent on the downwind side of the mountain, the cloud droplets evaporate as the air warms because of compression heating (which lowers the relative humidity). The cloud quickly dissipates after reaching the

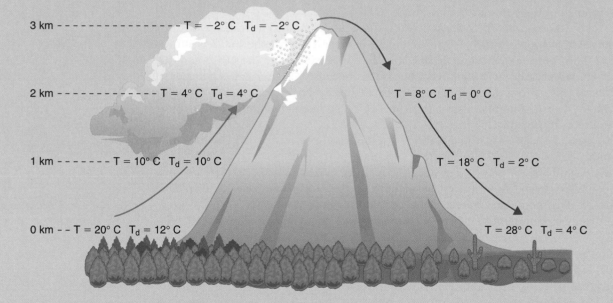

Our simple conceptual model consists of a mountain peak 3 kilometers high in the midlatitude westerlies. Suppose that the air temperature (T) of the wind approaching the mountain is 20° C with a dew point (T_d) of 12° C. Air whose temperature is higher than its dewpoint is unsaturated, or "dry." As the air is orographically lifted on the windward side it expands and cools at the dry adiabatic lapse rate—approximately 10° C for each kilometer it is lifted.

As the parcel cools, the relative humidity increases. As the parcel is forced to rise, the mixing ratio remains the same until a cloud forms. The parcel also expands as it rises, thus decreasing the vapor pressure. The air must therefore be

summit. The now-heated sinking air is depleted of moisture, resulting in a local minimum in rainfall on the downwind, or leeward, side. This local minimum of precipitation is called a *rain shadow*.

In addition, as the air moves down the mountainside, the temperature increases at the dry adiabatic lapse rate and the dew point increases at about 2° C per kilometer. When the air returns to sea level, its temperature is 28° C with a dew point temperature of 4° C. At the end of its trip across the mountain, the air has warmed and is drier than when it began. In this way, topography can dictate the weather and climate of a region.

There are also low-pressure regions in the vicinity of Iraq and the Southwest United States in July. These lows are associated with deserts and are referred to as *heat lows* or **thermal lows.** These heat lows develop because of intense surface heating caused by absorption of solar radiation that occurs in the absence of clouds. It is no accident that these lows form at about the same latitude as the subtropical highs, where sinking air leads to persistently clear skies. The lows form when surface heating warms the air above, causing it to expand both vertically and horizontally throughout a deep layer. This expansion reduces the air density, making the pressure near the surface lower than it would be if the air were cooler.

Monsoons

A **monsoon** is a weather feature driven by seasonal differences in the heating of land and ocean along with seasonal shifts in global-scale circulations. Our original model of atmospheric circulation assumed that the Earth was covered with water. To examine an important seasonal variation in precipitation, we will have to add land masses to our conceptual model. The distribution of land and especially large mountain ranges has a strong influence on weather patterns (Box 7.2). One good example of a place where this effect is prominent is the Indian sub-continent. The Indian monsoon that occurs there is a complex interaction of seasonal shifts in the global circulation patterns, thermal heating differences between land and sea, and the interaction of winds with the Himalayan mountains.

Blue Skies

**Moisture and Stability/
Adiabatic**

In spring and summer, the Sun heats the land and the air above it. With cooler air over the surrounding water, a horizontal pressure gradient, directed from ocean to land, is established near the surface, bringing humid air inland. The solar heating inland triggers convection. This causes hot, humid air to rise, expand, and cool, leading to condensation and rain. Orographic lifting by the Himalayas cools the air when it reaches the mountains, generating additional rain. Air rushing out of the tops of the thunderstorms flows out over the Arabian Sea and the Bay of Bengal where it sinks, completing the monsoon circulation (Figure 7.21). The air flowing over the water remains over the ocean a long time, which causes it to gain moisture.

During autumn and winter, the air above the land cools faster than the air over the water, establishing a pressure gradient force from land to sea. The winds are reversed from the summer monsoon flow—land to sea instead of from the sea to the land (Figure 7.22). Sinking air above the land suppresses cloud development and precipitation. As a result, the winter monsoon is a dry season and the summer monsoon is a wet season.

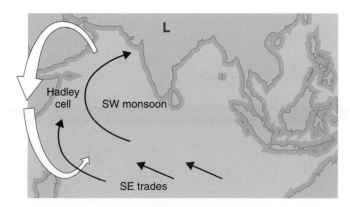

Figure 7.21

A depiction of the complete monsoon circulation during summer. Low-level winds travel over the warm Arabian Sea and supply the moisture for the torrential rains.

http://info.brookscole.com/ackerman

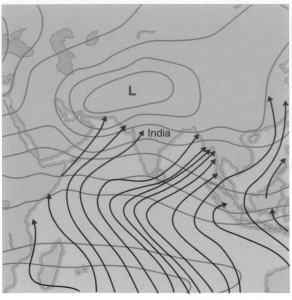
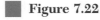

Summer monsoon

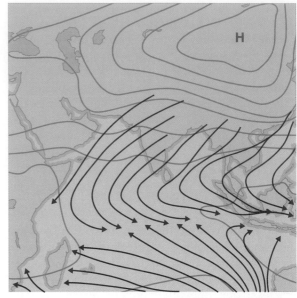

Winter monsoon

Figure 7.22

The surface wind (arrows) and pressure patterns (contours) associated with the summer and winter Indian monsoons. During summer the winds blow from sea to land. This pattern reverses in winter. (Source: Hastenrath, S., Climate and Circulation in the Tropics, *Kluwer, 1988.)*

 PUTTING IT ALL TOGETHER

 SUMMARY

In this chapter we developed a conceptual model to explain patterns of global circulation and their associated weather and climate.

Upward vertical motions near the equator explain the Intertropical Convergence Zone (ITCZ), which can be identified in global satellite imagery and precipitation maps. When rising air reaches the tropopause and moves poleward, it comes closer to the Earth's axis of rotation and increases speed to conserve angular momentum. This produces the subtropical jet stream.

The descending air of the Hadley Cell is compressed and warms, lowering the relative humidity. This sinking air explains the large deserts of Africa, Saudi Arabia, and Australia. The sinking air of the Hadley Cell results in calm winds at the surface, producing the "horse latitudes" dreaded by ancient mariners who used wind power to travel. Some descending air that reaches the surface moves equatorward to supply moisture to the ITCZ. As air moves equatorward, the Coriolis force acts to produce the steady northeast and southwest trade winds sought by sailors.

Some of the air associated with the descending branch of the Hadley Cell moves poleward and clashes with cold polar air masses that are moving equatorward. Fronts exist where the cold air meets the warm subtropical air, producing midlatitude storms that tend to move from west to the east. These storms are embedded in the midlatitude westerlies, causing them to move eastward. The polar jet stream exists in the vicinity of fronts.

In our simple conceptual model the globe is covered with water—a good approximation, since water covers 70% of the Earth's surface. However, the differences between land and water are important in weather and climate. The Indian monsoon provides an example of how ocean, land, and topography interact to form clouds and precipitation or to bring about a dry season. A monsoon is a seasonal reversal in wind patterns associated with changing regional energy budgets. The change in wind direction results in dry and wet seasons. Over the Indian subcontinent, the wet season occurs in summer and the dry season in winter.

The general circulation of the atmosphere transports heat from the equator toward the poles. The world's oceans also transport heat. How ocean circulations interact with the atmosphere to affect weather and climate is the topic of the next chapter.

KEY TERMS

You should understand all of the following terms. Use the glossary and this chapter to improve your understanding of these terms.

Angular momentum
Blocking pattern
Conservation of angular
  momentum
Doldrums
Hadley cell
Horse latitudes
Index cycle

Intertropical Convergence
  Zone
Jet stream
Meridional flow pattern
Midlatitude westerlies
Monsoon
Polar easterlies
Polar front

Polar front jet stream
Rossby wave
Short-wave
Split flow pattern
Subtropical highs

Subtropical jet stream
Thermal lows
Trade winds
Zonal flow pattern
Zonal index

REVIEW QUESTIONS

1. What might the global circulation of the atmosphere be if the Earth did not rotate and were covered with water? (Hint: start with an ITCZ.)
2. Explain the seasonal shift in the polar front jet over the continental United States from January to July.
3. Why are the subtropical highs referred to as semipermanent highs rather than permanent high pressure systems?
4. Draw a picture of the global air flow near the surface of the Earth.
5. Explain the existence of the Intertropical Convergence Zone.
6. Discuss the differences and similarities between a land or sea breeze and a monsoon.
7. Using the "Atmospheric Circulation/Global Atmosphere" section of the *Blue Skies* CD-ROM, locate the major features of the atmospheric circulation pattern in a week-long animation of weather satellite images.
8. Explain why the Coriolis force does not influence how your sink drains, but does affect the subtropical jet stream.
9. Based on your knowledge of how the Coriolis force affects the Hadley Cell, how far poleward do you think the Hadley Cell would extend if the Earth didn't rotate around its axis?
10. Just south of the Tropic of Cancer on the island of Kauai lies Mount Waialeale (elevation: 1.6 km), which receives approximately 1170 centimeters (460 inches) of rain a year. Explain why this region receives so much rain.
11. Why are the Southern Hemisphere westerlies slightly faster than the Northern Hemisphere westerlies?
12. The zonal wind at 500 mb is from the east at 20 meters per second at 35° latitude, and it is from the west at 20 meters per second at 55° latitude. Using just this information and the definition of the zonal index, can you tell if the Rossby wave pattern is high-amplitude or low-amplitude? Would you expect a cold air outbreak in the subtropics in this situation?
13. Use the interactive simulation in the "Moisture and Stability/Adiabatic" section of the *Blue Skies* CD-ROM to test the effect of crossing a mountain on an air parcel. Try the simulation with both moist (temperature and dew point close together) and dry (temperature and dew point far apart) parcels. Why does the mountain crossing affect the parcel's final temperature and dew point in the moist cases but not in the very dry cases? Relate this phenomenon to the typical weather conditions of the mountainous regions of western North America.

WEB ACTIVITIES

Choose Chapter 7 on the textbook's Web site:
www.brookscole.com/ackerman
and select from the following resources:

- Interactive Modules that illustrate and extend your understanding of key topics in this chapter
- Tutorial Quizzes to test your mastery of terms and concepts

For additional readings, go to the InfoTrac College Edition, your online library, at:
http://www.infotrac-college.com

Atmosphere–Ocean Interactions: El Niño and Tropical Cyclones

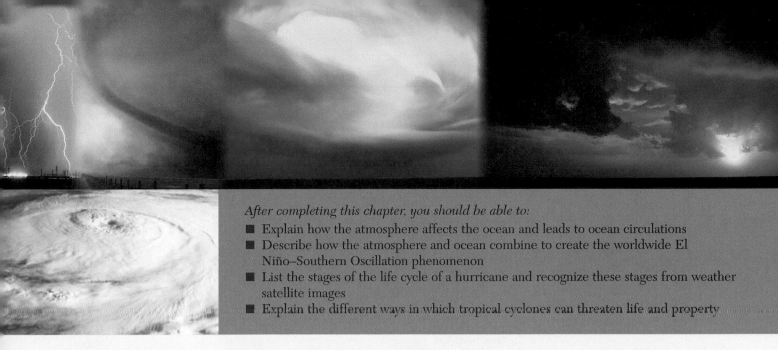

After completing this chapter, you should be able to:

- Explain how the atmosphere affects the ocean and leads to ocean circulations
- Describe how the atmosphere and ocean combine to create the worldwide El Niño–Southern Oscillation phenomenon
- List the stages of the life cycle of a hurricane and recognize these stages from weather satellite images
- Explain the different ways in which tropical cyclones can threaten life and property

Introduction

In the early 1990s, beachcombers along the Pacific coasts of Alaska, Canada, and the northwest United States made some surprising discoveries. Instead of seashells, they found hundreds of plastic bathtub toys and Nike tennis shoes washed up on the beach! Where did this flotilla of flotsam come from? Two separate accidents in the north Pacific in 1990 and 1992 caused 61,000 shoes and 29,000 toys to spill off of ships into the ocean. Once overboard, the shoes and the toys floated, in some cases for several years. However, they didn't just bob up and down in one place in the Pacific. Driven by strong winds at the surface, steady ocean currents carried the shoes and toys eastward toward the North American coastline.

The atmosphere interacts with an ocean across 70 percent of the Earth's surface. In this chapter we learn what the ocean is made of and how the force of wind on water creates ocean currents. Then we examine how cross-talk between the atmosphere and oceans creates oscillations in weather and climate such as the El Niño phenomenon. Finally, we investigate how air and ocean can combine to form nature's deadliest storm system, the tropical cyclone. These ferocious atmospheric storms, often called hurricanes, cannot survive without the moisture supplied by warm ocean water. Killer tropical cyclones are a far cry from plastic toys washing up on an Alaska shore, but both are signs of the strong linkages between the atmosphere and oceans.

Oceanography

Oceanography is the study of the oceans and is essential to an understanding of weather and climate. The oceans play three important roles in determining

weather and climate: (1) they are a source of atmospheric water vapor; (2) they exchange energy with the atmosphere; and (3) they transfer heat poleward.

Each year approximately 396,000 cubic kilometers (almost 95,000 cubic miles) of water vapor circulate through the atmosphere through the hydrologic cycle. Most of the water that recycles into the atmosphere comes from the oceans (see Chapter 1). About 334,000 cubic kilometers (almost 80,000 cubic miles) of water evaporate from the oceans every year. At any given time, the atmosphere contains approximately 15,300 cubic kilometers (about 3,500 cubic miles) of water. It takes approximately 2 weeks for all of the water in the atmosphere to be recycled. The oceans provide the majority of water needed to form precipitation.

Exchanges of heat and moisture occur at the interface between the atmosphere and the ocean. Figure 8.1 shows the net energy gains and losses at this interface for Northern Hemisphere winter and summer. On average, the ocean gains energy during summer and loses energy to the atmosphere in winter. Therefore, the oceans normally warm in the summer and cool during the winter. Energy budget maps like Figure 8.1 are a measure of the average interaction between the atmosphere and ocean. Maximum exchanges of energy occur in the Northern Hemisphere winter to the east of North America and Asia. Warm ocean currents flow northward in these regions (see Chapter 3) and supply energy to winter storms that are leaving these continents heading eastward. As we will learn later in this chapter, warm ocean waters also supply the energy that fuels hurricanes.

The rate of heat and moisture transfer, as discussed in Chapter 2, depends on the temperature difference between the air and the water, as well as the strength of the winds. Warm sea surface temperatures (SSTs) and strong winds are favorable for large heat exchanges between the ocean and atmosphere. It is therefore important to understand where the warm waters of the oceans

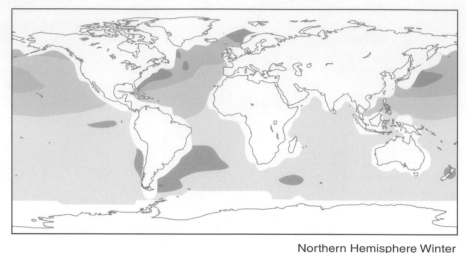

Northern Hemisphere Winter

Surface heat flux
(w/m²)

−100   100

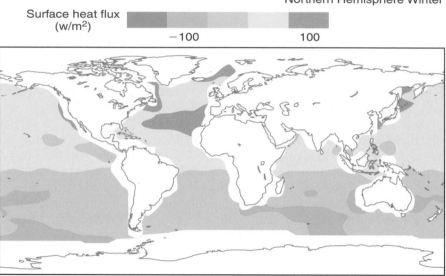

■ **Figure 8.1**

The energy gains and losses of the Earth's oceans. Positive (yellow and orange) values indicate energy gains by the ocean while negative (green) represent energy losses by the ocean.

Northern Hemisphere Summer

are and to know why they are there. We learned in Chapter 3 that warm currents of the middle and high latitudes border the eastern coasts of some landmasses. Countries such as Great Britain have mild winters and small annual temperature ranges. The western coasts of landmasses that border cool waters, such as northern California, have low average temperatures and small annual ranges of temperature (Chapter 3). This chapter will discuss the origins of these cool and warm currents.

Like the winds, the ocean currents transport heat from the tropics to higher latitudes. Figure 8.2 shows the poleward transport of heat in the Northern Hemisphere by the atmosphere and oceans required to balance the radiation budget, as discussed in Figure 2.20 (reproduced as an inset of Figure 8.2). The poleward transport of heat by the atmosphere is discussed in Chapter 7. At about 30° latitude, the atmosphere and ocean each transport about the same amount of heat. Equatorward of 30°, the ocean transfers the majority of heat required to maintain balance. The atmospheric transport has a broad peak at around 30° to 60° north latitude. The oceans and the atmosphere therefore work together to redistribute the Sun's energy toward the poles. This means that we cannot study the atmosphere without taking the oceans into account. For example, any changes in the ocean circulation patterns may change the heat transport by the oceans, which in turn may change heat transport by the atmosphere, and thus change the world's climate.

The teamwork between the atmosphere and oceans is a two-way street. Changes in the ocean and its circulation patterns affect the atmosphere and vice

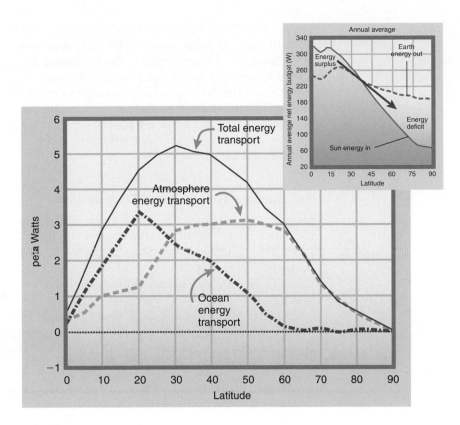

■ **Figure 8.2**
The yearly mean transport of energy by the atmosphere and ocean as a function of latitude for the Northern Hemisphere. The inset indicates the need for a poleward energy transport in order to maintain energy balance.

versa. The Sun is the primary source of energy that warms the oceans, and atmospheric conditions determine the amount of solar energy reaching the ocean. Cloud-free conditions allow more energy to enter the oceans while cloudy conditions reduce the amount of solar energy reaching the surface. Winds also play an important role in determining ocean temperatures. The solar energy that is absorbed at the surface of the water can be mixed to deeper layers by the winds, which stir the upper layer of the ocean. Next we investigate the temperature patterns of the world's oceans.

■ Ocean Temperature

Chapter 1 described the atmosphere in terms of vertical distribution of temperature. We can do the same with the oceans, based on observations of temperatures at various ocean depths at a given location. The Sun warms the atmosphere from below, but it heats the top of the oceans. Therefore, the vertical temperature profile of the oceans looks a little like an upside-down version of the atmosphere's temperature profile, with warm water at the top and colder water below, as seen in Figure 8.3. Based on measurements of temperature we can classify the ocean into three vertical zones.

The characteristic feature of the top 100 meters of the ocean at a given location is a constant temperature. The uniform layer is called the **surface zone** or *mixed layer.* Wind-driven waves and currents mix this layer so that the temperature is relatively uniform. Only about 2% of the world's ocean waters are within the surface zone. The bottom layer, or **deep zone,** lies below about 1000 meters and is composed of cold water at a temperature between $-1°$ to $3°$ C ($30.5°–37.5°$ F). The temperature in the deep zone is uniform. The transition zone between the surface and deep layers is the **thermocline** (*therme* means "heat" and *clinare* means "to lean"). In the thermocline, temperature decreases rapidly with depth down to a depth of about 1000 meters.

Figure 8.3 is a typical temperature distribution of the ocean with depth. This temperature variation with depth is a function of latitude. Figure 8.4 shows the

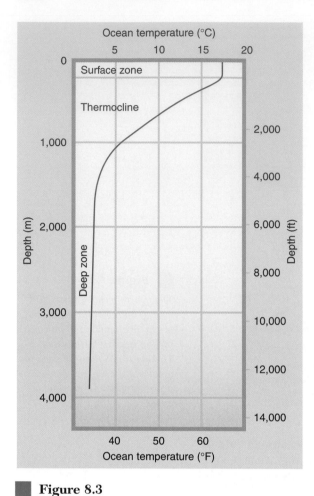

Figure 8.3

Vertical profile of temperature in the oceans, indicating the surface zone, the thermocline, and the deep zone.

temperature profile with depth for oceans typical of tropical, midlatitude, and polar regions. The deep zone temperatures are similar for all three regions. The tropical waters have the warmest surface zone temperatures and the polar regions have the coldest. This is because the polar regions receive less solar energy. The tropical and midlatitude regions have the steepest thermocline. For our purposes, the important point is that the world's oceans, even those that are warm at the surface, are cold *beneath* the surface.

Sea Surface Temperature Distributions

Interactions between the atmosphere and ocean occur at the surface and result in the transfer of heat and moisture. The distribution of SSTs is therefore important in meteorology. Indeed, as we shall see later in this chapter, the SST determines where tropical cyclones can develop and grow.

Figure 8.5 shows the SST distribution in the world's oceans, with red regions representing warm water and blue regions cold water. We can draw the following general conclusions from this figure:

1. Western coasts in the subtropics and middle latitudes are bordered by cool water.
2. Western coasts in tropical latitudes are bordered by warm water.
3. Eastern coasts in the middle latitudes are bordered by warm water.
4. Eastern coasts in polar regions are bordered by cool water.

To explain these temperature distribution patterns, we have to understand the circulation of the world's oceans.

Ocean Currents

An **ocean current** is a massive, ordered pattern of water flow. The world's ocean currents are shown in Figure 8.6. The global-scale wind patterns such as the trade winds and the midlatitude westerlies (Chapter 7) blow steadily over large regions of the ocean and push the ocean surface in the same general direction as the wind. As a result, the ocean current patterns in Figure 8.6 resemble the surface wind patterns discussed in Chapter 7.

The swirling ocean **gyres** that lie underneath the subtropical anticyclones are a striking example of how the atmosphere affects the oceans. These immense ocean circulations spin in the same direction as the anticyclones, causing warm water to flow poleward along the eastern coasts in strong currents such as the Gulf Stream (Figure 8.7). In addition, the winds of the subtropical highs cause cold water to flow equatorward near the western coasts, such as the Canary Current near Africa and the Peru Current off the Pacific coast of South America. Thus the effect of wind on the oceans helps explain many of the features of the temperature distribution in Figure 8.5. The fact that warm water flows poleward and cold water flows equatorward in these gyres also helps explain how oceans transport heat poleward, as we saw in Figure 8.2.

We can explain even more about ocean circulations by considering the balance of forces on an ocean current, just as we did for winds in Chapter 6. In Figure 8.8, the ocean currents that make up the North Atlantic gyre are shown with the winds that blow over them. Notice that these currents actually flow somewhat to the right of the wind direction. Why? As we learned in Chapter 6, moving air is subject to the Coriolis force that pushes to the right in the Northern Hemisphere and to the left in the Southern Hemisphere. The same holds true for moving ocean water. The frictional force created by the wind pushes the ocean in the direction of the wind, but in combination with the Coriolis force it leads to a current that is at an angle to the wind direction. Surface currents therefore flow at an angle of approximately 45° to the wind, to the right of the wind in the Northern Hemisphere and to the left in the Southern Hemisphere.

These surface currents also have a major impact on ocean flow beneath the surface, causing two key atmosphere–ocean interactions known as the **Ekman spiral** and **Ekman transport** in honor of their discoverer, who investigated why ice in the ocean drifted at an angle to the wind.

To explain the Ekman spiral, let's divide the ocean into slabs of water as in Figure 8.9, starting with the top layer and working our way down into the ocean. As we have noted, the combination of friction caused by the wind and the Coriolis force leads to a surface ocean current in the top layer that flows at about 45° to the right of the wind in the Northern Hemisphere. This is not the end of the story, however. The slab of water underneath the top layer feels the surface current as friction. This slab moves initially in the direction of the surface current, but it too is then affected by the Coriolis

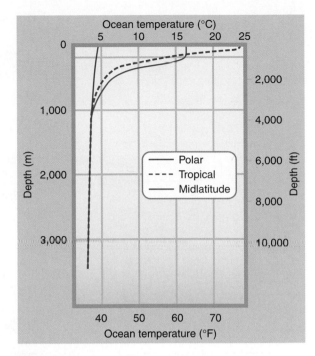

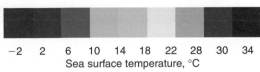
Figure 8.4

While all three zones—the deep zone, the thermocline, and the surface zone—appear at all ocean regions, their depths and the steepness of the thermocline are functions of latitude.

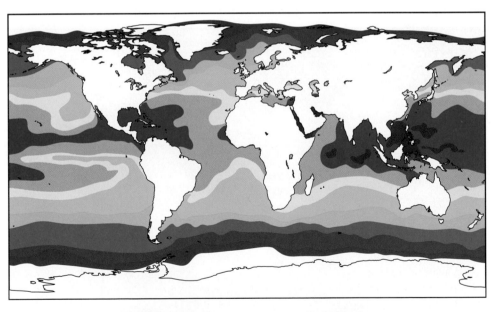

Sea surface temperature, °C

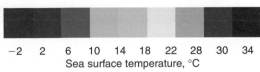
Figure 8.5

Sea surface temperature distributions across the globe. Western coasts in the subtropics and middle latitudes are bordered by cool water, and western coasts in tropical latitudes are bordered by warm water. Eastern coasts in the middle latitudes are bordered by warm-water currents, and eastern coasts in polar regions are bordered by cool water.

http://info.brookscole.com/ackerman

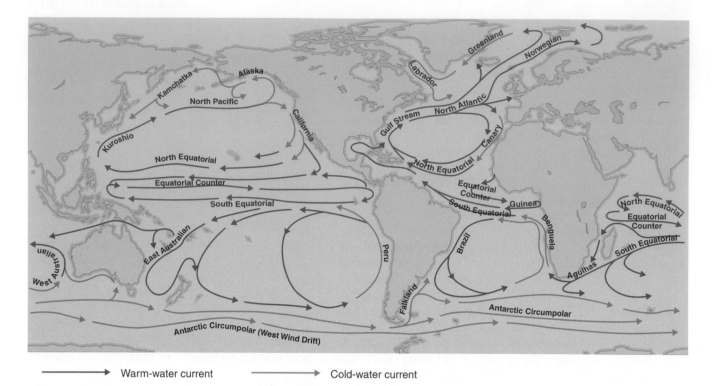

Warm-water current   Cold-water current

Figure 8.6

The major ocean currents. Warm currents are represented by red arrows and cold currents by blue arrows. In general, warm waters flow poleward or westward. Cold currents, except the Antarctic Circumpolar Current, tend to flow toward the equator.

Figure 8.7

A detailed infrared satellite view of the Gulf Stream (dark orange and red area extending from lower left). The undulations of this current, known as Gulf Stream eddies, are reminiscent of cut-off lows and highs that form in association with the atmospheric jet stream (Chapter 7).

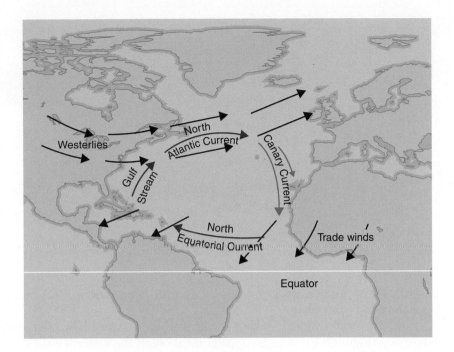

Figure 8.8

The North Atlantic gyre is comprised of four currents: The North Equatorial Current, the Gulf Stream, the North Atlantic Current, and the Canary Current. These surface currents are driven by the trade winds and the westerlies, but are deflected to the right of the wind direction by the Coriolis force.

force and its resultant motion is to the right of the surface current. The third slab feels the motion of the second slab as friction, and it too moves and is affected by the Coriolis force, resulting in a motion that is to the right of the slab above it. This clockwise turning of the ocean currents proceeds down to a depth of about 100 meters, with the current gradually reducing in strength from the surface to this depth.

The cumulative effect of the Ekman spiral, from the surface to 100 meters, is to move water at a 90° angle to the direction of the wind, as shown in Figure 8.9. This is called *Ekman transport* and occurs to the right of the wind in the Northern Hemisphere and to the left of the wind in the Southern Hemisphere. This horizontal

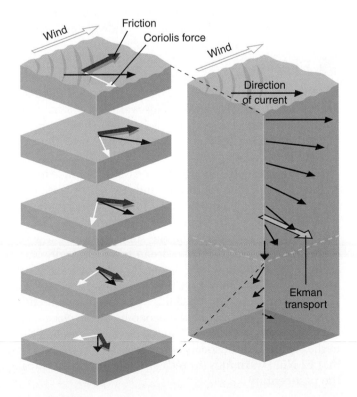

Figure 8.9

The Ekman spiral in the water is a change of ocean current direction and speed with depth. The body of water can be thought of as a set of layers. The top layer is driven by surface wind and each layer below is driven by frictional drag of the layer above. The direction that the water in each layer flows rotates to the right in the Northern Hemisphere.

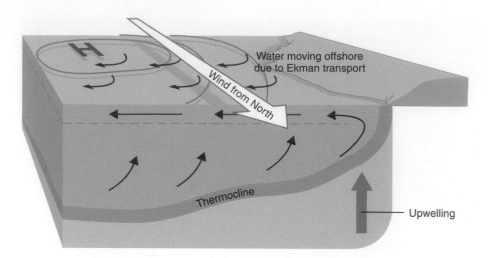

■ **Figure 8.10**

Surface winds along coastlines cause upwelling when they create ocean currents that move offshore.

movement of surface layer water generates an important vertical motion in the oceans called **upwelling.**

To understand why upwelling occurs, consider a wind blowing parallel to the shoreline (Figure 8.10). This is a typical situation off the western coast of continents because of the presence of the subtropical highs. Friction causes the surface waters to move and the Coriolis effect deflects the motion of the water. The resulting Ekman spiral leads to a transport of water away from the shoreline. As the surface water moves away from the coast it must be replaced by water rising from the thermocline.

Upwelling has a profound effect on ocean temperatures and life along coastlines. As we saw earlier, ocean temperatures decrease with depth. Therefore, upwelling brings cold water to the surface, and this water is rich in nutrients. These nutrients are an important food source for marine organisms living near the surface. These organisms, in turn, are sources of food for fish and birds, and millions of people across the world depend upon the abundance of life in and near regions of upwelling. When upwelling ceases because of a change in wind patterns, the result can be catastrophic for coastal regions. One example of this, which turns out to have implications for weather and climate worldwide, is El Niño.

■ El Niño

El Niño is a periodic warming of the equatorial Pacific Ocean between South America and the Date Line. This warming is a natural variation of the ocean and atmosphere and is not caused by human activities. El Niño is an excellent example of the interaction between the atmosphere and the ocean and how these interactions affect climate.

The waters off the western coast of South America typically experience upwelling, because the southerly winds associated with the subtropical high cause water to be transported offshore, to the left of the wind. (In the Southern Hemisphere, the Coriolis force and Ekman transport are to the left.) The upwelling water is cold and rich in nutrients, even in the summer. Every few years, the upwelling slows or ceases in this region and the ocean turns markedly warmer, sometimes 4° C (7° F) above normal. This warming phenomenon typically first appears off the coast of South America around the Christmas season and lasts for several months. The term El Niño, or "the Christ Child," was used by the fishermen along the coast of Ecuador and Peru, where the economy of the coastal communities depends on the cold, nutrient-rich waters. The warm El Niño waters disrupt the marine food chain in the region and the local economy suffers from the loss of fish. Figure 8.11 reveals that El Niño warmings (in red) have occurred periodically every 2 to 7 years during the past century.

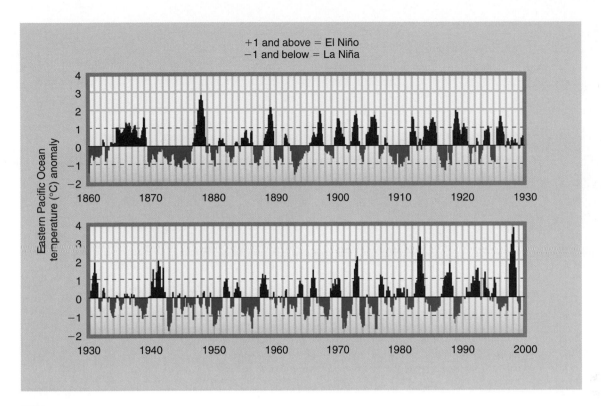

Temperature anomalies (red = warmer than normal, blue = colder than normal) across the eastern equatorial Pacific Ocean during 1860–1999. Red periods are El Niño episodes, and blue periods are La Niña episodes. (Source: Davis, M., Late Victorian Holocausts: El Niño Famines and the Making of the Third World, Verso Press, 2001, p. 235.)

Under normal conditions, the SSTs off Peru are cold because of coastal upwelling, and waters in the western equatorial Pacific are warm. SSTs off South America's western coast are 8° C (14° F) cooler than SSTs in the western Pacific. The steady easterly trade winds push the water towards Indonesia, raising the level of the ocean one-half meter higher there versus the South American coast. As the water flows westward, it warms as a result of absorption of solar energy and heat exchanges with the atmosphere. Over the western Pacific Ocean, the warm water evaporates and is accompanied by precipitation (Figure 8.12, A).

An El Niño event is triggered when, for reasons not fully understood, the trade winds weaken or reverse direction and blow from west to east. This allows the large mass of warm water that has piled up near Indonesia to move eastward along the equator until it reaches the coast of South America. During an El Niño event, two distinct changes occur in the equatorial Pacific Ocean: (1) Cold coastal waters are replaced by warm waters, and (2) the height of the ocean surface drops over Indonesia and rises in the eastern Pacific, forcing the thermocline lower near South America and preventing upwelling. Satellites can measure both of these features with an active sensor called an *altimeter*. Figure 8.13 shows global deviations in the height of the ocean surface as measured by a NASA satellite. Above-normal heights over the central and eastern equatorial Pacific, as seen in this figure, are the smoking gun indicating an El Niño is in progress.

El Niño Life Cycle

How Does El Niño Affect Global Weather Patterns?

During El Niño, the western Pacific experiences below-normal precipitation as the heavy rains follow the warm water and move eastward toward South America (Figure 8.12, B). The shift in the large tropical rain clouds alters the typical pattern of the subtropical jet stream. Air rising in these thunderstorms moves northward and

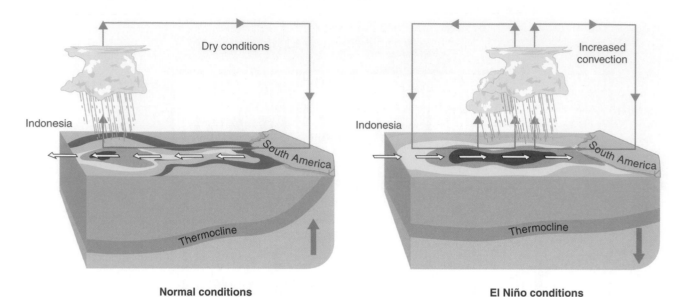

Figure 8.12

The trade winds normally cause the equatorial surface waters to move westward, piling up warm water in the western Pacific. During an El Niño event, the winds weaken and the warm water propagates eastward. Thunderstorms follow the movement of warm water into the central Pacific.

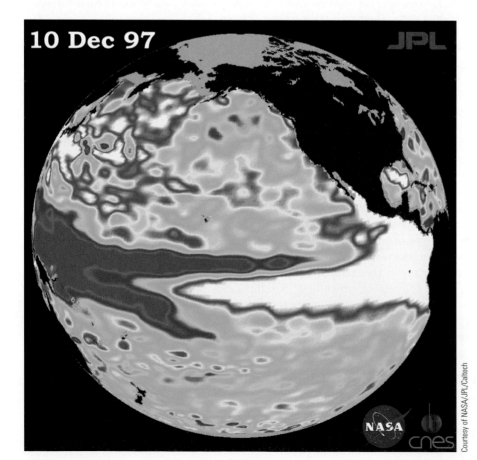

Figure 8.13

Satellite observations of changes in sea level height during an El Niño event. Tropical waters in the western Pacific are below normal, while those over the now-warm waters of the eastern tropical Pacific are above normal.

strengthens the subtropical jet over the southern United States. (See Figure 7.6 in Chapter 7 for a satellite image showing the connection between El Niño thunderstorms and the subtropical jet.) This change in the jet stream allows El Niño to affect the weather and climate of the midlatitudes as well as the tropics. To determine the impact of El Niño on global weather, meteorologists compare average weather conditions with weather conditions experienced during an El Niño year. There have been ten major El Niño events in the past half century: 1957–58, 1965, 1968–69, 1972–73, 1976–77, 1982–83, 1986–87, 1991–92, 1994, and 1997–98. The 1997–98 El Niño was the strongest recorded in the past century (see Figure 8.11). Figure 8.14 compares differences in temperature and precipitation between normal and El Niño years for winter and summer conditions. The impacts of El Niño upon climate in temperate latitudes are most evident during winter. A weak polar jet stream forms over eastern Canada and as a result a large part of North America is warmer than normal. During winter, the southeast United States experiences above-normal precipitation in connection with a strong subtropical jet stream. The increased rainfall across the southern United States and in coastal Peru, which is normally a desert, can cause destructive flooding. Drought in the western Pacific has also been

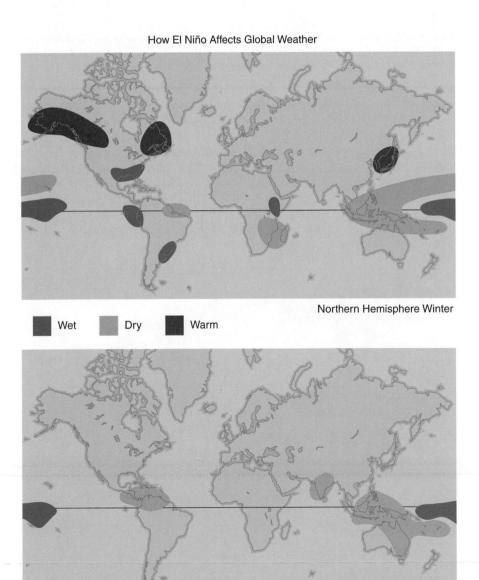

How El Niño Affects Global Weather

Northern Hemisphere Winter

Wet   Dry   Warm

Northern Hemisphere Summer

Figure 8.14

Changes in global weather patterns associated with an El Niño event.

TABLE 8.1 Global Impacts of Five Major El Niño Events

Region	1877–1878	1899–1900	1972–1973	1982–1983	1997–1998
India	D°	D†	D†	d	—
Philippines	d	D	d	D°	d
Australia	D	D°	D	D	D
North China	D°	D†	D	d	d
Yangzi River, China	F	—	—	F	F
South Africa	D	d	D	D	D
East Africa	d	d	D	D	D
Subsaharan Africa	d	D	D°	D	—
Northeast Brazil	D	d	D	D	D
South Brazil	?	?	—	F	—

D, Intense drought; *d,* moderate drought; *F,* intense flooding; *f,* moderate flooding; —, no weather events such as droughts or floods; ?, no data. An asterisk (°) indicates the worst drought or flooding for the region in the century. A dagger (†) indicates the second-worst drought or flooding for the region in the century.

Source: Adapted from Davis, M. *Late Victorian Holocausts: El Niño Famines and the Making of the Third World.* New York: Verso Press, 2001 (page 244). (Source: Davis, M., *Late Victorian Holocausts: El Niño Famines and the Making of the Third World,* Verso Press, 2001, p. 244.)

associated with devastating fires in Australia and Indonesia. Table 8.1 summarizes the effects of severe El Niños across the globe. Note that each El Niño event is unique.

Changes in precipitation and temperature patterns caused by El Niño affect snowfall in the United States (Figure 8.15). In the Southwest, there is a slight tendency toward cooler winters and a strong tendency toward wet winters, which makes higher-elevation snowpack deeper. In the Pacific Northwest, El Niño winters are warmer and drier than usual, so that at a given elevation there is less precipitation and the freezing level is at a higher altitude. The type of precipitation is more

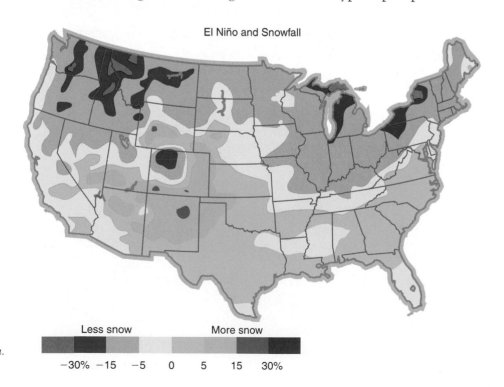

El Niño and Snowfall

■ Figure 8.15

Percent changes in snowfall in El Niño winters versus the average of all other winters. A decrease in snowfall during an El Niño event can be disastrous to winter recreation areas. (Copyright © www.sws.uiuc.edu. Used with permission.)

Less snow   More snow

−30% −15   −5   0   5   15   30%

likely to be rain, and the accumulation season is shorter. These factors produce a smaller snowpack accumulation by the end of winter in the Pacific Northwest. A significant reduction in total winter snowfall also occurs in the Midwest and New England regions during strong El Niño events.

Atmospheric scientists have discovered that El Niño is associated with a seesaw of atmospheric pressure between the eastern equatorial Pacific and Indo-Australian area. This seesaw in pressure is referred to as the **Southern Oscillation.** Scientists have been studying this oscillation since the 1890s. Generally, when pressure is high over the Pacific Ocean, it tends to be low over the eastern Indian Ocean, and vice versa. The Southern Oscillation is monitored by measuring sea-level pressure at Tahiti in the east, and at Darwin, Australia in the west. The difference in the pressure at these locations is called the Southern Oscillation Index (SOI). With a high positive SOI, pressure is high in the western Pacific, the trade winds are strong, and rainfall in the warm western tropical Pacific is plentiful.

The Southern Oscillation is closely linked to El Niño. High negative values of the SOI represent an El Niño, or "warm event" (Figure 8.16; compare to Figure 8.11). El Niño and the Southern Oscillation often occur together, but they can also happen separately. When an El Niño and Southern Oscillation occur together, the event is referred to as *ENSO.*

Blue Skies

Atmospheric Circulation/ Southern Oscillation

ERBE does ENSO

■ La Niña

La Niña is defined as cooler-than-normal sea surface temperatures in the eastern tropical Pacific Ocean. La Niña is the counterpart of El Niño. In a La Niña event, the SSTs in the equatorial eastern Pacific drop well below normal levels. Intense trade winds move warm surface waters of the Pacific westward, while increasing cold water upwelling in the eastern equatorial Pacific. La Niña years often, but certainly not always, follow El Niño years. La Niña conditions typically last for 9 to 12 months but on occasion may persist for as long as 2 years.

Both El Niño and La Niña impact the atmosphere globally. In some locations, especially in the tropics, La Niña produces weather departures opposite of those from El Niño. During La Niña, weather is drier than normal in the Southwest

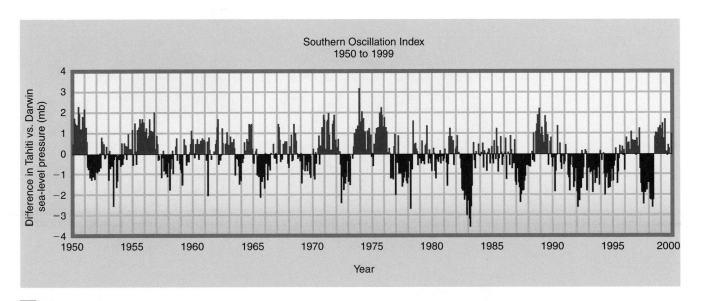

■ **Figure 8.16**

The Southern Oscillation Index (SOI), a measure of changes in atmospheric pressure patterns across the equatorial Pacific. Note that negative values of the SOI often occur during El Niño events (compare to Figure 8.11).

http://info.brookscole.com/ackerman

United States in late summer through the subsequent winter. Drier-than-normal conditions also typically occur in the Central Plains during autumn and in the Southeast during winter. With a well-established La Niña the Pacific Northwest is wetter than normal in the late fall and early winter. As for temperature distributions in the United States, La Niña winters are on average warmer than normal in the Southeast and colder than normal in the Northwest.

Other Oscillations

The El Niño–Southern Oscillation is a primary way in which the atmosphere and oceans interact with each other. In the past decade, scientists have uncovered other oscillations that can affect weather and climate. One is the **Pacific Decadal Oscillation** (PDO), so named because it is a seesaw of atmospheric pressure and SSTs (like ENSO) that occurs over the North Pacific over periods of several decades (Figure 8.17). The North American climate changes associated with the PDO are broadly similar to those connected with El Niño and La Niña, though generally not as extreme. Furthermore, the PDO and El Niño affect each other. When the PDO causes warm equatorial Pacific waters, El Niños can be more severe; this was the case during the 1980s and 1990s. The cause of the PDO remains a mystery, however.

The **Arctic Oscillation** is yet another atmosphere–ocean oscillation that changes from decade to decade. In this oscillation, a seesaw in atmospheric pressure between the North Pole and the central Atlantic Ocean affects Arctic Ocean temperatures as well as the strength and direction of the jet stream over the United States and Europe. Atmospheric scientist Mike Wallace of the University of Washington states that the Arctic Oscillation is the "embodiment of what earlier researchers have called the *Northern Hemisphere index cycle*" that we discussed in Chapter 7. The Arctic Oscillation demonstrates how decades-old problems in meteorology cannot be fully understood without studying the interactions between the atmosphere and the oceans.

The Pacific Decadal Oscillation

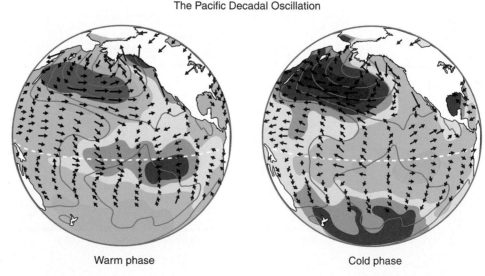

Warm phase   Cold phase

■ Figure 8.17

Surface atmospheric pressure patterns (contours), SSTs (shading), and winds (arrows) associated with the Pacific Decadal Oscillation (PDO). The warm phase of the PDO is shown at left, and the cold phase at right. (Source: http://tao.atmos.washington.edu/pdo/.)

Tropical Cyclones: Hurricanes and Typhoons

What Are They?

The Earth's atmosphere and oceans can interact in all sorts of ways. We've already seen that large-scale oscillations of weather and climate occur when the vast oceans and the air above them "talk" to each other over periods of several years. A different kind of weather feature develops when certain regions of the tropical oceans interact with the atmosphere during the summer and fall of each year. From space, they look like large circular swirls of clouds (Figure 8.18). They tend to be several hundred kilometers in diameter. These swirls are clearly much bigger than an individual thunderstorm, but are also much smaller than the global circulations induced by El Niño.

Because of the location of their birth and the pattern of their clouds, these swirls are given the generic name **tropical cyclones.** In the tropical regions of North and Central America, the most powerful of them are called **hurricanes.** Residents of the western Pacific call them **typhoons.** In most other parts of the world, such as the Indian Ocean, they are simply called "cyclones." In an average year, about five or six hurricanes form in the Atlantic and the Gulf of Mexico, nine form in the eastern Pacific off of Mexico, and sixteen typhoons form in the western Pacific Ocean.

Regardless of their name, these storms are the most highly organized and destructive weather patterns on Earth. A hurricane hit Galveston, Texas, in 1900 and killed approximately 8000 people. A cyclone ravaged Bangladesh in 1970 and killed

Figure 8.18

Computer-generated satellite view of Hurricane Mitch approaching Honduras on October 26, 1998, at 13:15 UTC.

Figure 8.19

Wind damage caused by Hurricane Andrew just south of Miami, Florida, in August 1992.

300,000 people. And in 1992, Hurricane Andrew blew through southern Florida and caused about $25 *billion* in damage (Figure 8.19). Table 8.2 lists the most damaging hurricanes to hit the United States, including many mentioned in this chapter. To save lives and property, we need to learn more about these remarkable storms.

What Do They Look Like?

As we learned in Chapter 6, a cyclone is a center of low pressure. Weather satellite photographs (Figure 8.20) indicate that the innermost part of a strong tropical cyclone's center is almost entirely clear of clouds. This region is known as the **eye.** The eye can be as small as 8 kilometers (5 miles) across, but can also be ten times that size or more. Immediately surrounding the eye is a narrow, circular, rotating region of intense thunderstorms called the **eye wall.** The eye wall winds thrust up as much as a million tons of air every *second*. Figure 8.21 shows you the cumulonimbus clouds of the eye wall from the perspective of an airplane flying inside the eye itself!

Perhaps the most remarkable aspect of the hurricane is the near-perfect circular pattern of the eye and the surrounding cloudiness, which is especially true for the most intense tropical cyclones. The whirling circles-within-circles of a strong hurricane are evidence of a rare combination of power and coordination. Next, we learn where the power of tropical cyclones originates, which is intimately connected to the areas where they develop.

How and Where Do They Form?

Tropical cyclones are essentially large weather engines, and any engine needs energy to run. In Chapter 7 we learned that the unequal heating of different parts of the Earth by the Sun drives the huge engine of global weather and climate. But the tropics are another matter, since temperature contrasts there are usually small. Warm air, warm water—it sounds more like an advertisement for a vacation cruise than a prelude to stormy weather.

The "secret" energy source of a tropical cyclone is the large latent heat of water, which we first discussed in Chapter 2. Air over the tropical oceans is drier than you might think. This is because the trade winds are partly made up of air that has

TABLE 8.2. The Most Damaging Hurricanes to Affect the United States since 1900

Rank	Hurricane	Year	Minimum Pressure at Landfall (mb)	Saffir–Simpson Category at Landfall	Deaths (U.S.)	Damage (1998 U.S. Dollars)
1	Southeast Florida/Alabama (unnamed)	1926	935	4	243	83,814,000,000
2	Andrew (Southeast Florida/Louisiana)	1992	922	4	58	38,362,000,000
3	Galveston, Texas (unnamed)	1900	931	4	8000+	30,856,000,000
4	Galveston, Texas (unnamed)	1915	945	4	275	26,144,000,000
5	Southwest Florida (unnamed)	1944	962	3	300	19,549,000,000
6	New England (unnamed)	1938	946	3	600	19,275,000,000
7	Southeast Florida/Lake Okeechobee (unnamed)	1928	929	4	1836	15,991,000,000
8	Betsy (Southeast Florida/Louisiana)	1965	948	3	75	14,413,000,000
9	Donna (Florida/Eastern United States)	1960	930	4	50	13,967,000,000
10	Camille (Mississippi/Louisiana/Virginia)	1969	909	5	256	12,711,000,000
11	Agnes (Northwest Florida/Northeast United States)	1972	980	1	122	12,408,000,000
12	Diane (Northeast United States)	1955	987	1	184	11,861,000,000
13	Hugo (South Carolina)	1989	934	4	26	10,872,000,000
14	Carol (Northeast United States)	1954	960	3	60	10,509,000,000
15	Southeast Florida/Louisiana/Alabama (unnamed)	1947	940	4	51	9,630,000,000
16	Carla (North and central Texas)	1961	931	4	46	8,194,000,000
17	Hazel (South Carolina/North Carolina)	1954	938	4	95	8,160,000,000
18	Northeast United States (unnamed)	1944	947	3	390	7,490,000,000
19	Southeast Florida (unnamed)	1945	951	3	4	7,318,000,000
20	Frederic (Alabama/Mississippi)	1979	946	3	5	7,295,000,000
21	Southeast Florida (unnamed)	1949	954	3	2	6,767,000,000
22	Southern Texas (unnamed)	1919	927	4	600+	6,200,000,000
23	Alicia (Northern Texas)	1983	967	3	21	4,702,000,000
24	Floyd (North Carolina)	1999	956	2	56	4,500,000,000
25	Celia (Southern Texas)	1970	945	3	11	3,869,000,000
26	Dora (Northeast Florida)	1964	966	2	5	3,603,000,000
27	Fran (North Carolina)	1996	961	3	37	3,591,000,000
28	Opal (Northwest Florida/Alabama)	1995	942	3	9	3,478,000,000
29	Georges (Southwest Florida/Mississippi)	1998	965	2	4	3,073,000,000
30	Cleo (Southeast Florida)	1964	971	2	0	2,823,000,000

Dollar amounts are normalized to 1998 standards. Notice that earlier storms caused many more deaths than comparable, but more recent, storms. This is a measure of the effectiveness of today's hurricane detection and warning systems. In some cases, intense hurricanes caused little damage and few deaths because sparsely populated regions were hit.

Source: Adapted from information provided by Dr. Chris Landsea, Atlantic Oceanographic and Meteorological Laboratory.

descended from above—part of the downward branch of the Hadley Cell we studied in the last chapter. Although both the air and water may be warm and calm, evaporation can take place because the air is not at 100% relative humidity. Silently and invisibly, water changes from liquid to vapor and enters the atmosphere. The energy required to make this change comes from the Sun, and this energy is lying in wait—latent—ready to be released when the vapor is condensed into liquid again. This happens in rising air in a cloud or thunderstorm.

http://info.brookscole.com/ackerman

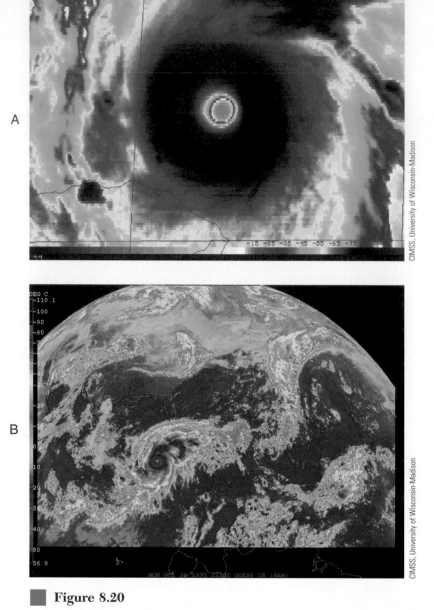

Figure 8.20

*Infrared satellite image of Hurricane Mitch in the western Caribbean Sea on October 26, 1998, showing (**A**) a close-up view of the hurricane's eye, and (**B**) a view of the cyclone (lower left) and other weather patterns at the time. Note the clear eye, the fierce eye-wall ring of thunderstorms, and the difference in the hurricane's appearance compared to other tropical and extratropical weather systems. Mitch later moved inland over Central America; its flooding rains there caused one of the worst disasters in Western Hemisphere history, killing tens of thousands of people.*

However, this process alone is not enough to power a tropical cyclone. A tropical cyclone adds fuel to its own fire by drawing surface air toward its low-pressure center (see Chapter 6). The tight pressure gradient nearer the center means that the winds grow stronger as the air approaches the eye. The faster the wind blows, the more evaporation takes place (this is why you blow-dry your hair or hands instead of merely warming them). Increased evaporation means more water vapor in the air and more energy ready to be liberated in the tropical cyclone's thunderstorms as water vapor condenses. In short, evaporation and condensation of water are the keys to understanding the power of tropical cyclones. Figure 8.22 summarizes the atmosphere–ocean interactions that fuel these cyclones.

Robin Moyer

Figure 8.21

Photograph of the eye wall of Typhoon Vera taken from a "hurricane hunter" aircraft.

How strong is the engine that powers a tropical cyclone? The energy released by condensation in a single day in an average hurricane is at least 200 times the entire world's electrical energy production capacity! Part of this energy is expended reducing the central pressure of the storm and strengthening the winds. Figure 8.23 shows the minimum possible central sea-level pressure for a tropical cyclone based on SSTs

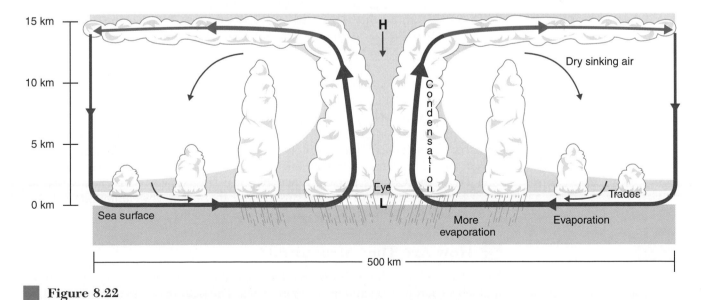

Figure 8.22

Cross-sectional view of the processes involved in fueling a hurricane. (Source: From "Toward a General Theory of Hurricanes," by Kerry Emanuel. American Scientist, July–August 1988, p. 376. Adapted by permission.)

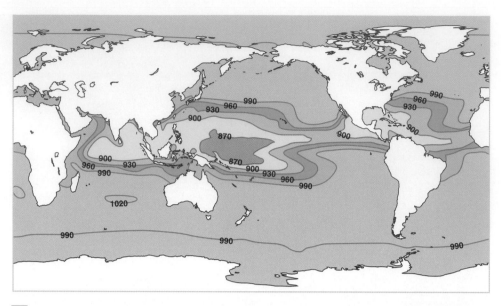

■ **Figure 8.23**

The minimum possible central pressure in a "perfect" tropical cyclone for average September conditions, based on the hypothesis that hurricane formation and intensity are tied directly to sea surface temperatures. Regions with minimum pressures less than 990 mb are the birthplaces of tropical cyclones. (Source: www-paoc-mit.edu/~emanuel/pcmin/climo.html)

in an average September. Notice how low the pressures can go—below 900 mb in some regions! Tropical cyclones can create the lowest sea-level pressures observed on Earth, the record being 870 mb in the Pacific, about the same surface pressure as at Denver, Colorado, the "Mile-High City."

Figure 8.23 also shows us where tropical cyclones develop. The very lowest pressures, and therefore the strongest possible tropical cyclones, are located in the regions of warmest ocean temperatures (compare to Figure 8.5). In regions with SSTs below 26.5° C (80° F), tropical cyclones cannot form. This is the fundamental reason why they are *tropical* cyclones. But why only in the tropics? Evaporation rates increase very quickly as temperatures increase, and evaporated water is the fuel for these storms. In other words, a few degrees' difference in SSTs can be the difference between no storm, a run-of-the-mill tropical cyclone, and a record-setting hurricane.

Some low-latitude regions are not conducive to tropical cyclone development. The far eastern boundaries of the tropical oceans are home to cool ocean currents, as we learned earlier in this chapter. The ocean temperatures are below 26.5° C, and therefore tropical cyclones do not form there.

Tropical cyclones also do not form within about five degrees latitude of the equator. The SSTs there are definitely warm enough for tropical cyclones. But tropical cyclones never form near the equator. Why is this so? Because, as we learned in Chapter 6, the Coriolis force is negligible near the equator. This means that air tends to flow straight into low-pressure centers there. The low cannot intensify because it "fills up" before the pressure can drop very much.

■ **How Are They Structured?**

Now that we know the power source for tropical cyclones, we can explore their remarkably coordinated structure in more detail. The pattern of winds in a tropical cyclone is fascinating, and we can explain it fairly easily. Figure 8.24 is a time series of wind speeds recorded when powerful Hurricane Celia passed directly over a weather station in Texas. The winds of a tropical cyclone are strong, and consistently

Blue Skies

Hurricane/Virtual Hurricane

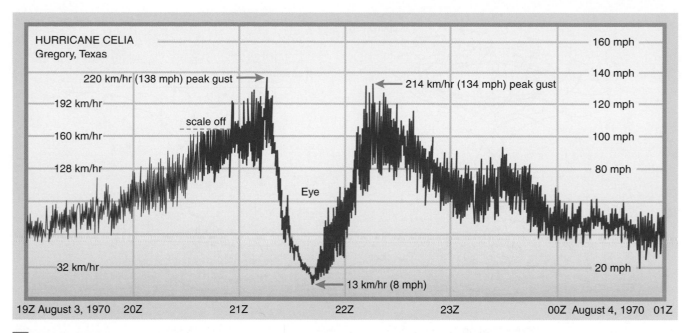

Figure 8.24

Wind speeds recorded near Corpus Christi, Texas, during the passage of Hurricane Celia directly over the anemometer on August 3, 1970. (From The Hurricane and Its Impact *by Robert H. Simpson and Herbert Riehl. Copyright © 1981. Reprinted by permission of Louisiana State University Press.)*

strong for many hours. The bull's-eye of isobars around the eye of the tropical cyclone ensures that it is a big blow; as we learned in Chapter 6, a strong pressure gradient means a strong wind.

The opposite is true *inside* the eye of the tropical cyclone. As the eye passes over a location it causes a nearly calm "halftime" between the front and back sides of the storm. Why? The air spiraling into a strong tropical cyclone rises in the eye wall, and it is there that the strong pressure gradient ceases. Instead, air *sinks* gently from above. We learned in earlier chapters that sinking air warms adiabatically and is very stable. This explains why the eye in a strong storm is cloudless (except for some harmless cumulus and cirrus clouds), calm, and warm.

Sinking air suggests an anticyclone (Chapter 6). Indeed, one of the hallmarks of a strong tropical cyclone is an intense low at the surface and a high-pressure center or ridge near the tropopause. Generally, the stronger the cyclone, the stronger the upper-level ridge or high. This high is caused in part by the cyclone itself. Without it, the cyclone cannot remove air from the eye quickly enough to keep its pressure low. Therefore, the nearby presence of an upper-level low or trough, or even a high-speed jet stream, is enough to weaken or destroy a tropical cyclone. The weakening in these cases is attributed to **wind shear,** meaning that the winds are changing fast enough in the vertical to disrupt the cyclone's own high-building process at those levels. Figure 8.25 is a schematic view of how the winds in a hurricane differ in the lower and upper levels in both strengthening and weakening cases.

The radar image of Hurricane Hugo in Figure 8.26 illustrates two more important features of tropical cyclones. First, notice that the radar-reflecting rain in this hurricane, like most, is not symmetrically distributed as the winds are. Instead, the heaviest rain is contained in spiral **rainbands** that curve into and blend with the eye wall. This characteristic of tropical cyclones, so fundamental that it is incorporated into the weather-map symbol for them, is still not well understood by meteorologists. This is one of many reasons that "hurricane hunters" still fly planes into these storms to gather weather data such as that shown in Figure 8.26; Box 8.1 explains the work of hurricane hunters in more detail.

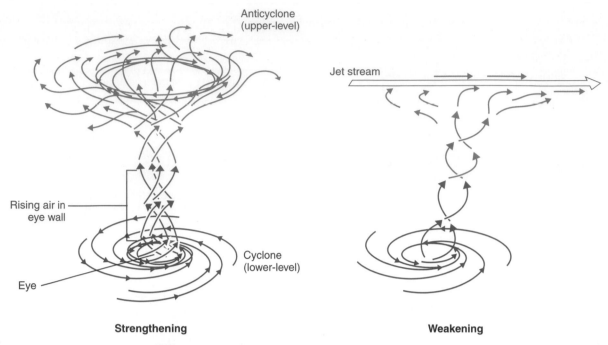

Figure 8.25

The lower-level and upper-level wind flows in a tropical cyclone in conditions (left) favorable and (right) unfavorable for growth of the storm. (Source: Adapted from Nese, J. and Grenci, L., A World of Weather: Fundamentals of Meteorology, Kendall/Hunt, 1998, Figure 11.10.)

The second feature can be seen in the wind measurements from aircraft that are overlaid on the radar image in Figure 8.26. The winds are strongest on the north and east sides of the cyclone. Surprisingly, this is not because the pressure gradient is different on one side of the storm than on the other. It is because the storm's forward motion adds or subtracts from the winds. In the same way, if you throw a baseball out of a moving car, the baseball's velocity is the combination of the velocity of

Figure 8.26

Airborne radar view of Hurricane Hugo at peak intensity over the northeast Caribbean in September 1989. The white lines indicate the flight paths of the "hurricane hunter" aircraft (see Box 8.1), which measured total storm winds (shown in white) at various points along the flight paths. The brightest colors indicate the most intense thunderstorms in the eye wall.

BOX 8.1

The "Hurricane Hunters"

Why would anyone in their right mind *intentionally* fly into the middle of a hurricane? During every hurricane season, U.S. Air Force (USAF) and National Oceanographic and Atmospheric Administration (NOAA) pilots and scientists do just that, time and again. Their goal: to obtain up-to-the-minute weather data on storms that cannot be obtained by any other means. They are the "hurricane hunters."

The first exploration of a major hurricane by a "reconnaissance aircraft," as they are called, occurred in 1944 during the Great Atlantic Hurricane. Since 1955, no planes have been lost in the Atlantic while penetrating these deadly storms. (There was one close call during a rough flight into Hurricane Hugo in 1989 when three engines failed and the plane was almost submerged by a high wave!) The pilots carefully choose their altitude (high, away from the roiling ocean) and path (straight through to the eye—no dilly-dallying in the eye wall) to maximize their safety while gathering crucial weather data.

Today, hurricane hunting is a highly coordinated effort between the National Hurricane Center (NHC), USAF and NOAA. When the NHC needs to know if a tropical disturbance far out in the Atlantic is getting better organized, the USAF pilots fly long missions to search for signs of a closed circulation. They use weather instruments that have been specially installed on their large cargo airplanes. In addition,

the planes launch "dropsondes"—the reverse of radiosondes—that drop out of the plane and into the storm from above. These dropsondes provide information about the weather conditions in the very center of the disturbance, below where the planes normally fly. Finally, the pilots and scientists use their eyes to see, literally, if the storm clouds are becoming better organized. This information is relayed to NHC, where meteorologists use it to decide whether or not a disturbance has achieved tropical depression or tropical storm status.

If a mature hurricane comes closer to the United States, the NOAA hurricane hunters take over. NOAA's specially instrumented P-3 aircraft (see the photo below) include on-board radars. Figure 8.26 is a radar image from one of the NOAA flights into Hugo. The measurements of winds and location of the hurricane's center, as shown on Figure 8.26, are extremely important pieces of information for hurricane forecasters. They are much more accurate than satellite estimates.

For this reason, hurricane hunters will continue to brave the bumpy flights into these storms for the foreseeable future. There is also an added bonus: hurricane hunters often say that a strong hurricane's eye wall, viewed from an airplane circling in its eye (see Figure 8.21) is one of the most beautiful and awe-inspiring sights in all of nature.

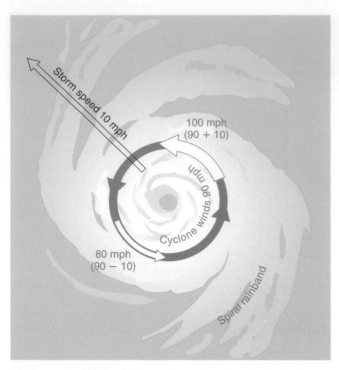

■ **Figure 8.27**

Schematic view of how a tropical cyclone's forward motion affects the speed of the total wind associated with the storm. (Source: Adapted from: www.aoml.noaa.gov/hrd/tcfaq/tcfaqD. html#D6.)

Radar Loop of a Gulf Hurricane

the ball and the velocity of the car. Figure 8.27 shows that the storm's forward speed adds to the wind on the right side of the eye, but subtracts from the wind on the left side of the eye. As a result, the difference in wind on the right versus the left side of the eye is approximately twice the cyclone's forward speed.

We now know quite a bit about the organization of a tropical cyclone. The fact that these storms are given names (Box 8.2) suggests that, like people, they are unique and have individual "lives." Next, we examine the life cycle of a typical tropical cyclone.

■ What Are the Different Stages of Their "Lives"?

Tropical cyclones usually begin small, as **tropical disturbances.** These disturbances are not even low-pressure centers, just disorganized clumps of thunderstorms. In the Atlantic Ocean, these disturbances take the form of **easterly waves** in the wind patterns that move along with the easterly jet stream across Africa and into the extreme eastern Atlantic. Easterly waves are also seen along the Pacific coast of Mexico. Figure 8.28 shows an example of an easterly wave in the Atlantic Ocean east of Puerto Rico. As the satellite-observed winds and clouds indicate, winds converge into the wave on its east side, leading to clouds and precipitation.

The vast majority of tropical disturbances die without growing any stronger. However, about one in every ten develop a weak low-pressure center of roughly 1010 mb. Once the low-pressure center is identified and cyclonic rotation is noticed in the winds as in the lower right of Figure 8.28, it is called a **tropical depression.**

Some, but not all, tropical depressions gain more strength. Their central pressures drop below about 1000 mb, the pressure gradient between the center and the edge of the storm increases, and the winds strengthen. When the winds in a depression reach consistent speeds of 35 knots (39 mph), a **tropical storm** is "born" and given a name (see Box 8.2).

Roughly half of tropical storms intensify further. Their central pressures drop below about 990 mb and their winds keep strengthening because of the increasing pressure gradient as shown by the tightening of isobars around the center. An eye forms that can be seen in visible or infrared satellite pictures. When the highest sustained winds in a tropical storm reach 65 knots (74 mph), the storm is reclassified as a hurricane or a typhoon depending on its location. A few typhoons develop with winds above 240 km/hr (150 mph) and are called, appropriately enough, **supertyphoons.**

After one or two weeks (in rare cases as long as a month), even the strongest of hurricanes and typhoons lose their strength, or "dissipate," because they move over colder water or land, away from their energy source. On rare occasions a hurricane will make the transition from a tropical, warm-water cyclone to an extratropical cyclone (Chapter 10) with cold air and fronts near its center.

What does the mature portion of a hurricane's or a typhoon's life cycle look like? Figure 8.29 shows the evolution of Hurricane Georges over 10 days in September 1998, all superimposed on the same map. Notice how the appearance of the eye changes with time. Meteorologists routinely use the appearance of the eye in satellite images to estimate the strength of a hurricane. Using this figure, we can learn how they do this while studying Georges' growth and decay.

Figure 8.29 shows an eye forming in the center of Georges on September 18. The peak winds in Georges at that time are around 153 km/hr (95 mph). This

BOX 8.2

Naming Hurricanes

Hurricanes come in large and small sizes; some have big eyes, and some have squinty eyes. They are born, grow, and then die. It's not surprising, then, that human names have been assigned to hurricanes for centuries.

The earliest instance of hurricane naming was in the Caribbean islands, where hurricanes were associated with the Roman Catholic saint's holiday when they hit the islands. For example, Hurricane "Santa Ana" struck Puerto Rico on July 26, 1825.

The first known scientific use of hurricane naming arose in the Pacific during World War II. It was an easy and effective way to distinguish one tropical cyclone from another on the weather maps. The system was simple and alphabetical: the name of the first storm of the season would begin with A; the second, B; the third, C; and so on.

Naming hurricanes also gave homesick soldiers a way to recall loved ones. Thus began the practice by predominantly male meteorologists of giving female names to hurricanes. This practice persisted at the National Hurricane Center in Miami until the late 1970s. One retired NHC meteorologist ironically commented, "It is surely only a coincidence that many names of NHC employees, their wives, and other female relatives appear on the [name] list." This may explain the distinctly Southern flavor of the names of past hurricanes.

Beginning in 1979, the list of names for each year's tropical storms and hurricanes was broadened to include male and Spanish and French names. The alphabetical naming system continues, now overseen by the World Meteorological Organization (WMO). The WMO composes different alphabetical lists for different ocean basins (for example, starting in 2000, northwest Pacific typhoons now have Asian names). The names of severe and/or notable hurricanes are "retired" and taken off the lists. Horror sequels are popular in the movies, but no meteorologist wants to tempt fate by naming a new hurricane Andrew, Georges, or Mitch, to cite a few of the recently retired names.

Following is the list of names for Atlantic Ocean, Gulf of Mexico, and Caribbean Sea tropical storms and hurricanes for the years 2001–2006. You may be wondering what happens if there are more storms than names. That has never happened in the Atlantic—although in 1995 it was a close call, with Tanya lasting until November 1.

Atlantic Hurricane Names by Year

2001	2002	2003	2004	2005	2006
Allison	Arthur	Ana	Alex	Arlene	Alberto
Barry	Bertha	Bill	Bonnie	Bret	Beryl
Chantal	Cristobal	Claudette	Charley	Cindy	Chris
Dean	Dolly	Danny	Danielle	Dennis	Debby
Erin	Edouard	Erika	Earl	Emily	Ernesto
Felix	Fay	Fabian	Frances	Franklin	Florence
Gabrielle	Gustav	Grace	Gaston	Gert	Gordon
Humberto	Hanna	Henri	Hermine	Harvey	Helene
Iris	Isidore	Isabel	Ivan	Irene	Isaac
Jerry	Josephine	Juan	Jeanne	Jose	Joyce
Karen	Kyle	Kate	Karl	Katrina	Keith
Lorenzo	Lili	Larry	Lisa	Lee	Leslie
Michelle	Marco	Mindy	Matthew	Maria	Michael
Noel	Nana	Nicholas	Nicole	Nate	Nadine
Olga	Omar	Odette	Otto	Ophelia	Oscar
Pablo	Paloma	Peter	Paula	Philippe	Patty
Rebekah	Rene	Rose	Richard	Rita	Rafael
Sebastien	Sally	Sam	Shary	Stan	Sandy
Tanya	Teddy	Teresa	Tomas	Tammy	Tony
Van	Vicky	Victor	Virginie	Vince	Valerie
Wendy	Wilfred	Wanda	Walter	Wilma	William

■ **Figure 8.28**

Visible satellite picture of two tropical disturbances in the Atlantic Ocean on August 14, 2001. Clouds are white, and winds derived from the satellite observations are depicted using the flag/flagpole symbols explained in Chapters 1 and 6. White arrows indicate the general direction of the wind flow. The disturbance in the lower center of the image is a tropical wave, so named for the hump in the easterly wind pattern. The disturbance at lower right is a tropical depression, which is distinguished from a tropical wave by its closed cyclonic wind circulation. This depression later intensified to become Tropical Storm Chantal and hit the Yucatan Peninsula of Mexico.

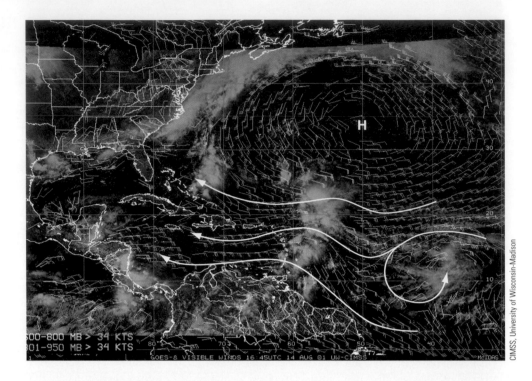

matches our expectations; as we learned earlier, eyes generally form in tropical cyclones when they strengthen to hurricane status. As the storm strengthens to 232 km/hr (145 mph) on September 20, the eye becomes better defined and then shrinks into a tight, narrow pinhole only a few miles across. This is a fairly good rule of thumb: the stronger the hurricane or typhoon, the smaller the eye and the more circular the storm appears on satellite pictures.

Georges's eye is hard to identify when the cyclone moves over land during September 21–25, even over small Caribbean islands. This is because the mountains of these islands, combined with the lack of warm water underneath, disrupt the

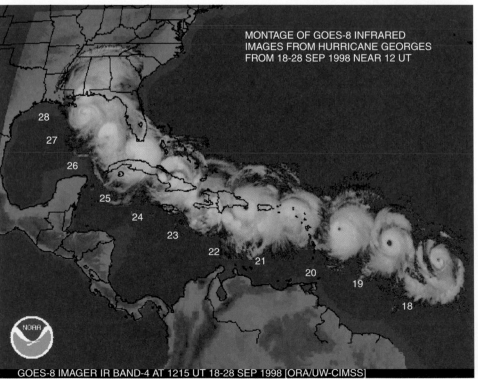

■ **Figure 8.29**

Hurricane Georges at different stages in its life cycle, as seen in a series of daily infrared satellite images.

hurricane's circulation and cause a weakening of winds and the disappearance of the eye. During this period, Georges is strongest on September 22, with winds of 176 km/hr (110 mph). Not coincidentally, this is when the eye is most easily seen and when Georges has temporarily moved over water again.

Judging from the hurricane's appearance, Georges never regains its strength after passing over Cuba. Just a hint of an eye can be made out when Georges crosses the Gulf of Mexico on the 26th and 27th with top winds of 160 km/hr (100 mph). In the final image from September 28, the cyclone no longer has an eye or even a clearly circular shape. It is over land along the coast of Mississippi, and its wind speeds rapidly fall below hurricane and even tropical storm status during the next day. Georges, over land and away from warm tropical waters, dies.

Meteorologists have recently discovered that the life cycle of a tropical cyclone over water is strongly affected on a *day-to-day* basis by the SSTs beneath it. In Figure 8.30, the evolution of Hurricane Floyd in 1999 is shown as a succession of circles whose color and size indicate the maximum sustained wind speeds at different

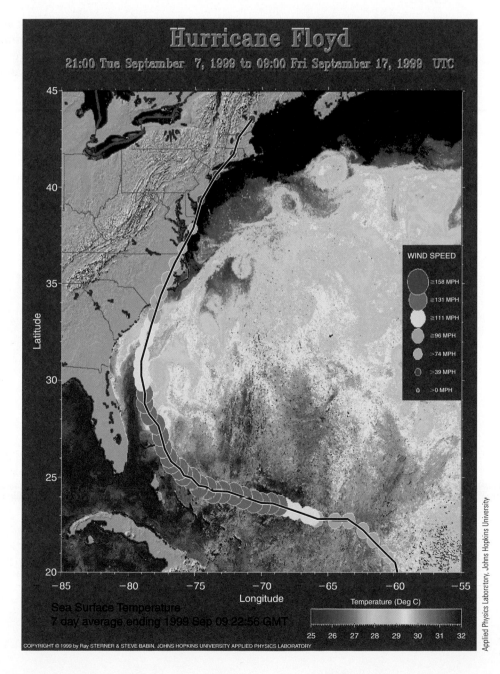

Applied Physics Laboratory, Johns Hopkins University

■ **Figure 8.30**

The relationship between Hurricane Floyd in September 1999 and the sea surface temperatures beneath it, as determined by satellite observations of SSTs. Circles denote Floyd's position and wind speed; colors indicate the ocean temperatures. (Source: http://fermi. jhuapl.edu/hurr/99/floyd/ floyd09sep7dsst_25_32.gif.)

times. Underlying the dots, satellite observations of ocean temperatures are shown for the same period. Floyd increases from a minimal hurricane to one packing winds over 208 km/hr (130 mph) at the same time that the storm moves over waters that were significantly warmer than the 26.5° C threshold for cyclone formation and growth.

Later, Floyd weakens as it nears land because it passes over cooler waters to the east and west of the warm Gulf Stream. These waters cooled because the winds of an earlier hurricane, Dennis, stirred up the ocean near the North Carolina coast and brought cooler water to the surface via upwelling as explained earlier in this chapter. Finally, Floyd, like Georges, diminishes to tropical storm status after its circulation leaves the ocean and moves over land. The presence of warm water beneath the storm's circulation is a major key to understanding tropical cyclone strength.

■ What Does a Year's Worth of Tropical Cyclones Look Like?

Now that we have a sense of the life cycle of a single tropical cyclone, let's look at an entire season's worth of cyclones to see how different storms evolve at different places and times in the year. Figure 8.31 shows the paths and stages of the different tropical storms and hurricanes that affected the Atlantic Ocean, Gulf of Mexico, and Caribbean Sea in 1995—all nineteen of them! We can use this modern record-setting year for tropical cyclones to illustrate the seasonal cycle of their development.

Hurricane "season" carefully follows the seasonal cycle of SSTs. However, this is not the same as saying "hurricanes form in summer." Because it takes time for the oceans to warm up, the Atlantic hurricane season doesn't officially start until June 1

The 1995 Atlantic Hurricane Season

1	Allison	Jun 03-06
2	Barry	Jul 06-10
3	Chantal	Jul 12-20
4	Dean	Jul 24-Aug 02
5	Erin	Jul 31-Aug 06
6	Felix	Aug 06-22
7	Gabrielle	Aug 09-12
8	Humberto	Aug 22-Sep 01
9	Iris	Aug 22-Sep 04
10	Jerry	Aug 22-25
11	Karen	Aug 26-Sep 03
12	Luis	Aug 27-Sep 11
13	Marilyn	Sep 12-22
14	Noel	Sep 26-Oct 07
15	Opal	Sep 27-Oct 05
16	Pablo	Oct 04-06
17	Roxanne	Oct 07-21
18	Sebastien	Oct 20-25
19	Tanya	Oct 27-Nov 01

——— Hurricane
——— Tropical storm
——— Tropical depression
+++++++++++ Extratropical

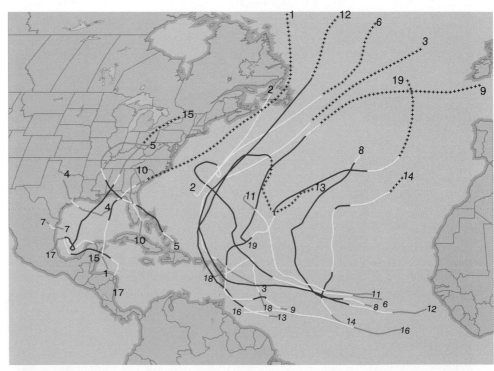

■ Figure 8.31

The names, dates, paths, and life-cycle stages of all tropical storms and hurricanes in the Atlantic Ocean basin in 1995, a year that set a modern record for number of storms.

and lasts until November 30. In rare instances, hurricanes can form even in December; the calendar isn't the important thing—the SSTs and the lack of wind shear are. The peak of hurricane activity in the tropical waters south and southeast of the United States is typically in early to mid-September.

In Figure 8.31, we see that the early-season cyclones tend to form in the Gulf and the western Caribbean. Why is this the case? Largely because these smaller bodies of water are able to warm up more quickly than the Atlantic and reach the magic 26.5° C threshold sooner.

By late August and September, attention shifts to the eastern Atlantic. Around this time, potentially dangerous **Cape Verde hurricanes** develop from easterly waves off the coast of Africa and travel thousands of miles across open water, gathering strength each day. Even though they begin life far from land, Cape Verde cyclones are cause for worry: Approximately 85% of intense hurricanes in the Atlantic begin as easterly waves. In 1995, Hurricane Luis developed south of the Cape Verde Islands, intensified to hurricane strength by August 30, and roared northwest and then northeast for the next 12 days before becoming an extratropical cyclone east of Canada!

In October, the Atlantic is cooling and the midlatitude weather patterns cause enough wind shear to reduce the threat of Cape Verde hurricanes. At this time, the still-warm Caribbean and Gulf of Mexico can give birth to late-season cyclones. For example, in 1995, Hurricane Opal developed along the Caribbean coast of the Yucatan Peninsula, wandered into the southern Gulf of Mexico, and then rapidly intensified over especially warm Gulf waters as it moved toward the north-central Gulf Coast on October 4. Fortunately, it weakened just as quickly, sparing Alabama and the Florida Panhandle from terrible damage.

Another aspect of the seasonal cycle of tropical cyclones is seen in Figure 8.31. Notice that the paths of most of the Atlantic storms tend to follow the same arc-shaped track, first moving west or northwest and then "recurving" toward the northeast. Why do they do that? Tropical cyclones are low-pressure centers near the surface, and they tend to move around the edge of the large surface **Bermuda high** that sits over the Atlantic in summer. (This high is part of the belt of subtropical anticyclones that are associated with sinking air in the Hadley Cell that we studied in the last chapter.) So, the early-stage paths toward the west or northwest mark the southern and western edges of this high. In 1995, the high did not extend very far west; this was a key reason that this bumper crop of Atlantic tropical cyclones did *not* cause record-setting damage to the U.S. East Coast.

The **recurvature** of Atlantic tropical cyclones to the north and northeast is partly caused by the storms' continuing to skirt the edge of the Bermuda high. However, another factor is involved: the wind patterns of the middle latitudes. By August and September, it is not uncommon to have extratropical cyclones push southeast out of Canada and bring with them strong winds at high altitudes. These jet stream winds help sweep tropical cyclones ahead of them, toward the northeast, while disrupting their circulation and weakening them as the cyclones head out over cold Atlantic waters.

Recurving tropical cyclones often accelerate rapidly. The Great New England Hurricane of 1938 roared across Long Island at an estimated forward speed of 60 mph! Based on our earlier discussion of the effect of forward motion on overall storm winds, this means that people to the right of the hurricane's eye in Rhode Island experienced winds 192 km/hr (120 mph) faster than did luckier residents (including the co-author's father) near New York City who were to the left of the eye.

It should be stressed that every hurricane season is different. In 1995 there were nineteen hurricanes, but none did significant damage. But 1992 is remembered as a terrible year for hurricanes in the United States because Hurricane Andrew ravaged Miami in August. However, Andrew was the only major hurricane that entire year!

TABLE 8.3.   The Saffir–Simpson Scale for Hurricanes

Saffir–Simpson Category	Maximum Sustained Wind Speed			Minimum Surface Pressure	Storm Surge		Damage Level
	mph	m/s	knots	mb	ft	m	
1	74–95	33–42	64–82	greater than 980	3–5	1.0–1.7	Minimal

Description: Damage primarily to shrubbery, trees, foliage, and unanchored homes. No real damage to other structures. Some damage to poorly constructed signs. Low-lying coastal roads inundated, minor pier damage, some small craft in exposed anchorage torn from moorings.
Example: Hurricane Agnes (1972)

| 2 | 96–110 | 43–49 | 83–95 | 979–965 | 6–8 | 1.8–2.6 | Moderate |

Description: Considerable damage to shrubbery and tree foliage; some trees blown down. Major damage to exposed mobile homes. Extensive damage to poorly constructed signs. Some damage to roofing materials of buildings; some window and door damage. No major damage to buildings. Coastal roads and low-lying escape routes inland cut by rising water 2 to 4 hours before arrival of hurricane center. Considerable damage to piers. Marinas flooded. Small craft in unprotected anchorages torn from moorings. Evacuation of some shoreline residences and low-lying areas required.
Example: Hurricane Georges (1998)

| 3 | 111–130 | 50–58 | 96–113 | 964–945 | 9–12 | 2.7–3.8 | Extensive |

Description: Foliage torn from trees; large trees blown down. Practically all poorly constructed signs blown down. Some damage to roofing materials of buildings; some wind and door damage. Some structural damage to small buildings. Mobile homes destroyed. Serious flooding at coast and many smaller structures near coast destroyed; larger structures near coast damaged by battering waves and floating debris. Low-lying escape routes inland cut by rising water 3 to 5 hours before hurricane center arrives. Flat terrain 5 feet or less above sea level flooded inland 8 miles or more. Evacuation of low-lying residences within several blocks of shoreline possibly required.
Example: Hurricane Celia (1970)

| 4 | 131–155 | 59–69 | 114–135 | 944–920 | 13–18 | 3.9–5.6 | Extreme |

Description: Shrubs and trees blown down; all signs down. Extensive damage to roofing materials, windows and doors. Complete failures of roofs on many small residences. Complete destruction of mobile homes. Flat terrain 10 feet or less above sea level flooded inland as far as 6 miles. Major damage to lower floors of structures near shore due to flooding and battering by waves and floating debris. Low-lying escape routes inland cut by rising water 3 to 5 hours before hurricane center arrives. Major erosion of beaches. Massive evacuation of all residences within 500 yards of shore, and of single-story residences within 2 miles of shore, possibly required.
Example: Hurricane Andrew (1992)

| 5 | 156+ | 70+ | 136+ | less than 920 | 19+ | 5.7+ | Catastrophic |

Description: Shrubs and trees blown down; considerable damage to roofs of buildings; all signs down. Very severe and extensive damage to windows and doors. Complete failure of roofs on many residences and industrial buildings. Extensive shattering of glass in windows and doors. Some complete building failures. Small buildings overturned or blown away. Complete destruction of mobile homes. Major damage to lower floors of all structures less than 15 feet above sea level within 500 yards of shore. Low-lying escape routes inland cut by rising water 3 to 5 hours before hurricane center arrives. Massive evacuation of residential areas on low ground within 5 to 10 miles of shore possibly required.
Example: Hurricane Camille (1969)

Source: Adapted from information provided by Dr. Chris Landsea, Atlantic Oceanographic and Meteorological Laboratory.

■ How Do They Cause Destruction?

Tropical cyclones can create havoc in a variety of ways. They can damage and kill with wind, seawater, and even with rainwater. To preserve life and property, we have to understand how to defend ourselves against such a variety of threats. Below, we examine the destructive capacity of tropical cyclones in more detail.

Wind

The most obvious threat of a tropical cyclone is the powerful wind associated with its incredibly tight pressure gradient, which as we have seen can blow in a single spot for many hours. Damage is inevitable; a 200-mph wind gust in the most severe cyclones can exert a weight of *30 tons* against the wall of a house! Wind damage is such a hallmark of hurricanes that they are classified by meteorologists using the **Saffir–Simpson scale,** which rates hurricanes on a scale of 1 to 5 based on the damage their winds (and storm surge, see below) would cause upon landfall. Table 8.3 explains the Saffir–Simpson scale and relates the observed damage to estimated wind speeds. Figure 8.32 is a unique before-and-after photograph of wind damage due to Hurricane Andrew—a Category 4 hurricane on the Saffir–Simpson scale.

Fortunately, most hurricanes do not produce the extreme winds linked with the highest category on the Saffir–Simpson scale. In the past century, only three of these "Cat 5" hurricanes have crossed the United States coastline: the Labor Day Storm of 1935 in Florida, Hurricane Camille in Mississippi in 1969, and Hurricane Allen at the southern tip of Texas in 1980. While Camille caused major damage along the Gulf Coast, the other two hurricanes caused much less damage.

Why can the damage due to similarly intense storms differ so much? The Labor Day storm was intense but very small; it caused extensive damage in the Florida Keys but Miami was left unscathed. Hurricane Allen diminished in intensity just prior to landfall and struck one of the few stretches of the Gulf Coast that was not covered with expensive beachfront buildings. The amount of damage depends not only on the size and strength of the storm, but also on what is in its way. This is particularly relevant to the United States, where soon over 50% of the population will live within 50 miles of the ocean. This boom in coastline development means that damage caused by tropical cyclones will continue to increase, even as meteorologists learn more about the storms and how to protect against them.

Hurricane Winds

South Florida Sun-Sentinel, Tribune Publishing

■ Figure 8.32

Before-and-after photograph illustrating Hurricane Andrew's wind damage at Fairchild Tropical Garden, south of Miami. The hand-held postcard shows what the scene in the larger photo looked like before the hurricane.

To add insult to injury, hurricanes also contain smaller whirlwinds inside them that can cause additional damage. Upon landfall, nearly every hurricane produces at least one tornado (Chapter 11) that wreaks localized havoc in a narrow path for a few miles at most. In Hurricane Andrew, meteorologists identified tornado-sized "mini-swirls" that tacked on more speed to Andrew's already powerful winds. These mini-swirls helped cause the worst damage attributed to Andrew, as seen in Figure 8.19. Houses just a few blocks away escaped with much less damage because of the local nature of the mini-swirls.

Seawater

The winds in a tropical cyclone push ocean water in front of them. The stronger the wind, the more water is "piled up" by the winds. As the cyclone nears shore, the winds to the right of the eye, which blow onshore, push this water inland. This dome of water is usually about 80 kilometers (50 miles) wide and causes massive flooding near and to the right of the eye where it makes landfall. This process of wind-induced seawater flooding is called **storm surge.** It is extremely rapid; in Galveston in 1900, two different observers reported that the sea level rose four feet in just a few *seconds*.

Storm surge is by far the deadliest weapon in the tropical cyclone's arsenal, historically causing as much as 90% of all hurricane-related deaths. Simply put, this is because water is heavier than air. The brute force of even a small storm surge dwarfs the power of the strongest wind gust. Anyone dragged underwater during a storm surge is at great risk of drowning. Hundreds of thousands of people in low-lying Bangladesh have died by drowning in storm surges. The 8,000 or more deaths in Galveston in 1900 were primarily due to the 5-meter storm surge there (Figure 8.33). Thanks to warnings, Hurricane Andrew's storm surge of nearly 6 meters was not a major killer—but it did severely damage the world headquarters of Burger King, located on the coast just south of Miami.

Rainwater

A tropical cyclone that has moved over land may seem to be dying, but it still has one last knockout punch left: flooding caused by heavy rains. Tropical cyclones, even relatively weak storms that never attain hurricane status, are capable of causing record-setting amounts of rainfall over land. In June of 2001, weak Tropical Storm Allison dropped close to 1 meter (3 feet) of rain on metropolitan Houston, Texas, much of it in just 24 hours (Figure 8.34). At least 20 people were killed in the

■ **Figure 8.33**

Storm-surge damage in Galveston in September, 1900. Little is left but timbers as far as the eye can see.

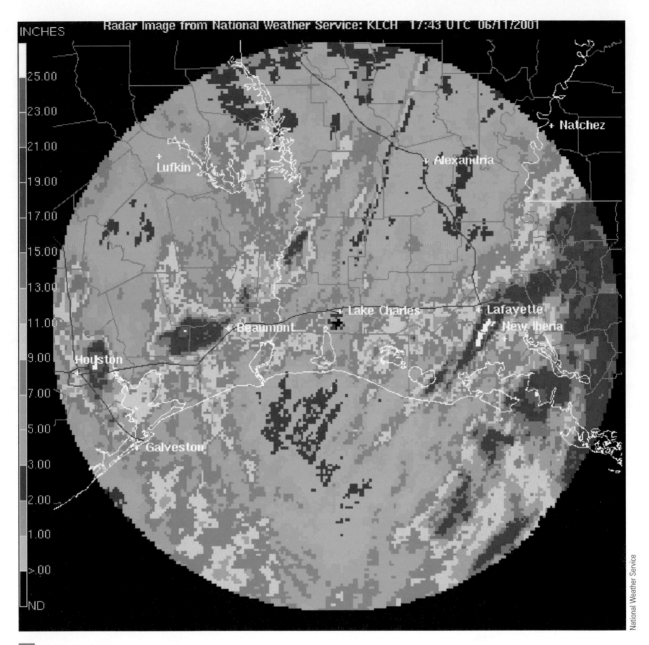

National Weather Service

■ **Figure 8.34**

A Doppler radar estimate of the total rainfall resulting from Tropical Storm Allison, as calculated by the National Weather Service radar at Lake Charles, Louisiana. Purple and white regions indicate areas where more than 20 inches fell, according to radar estimates. In the white area, between the u and s of Houston, on-site gauges measured nearly 36 inches of rain. Allison caused rainfall in every Gulf and Atlantic Coast state, from Texas to Maine.

massive urban flooding that resulted; over 40,000 homes were damaged, with costs estimated near $5 billion. Hurricane Camille is said to have dumped 76 centimeters (30 inches) of rain in 6 hours on the James River Valley of Virginia 3 days after its 320-km/hr (200-mph) winds raked Mississippi; the resulting floods and mudslides killed 109 people.

In the past thirty years, about 60% of hurricane-related deaths in the United States have occurred as a result of rain-fed floods. Hurricane Floyd (see Figure 8.30) illustrates the damage that a tropical cyclone's rainwater can cause. By the time Floyd reached the coastline of North Carolina, its winds had weakened to Category 2 on the Saffir–Simpson scale. Its ability to do wind-related damage had

Dave Saville, FEMA News Photo

 Figure 8.35

A flooded subdivision of Greenville, North Carolina, following the heavy rains of Hurricane Floyd in 1999.

diminished, but its capacity to cause destruction with flooding rains had not. The floods that ensued along the U.S. East Coast (Figure 8.35) accounted for the majority of Floyd's death toll of 57—more fatalities than were attributed to the stronger Andrew. Over half of the victims died in, or while attempting to abandon, their vehicles. Floyd was the deadliest hurricane in the United States since another flood-producer, Agnes, in Pennsylvania in 1972. Damage caused by Floyd was estimated at up to $6 billion.

Tropical cyclones are part of the overall pageantry of weather, however, and they do good as well as harm. Their winds blow down forests and make way for new growth. Storm surges carve new channels and cleanse wetlands. The rainfall from many a tropical storm, or even a wave or disturbance, is enough to save crops from summer and autumn droughts. From a human perspective, however, tropical cyclones are generally far too much of a good thing in all respects, and must be anticipated and avoided if at all possible.

How Do We Observe and Forecast Tropical Cyclones? Past, Present, and Future

A Category 4 Cape Verde hurricane bears down on a major American city from the open seas. In Galveston in 1900, almost 10,000 people die as a result. In Miami in 1992, the storm directly kills only 15. The enormous difference in death tolls is explained by the vast improvements in hurricane detection, forecasting, and warning made during the 20th century.

In 1900, weather satellites did not exist, the global network of weather observations was in its infancy, and countries did not routinely share weather data. The U.S. Weather Bureau was even afraid to use the word "hurricane" in its statements, to avoid undue concern! In the case of the Galveston hurricane (Figure 8.36), the Weather Bureau ignored reports from Cuban meteorologists and expected the storm to recurve as usual to the northeast along the U.S. East Coast. Assumption became "fact" as the official government reports stated, wrongly, that the storm was traveling northeast in the Atlantic. Instead, a high-pressure system shunted the

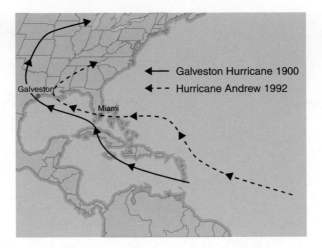

Galveston Hurricane 1900

Hurricane Andrew 1992

Figure 8.36

Tracks of the Galveston Hurricane of 1900 and Hurricane Andrew in 1992. The storms were similar in location of origin, strength, and severity, but took different paths because of the different wind patterns that steered them.

hurricane far to the west through the Gulf of Mexico. The hurricane did not recurve until it hit an unprepared Galveston. Even Galveston's chief Weather Bureau meteorologist was forced to ride out the storm in his floating house, so unexpected was the storm's fury. Chicago was surprised by hurricane-force wind gusts by the recurving storm! Failed forecasts of "fair, fresh" weather based on a lack of scientific understanding and observations cost thousands of lives.

In 1992, Hurricane Andrew was tracked from its earliest stages by weather satellite—the biggest single advance in hurricane detection and forecasting in the 20th century. When Andrew made a sudden course change and intensified rapidly, the National Hurricane Center (NHC) in Miami knew about it within hours and immediately changed their expectations for when and where the storm would make landfall, and how strong it might be when it did. "Hurricane hunter" aircraft flew into the storm to get precise observations of the storm that were more accurate than satellite-based estimates. Hurricane watches were issued for southern Florida, as they always are when a hurricane may threaten a region within the next 36 hours. The watches were upgraded to warnings when it became likely that Andrew would hit the Miami vicinity within 24 hours. All of this information was widely disseminated to the public through government and media sources (Box 8.3). Finally, weather radar followed the storm all the way into the NHC's backyard of Miami. Figure 8.37 is the last scan of the NHC's radar before Andrew's 265-km/hr (164-mph) winds blew the radar dish off the building's roof! Detection and surveillance are conducted in this detailed manner for each and every hurricane that threatens the United States. For this reason, deaths resulting from hurricanes have decreased drastically in the United States in the past century.

Accurate observation of each tropical cyclone is one important advance in forecasting. Recently, computer models of hurricanes have improved to the point that they can be used to anticipate, several hours or a day in advance, what a particular hurricane will do. These models are still not good at capturing the quirky changes in direction and intensity of hurricanes, but are getting better each year. In Chapter 13 the science of numerical weather forecasting is covered in detail.

The most exciting advance in tropical cyclone prediction is based on a growing understanding of the structure of these storms. During the 1980s and 1990s, Colorado State University meteorologist Professor Bill Gray developed and refined a method for forecasting tropical cyclones in a *statistical, seasonal* sense. His method wouldn't have told you if Andrew would hit Miami on August 24, but it can tell you up to a year ahead of time if the next hurricane season will probably see more than its usual share of hurricanes in the Atlantic.

Professor Gray's forecasts are based primarily on the El Niño cycle, rainfall in the sub-Saharan Sahel region of Africa, and the direction of winds in the lower strat-

Blue Skies

Hurricane/Hurricane Forecasting

BOX 8.3

Bryan Norcross, TV Meteorologist and Hero

When most people think of meteorologists, they think of TV weather personalities. This is a misconception, because the majority of meteorologists are not on TV, and many TV weathercasters are not meteorologists. Most TV weathercasters, meteorologists or not, don't make their own forecasts. They rely instead on the official National Weather Service forecasts. This doesn't mean that TV weatherpersons are not serious about forecasting however. They perform an important function by using their wide exposure and their personalities to convey important weather information to the public. One of the best examples is Miami TV meteorologist Bryan Norcross. When Hurricane Andrew blasted Miami, Norcross became a national hero.

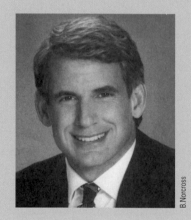

Norcross (pictured here) is part newsman, part meteorologist. He has a bachelor's degree in math and physics and master's degrees in communications and meteorology. Before coming to Miami, he worked as a news producer in Denver and a news director in Louisville. He also oversaw reporting on political campaigns and the Olympics. In Miami, Norcross returned to TV meteorology, work he had done previously at CNN. His background in news reporting helped him explain complicated aspects of meteorology to a wide audience. In November 1991 he even appeared on national television in a two-part episode of the sitcom

"Golden Girls." The plot: a terrible hurricane hits Miami, and Norcross spreads the news.

Life imitated television just a few months later, on August 23, 1992. Hurricane Andrew was bearing down on a city that had not experienced a hurricane in nearly thirty years. Untold thousands of residents, newcomers to the region, had no idea what to expect or how to protect themselves. Worse yet, Andrew came ashore just south of Miami shortly after 5:00 AM, when most sources of information were off the air, either intentionally or as a result of the storm's winds. Into this breach stepped Bryan Norcross.

Norcross provided frequent updates on the predicted path and severity of Andrew throughout the evening of the 23rd, and then stayed on all night with continuous coverage of the storm. He didn't make the forecasts; the National Hurricane Center did. But Norcross did broadcast these forecasts with live interviews with the NHC's director. He also provided numerous safety tips to residents who had little knowledge of how to protect themselves from 240 km/hr (150-mph) wind gusts. To emphasize the risks, at 5:00 AM Norcross and his overnight crew took shelter in their own studio, broadcasting from a closet. A resident from a hard-hit section of Miami called him, on the air, and said her roof was coming off. What should she do? Norcross replied, get away from all the windows, crouch inside a bathtub, and pull a mattress over her head. *Now.*

For his efforts during the Hurricane Andrew crisis, Bryan Norcross was given national journalism and meteorology awards. (He even was the subject of a 1993 TV-movie, which won no awards whatsoever.) Norcross is one of many TV weather personalities who save lives with timely, useful information.

osphere. A novice to tropical cyclones might wonder why these parameters could help one predict the number of storms in a given year. Based on our discussion of atmosphere–ocean interactions, they make sense. Earlier in this chapter we saw that El Niño leads to a strong jet stream over the subtropics. This jet destroys the carefully organized circulation of hurricanes. On average, then, El Niño years are low-hurricane years, and La Niña years are high-hurricane years. High rainfall in the Sahel means lots of thunderstorm "seedlings" that move east and develop into easterly waves and, sometimes, into Cape Verde hurricanes. Finally, strong winds in the lower stratosphere, like upper-tropospheric jet streams, lead to wind shear that disrupts the upper-level winds of a growing hurricane.

The accuracy of Professor Gray's forecasts is illustrated in Figure 8.38. In about seven of every ten years, his methods can accurately predict whether a given hurricane season will be more active or less active than normal. This is not the same as predicting a specific hurricane in a specific location in advance—that is a task for computer models (Chapter 13). However, it is a vast improvement over expectations

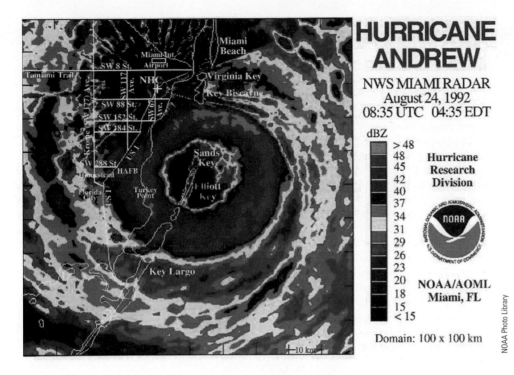

■ **Figure 8.37**

The last image of Hurricane Andrew from the National Hurricane Center's weather radar before it was blown off the roof of the NHC's headquarters in Coral Gables, Florida (just left of center near top), at 4:35 EDT on August 24, 1992. The circular band of red around Sands Key and Elliott Key is the eye wall. (Source: www.aoml.noaa. gov/hrd/tcfaq/tcfaqF.html#F3.)

based on climatological records, and is helping coastal residents as well as businesses and insurance companies prepare months in advance for the chance of damaging storms. As we will see later in Chapter 13, these kinds of long-range statistical predictions are the future of weather forecasting.

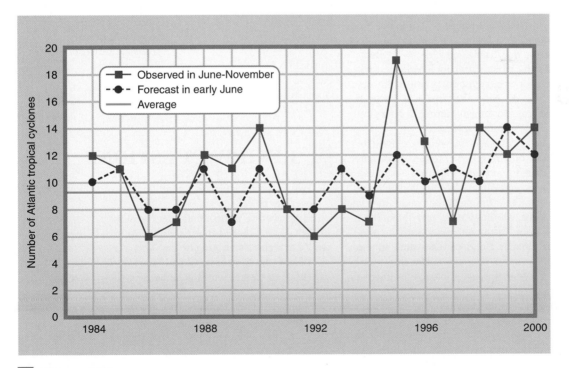

■ **Figure 8.38**

A comparison of the number of observed tropical storms and hurricanes versus the number predicted by Professor Bill Gray's statistical methods at the beginning of hurricane season. The green line represents the number of hurricanes expected based on a long-term average that, unlike Gray's method, does not take into account El Niño cycles and wind shear.

PUTTING IT ALL TOGETHER

 SUMMARY

Based on temperature, the vertical structure of the ocean can be classified into three basic layers: the surface zone, a deep-water zone, and a transition zone. In the bottom layer, or the deep zone, the temperature is uniform with depth. The surface zone is the warmest layer, and is warmest in the tropical waters. Ocean temperature decreases toward the poles and with increasing depth.

The atmosphere and ocean interact in a variety of ways, through exchanges of heat and moisture and also by the creation of surface ocean currents. These currents are driven by the winds. Some, like the Gulf Stream, move in well-defined boundaries like a river, while others are broad and diffuse. Surface currents generated by the winds can induce vertical circulations in the oceans. This can cause upwelling of nutrient-rich waters. For example, over the eastern tropical Pacific Ocean, winds typically rotate counterclockwise around a high-pressure system, producing along-coast winds off Peru and easterly winds along the equator in the eastern Pacific. The combined effect of wind and the Coriolis force produces offshore flow off Peru, causing upwelling in which cold, deeper water comes to the surface. This produces the cold SSTs off Peru and along the equator in the eastern Pacific. In the western Pacific on the equator this upwelling does not occur, and as a result the SST is higher there than in the east.

El Niño and La Niña are extreme phases of a naturally occurring climate cycle, and describe large-scale changes in SST across the tropical Pacific. El Niño and La Niña result from interaction between the surface of the ocean and the atmosphere in the tropical Pacific. During El Niño, warm SSTs replace the cold upwelling waters in the eastern Pacific. This shifts thunderstorm activity eastward across the Pacific and ultimately affects weather and climate across the globe. In many locations, especially in the tropics, La Niña (or a cold episode) produces the opposite climate variations from El Niño. For instance, parts of Australia and Indonesia are prone to drought during El Niño, but are typically wetter than normal during La Niña.

The oceanic patterns of El Niño and La Niña are closely related to a Pacific-wide seesaw of atmospheric pressure known as the Southern Oscillation. Recently similar oscillations have been discovered in the north Pacific and Atlantic oceans, highlighting the importance of atmospheric-ocean interactions in meteorology.

Tropical cyclones are large, whirling storms that obtain their energy from warm ocean waters. They stand out on satellite photographs because of their circular cloud patterns and, in the stronger storms, a nearly clear eye at the center. The clarity and size of the eye on satellite images helps meteorologists estimate a cyclone's strength.

A tropical cyclone begins as a disorganized tropical disturbance. A few grow to hurricane or typhoon strength with winds up to 200 mph. Most weaken within a week or two. They typically move west or northwest and then recurve toward the northeast, but each storm is unique. Hurricanes affecting the United States often form in the Gulf of Mexico and Caribbean Sea in early summer and late fall, with powerful Cape Verde hurricanes dominating attention in August and September. The location of warm ocean waters largely determines the birthplaces and intensities of tropical cyclones.

Tropical cyclones cause destruction with extremely high winds, storm surges of seawater, and flooding rainfall. Of these, storm surge has been the most deadly and devastating in the past, but recently flooding rains have killed more U.S. residents than wind or storm surge. Each hurricane packs its own unique combination of these weapons.

Weather satellites have revolutionized the forecasting of tropical cyclones. Today no tropical cyclones go undetected, a vast improvement upon the situation a century ago. Thousands of lives are saved as a result. Statistical forecasts now provide long-range estimates of the activity of hurricane seasons as much as a year in advance.

 KEY TERMS

You should understand all of the following terms. Use the glossary and this chapter to improve your understanding of these terms.

Arctic Oscillation	Pacific Decadal Oscillation
Bermuda high	Rainbands
Cape Verde hurricanes	Recurvature
Deep zone	Saffir–Simpson scale
Easterly wave	Southern Oscillation
Ekman spiral	Storm surge
Ekman transport	Supertyphoon
El Niño	Surface zone
Extratropical cyclone	Thermocline
Eye	Tropical cyclones
Eye wall	Tropical depression
Gyre	Tropical disturbance
Hurricanes	Tropical storm
La Niña	Typhoon
Ocean current	Upwelling
Oceanography	Wind shear

REVIEW QUESTIONS

1. Why is the ocean surface zone thicker over the tropics than over the poles?
2. How can SST distribution affect weather?
3. What is the thermocline?
4. What is the North Atlantic gyre?
5. What is the Ekman spiral and how is it formed?
6. How can surface winds generate upwelling along coastal regions?
7. What is El Niño and how does it impact the weather where you live?
8. What is La Niña and how does it impact the weather where you live?
9. Is a typical tropical cyclone bigger or smaller than the following weather phenomena:
   a. Hadley Cell
   b. Air parcel
   c. Thunderstorm
   d. Southern Oscillation?

10. Using the "Atmospheric Circulation/Southern Oscillation" section of the *Blue Skies* CD-ROM, estimate how long in months it takes for (a) warm water to move across the Pacific from Indonesia to South America, and (b) changes in SSTs to occur in the equatorial Pacific precipitation patterns in North and South America. Why do you think it takes this long for the atmosphere to react to an El Niño?

11. When a tropical cyclone "opens its eye" for the first time, what can you say about its approximate wind speed? Is it strengthening or weakening? When a hurricane narrows its eye and becomes very circular, is it strengthening or weakening?

12. If you could cover the Gulf of Mexico with water-impermeable plastic wrap, what would the effect be on the development and intensity of tropical cyclones over the Gulf?

13. In some El Niño years, the ocean waters off the west coast of Mexico also become abnormally warm, and tropical fish are sighted unusually far north. How might this unusual warming affect the intensity of tropical cyclones that develop in this region?

14. Only one tropical cyclone has developed in the South Atlantic Ocean off the coasts of South America and Africa in the last *century*. What ocean temperature and wind patterns might be responsible for this amazing lack of tropical cyclone activity?

15. Why is it that hurricanes never form off the West Coast of the United States, but can form off the East Coast?

16. Global warming, if occurring, should lead to warmer ocean temperatures and less wind shear because of a reduced latitudinal temperature gradient, but perhaps more wind shear because of "perpetual El Niño" conditions. Discuss how each of these three changes might affect future tropical cyclone activity and intensity.

17. Using the "Hurricane/Virtual Hurricane" section of the *Blue Skies* CD-ROM, explore the structure of Hurricane Fran in a simulation of a hurricane hunter aircraft. What Saffir–Simpson category is Fran? Given the wind information provided on the simulation, what do you think the forward speed of Fran is? Why?

18. Using the "Hurricane/Hurricane Forecasting" section of the *Blue Skies* CD-ROM, try your hand at forecasting the movement of the archived storm Zarathustra. Does this hurricane move in the same direction as the upper-level winds? How do SSTs affect the strength of the hurricane? What happens to the central pressure and wind speed of the storm when it moves over land? Explain why this happens with reference to important points raised in this chapter.

19. In 1985, Hurricane Gloria recurved northeastward along the U.S. East Coast and traveled quickly just east of the coastline for hundreds of miles. East Coast residents braced for a fearful storm; a Boston radio station even played a marathon of songs titled "Gloria" in anticipation. But in the end, most inland residents were disappointed, saying that the winds were much less than they had expected. Using a map of the United States and your understanding of the relationship between total storm winds and the forward motion of the storm, explain why the inland residents experienced relatively weak winds from this strong hurricane.

20. If you lived or ran a business on the U.S. East Coast and you heard that Professor Gray's forecast was for a very active and damaging hurricane season in the near future, what actions would you take to protect your life and property?

21. True or false: In a hurricane, you want to drive to a safe location in order to stay out of the high winds. Explain your answer.

22. Why don't the damage totals and death tolls from U.S. hurricanes always increase with increasing Saffir–Simpson category number? Discuss the different variables that can make a Category 1 storm more of a killer and destroyer of property than a Category 5 storm.

23. Can tropical cyclones exist on other planets in our solar system? Explain why or why not.

24. Using information in this chapter, explain how El Niño could prevent the formation of hurricanes in the Atlantic by a mechanism other than subtropical jet strengthening. (Hint: See Table 8.1 and refer to Professor Gray's forecast methods.)

■ WEB ACTIVITIES

Choose Chapter 8 on the textbook's Web site:
http://info.brookscole.com/ackerman
and select from the following resources:
- Interactive Modules that illustrate and extend your understanding of key topics in this chapter
- Tutorial Quizzes to test your mastery of terms and concepts

 For additional readings, go to the InfoTrac College Edition, your online library, at:
http://www.infotrac-college.com

Air Masses and Fronts

After completing this chapter, you should be able to:

■ Explain what an air mass is and identify the regions in which different air masses develop

■ Explain how and why air masses change once they leave their "birthplaces"

■ Characterize the different ways in which clashes of air masses create fronts

■ Contrast the varying types of weather that are associated with cold, warm, stationary, and occluded fronts

Introduction

Lightning bolts around dawn soon are gone. Later in this chapter we'll explain the science behind this bit of weather folklore, but for the moment just *say* it out loud. Then have a friend from another part of the country say it to you. Depending on where you grew up, your pronunciation of these words may differ from your friend's. We call this difference an accent, and it arises because your pronunciation is influenced by the way people around you as a child said them. For example, people who grew up in the rural South might begin by saying "lahtning boats." Residents of Canada and the far northern United States sometimes pronounce "around" as "aroond." Native New Englanders could finish the sentence by saying "dahn soon ah gahn." The words are the same, but how they sound differs from one region to another.

We can draw an analogy between accents and the atmosphere. The composition of the atmosphere is the same worldwide, but where the air "grows up" determines its temperature and humidity. In other words, air develops an "accent" and takes on the characteristics of its surroundings. Just as we can identify a vocal accent and label it with its region of origin, we can do the same with air. This leads to the concept of air masses, the first topic of this chapter.

People sometimes lose or modify their accents when they move away from where they grew up. A New Englander who moves to the South may, over time, find herself saying "dawwn" instead of "dahn." Air masses are modified in a similar way; they move around and in the process become cooler or warmer, drier or moister, depending on where they have traveled. Along the way, these air masses can lead to record cold outbreaks and

lake-effect snows in winter and deadly heat waves in summer.

Finally, different air masses, like different accents, don't mix together easily. In the atmosphere, air masses with differing characteristics clash in regions called *fronts.* Much of the stormy weather observed in the middle latitudes, from clouds and rain to thunderstorms and tornadoes, occurs in frontal regions as a result of air masses interacting with each other. In this chapter, we will also explore the different types of fronts and the typical weather associated with them.

What is an Air Mass?

An **air mass** is an extremely large body of air whose properties of temperature and moisture content (humidity) are similar in any horizontal direction. Air masses can cover hundreds of thousands of square miles. Air masses are formed when air stagnates for long periods of time over a uniform surface. The characteristic weather features (temperature and moisture) of air masses are therefore determined by the surface over which they form. An air mass acquires these attributes through heat and moisture exchanges with the surface. As a result, an air mass can be warm or cold, and moist or dry.

Observations

In a typical year, air mass weather kills more people in the United States than all other weather phenomena combined. Heat waves in the summer can bring perilous weather to cities not accustomed to handling such hazardous conditions (Box 9.1). These heat waves are caused by very hot, stagnant air masses. For example, between July 13 and 14 of 1995 the deaths of over 700 people, and over 500 just in Chicago, were attributed to a hot and humid air mass that occupied much of the central and eastern United States. High temperatures exceeded 40° C (104° F) in Illinois and Wisconsin, and even the low temperatures hovered just below a sultry 30° C (86° F) (Figure 9.1). Record-high dew points of 26° to 28° C (79° to 82° F) contributed to the excruciating conditions. A heat wave in the Midwest in late July 1999 resulted in average daily temperatures of greater than 33°C (90° F)! More than 200 deaths in the Midwest were attributed to this 1999 heat wave (Figure 9.2).

Problems and hazards are also caused when air masses move. For example, between February 12 and 16 of 1990, a large mass of cold air spread southward from northern Canada into the central United States. The associated weather, which included freezing rain, tornadoes, severe thunderstorms, and flooding, caused an estimated $120 million of damage. Cold air outbreaks such as these have cost American homeowners at least $500 million per year when frozen water pipes burst and inflict water damage.

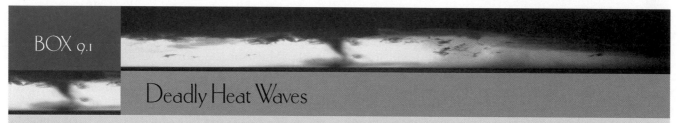

BOX 9.1

Deadly Heat Waves

Each summer in the United States approximately 175 to 200 deaths are attributable to heat waves. Most of these deaths occur in cities, particularly northern cities. Regions that suffer under intense hot spells are usually dominated by a surface high-pressure system with a midtropospheric ridge aloft. Dew points are also high, and to compound matters, wind speeds are often low. Clear or partly cloudy skies allow intense solar radiation to further heat the ground and the air mass. High humidity and stagnant air reduce the body's ability to cool down through sweating (see Boxes 3.3 and 4.1). When these conditions persist day and night for several days, lives are endangered.

Overexposure to heat leads to giddiness and nausea. Heat cramps are generated when the body loses too much water and salt through perspiration. Symptoms of heat cramps are muscle spasms, fever, and nausea. Heat exhaustion occurs when the body temperature and blood pressure drop and the skin turns cool and clammy. Perspiration is profuse as the body continues trying to lose energy. More dangerous than heat exhaustion is heat stroke, which poses an immediate threat to life. Most victims of heat stroke are people over 60, although infants are also susceptible. The first signs of a heat stroke are that the skin becomes dry and hot as sweating ceases, the face becomes flushed, and the pulse accelerates. Once the body loses its ability to cool down, death can occur within a few hours. A victim of heat stroke needs immediate medical attention. Sponging the body with water will help cool the body.

During a heat wave, the threat to health is increased in large cities by what is called the "urban heat island" effect, which amplifies the heat by 1° to 4° C (2° to 7° F) or more (see figure). During the heat of the day, poorly ventilated urban houses and apartments that do not have air conditioning can become like brick ovens. To avoid the dangers of heat exhaustion it is important to drink water, stay out of direct sunlight, and minimize physical activity. Reduction in the loss of life can be accomplished by monitoring the health of the urban elderly. The news media helps by not only providing useful information about the weather but also by explaining procedures that lessen heat stress and giving the locations of places where relief is available.

Heat waves also have a strong economic impact even if there is no loss of human life. A prolonged heat wave can lead to the widespread use of air conditioning, leading to increased demands for power that stress gas and electric utilities. During the July 1995 heat wave, Chicago residents set a record by using almost 20 megawatts of energy in just one day. Excessive demand can leave people without power for hours and sometimes days. Transportation can be stymied when highway surfaces and railways buckle and warp in the heat. All types of outdoor work, such as landscaping and construction, experience reduced productivity. Agriculture is especially vulnerable. Heat waves stunt crops and kill livestock; on July 14, 1995, 850 cows died from heat exposure in Wisconsin.

Intense heat waves are often broken with the invasion of a polar air mass. Violent storms can be associated with cold fronts, resulting in additional economic losses and the potential for further loss of life.

Air Mass Types

Air masses are classified according to the temperature and moisture characteristics where they develop. Cold air masses originate in polar regions and are therefore called polar air masses. Warm air masses typically form in tropical or subtropical regions and are called tropical air masses. Moist air masses form over oceans and are referred to as maritime air masses. Dry air masses that form over land surfaces are called continental air masses. The five primary air mass regions are summarized as follows:

- Polar (P): formed poleward of 60° north or south—cold
- Tropical (T): formed within about 30° of the equator—warm
- Arctic (A): formed over the Arctic—very cold
- Continental (c): formed over large land masses—dry
- Maritime (m): formed over the oceans—moist

These regions overlap. For example, a tropical air mass can be warm and dry or warm and moist, depending on whether an ocean or a continent is

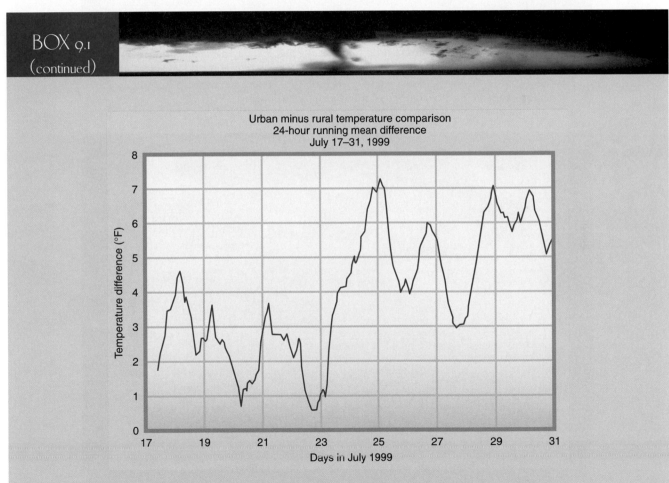

BOX 9.1
(continued)

Urban minus rural temperature comparison
24-hour running mean difference
July 17–31, 1999

The urban heat island is demonstrated by a comparison of the differences in hourly temperatures between Chicago, IL and Aurora, IL (a nearby suburban station). During the late afternoon and early evening of July 25 to 31, 1999, the average temperatures were about 3° to 5° C (5° to 7° F) higher in Chicago. During this heat wave, most of the people who died on July 29 and 30 lived in large cities.

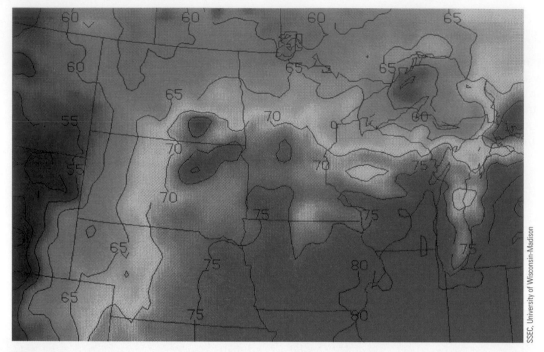

Figure 9.1

Surface temperatures across the north-central United States and southern Canada on the morning of July 14, 1995. The red region is a hot air mass in which temperatures exceed 24° C (75° F). Hot conditions during the night and morning can be as life-threatening as record highs in the afternoon because of the 24-hour-a-day stress caused by the heat.

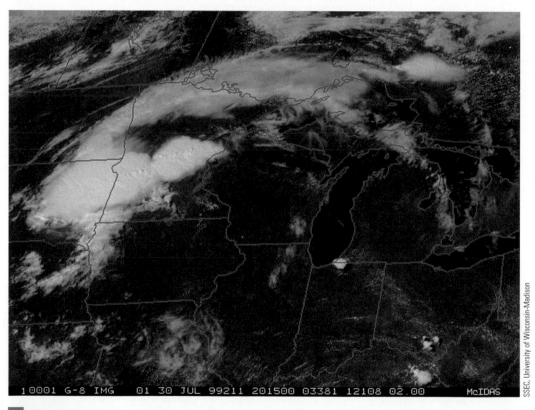

Figure 9.2

The thunderstorms in this visible satellite image taken on July 30, 1999, do not pose the major weather crisis occurring during this time. The clear area in the middle of the image is a hot, cloud-free air mass. An intense heat wave caused by this air mass killed over 200 people in the Midwest.

TABLE 9.1   Temperature and Moisture Characteristics of Air Masses

Air Mass	Winter Characteristics	Summer Characteristics
Continental polar (cP)	Very cold and dry	Cool and dry
Maritime polar (mP)	Cool and humid	Mild and humid
Continental tropical (cT)	Cool and dry	Very hot and dry
Maritime tropical (mT)	Warm and humid	Warm and humid
Arctic (A)	Bitter cold and dry	—

Identifying Air Masses

beneath it. This suggests a two-letter classification scheme for air masses, with the first letter denoting land versus ocean, and the second letter indicating the latitude band of origin.

Four main air mass types are dominant: **continental polar (cP), continental tropical (cT), maritime polar (mP)**, and **maritime tropical (mT).** An air mass's temperature and humidity are a function of the time of year when it forms. Winter air masses will be colder than those that form during the summer. Some classification schemes differentiate cold air masses into polar and arctic air masses. Arctic air masses (A) are much colder than polar air masses and form in winter over snow-covered surfaces in Siberia, the Arctic Basin, North America, and Greenland. The temperature and moisture properties of these air masses as a function of season during which they originate are shown in Table 9.1.

Air Mass Source Regions

Air mass source regions are the "birthplaces" of air masses. A source region must have light winds or no winds, so the air has time to acquire the temperature and moisture properties of the region's surface. Strong winds would move the air before it could acquire the characteristics related to the surface. Given the requirements of a uniform surface and light or no winds, not all parts of the world can generate air masses. Good source regions for air masses are the subtropical belts, which have light winds and thus favor the development of air masses with uniform temperature and moisture (Figure 9.3).

A source region must have an extensive and homogeneous surface. Large water bodies or flat uniform lands are potentially good source regions for air masses. Coastlines are not very good source regions of air masses because these locations have both land and water.

Once formed, air masses are not confined to their source regions. They move on to other regions. How they move is determined by upper air patterns and is discussed in the next chapter. Before we look at the air masses that affect North America, we need to revisit the concept of atmospheric stability (Chapter 3).

Atmospheric Stability Revisited

The **static stability** of the atmosphere is determined by comparing the temperature of a rising parcel with the temperature of the atmosphere at the same altitude as the parcel. This is known as the parcel's "environment." At a given pressure, the colder the air parcel, the greater its density. If an air parcel is colder than its environment, it is more dense than its surroundings and will descend. The atmosphere is stable when an ascending parcel, which cools as a result of expansion, becomes cooler than its environment. Therefore, as discussed in Chapter 3, any region of the atmosphere that has a temperature inversion is stable.

http://info.brookscole.com/ackerman

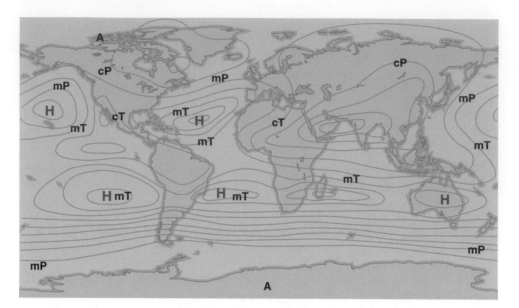

 Figure 9.3

Major air mass sources of the world. Observe the correlation between the source regions and surface high-pressure systems. Blue lines are surface isobars.

If an air parcel is warmer than the surrounding air, it will rise. If an ascending parcel within an air mass becomes warmer than its environment, the air mass is said to be unstable. An unstable atmosphere has a temperature and moisture structure favorable for lifting air parcels and thus forming clouds and possibly precipitation. A stable atmosphere inhibits rising motions and is usually less cloudy and drier.

Another important concept is the means by which the stability of the atmosphere changes. Warm air overlying cold air results in a stable atmosphere. So, when the lower regions of the troposphere are cooled, the air mass becomes more stable. This explains why polar air masses are generally stable. However, when cold air moves over a warm surface (for example, a warm body of water), the air near the surface warms. Warming the air near the ground makes the atmosphere less stable. This is why tropical air masses are, in general, less stable than polar air masses.

To make an air mass unstable, the environment must be more conducive to vertical motions. This happens when the lower layers of the atmosphere warm up through energy exchanges with the surface under the air mass.

Air Masses Affecting North America

Figure 9.4 shows the major air mass source regions that affect North American weather and climate. The arrows show the typical paths the air masses take once they begin to move. Below, we examine each type of air mass.

Maritime Polar Air Masses

Maritime polar air is formed over the oceans at high latitudes. The North Pacific is a good source region of mP air masses. These North Pacific air masses often move into the Gulf of Alaska and then to the West Coast of North America. During winter, they can influence the weather as far south as California. The East Coast of North America is also affected by North Atlantic air masses, which approach the coast from the northeast. The stormy weather pattern associated with these advancing air masses is referred to as a *northeaster* (typically pronounced and spelled *nor'easter*). Low-pressure systems that comprise a nor'easter typically move up the East Coast of North America between October and April. The counterclockwise winds around the low draw moist mP air over the Atlantic Ocean inland. At the same time, cold cP air moves eastward toward the East Coast and meets with

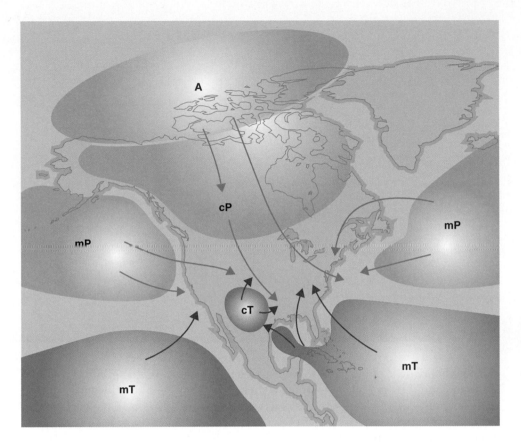

■ **Figure 9.4**

The major air mass source regions that affect North American weather. The arrows show the typical paths these air masses take once they begin to move. Which air masses affect the weather where you live?

northward-moving mT air from the Gulf of Mexico. This clash of warm and cold has produced some of the heaviest snowfalls on record along the East Coast (Figure 9.5).

The weather associated with mP air masses is quite variable. Winter mP air masses are formed in polar regions over water. Cold mP air that moves over a warm surface is unstable and can bring rain showers. When the mP air moves over a surface that is only slightly warmer, the atmosphere is less unstable and may produce stratus clouds and drizzle.

■ Continental Polar Air Masses

Continental air masses affecting North America are formed over the interior high-latitude regions of the continent, such as Alaska and Canada. The air masses formed in winter are very cold and dry. They require long, clear nights to form, during which strong radiational cooling causes the temperature of air near the surface to plummet, especially over snow-covered ground. Because the air above the surface is not cooled as strongly, temperatures in the lower troposphere in a cP air mass can increase with altitude. These surface temperature inversions are often observed, indicating the stable nature of these air masses.

In winter, the weather associated with cP air masses is cloud-free and frigid. From northern Canada cP air masses move southward and can result in bitter cold temperatures as far south as Florida. This can lead to crop damage and cause economic hardship (Figure 9.6). The wintry temperatures and surface temperature inversion can also lead to reduced air quality and poor visibility as pollution levels increase with the burning of heating fuels such as coal, fuel oil, and wood.

When formed in summer, cP air masses have more moderate temperatures. The snow that would cool this air in winter has melted because of the increased warmth from the Sun. Surface temperature inversions are usually absent in cP air masses formed in summer. The air mass has formed over a relatively dry surface, so

SSEC, University of Wisconsin-Madison

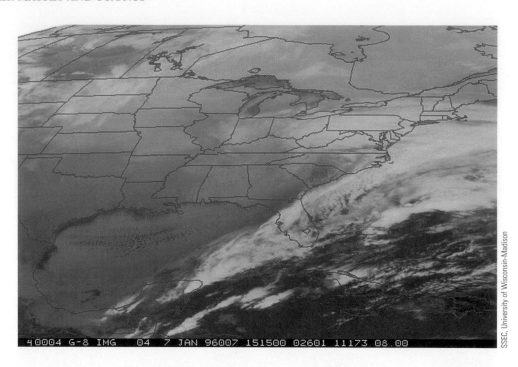

Figure 9.5

Weather associated with the combination of warm and cold can produce a nor'easter. This infrared satellite image shows a nor'easter that ravaged the East Coast during January 7–8 in 1996. By the time the storm was over, 3 feet of snow had accumulated in Virginia and North Carolina. At least 86 people died because of the weather associated with the storm.

the characteristic weather is cool, dry, and clear—a pleasant change from a humid summer caused by mT air masses. Daytime heating, however, may lead to puffy cumulus clouds and even showers as the air mass becomes unstable, but clear skies return soon after sunset.

Arctic Air Masses

Arctic air masses are formed over the frozen Arctic and are much colder than cP air masses (Figure 9.7). This air mass type is confined to a shallow layer near the surface. Because the layer is not very deep (less than 0.6 kilometer [1 mile]), there is little vertical lifting associated with its movement, and thus little precipitation. Arctic air can move south and occasionally reach into the United States. When it does, record-setting cold outbreaks are the consequence. Figure 9.8 shows the hourly temperature at Madison, Wisconsin, during a prolonged Arctic outbreak in 1996.

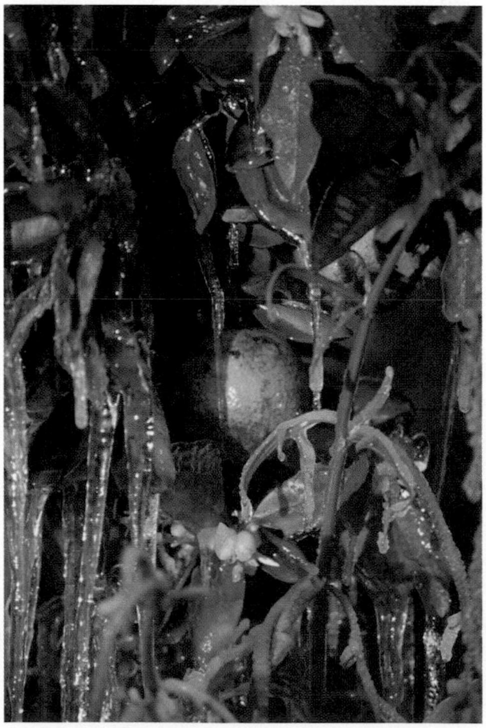

The Photographer's Window

Figure 9.6

Outbreaks of Canadian continental polar air masses during early spring can result in bitter cold temperatures as far south as the orange groves of Florida.

Continental Tropical Air Masses

Continental tropical air masses are hot and dry and typically form over tropical and subtropical deserts and plateaus. The hot surface temperatures and the lack of vegetation make the cT air mass hot and dry. The southwest United States and northern Mexico are summer source regions in North America.

Continental tropical air influences the southwest and central United

Figure 9.7

Arctic air masses are extremely cold.

States in summer as warm, dry air arrives from the Mexican Plateau. The hot temperatures near the ground make the air mass unstable, but the dryness limits cloud formation. The contrast between dry cT air and moist mT air over the south-central United States can be so extreme that meteorologists label it as a special kind of front known as a "dryline." Later in this chapter we will discuss an example of the dryline.

When cT air from the Mexican Plateau moves aloft over the southern Great Plains, it can create a thin but strong inversion known as a "capping inversion" or "the cap" to storm chasers. We discuss this type of inversion in Chapter 11.

◼ Maritime Tropical Air Masses

The sultry summer weather of the eastern United States is strongly influenced by mT air masses that form over the Gulf of Mexico, the subtropical western Atlantic Ocean, and the Caribbean Sea. The moisture source of precipitation for the midwestern United States is the mT air masses, particularly those originating over the Gulf of Mexico. When the air mass is stable, however, an oppressively humid heat wave is the result (Figure 9.9). Summertime weather in the southwestern United

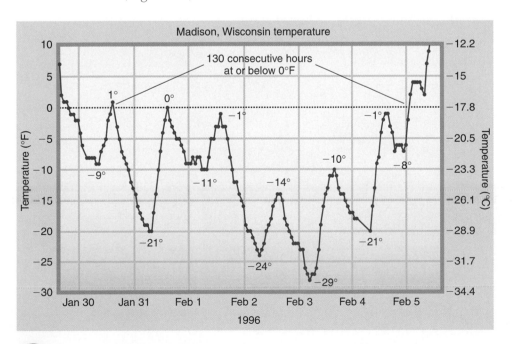

Figure 9.8

The hourly temperature observations at Madison, Wisconsin, during the cold air outbreak of January 30–February 5, 1996. Numbers indicate daily high and low temperatures, which may be a degree different from the highest and lowest hourly temperatures in some cases. For nearly a week, the daily high and low temperatures both registered 11° to 22° C (20° to 40° F) below the normal values for a Wisconsin winter!

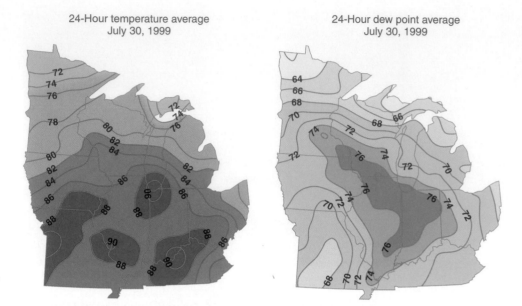

24-Hour temperature average
July 30, 1999

24-Hour dew point average
July 30, 1999

Figure 9.9

The high temperatures and dew points of the heat wave in the Midwest during the last two weeks of July in 1999 (shown at right in °F) were responsible for some 232 deaths.

States is influenced by Pacific mT air masses. Moisture supplied by these mT air masses to the Southwest generates rains referred to as the Arizona Monsoon (Figure 9.10).

Air Mass Modification

As air masses move from one place to another, their properties, such as temperature, moisture, and stability change as the air masses exchange heat and moisture with the underlying surface. This process is called **air mass modification.** There are two primary mechanisms that modify an air mass: heat exchanges with the surface and mechanical lifting.

Heat exchanges with the surface primarily affect the lowest regions of the air mass. The rate at which the air mass temperature and moisture change is determined

Figure 9.10

Maritime tropical air masses supply the moisture for the summertime rains of the southwestern United States. This seasonal precipitation cycle is referred to as the Arizona Monsoon.

PictureQuest International

by the rate of heat and moisture exchanges. The greater the temperature difference between the air mass and the surface over which it is moving, the greater the heat exchange. Exchanges of moisture are at a maximum when the relative humidity is low and the surface is wet (Figure 9.11).

As a cold cP air mass moves over a warm body of water, there is a rapid exchange of heat and moisture. The lowest layer of the air mass warms and moistens, increasing the instability of the air mass. If the temperature difference between the air and water is large, rapid evaporation causes the air to saturate and form a fog. These are *steam fogs* (see Chapter 4). Steam fog is common over the Great Lakes and the north Atlantic in fall and winter when the cP air flows over the warm waters (Figure 9.12). This fog can be as deep as 1500 meters, with swirling columns of fog called steam devils. The movement of cP air masses over the Great Lakes can increase snowfall downwind. The increased snowfall is referred to as *lake-effect snow* (Box 9.2).

Fogs over the Great Lakes also form during summer when mT air moves over the cooler lake waters. This is an example of an advection fog (see Chapter 4). In this weather situation, the mT air near the surface cools, increasing the relative humidity and causing a fog to develop. The cooling of the air occurs rapidly in the lowest 50 meters. In this case the lower layers of the air mass are cooling, causing the air mass to become more stable. This suppresses vertical mixing, limiting how deep the fog will form.

An extreme example of air mass modification occurs over the Gulf of Mexico during the cool season. A cold cP air mass plunges south across the Great Plains and stalls over the warm Gulf of Mexico. During the next few days, this air mass modifies rapidly while the large-scale weather pattern draws the air mass northward again across the Gulf Coast of Texas. However, by the time the air mass returns to the mainland, it is considerably warmer and moister than before. In fact, it may even be reclassified as an mT air mass! This remarkable transformation is called a "return flow event" and can happen as many as 25 times during the months of November through April. These events supply the southern United States with warm, moist air

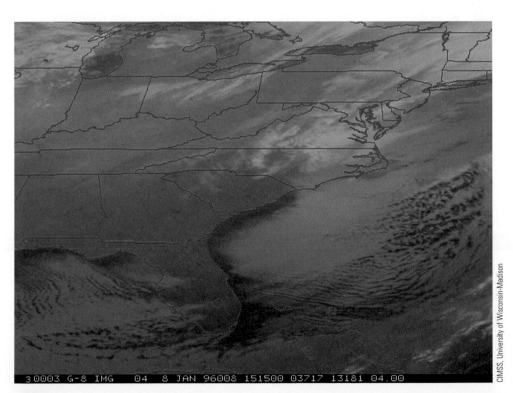

30003 G-8 IMG   04   8 JAN 96008 151500 03717 13181 04.00

CIMSS, University of Wisconsin-Madison

Figure 9.11

This infrared satellite image provides a view of the East Coast after the passage of a cold front associated with the nor'easter shown in Figure 9.5. As the cold and dry cP air mass moves over the warm Gulf Stream, the air mass warms quickly and becomes unstable. Mixing causes the low-level clouds to form over the water.

 Figure 9.12

A winter cP air mass that moves over a warmer body of water becomes unstable as energy and moisture exchanges with the surface warm and moisten the lower levels of the air mass.

during the winter and help contribute to the secondary maximum in severe weather activity across the Southeast in November (see Chapter 11).

Air masses can also be modified when lifted or forced to descend by topography. As an mP air mass from the Pacific Ocean moves over the northwest coast of the United States, the mountains lift it. This causes the air to cool, increasing the relative humidity, and leading to cloud formation and snowfall in the mountains. A cP air mass formed over the Greenland Plateau will descend as it moves off the high plateau, warming along the way.

Fronts

Although an individual air mass can change over time, different air masses do not mix readily. Therefore, when two air masses come in close contact they retain their separate identities for several days. The transition zone between two different air masses is called a **front.** Fronts can be hundreds of miles long and exist as long as the air masses they separate remain distinct.

Norwegian meteorologists around the time of World War I laid the foundation for our concepts of fronts and their movements. They observed large air masses with different temperature and moisture properties that advanced and retreated versus one another. Clashing air masses often led to disruptive weather conditions. The boundary between air masses was called a "front," analogous to the boundaries used on military maps to separate battling armies.

A frontal zone is a sloping surface that separates two air masses. Figure 9.13 is a three-dimensional sketch of a front. The area where the front meets the ground is called the *frontal zone.* The frontal zone is typically featured on surface weather maps because this is where the air mass contrasts are usually most prominent. While fronts are represented on weather maps as lines, the front is actually a zone where weather conditions rapidly change across distances of a few miles as one goes from one air mass to the other.

Fronts are classified by the temperature changes that result after an air mass passes over a given location. A **cold front** indicates that colder air will follow the front's passage; a **warm front** means warmer air will follow. But there are more

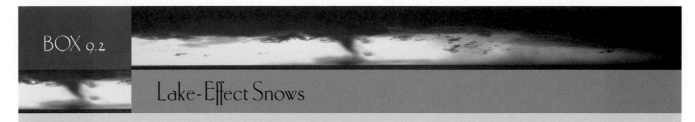

BOX 9.2

Lake-Effect Snows

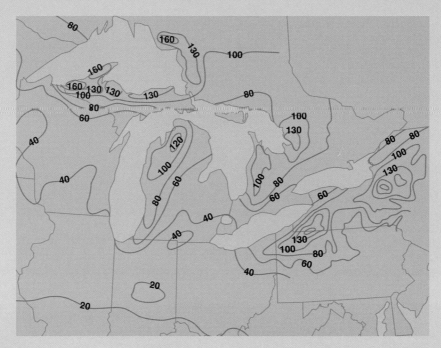

The figure above shows the average annual snowfall for the Great Lakes region. Two distinct patterns are discernible. The first is that, in general, snow depth increases northward. This is expected, because temperature usually decreases poleward. The other distinct feature is the difference in the amount of snow along the shoreline. Regions with the localized maximum in snowfall are on the southerly and easterly sides of the Lakes and are referred to as *snowbelts*. Because of this, some regions along the coast have more snow than regions to the north. Why does the geographic location of the snowbelts vary for each lake?

The Great Lakes modify the weather and climate in the region by modifying air masses that move over them. As the cold air moves over the water, the lower layers are warmed and moistened by the underlying lakes. This makes the air mass unstable. Evaporation increases the moisture content of the air mass, which is then precipitated in the form of snow on the land downwind. Maximum heat and moisture exchanges occur when the air is cold and the temperature difference between the air and the water is large. This condition tends to occur during early winter; this is when the most lake-effect snow is produced. A long path across the warm water by the air mass results in heavy precipitation over the land (see figure at left). The longer the path or "fetch," the more evaporation and the greater the potential for large snowfall amounts over the land on the downwind side of the lake. Hills can amplify the snowfall amounts by providing additional lifting. The location of a snowbelt along a particular lake is a function of the temperature difference between the air mass and the water, the fetch, and the terrain on the leeward side of the lake.

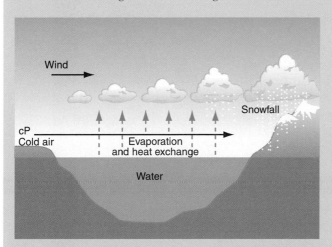

As a cold polar air mass moves over a warmer body of water, it becomes unstable. This results in increased snowfall on the downwind side of the lake.

Lake-effect snows are good for the economy of a region, particularly ski resorts. They also provide water for reservoirs and rivers. Too much lake-effect snow can be hazardous, as during the winter of 1976–1977 (see photo at right). During that winter, lake-effect snows helped to produce 40 straight days of snowfall in Buffalo, New York. A blizzard during this time generated 9-meter (30-foot) snowdrifts and resulted in the deaths of 29 people. More recently, on November 20, 2000 Buffalo was blitzed with 63 centimeters (24.8 inches) of snow in one day, with most of it falling in just seven hours! A radar image from that snowstorm (see figure below) shows a streak of precipitation created by cold southwesterly winds blowing the length of warmer Lake Erie toward Buffalo (*BUF* in the figure). Lake-effect snow can bombard a location as long as all the ingredients–cold winds, warm water, and a long fetch–are present. This was demonstrated when 81.5 inches of snow—over 2 meters, or nearly 7 feet—fell on Buffalo during December 24–28, 2001.

Red Cross workers search for victims buried in cars following heavy lake-effect snowfall near Buffalo, New York, in February of 1977.

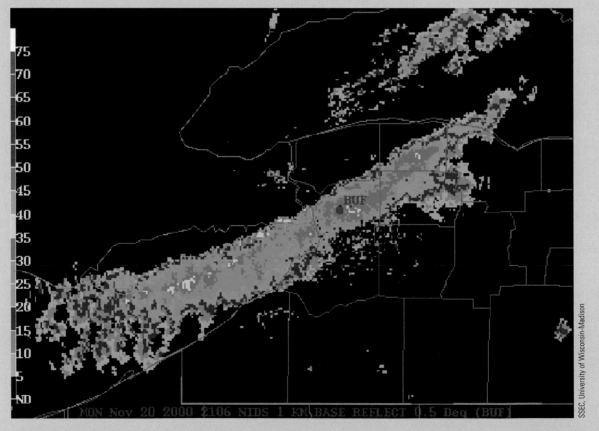

Radar image of lake-effect snow in Buffalo on November 20, 2000.

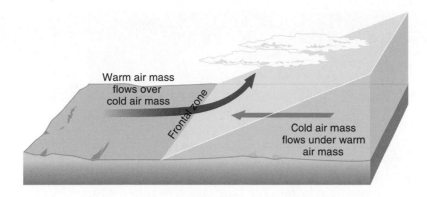

Warm air mass
flows over
cold air mass

Frontal zone

Cold air mass
flows under warm
air mass

Figure 9.13

A *schematic three-dimensional representation of a generic frontal zone.*

weather changes than just temperature during a frontal passage. In this section, we will investigate the changes in temperature, pressure, wind direction, and cloudiness of an idealized, well-developed cold front. Then we will do the same for a warm front. We will also discuss **stationary** and **occluded** fronts.

This textbook presents an idealized example of the type of weather associated with cold and warm fronts. Nature is often much more complex, but these examples should help you to identify the general features of fronts. Chapter 10 presents examples of real-life fronts associated with a severe storm; look for similarities and differences between the idealized cases below and the observations in Chapter 10.

Frontal Analysis

Cold Fronts

Figure 9.14 shows an idealized structure of temperature, pressure, wind direction, and precipitation associated with a cold front. The blue triangles point in the direction of the movement of the front. To imagine how temperature changes as a cold front approaches, observe how the isotherms change along the line in the figure that is perpendicular to the front. As a cold front approaches, the temperature remains steady or rises under the influence of warm southwesterly winds (Figure 9.14, *A* and *C*). The temperature then quickly drops in the frontal zone region as the cold air mass moves into the area.

As the cold front approaches, a decrease in pressure is observed (Figure 9.14, *B*). Fronts develop in regions of lower pressure because, as we learned in Chapter 6, air at the surface tends to blow toward lower pressure. This convergence of air helps intensify temperature and moisture contrasts, creating and sustaining fronts. Changing frontal weather is therefore a good bet when the barometer shows a consistent drop in pressure.

After cold frontal passage the atmospheric pressure begins to increase as a cold, high-pressure air mass moves into the region (Figure 9.14, *B*). The largest pressure changes usually occur in the frontal zone.

Strong wind shifts often accompany fronts. Air flows into the low-pressure area in a counterclockwise direction in the Northern Hemisphere, and because of friction crosses the isobars. Thus, ahead of a cold front the winds tend to be from the southwest (Figure 9.14, *C*). Wind direction is variable in a frontal zone as the winds shift to a northwesterly direction after the cold front passes. These winds advect the cold air into the region. If a cP air mass is behind the front, it has low humidity and the visibility is usually good.

Precipitation associated with our ideal cold front is shown in Figure 9.14, *D*. Precipitation is confined to the frontal zone, usually in the form of scattered showers and thunderstorms. When temperature and moisture contrasts are strong, a squall line is sometimes observed in advance of the cold front. The reasons for this are discussed in Chapter 11. After the front passes, the sky clears. Scattered cumulus or stratus clouds sometimes accompany the cold air mass behind the front.

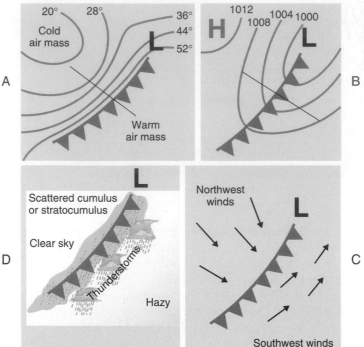

■ **Figure 9.14**

*Surface weather associated with a cold front. Clockwise from upper left: **A,** Temperature in degrees Fahrenheit. **B,** Pressure in mb. **C,** Wind direction. **D,** Clouds and precipitation. The L marks the location of the lowest pressure. The thin line in **A** and **B** is the cross-section used in Figure 9.15.*

Showers behind the cold front can occur when a cP air mass moves over a warm surface (Figure 9.14, *D*).

Fronts extend above the surface, because air masses are three-dimensional phenomena. Figure 9.15 is a vertical slice through our idealized cold front. The front slopes upward, toward the pool of cold air. The altitude of the cold frontal zone typically increases 1 kilometer for each 50 to 100 kilometers of horizontal distance along the surface. The slope is steepest where the air mass meets the surface, sloping more gently behind the surface front. The cold air often moves fastest near the surface for a variety of reasons, including the fact that the cross-frontal winds that push the cold front along are greatest near the surface. This gives the characteristic curved profile of the cold front (see Figure 9.15).

The steep slope near the surface frontal zone forces the warmer air it is replacing to rise rapidly. The abrupt vertical motion tends to generate thunderstorms. Thus, precipitation associated with the cold front is often intense, but brief and restricted to the immediate vicinity of the cold frontal zone near the surface.

The duration of the precipitation depends on the horizontal extent of the vertical lifting and the speed of the front. The precipitation band is narrow because of the steep cold front. Unlike daytime showers, frontal precipitation can occur 24 hours a day and remain intense even at night. The speed of a cold front is observed to vary from nearly stationary to 50 km/hr (31 mph) or greater, especially over the western Great Plains where flat ground and the presence of the Rocky Mountains just to the west help funnel cold air rapidly toward Mexico. The fairly rapid speed of cold fronts and their ability to generate stormy weather without the added "juice" of daytime heating together explain why "lightning bolts at dawn soon are gone."

■ **Warm Fronts**

Figure 9.16 depicts the temperature, pressure, wind direction, and cloud patterns typically observed in association with a warm front. The red half-circles point toward the direction of movement of the warm air mass. After the passage of a warm front, the air is warmer and often more humid. As a warm front approaches, the

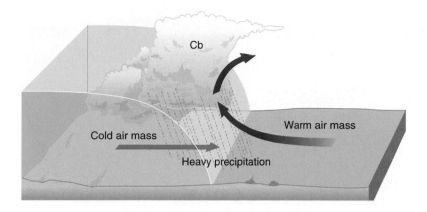

■ **Figure 9.15**
Vertical slice through a cold front. The steepness of the cold front favors the development of thunderstorms.

temperature initially remains steady or climbs slowly, and then rapidly increases during the frontal passage (Figure 9.16, *A*). As with the cold front, the warm front is located in a region of lower pressure relative to regions ahead and behind the front (Figure 9.16, *B*). Therefore, the pressure drops as the warm front approaches and then increases as it passes by. The changes in pressure associated with the warm front are not as rapid as those associated with the cold front.

The winds ahead of the warm front are typically observed to be easterly or northeasterly (Figure 9.16, *C*). Again, this is associated with the counterclockwise convergence of surface air in toward low pressure. Behind the front, the winds are typically from a southerly direction.

The vertical slope of the warm frontal zone is gentler than that of the cold front. Its altitude increases 1 kilometer for every 200 kilometers or so of horizontal distance perpendicular to the surface front. As a result, the warm frontal air slides upward gradually over the cooler air ahead of the front. This upgliding warm air is a form of **overrunning.** Overrunning occurs when an air mass moves over a colder air mass that is denser. Deep layers of stratiform clouds (cirrostratus, altostratus, and nimbostratus) are often the result of overrunning. The gentle rise of the warm

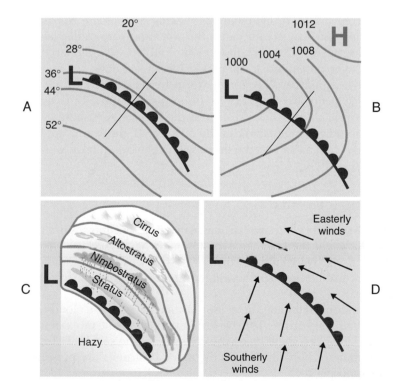

■ **Figure 9.16**
Surface weather associated with a warm front. Clockwise from upper left: **A,** *Temperature in degrees Fahrenheit.* **B,** *Pressure in mb.* **C,** *Wind direction.* **D,** *Clouds and precipitation. The* L *marks the location of the lowest pressure. The thin line in* **A** *and* **B** *is the cross-section used in Figure 9.17.*

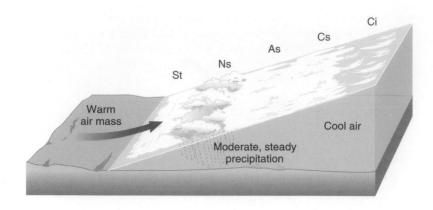

■ Figure 9.17

Vertical structure along an idealized warm front. As the front approaches a location, cirrus (Ci) clouds appear first, followed by cirrostratus (Cs), altostratus (As), nimbostratus (Ns), and stratus (St) clouds.

Blue Skies

Weather Analysis/Find the Front

frontal slope promotes condensation and clouds, but not usually the thunderstorms that typify cold frontal precipitation.

Steady precipitation falls from nimbostratus clouds that result from the overrunning (Figure 9.16, *D*). The precipitation can be far in advance of the frontal zone. The frontal slope also tends to generate stratus-type clouds far in advance of the frontal zone. So, ideally, we could predict the approach of a warm front by watching the changing cloud conditions (see Figure 9.16, *D*). Cirrus appear first, followed by cirrostratus, which may result in a halo around the Sun (see Chapter 5). Cirrostratus are followed by altostratus clouds, which make the Sun appear milky or watery. Stratus and nimbostratus follow the middle-level clouds. As discussed in Chapter 4, nimbostratus clouds produce steady precipitation. If the warm air is unstable, cumuliform clouds may also appear.

A slice through a warm front revealing its vertical structure is shown in Figure 9.17. The gentle slope generates stratus-type clouds, starting with cirrus, cirrostratus, altostratus, nimbostratus, and stratus nearest the front. The cirrus that first heralds the approach of the front can be 1000 kilometers (600 miles) ahead of the surface front. Precipitation is steady and moderate, and can last for several days, depending on how fast the front is moving. Warm fronts typically move at about half the speed of cold fronts. In mountainous regions such as the Appalachian Mountains, the warm front may stall because the warm air cannot displace cold air entrenched in the valleys. The slow movement coupled with the large-scale vertical lifting explains why precipitation can persist for several days.

Hazardous weather can accompany warm fronts. If the air near the ground is below freezing, sleet or freezing rain can occur (see Chapter 4). In addition, as rain falls from the warm air into the cold air near the surface, it evaporates and increases the relative humidity of the air near the surface. If the warm air is significantly warmer than the cold air below, the drops will evaporate quickly and form a fog referred to as a *frontal fog.*

■ Stationary Fronts

Stationary fronts occur when two air masses collide, but move little or not at all at the surface. Although the front appears stationary at the surface, the air above can be moving, causing overrunning. This is why the weather conditions along a stationary front are sometimes similar, though milder, to a warm front with a stable, warm air mass.

The warm air overrunning the colder air mass causes the clouds and precipitation along a stationary front. This overrunning can result in extended periods of cloudiness and light precipitation on the cold side of the stationary front. As with all fronts, the amount of rain depends on the amount of moisture present and the stability of the atmosphere.

Just because a stationary front is not moving, don't assume that it is a weakling among fronts. A stationary front may represent a hard-fought stalemate between clashing air masses with sharply contrasting temperatures. By the thermal wind relationship (Chapter 6), this means that mid-tropospheric winds above a stationary front can be strong. This is an important factor in the development of thunderstorm-related windstorms known as "derechos," which are covered in Chapter 12.

◼ Occluded Fronts

The Norwegian meteorologists who began the study of fronts hypothesized that fast-moving cold fronts could "catch up" to slower warm fronts to form occluded fronts, or occlusions. The occluded front is represented on weather maps by a purple line with alternating triangles and half-circles. The direction of movement is indicated by the direction the triangles and half-circles point.

While warm and cold fronts separate cold and warm air masses, an occluded front marks the surface boundary between two polar air masses. The warm air mass is aloft, lifted by its interaction with the polar air masses. There are two basic types of occluded front: the cold-type occlusion and the warm-type occlusion. In the cold-type occlusion, the cold cP air behind the cold front digs underneath the warm and the cool air masses ahead of it. In the warm-type occlusion, the cool mP air behind the cold front overruns the colder cP air, as does the warm air aloft (Figure 9.18). The weather ahead of the occluded front is similar to that ahead of a warm front. The weather behind the occluded front is similar to that of a cold front.

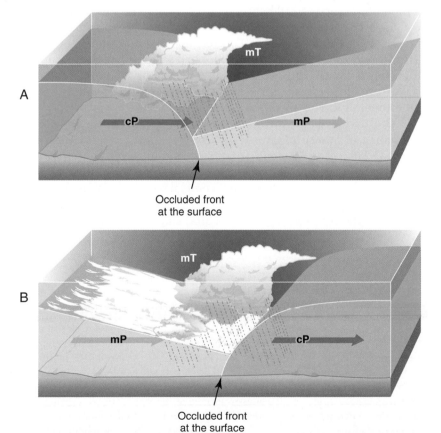

Occluded front
at the surface

Occluded front
at the surface

◼ **Figure 9.18**

The air mass behind the cold front can be colder or warmer than the air mass ahead of the warm front. As a result, there are two idealized types: **A,** *Cold-type occluded front.* **B,** *Warm-type occluded front.*

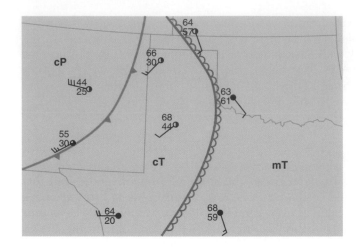

 Figure 9.19

Surface weather map showing a dryline (brown line with open semicircles) over the Texas Panhandle and a cold front over New Mexico. Notice the difference in dew point and wind direction in the three different air masses. (Source: From Vasquez, T., Weather Forecasting Handbook, Weather Graphic Technologies, Garland, Texas, 2000.)

Although conditions along an occluded front are similar to a combination of warm and cold fronts, recent research has substantially modified this traditional view of occluded fronts. Cold fronts generally do not "catch up" to warm fronts after all, although the weather associated with them may resemble that shown in Figure 9.18. Instead, occlusion is part of the maturing process of the extratropical cyclone, which is examined in detail in the next chapter.

Drylines

In late springtime, dry cT air from the plateaus of Mexico is drawn northeastward into the south-central United States by the counterclockwise circulation of extratropical cyclones in the lee of the Rocky Mountains (Chapter 10). The difference in temperature can be small between the cT air and the southerly winds bringing moist mT air over eastern Texas. However, the clash in dew points between the two air masses is breathtaking: 5° C (9° F) or more over a distance of a kilometer or two. This frontal zone defined by moisture and wind rather than temperature contrasts is known as the **dryline.** Figure 9.19 depicts a dryline on a weather map. It is analyzed with clear semicircles in a continuous line that point in the direction of the movement of the dryline.

The dryline helps provide a focus for thunderstorm activity over Texas and Oklahoma in late spring, to the delight of storm chasers. Air converges and lifts about 15 kilometers (10 miles) east of the dryline, causing thunderstorms just like a cold front. In a region that averages only about 25 centimeters (10 inches) of rain each year, these thunderstorms are a crucial part of the annual water supply. Therefore, the dry parching wind behind the dryline actually plays a key role in the agriculture of the semiarid regions of central and west Texas.

PUTTING IT ALL TOGETHER

SUMMARY

An air mass is a large body of air with similar temperature and moisture properties. The movement of air masses generates fronts and causes changing weather conditions. These changes are sometimes welcomed while at other times they are dreaded because of the weather hazards they bring.

Air masses adopt the characteristics of the source regions in which they form. Cold air masses are referred to as polar air masses (P) because they typically form in the polar regions where the surface is cold. Warm air masses are of subtropical or tropical origin; both are referred to as tropical air masses (T). Air masses that form over water are referred to as maritime (m) while those generated over continents are referred to as continental (c). Maritime air masses are usually cooler and moister than continental air masses formed at the same latitude. Mixing and matching these regional categories leads to the four basic air mass types: cP, cT, mP, and mT. Some classification schemes also use the letter A to denote bitter cold Arctic air masses.

An air mass eventually moves and exchanges heat and moisture with the ground it migrates over in a process known as air mass modification. When a cold air mass moves over a warm surface, lower layers of the troposphere warm, increasing the instability of the air mass. This favors rising motion, which increases the possibility of condensation and precipitation. Conversely, when the warmer air mass moves over a cold surface, it is cooled, increasing the stability of the air mass and opposing the formation of clouds and precipitation.

Fronts form when air masses collide. The colder air mass is pushed under the warmer air mass. When a cold air mass replaces a warm air mass, the boundary between the two air masses is called a cold front. Cold fronts are often associated with a narrow band of clouds and intense precipitation. A warm front occurs when the warm air mass replaces a cooler air mass. Warm fronts herald their approach with a large deck of steadily lowering and thickening clouds. Moderate precipitation occurs as a warm front nears. A stationary front occurs when neither air mass is advancing. Occluded fronts are formed when a cold front catches up with and overtakes a warm front, although this explanation is oversimplified. Drylines are moisture fronts that help trigger thunderstorms over the southwest United States.

Several general conclusions can be drawn from our study of fronts:

1. Fronts form at the boundaries between air masses of different temperatures and moisture levels.
2. Warmer air always slopes upward over colder air.
3. Clouds and precipitation form as a warm air mass rises over more dense colder air.
4. The front always slopes upward over the cold air.
5. Pressure drops as a front approaches.
6. In the Northern Hemisphere, wind direction near the ground shifts clockwise as the front passes.

 # KEY TERMS

You should understand all of the following terms. Use the glossary and this chapter to improve your understanding of these terms.

Air mass	Maritime polar (mP)
Air mass modification	Maritime tropical (mT)
Cold front	Occluded front
Continental polar (cP)	Overrunning
Continental tropical (cT)	Static stability
Dryline	Stationary front
Front	Warm front

 # REVIEW QUESTIONS

1. What kind of surface pressure pattern is associated with air masses?
2. Why is mountainous terrain a poor source region for air masses?
3. Explain the requirements for a geographic region to serve as a source region for an air mass.

4. Describe the characteristics of each of the four main types of air mass.
5. Use the text's Web site to try your hand at identifying different types of air masses on an actual weather map. What weather variables distinguish one air mass from another?
6. Describe the type of clouds and precipitation associated with an approaching warm front.
7. At the beginning of the chapter, we made an analogy between accents and air masses. Explain why there is no "O'Hare Airport" accent despite the fact that millions of people go through this Chicago airport every year. Then extend your reasoning to explain why there are no midlatitude air mass source regions even though air travels through the middle latitudes.
8. Discuss the differences and similarities between a warm front and a cold front.
9. Explain why a cold cP winter air mass would make it difficult for firefighters to do their job.
10. Would freezing rain be more likely with the passage of a warm front or a cold front? Why?
11. Before the advent of satellites and advanced systems for tracking fronts, weather predictions were often made by observing changes in wind direction, pressure, and cloud types. These predictions drew on knowledge accumulated over years of weather observations. The predictive nature of these observations is sometimes revealed in weather lore. The following weather lore predicts precipitation:

    Ring around the Moon or Sun,
    Rain before the day is done.

    Is there a physical basis for this statement?
12. Use the text's Web site or the "Weather Analysis/Find the Front" section of the *Blue Skies* CD-ROM to locate a cold front and a warm front on an actual weather map. What weather variables do you focus on when trying to identify the location of a front?

 # WEB ACTIVITIES

Choose Chapter 9 on the textbook's Web site:
http://info.brookscole.com/ackerman
and select from the following resources:
- Interactive Modules that illustrate and extend your understanding of key topics in this chapter
- Tutorial Quizzes to test your mastery of terms and concepts

 For additional readings, go to the InfoTrac College Edition, your online library, at:
http://www.infotrac-college.com

Extratropical Cyclones and Anticyclones

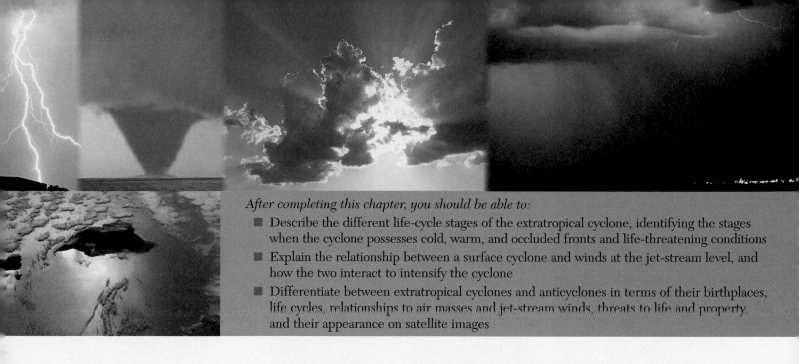

After completing this chapter, you should be able to:

■ Describe the different life-cycle stages of the extratropical cyclone, identifying the stages when the cyclone possesses cold, warm, and occluded fronts and life-threatening conditions

■ Explain the relationship between a surface cyclone and winds at the jet-stream level, and how the two interact to intensify the cyclone

■ Differentiate between extratropical cyclones and anticyclones in terms of their birthplaces, life cycles, relationships to air masses and jet-stream winds, threats to life and property, and their appearance on satellite images

Introduction

What do you see in the diagram to the right: a vase, or two faces? This classic psychology experiment exploits our amazing ability to recognize visual patterns. Meteorologists look for patterns in weather observations. Instead of pondering, "Is that a nose, or the narrow part of a vase?" the meteorologist asks: "Is this a line of rain showers? Could that be a wind shift? Should I be focusing on the areas of warm and cold air or the regions in between?" Finding the "face" in the weather is a lot more confusing than in a textbook diagram!

Early 20th-century weather forecasting reflected this confusion. Some meteorologists focused on the location and shape of low-pressure areas; others scrutinized temperature patterns; still others looked at cloud types. No one yet saw the overall "face," only these individual weather features.

American Meteorological Society

In 1918 twenty-year-old meteorologist Jacob "Jack" Bjerknes (right; pronounced BYURK-nizz) discovered the "face in the clouds" of his native Norway: the extratropical cyclone or low-pressure system. His conceptual model of this type of cyclone emphasized the importance of the fronts that extend like a moustache from the center of midlatitude lows on a weather map. Jack and his father Vilhelm's "Bergen School" of meteorology fleshed out this face using their intuition and observations of the sky. They realized that the extratropical cyclone, like a face, has a distinctive three-dimensional shape. Furthermore, like a person the cyclone matures from youth to old age in a recognizable and predictable way. At last, meteorology knew what face to look for! Weather maps and forecasts have never been the same since.

In this chapter we learn how to recognize the face of the extratropical cyclone. To make this face come alive for you, we look at weather data leading up to a tragic day in American history: November 10, 1975. On that day, a cyclone fitting Bjerknes's description ravaged the midwestern United States and contributed to the wreck of the Great Lakes iron ore freighter *Edmund Fitzgerald*. Twenty-nine sailors perished on the *Fitzgerald* without a single "mayday." Over a quarter-century later, this tragedy lives on through TV documentaries, books, and folk songs. We learn here what the face of a classic cyclone looks like from a variety of angles, and why it looks that way.

Some faces aren't memorable, but the people behind them are. This is also the case with the cyclone's counterpart, the anticyclone or high-pressure system. Highs look like bland and boring blobs on the weather map, but they can also harbor potentially tragic weather conditions. This chapter ends with an examination of sultry, murky high in mid-July of 1999 that claimed the life of one of the most recognized faces in America: John F. Kennedy, Jr.

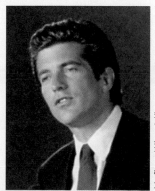

John Fitzgerald Kennedy Library

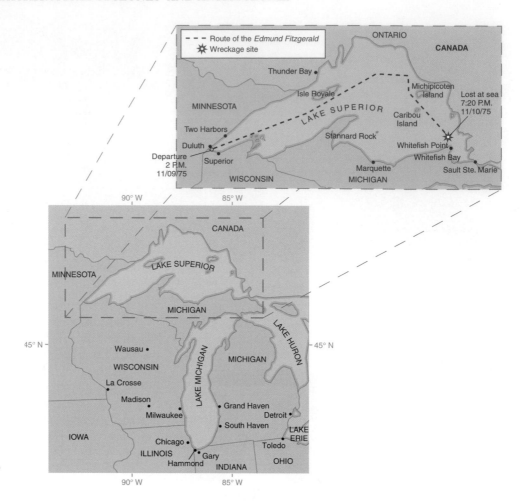

 Figure 10.1

Map of the Great Lakes region, showing the path of the Edmund Fitzgerald's *final voyage and locations referred to in this chapter.*

A Time and Place of Tragedy*

1975. The Vietnam War ends. A hand-held calculator costs more than $500. Unknown actor Sylvester Stallone stars in a low-budget movie called *Rocky*. Better known is NFL running back O.J. Simpson, on his way to a record 23 touchdowns for the Buffalo Bills. Meanwhile, teenagers are starting to buy platform shoes and boogie to the newest pop music fad, "disco." At the end of the year, baby Eldrick Woods is born near Long Beach, California. Future generations of golf fans will instantly recognize him by his nickname—"Tiger."

On the Great Lakes (Figure 10.1) in 1975, large boats deliver iron ore from Minnesota to the steel mills and car factories of Indiana, Michigan, and Ohio. Lake Superior, Earth's broadest lake, serves as a highway for these boats. The pride of the American ore freighters is the 222 meter (729 foot)-long *Edmund Fitzgerald* (Figure 10.2). Her respected captain, Ernest McSorley, has weathered 44 years on the lakes. His crew of 29 includes six sailors under the age of thirty, including 22-year-old Bruce Hudson and 21-year-old Mark Thomas (Figure 10.3).

In November of 1975, "Big Fitz" is completing just its seventeenth full shipping season on Lake Superior. It's a young boat by Great Lakes standards. The end of the shipping season comes when the "gales of November" howl across the Lake and

---

*Details of the *Edmund Fitzgerald's* last voyage used in this chapter are primarily derived from the following books: *The Wreck of the Edmund Fitzgerald* by Frederick Stonehouse (Avery Color Studios, Marquette, Michigan, reprinted 1997), *Gales of November* by Robert J. Hemming (Thunder Bay Press, Holt, Michigan, 1981), and the words of Captain Bernie Cooper in *The Night the Fitz Went Down* by Hugh E. Bishop (Lake Superior, Port Cities, Inc., 2000).

Figure 10.2

The last known picture of the ore freighter Edmund Fitzgerald, *taken on October 26, 1975, as it is unloading at Great Lakes Steel, Detroit River, Michigan.*

winter's icy cold freezes some of the lake surface solid (Figure 10.4, a view from space). One trip too many, during a fierce extratropical cyclone, will sink the *Fitzgerald* and its crew in 160 meters (530 feet) of churning Lake Superior water on November 10, 1975.

A Life Cycle of Growth and Death

In the 1920s, Jack Bjerknes helped discover that cyclones such as the one that would sink the *Fitzgerald* often follow a stepwise evolution of development. The left-hand side of Table 10.1 shows how Bjerknes himself depicted the **Norwegian cyclone model** life cycle. The cyclone arises as a **frontal wave** along a stationary front separating cold, dry cP air from warm, moist mT air. It is called a "wave" because the **warm sector** region between the cold and warm fronts resembles a gradually steepening ocean wave. In adolescence, the **open wave** develops strong cold and warm fronts with obvious wind shifts as the whole system moves to the east or northeast. Precipitation (in green) falls in a broad area ahead of the warm front and in a narrow line in the vicinity of the cold front, just as described in Chapter 9.

At full maturity, the **occluded cyclone** sprouts an occluded front, which Bjerknes conceived of as the result of the cold front outrunning the warm front (see Chapter 9). Typically, the barometric pressure at the center of the cyclone reaches its minimum during this stage, sometimes plummeting to 960 mb in a few intense cyclones—as low as in the eye of a Category 3 hurricane! Because of the strong gradient of pressure near its center, the cyclone's winds are usually strongest during this stage (Chapter 6). The accompanying satellite images in Table 10.1 reveal how the cloudiness associated with the fronts progressively wraps poleward and around the back side of the

Figure 10.3

Edmund Fitzgerald *sailors Bruce Hudson (left) and Mark Thomas on the main deck of the* Fitz.

Michipicoten
Island

Whitefish
Point

Whitefish
Bay

National Geographic Society

■ **Figure 10.4**

Looking down the length of Lake Superior from the Space Shuttle Atlantis *on March 27, 1992. In the foreground, ice (white areas) covers most of the lake between Michipicoten Island and Whitefish Point, and all of Whitefish Bay is ice covered. On November 10, 1975, these portions of the lake were ice free but were churned into huge waves by a midlatitude cyclone.*

Blue Skies

**Weather Forecasting/
Cyclogenesis**

cyclone. In the final stage, the **cut-off cyclone** (see Chapter 7) slowly dies a frontless death, as clouds and precipitation around the low's center dissipate.

The life cycle shown in Table 10.1 is an idealized conceptual model. Some real-life cyclones do not follow this lifecycle in every detail. For example, some cyclones occlude soon after birth, and many others never reach the cut-off stage. The cyclone that hits the *Edmund Fitzgerald* is an actual storm brought about by a unique combination of circumstances. Nevertheless, the conceptual model provides a remarkably accurate guide to the life cycle of the *Fitzgerald* cyclone from its birth on November 8, 1975 until its death several days later, as we see below.

■ Day One: Birth of an Extratropical Cyclone

Saturday, November 8, 1975, is a gorgeous day across much of the United States. Calm and almost summer-like conditions predominate. High temperatures above 26° C (80° F) bathe the South and Southwest, with seventies as far north as Indianapolis and Burlington, Vermont. The *Edmund Fitzgerald* glides on smooth Lake Superior waters toward its next load of iron ore at the Duluth (Minnesota)/Superior (Wisconsin) harbor.

Weather maps and satellite pictures (Figure 10.5) show some cloudiness associated with weak fronts extending from Iowa to Colorado and from Ontario, Canada, to Kentucky. Very little precipitation is falling, however. The fronts will soon stall

TABLE 10.1 The Life Cycle of the Extratropical Cyclone, based on the Bergen School model.

Stage	Weather Map Depiction of Norwegian Cyclone Model	Typical Satellite Image of Life-Cycle Stage	Typical Sea-Level Pressure at Cyclone Center	Corresponding Dates of Edmund Fitzgerald Cyclone
Birth (frontal wave)			1000–1010 mb	November 8, 1975
Young adult (open wave)			990–1000 mb	November 9, 1975
Mature (occluded cyclone)			960–990 mb	November 10–11, 1975
Death (cut-off cyclone)			Slowly rising from 960–990 mb up to 1010 mb	November 11–15, 1975

CIMSS, University of Wisconsin–Madison

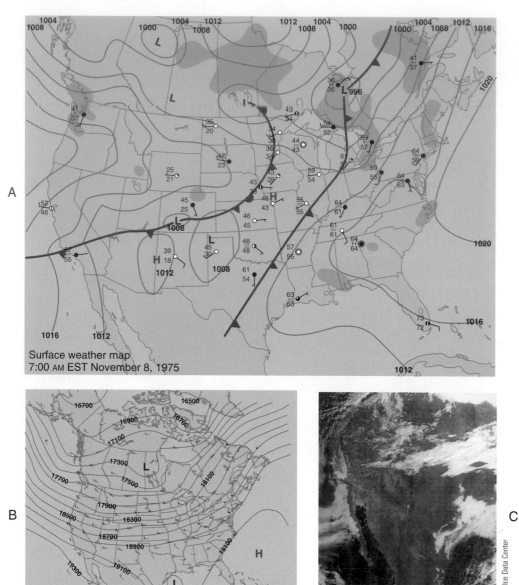

■ **Figure 10.5**

*Surface weather (**A**) and 500-mb conditions (**B**) for 7:00 AM (EST), Saturday, November 8, 1975. The heights in **B** are in feet, but the rules relating wind to height patterns are unaffected. The Defense Meteorological Satellite Program visible satellite picture from 10:30 AM (EST) is at the lower right (**C**). The satellite picture is centered over extreme western Texas; clouds are visible along the west coast of Baja California at the left of the image.*

and will combine into a stationary front lying southwest-to-northeast from New Mexico to New England. Along this stalling front a very weak (1006 mb) low is forming just northwest of Amarillo, Texas, in the northern Texas Panhandle. The low is so weak, and the air is so dry (note the low dew point of 30° F [–1° C] at Amarillo), that there is not a cloud in the sky associated with it (Figure 10.5, *C*).

The presence of a low just east, or "downstream," of the Rocky Mountains is a natural consequence of upper-level winds blowing across high mountain ranges. Figure 10.5, *B*, shows the observed winds and altitudes (in feet) at the 500-mb level, about midway between the ground and the tropopause. The strong west-to-east winds over the Rocky Mountains are occurring where the solid lines of equal height are closest together, just as we learned in Chapter 6.

During Saturday the 8th, the counterclockwise circulation around this weak low drags colder, drier air southward through the Rockies and draws warmer, moister air northward from the Gulf of Mexico. As a result, temperature and moisture gradients near the low become stronger throughout the day. The occluded front will turn into a "leading" warm front just northeast of the low and a "trailing" cold front that will extend to the southwest of the low.

At the same time, winds above the low intensify for two reasons. First, strong upper-troposphere jet-stream winds are moving into this region. Second, as we learned in Chapter 7, strong winds are usually located above strong surface temperature gradients, and the region around the low is becoming an area of strong temperature gradients. As we'll see shortly, this upper-level wind pattern helps the low grow, which in turn helps the fronts intensify, which intensifies the upper-level winds even more and makes the low grow even faster . . . it's a cycle of growth!

Meteorologists give this cycle of cyclone growth a special name: **cyclogenesis.** Why cyclogenesis happens at one place and time, but not another, is almost as complicated as explaining how human babies are born. Key ingredients for cyclogenesis include *surface temperature gradients, a strong jet stream,* and *the presence of mountains or other surface boundaries* (for example, a coastline near a warm ocean current). Wherever winds blow across temperature gradients, warm air can glide upward and cold air can dive and sink, leading to clouds and precipitation; we saw this with both cold and warm fronts in Chapter 9. This tilted pattern of rising and sinking air liberates energy for a cyclone and is called **baroclinic instability.**

The Norwegian cyclone model emphasized the importance of fronts and mountains in cyclogenesis. More recently, meteorologists who unlike the Bjerkneses, have access to upper-level data from radiosondes have identified short-waves (Chapter 7) in the jet stream as a main triggering mechanism. On November 8, 1975, all three of these ingredients—fronts, a strong jet stream, and mountains—give birth to a memorable extratropical cyclone, a low-pressure system structured completely differently than a hurricane but with the same ability to cause harm.

◼ Typical Extratropical Cyclone Paths

For residents of the Great Lakes region, a strengthening cyclone in the Texas Panhandle in November is cause for concern—even though the two locations are a thousand miles apart! This is because extratropical cyclones grow and move quickly away from their places of birth, often to the east or northeast.

Figure 10.6 shows the typical regions of cyclogenesis (shaded) and typical cyclone paths (arrows) for storms that hit the state of Wisconsin in each season of the year. Notice that the birthplaces of cyclones vary from season to season, moving north in the summer. Why? Because temperature gradients and jet streams, key ingredients for cyclogenesis, move north with the Sun during the summer.

Figure 10.6 indicates that during the fall and winter months the region from southern Colorado to the Oklahoma and Texas Panhandles is a prime breeding ground for Wisconsin-bound cyclones. In fact, this path is so common that these cyclones have a generic name attached to them: **Panhandle Hooks.** The "hook" describes the curved path that these cyclones often take, first bending to the southeast and then curving northeast toward the Great Lakes. (Box 10.1 explains this bending and curving.)

Other parts of the United States are hit by extratropical cyclones that form in regions far away from the Rockies. For example, the Northeast is pummeled by winter cyclones—the **nor'easters** discussed in Chapter 9—that develop along the East Coast over the Gulf Stream near Cape Hatteras, North Carolina. Especially during El Niño years, the West Coast is drenched by extratropical cyclones riding the **Pineapple Express** jet stream blowing northeast from Hawaii. The Pacific Northwest is often soaked and windblown by cyclones that spin themselves out in the Gulf of Alaska. And, as seen in Figure 10.6, the Great Lakes region is also visited by cyclones born in western Canada that move southeastward. These storms are called **Alberta Clippers** due to their birthplace and their fast forward speed.

Life Cycle of an Extratropical Cyclone

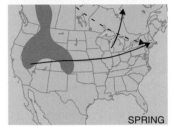

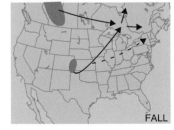

◼ **Figure 10.6**

Typical regions of cyclogenesis (shaded) and paths of cyclones (arrows; likeliest paths in solid lines, other possible paths in dashed lines) that affect the state of Wisconsin on a seasonal basis. (Source: Courtesy Pam Naber Knox, former Wisconsin State Climatologist.)

BOX 10.1

Making Cyclones and Waves

Extratropical cyclones, mountain ranges, and upper-level wave patterns in the wind—what's the connection? It's simple if you know a little about the atmosphere and ice skating.

We learned in Chapter 7 that the closer a skater's body is "tucked in" to the axis of rotation, the faster he or she spins. This is why skaters bring their arms and legs in tight against their bodies during fast spins.

What if we could alter the shape, but not the mass, of our skater? Imagine stretching and squashing the skater like a lump of clay. Stretching the skater (see right) has the same effect as pulling the arms and legs in, and so the skater spins faster.

The same principle, with a twist, works for a spinning cyclone. The tropopause acts as a "ceiling" on cyclones and most other weather. The

Conservation of angular momentum and skaters.

tropopause is, to a first approximation, at about the same altitude throughout the middle latitudes. So a cyclone can be stretched or squashed depending on the altitude of the surface that is beneath it. High mountains squash cyclones; lower oceans and plains help stretch them.

What does this imply about the spin of our cyclones? Just like our skater, a squashed cyclone spins more slowly and a stretched cyclone spins more rapidly. The rate of spin increases as the height of the cyclone increases; spin decreases as cyclone height decreases. And so a meteorological form of the Conservation of Angular Momentum is this: *spin divided by the cyclone height must be a constant.* (This works for anticyclones too.)

Now, the twist. A skater spins for a minute or so. This is far too short a time for the Earth's spin to affect the skater via the Coriolis force. But a cyclone spins for days and is strongly affected by the Coriolis force. Because of this, the cyclone, but not the skater, feels the effect of the Earth's spin. So the cyclone has a double dose of spinning. Therefore, the meteorological form of Conservation of Angular Momentum is:

$$\text{total spin} \div \text{cyclone height} = \text{constant.}$$

In meteorological terminology, spin around a vertical axis is called "vertical vorticity" or just "vorticity" for short. The spin of winds in a weather system, such as a cyclone, is called "relative vorticity" because it's measured relative to the ground. But the air and ground are all spinning due to the Earth's rotation too. The Earth's spin is called "planetary vorticity." Vorticity is defined as positive if the direction of rotation is in the same direction as the Earth's spin. Looking down from above the North Pole, the Earth appears to rotate counterclockwise, and so do cyclones; therefore, the relative vorticity of a cyclone in the Northern Hemisphere is positive.

With these definitions, we can put everything together and write a special version of the Conservation of Angular Momentum used in meteorology, known as the Conservation of Potential Vorticity:

$$\underbrace{(\text{planetary vorticity} + \text{relative vorticity}) \div \text{cyclone height}}_{\text{Potential Vorticity}} = \text{constant}$$

Conservation of potential vorticity lets us link up cyclones, mountains, and upper-level waves in the wind. Imagine a cyclone moving more or less eastward from the Pacific Coast to the Great Plains as depicted in the illustration on the facing page.

Over the Pacific, the cyclone is strong. Lows are associated with upper-level troughs. The trough's winds steer the cyclone southeast and then northeast in a cyclonically curved path.

As the cyclone moves over the high Rockies, however, it gets squashed. Conservation of potential vorticity says that in

---

The Panhandle Hook of November 1975 will also race toward the Great Lakes at a high rate of speed, gathering more energy with every passing hour. The next two days will make history, and tragedy, on Lake Superior.

Day Two: With the *Fitz*

Early on Sunday morning November 9th, 1975, workers at the Duluth/Superior docks begin the task of loading the *Edmund Fitzgerald*. On this trip, the *Fitz* will carry enough iron ore to make 7500 automobiles! The iron ore pellets slide down

BOX 10.1
(continued)

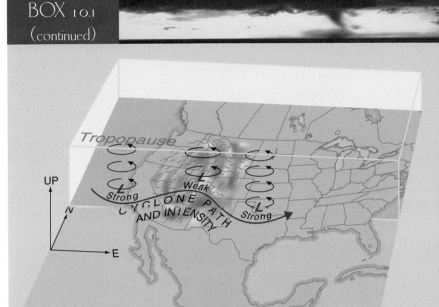

Cyclone strength and path changes due to conservation of potential vorticity over the Rocky Mountains.

this circumstance the total spin of the cyclone must decrease too. The once-strong Pacific low appears to decrease in intensity as it moves over the Rockies and its path curves anticyclonically over the mountains, heading southeast as the weather system crosses the Rockies.

Then suddenly the bottom drops out of the cyclone as it crosses the high Rockies and moves over the lower Great Plains. Over the Plains, the cyclone is stretched. By conservation of potential vorticity, this means the sum of the vorticities needs to increase too. The cyclone does this in two ways: by "spinning up" (intensifying rapidly), which increases its relative vorticity; and by heading toward the northeast, which increases its planetary vorticity. The northeast path is consistent with the steering winds in the upper-level trough that develops in tandem with the strengthening cyclone.

Because the cyclone seemed to disappear over the Rockies, its reappearance and strengthening just east of the Rockies seem to come out of nowhere. However, they are natural results of potential vorticity conservation, as is the characteristic hook-shaped path of the cyclone (southeast over the Front Range of the Rockies, northeast over the Plains).

Obviously, the cyclone is closely related to the upper-level winds. The wavy path of the cyclone is mirrored in the waviness of the jet-stream winds that guide the storm. These large-scale waves of wind are the Rossby waves we studied in Chapter 7. Rossby waves move horizontally, like ocean waves. We noted this in Chapter 7, but now we can be more precise about how quickly they move. Using the concept of potential vorticity conservation, famed 20th century meteorologist C.-G. Rossby found that these waves' west-to-east (W–E) speed is dependent on how fast the jet stream is and how big the wave is (size matters):

Rossby wave W–E speed =
500-mb W–E wind speed –
Rossby wave W–E wavelength × a constant

Since the two terms on the right-hand side of the equation are subtracted from each other, this means that the speed of a Rossby wave is a push–pull of wind speed versus size. Short-waves (the size of the Great Lakes region) move eastward almost as fast as the 500-mb wind because the wind speed factor dominates. However, long-waves (the size of a continent) travel very slowly eastward, stop, or even move backwards toward the west. This is because the wavelength term is almost as big, as big, or bigger, than the wind speed term.

As you read this chapter, compare the forward speed of the *Fitzgerald* cyclone to Rossby's simple formula: does the storm move fastest when it is young and small? Does it slow down as it matures and grows bigger? Try it and see!

huge chutes like marbles into the 21 hatches in the middle of the boat. The crew then anchors each of the 6350-kilogram (7-ton) hatch covers using 68 special clamps. Shortly before 2:00 PM, the *Fitzgerald* departs into the open waters of Lake Superior.

Less than 2 hours later, the *Arthur M. Anderson,* sister ship to the *Fitzgerald,* leaves Two Harbors, Minnesota, with a load of iron ore for the Gary, Indiana, steel mills near Chicago. The *Anderson's* captain, "Bernie" Cooper, recalled years later that this Sunday "was one of the special days on Lake Superior—just ripples on the water, sunny and warm for November. As we departed we could see the *Edmund Fitzgerald. . . .*" The two ships will sail together for the rest of the *Fitzgerald's* last journey.

http://info.brookscole.com/ackerman

Portrait of the Cyclone as a Young Adult

The baby cyclone of November 8 has matured overnight—literally. The surface weather map for Sunday morning on the 9th (Figure 10.7, *A*) shows a 999-mb low over Wichita, Kansas, a pronounced cold front digging southward into Texas, and a warm front pushing north toward Iowa. Temperature contrasts across the cold front are now 11° C (20° F), and 5.5° C (10° F) across the warm front. The 500-mb map (Figure 10.7, *B*) shows a deepening short-wave trough to the west of the surface low, part of the cycle of growth of the storm. The baby is now a young adult, teeming with energy.

A high-resolution satellite picture taken just three hours later (Figure 10.7, *C*, and enlarged in Figure 10.8) reveals that this cyclone has taken on a classic form at an early age—it's a meteorological Tiger Woods. Seen from space, it has the shape of a **comma cloud,** which is characteristic of mature extratropical cyclones and is quite different from the circular tropical cyclone. The tail of the comma is cloudiness along

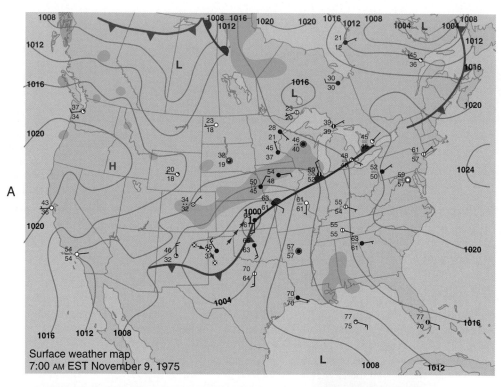

A

Surface weather map
7:00 AM EST November 9, 1975

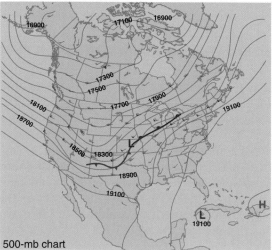

B

500-mb chart

C

National Snow and Ice Data Center

 Figure 10.7

*Surface weather (**A**) and 500-mb conditions (**B**) for 7:00 AM (EST), Sunday, November 9, 1975. The red-and-white boxes in **A** denote the location of the surface low 6, 12, and 18 hours earlier. The surface low and fronts are superimposed in **B** for reference. The Defense Meteorological Satellite Program visible satellite picture at the lower right (**C**) is from 10:12 AM (EST) that morning.*

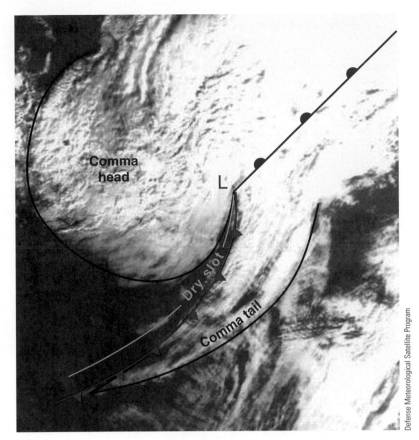

■ **Figure 10.8**

A close-up of Figure 10.7, **C.** *The center of the cyclone, which is underneath the "comma head" clouds, is just northwest of Wichita, Kansas, at this time.*

the trailing cold front, which a tropical cyclone lacks. The comma head is made up of clouds and light precipitation circling counterclockwise around the low's center. The cloudless region between the comma head and tail is the **dry slot,** a feature usually seen only in more mature extratropical cyclones and never in hurricanes—we discuss it in more detail shortly. The lumpy areas in and just to the east of the comma head are taller convective clouds casting shadows on the lower clouds around them (Omaha, Nebraska, just north of the warm front, reports a thunderstorm at 6:00 AM). All in all, this picture tells a simple story: this cyclone is mature beyond its tender age. The weather map provides an even more ominous fact: the cyclone is now racing northeastward toward the Great Lakes at up to 64 km/hr (40 mph).

■ Cyclones and Fronts: On the Ground

Madison, Wisconsin, lies just to the east of the cyclone's path on November 9. The changes in temperature, dew point, winds, pressure, and clouds at Madison on this day illustrate that an extratropical cyclone is a *frontal* cyclone. In other words, much of the exciting weather happens along the fronts connected to the low, just as Jack Bjerknes envisioned it.

Figure 10.9 tells the story as a timeline of weather at Madison. Before dawn on the 9th, Madison reports altostratus clouds and east winds. Temperatures hover around 10° C (50° F) and dew points are slowly rising into the forties. The warm front is approaching. Altostratus (As) clouds give way to stratus (St) clouds as the front nears, and light rain falls on Wisconsin's capital. The pressure falls slowly but steadily. In the afternoon, the warm front passes and winds shift to southeasterly. Despite fog (F) and the lack of sunshine, Madison temperatures soar into the sixties by nightfall. At 9:00 PM, Madison reports a temperature of 18° C (65° F) with a dew point of 14° C (58° F)—unusually balmy November nighttime weather in

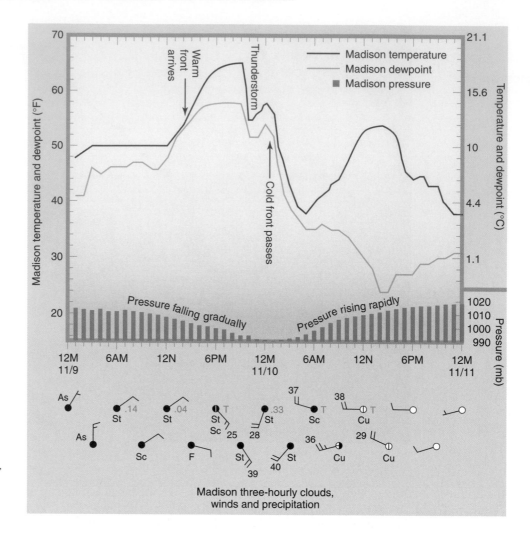

Figure 10.9

Madison, Wisconsin weather during the passage of the Edmund Fitzgerald cyclone. Cloud types, wind gusts (in knots), and precipitation (green, in inches) are shown at bottom, along with cloud cover and wind speed and direction.

Wisconsin. The wind circulation around the strengthening low has advected warm, moist mT air from the Gulf of Mexico all the way to the Great Lakes region.

Because Madison is close to the path of the cyclone, the warm sector is narrow and the cold front arrives soon after the warm front. Between 9:00 and 11:00 PM on the 9th, a thunderstorm ahead of the cold front hits the city, dropping about 0.85 cm (0.33 inch) of rain in a short time. The temperature drops 5.5° C (10° F) due to evaporational cooling, but the warm Gulf air pushes the thermometer back up to 14° C (58° F) by midnight. Then the cold front rushes through. The winds veer to a more westerly direction and strengthen, peaking at 46 knots (53 mph). The temperature and dew point drop into the thirties by sunrise on the 10th. Stratus clouds break up and are replaced by a few puffy cumulus clouds in the drier air, and the pressure rises rapidly. By nightfall on the 10th, the pressure is high, winds are slackening, the clouds have dissipated, and the cyclone is past—at Madison.

Cyclones and Fronts: In the Sky

Cyclones and fronts are three-dimensional. Jack Bjerknes and the Bergen School used observations from the surface, including cloud types, to indirectly determine the upper-level patterns. We can do the same with the *Fitzgerald* cyclone.

One technique for visualizing a cyclone is to take a vertical slice through it, as if it were a cherry pie. Let's make a cut across the center of the United States so we can slice through both the fronts and the center of the *Fitzgerald* cyclone itself. Figure 10.10, *A*, depicts this cut, and in Figure 10.10, *B*, the clouds and weather along

Airport city codes

SSM (Sault Ste. Marie, MI) COU (Columbia, MO)
DLH (Duluth, MN) MCI (Kansas City, MO)
MKE (Milwaukee, WI) ICT (Wichita, KS)
MDW (Chicago, IL) TUL (Tulsa, OK)
DSM (Des Moines, IA) AMA (Amarillo, TX)

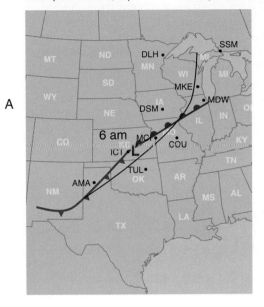
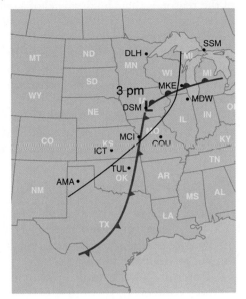

A

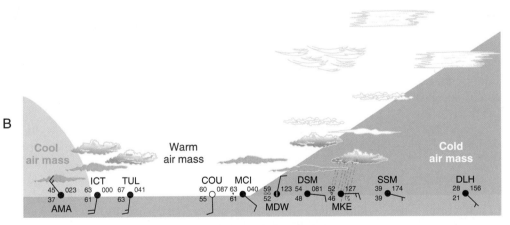

B

6AM CST November 9, 1975

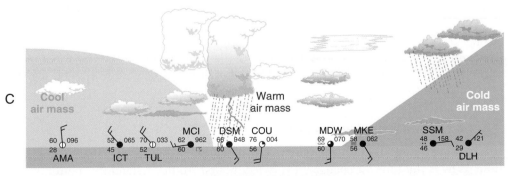

C

3PM CST November 9, 1975

Figure 10.10

A, *The approximate positions of the* Fitzgerald *cyclone and its fronts at 6:00 AM and 3:00 PM Central Standard Time, November 9, 1975. The line from Texas to the Great Lakes is used in constructing* **B** *and* **C,** *cross-sections of weather observations from the surface through the cyclone at 6:00 AM and 3:00 PM, respectively.*

this line from Amarillo, Texas, to the Great Lakes is shown graphically for 6:00 AM Central time on the 9th. Notice how well the real-life data compares to our idealized discussion of fronts in the last chapter. The transitions in clouds, precipitation, temperatures, moisture, and winds across the fronts are almost exactly what you would expect from reading Chapter 9. The cold front is less textbook-case, but part of the reason is that clouds ahead of a cold front often need a boost from the Sun in order to grow. At 6:00 AM in November the eastern sky is just beginning to glow over the Great Plains, so only stratocumulus clouds hover over Tulsa, Oklahoma.

By 3:00 PM on the 9th (Figure 10.10, C) the strengthening low and the Sun have combined to create severe weather. The Tulsa weather observer can see towering cumulonimbus clouds as the cold front pushes past; so can the observer at Des Moines, Iowa, in the immediate vicinity of the center of the cyclone. Severe thunderstorms are breaking out all over Iowa; winds shred the inflatable stadium dome at the University of Northern Iowa. Meanwhile, the lowering clouds and rising winds on Lake Superior concern the captain of the *Edmund Fitzgerald*.

◼ Back with the *Fitz:* A Fateful Course Correction

Captains McSorley and Cooper pay careful attention to the National Weather Service (NWS) forecasts on the 9th. Cooper recalls, "When the meteorologists become nervous we then start our own weather plots. It did show a low pressure to the south . . . normal November low pressure."

Unfortunately, the rapid growth of the cyclone tests the ability of the NWS and the boat captains to keep up with it. At first the severe weather experts in Kansas City call it a "typical November storm" headed south of Lake Superior. Because of the counterclockwise circulation around a low, this means strong east and northeast winds and waves of 1 to 2.5 meters (3 to 8 feet) on Lake Superior on Monday. But by mid-afternoon on Sunday the forecast is revised. Gale warnings are issued for winds up to 38 knots (44 mph)—not too unusual, but no longer typical.

The deteriorating weather forecasts prompt Cooper and McSorley to discuss over the radio the first of several weather-related course changes. The "fall north route" the two captains eventually choose to take (see Figure 10.1) will add many miles to the usual shipping path. However, by staying close to land and minimizing fetch (see Box 6.2), this route protects boats from high seas caused by north and northeast winds.

By 10:39 PM the cyclone compels the NWS to revise its forecast again. Now, the winds are supposed to howl at more than 40 knots (46 mph) across Lake Superior on Monday. More significantly, these winds will be coming from the west and southwest, generating waves of 2.5 to 5 meters (8 to 15 feet). The forecast of a storm near the Great Lakes has been consistent, but in just twelve hours the wave heights forecast for Monday have doubled and the wind direction has turned 180°.

At midnight, the *Fitzgerald* is 37 kilometers (23 miles) south of Isle Royale and reports 52-knot (60 mph) north–northeast winds, heavy rain, and 3.5-meter (10-foot) waves. Shortly the NWS will upgrade the gale warning to a storm warning. The cyclone is strengthening rapidly and aiming northward across Lake Superior. The strongest winds over the lake will soon be coming out of the west, not the northeast. This and later course corrections, based on the best available weather information at the time, will leave the freighter exposed to hurricane-force west winds and high seas on the 10th.

◼ Cyclones and Jet Streams

Why is the cyclone strengthening? Why is it moving so quickly and changing direction? Why is the forecast changing every few hours? To understand this better, we need to look closely at what is happening *above* the surface low.

During the 9th, the upper-troposphere wind pattern has been affected by two factors: the jet-stream winds moving east across the Rockies, and the increasing temperature gradients along the surface fronts. Cold air moving southward near the surface helps the upper-level trough "dig" or extend southward. Similarly, warm air moving northward leads to a "building," upper-level ridge (i.e., extending north and northwestward). These upper-level changes, in turn, affect three characteristics of the surface low: its speed, direction, and intensity.

Surface lows tend to move at about half the speed of the 500-mb winds above them. During the 9th, 85-knot (98 mph) winds over New Mexico increased and moved east toward the Great Lakes. This helps explain the rapid movement of this storm.

Surface lows and fronts also generally move in the same direction as the "steering winds" at the 500-millibar level. During the 9th (see Figure 10.7) the combination of the digging trough and the building ridge bends the wind pattern, causing the direction of these steering winds to shift from northeastward to a more northerly direction, across Lake Superior. In other words, the storm's steering wheel keeps turning on the 9th, and this is why the storm's future path is hard to forecast precisely.

A better understanding of cyclone–jet-stream relationship requires us to delve more deeply into concepts from Chapter 6 and the Norwegian cyclone model. Most important is the relationship between the surface low's strength and the jet-stream winds. A simple law of meteorology explains the connection: *Surface pressure drops when there is divergence of the wind in the column of air above the low.* Ordinarily, winds converge into a low near the surface, as explained in Chapter 6. So strong upper-level divergence must exist in order that the net effect is divergence above the low. If this happens, then the low will intensify, because there is less air over it to exert pressure on the surface (Figure 10.11).

Upper-level divergence can occur in two different ways. Either the winds can accelerate in a straight line, or else they can spread out rapidly in a variety of directions. The straight-line acceleration is called **speed divergence** and the spreading out is called **diffluence,** as illustrated in Figure 10.11.

The intertwined connections between the surface cyclone and the jet-stream pattern are summarized for the first three stages of the idealized Norwegian model in Figure 10.12. Upper-level divergence exists above and ahead of the low. (Remember from Chapter 6 that for equal spacing of height lines, the winds in a trough are slower than winds in a ridge. Therefore, divergence commonly occurs east of a trough. Figure 10.12 also shows some speed divergence and diffluence, especially for the mature cyclone.) This divergence helps the cyclone deepen and propagate northeastward, because the surface pressure drops along and ahead of the low. Upward motion brought about by this divergence helps in the formation of clouds and precipitation. The upper-level convergence behind the cyclone causes sinking air, which suppresses clouds. This sinking air also drags down warm air from the stratosphere above and behind the cyclone (discussed later in Box 10.3). These interrelationships between surface and jet-stream patterns make it clear that the extratropical cyclone is a complex three-dimensional phenomenon. This explains why radiosonde observations of upper-level winds are so crucial to weather analysis and forecasting in the middle latitudes.

Figure 10.12 is an idealized model, however. Figure 10.13 shows the actual winds measured by radiosondes at 300 mb at 6:00 PM Central time on November 9, 1975. Following the lines of constant height from Nebraska to Minnesota, the winds double in speed going from the trough to the ridge. As a result, on the 9th speed divergence exists over Iowa and Minnesota. Now notice that the winds over Minnesota are southerly, while the winds over Michigan are southwesterly. This spreading-out of the wind is an example of diffluence.

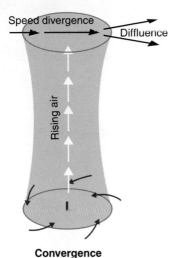

Figure 10.11

The relationship between the two types of divergence (speed divergence and diffluence) at upper levels and the development of a surface cyclone. Length of arrows is proportional to wind speed.

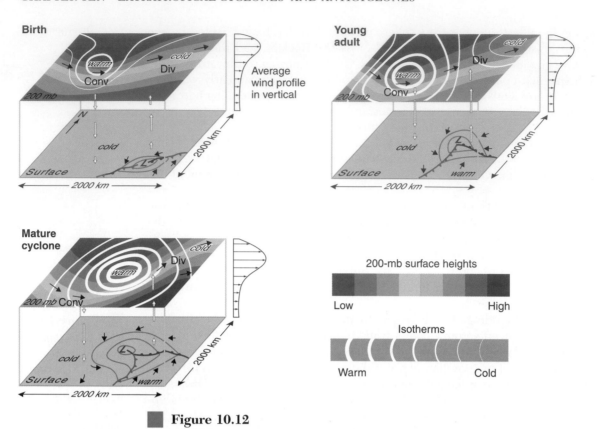

Figure 10.12

The relationship between upper-level (200 mb) and surface conditions during the first three stages of the Norwegian cyclone model. (Source: Adapted from Carlson, N., Mid-Latitude Weather Systems, *American Meteorological Society, 1998, p. 242.)*

The combination of these two types of divergence leads to sudden pressure drops over the upper Midwest. Therefore, this cyclone accelerates to the northeast and deepens rather rapidly, from a central minimum pressure of 999 mb at 6:00 AM (CST) to 993 mb 12 hours later. The lowest pressure of the cyclone is dropping 0.5 mb per hour. This cyclone is on its way to becoming the "Storm of the Year" on the Great Lakes.

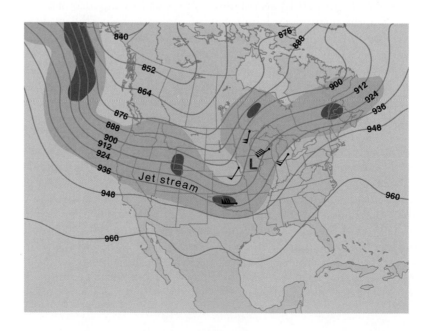

Figure 10.13

Winds and heights (tens of meters) at the 300-mb level for 6:00 PM (CST) on November 9, 1975. The pattern of winds here helps the surface cyclone (the L over Iowa) to strengthen rapidly.

Day Three: The Mature Cyclone

Early on November 10, the *Fitzgerald* sails into the Eastern time zone and heads for an unexpected rendezvous with a dangerous, mature, extratropical cyclone. The 7:00 AM (EST) surface weather map (Figure 10.14) reveals that this cyclone (and its associated fronts) now dominate the eastern half of the United States! Just 48 hours ago, it was a cloudless swirl over Texas.

The cyclone's power and scope of influence indicate that it has reached full maturity. The low itself is centered directly over Marquette, Michigan, in that state's Upper Peninsula. Its lowest pressure has plummeted overnight to 982 mb, a drop of nearly 1 mb per hour. To the west of the low, cold air spiraling in toward the storm across tightly packed isobars is causing blizzard conditions—heavy wet snow blown by 60-knot (69 mph) winds is tearing down power lines and piling up as much as

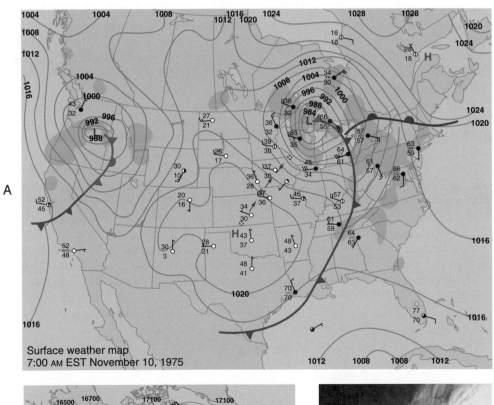

Surface weather map
7:00 AM EST November 10, 1975

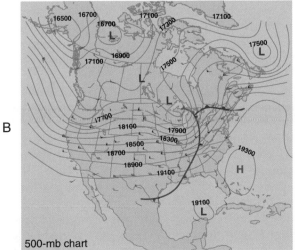

500-mb chart

National Snow and Ice Data Center

C

Figure 10.14

*Surface weather (**A**) and 500-mb conditions (**B**) weather for 7:00 AM (EST) on Monday, November 10, 1975. The Defense Meteorological Satellite Program visible satellite image in the lower right (**C**) is from 9:55 AM that morning and is centered over Kansas.*

36 centimeters (14 inches) of snow in parts of northern Wisconsin. Meanwhile, the cyclone's warm front has crossed over into Canada and its cold front has barreled south to the Gulf of Mexico. From Iowa to Tennessee, residents are combing through debris and tending to casualties after 15 tornadoes spawned ahead of the cold front injured 25 people. In the life cycle of this extratropical cyclone, Saturday's baby and Sunday's adolescent is Monday's ferocious adult.

■ Bittersweet Badge of Adulthood: The Occlusion Process

If you look closely at Monday's weather map in Figure 10.14, you will notice that the cold and warm fronts no longer join like halves of a moustache at the center of the cyclone. Instead, an occluded front joins the low and the cold and warm fronts.

What is an occluded front? Chapter 9 described the occluded front as the Bergen School saw it: one front "catches up" with another to form an occluded front.

Recent research indicates that occlusion is perhaps better understood as a process. During **occlusion,** the surface low gradually retreats from the zone of sharp temperature contrasts and "secludes" itself in the cold air to the north and west of the intersection of the cold and warm fronts. The surface low ultimately ends up directly underneath the upper-level trough—far removed from the tilted "baroclinic" situation of its youth. In an extratropical cyclone over a large body of water, the occlusion process can lend it a few of the characteristics of its tropical cousin, the hurricane (Box 10.2).

The irony is that the occlusion process isolates the mature extratropical cyclone from its fuel source, the fronts with their strong temperature gradients. By analogy, it's like a pop music star who makes it big, surrounds himself with bodyguards and secludes himself in a big mansion, and then loses touch with his fans. In both cases, success is bittersweet. The pop star is cut off from his inspiration, and the hits quit coming. Similarly, the mature cyclone's growth cuts it off from its energy source. The cyclone gradually turns into a frontless cut-off low and slowly dies. The occlusion process is therefore not only a badge of adulthood, but is also the beginning of the end for an extratropical cyclone.

An occluding cyclone is still at the peak of its powers initially, however, as the crew of the *Fitzgerald* learns on the 10th. As the cyclone races toward Lake Superior before dawn on the 10th, captains Cooper and McSorley adjust course again, "hauling up" near the eastern Canadian shore of the lake. There the boats will be less vulnerable to northeast winds because the winds will be blowing over only a short distance or "fetch" of water.

Unfortunately, the cyclone outmaneuvers the ore freighters. The cyclone is now moving northeast at 30 mph; even the powerful "Big Fitz" cannot match this speed. The low and the boats cross paths in Canadian waters in northern Lake Superior.

"The low had reached Lake Superior and was intensifying dramatically," Cooper recalls. "Shortly after noon, we were in the eye of the storm. The Sun was out . . . light winds, no sea." The *Anderson's* barometer, adjusted for sea level, bottoms out at 28.84 inches of mercury or 976.6 mb, as low a pressure as will be recorded for this cyclone.

The good weather cannot last because of the raging contrasts near the heart of an extratropical cyclone. As the low and its occluded front race past, strong winds from the west will soon lash at the *Anderson* and *Fitzgerald.* At 1:40 PM, Cooper and McSorley agree to yet another weather-related course correction, cutting between woody Michipicoten Island and Caribou Island on a southeastward heading toward safe haven in Whitefish Bay and Sault Ste. Marie, Michigan. But it's too late.

By 2:45 PM, the *Anderson* reports snow with northwest winds at a steady 42 knots (48 mph), a 180° reversal of direction and a doubling of speed in less than 2 hours. Instead of being sheltered from northeast winds, they are now fleeing rising winds and waves that are rushing at them from behind across the vast expanse of Lake Superior.

BOX 10.2

Cyclones and Water: Bomb and Bust

Is an extratropical cyclone the same thing as a hurricane? *No.* Extratropical cyclones thrive on ingredients such as fronts and strong jet streams, which we learned in Chapter 8 would kill off a hurricane by cutting off its fuel sources and disrupting its circulation.

Even so, it's true that the same fuel sources that feed the engines of hurricanes—warm surface waters and lots of water vapor—can also stoke the fires of their extratropical cousins. As a result, extratropical cyclones that are over oceans or that have access to lots of warm, moist air often become more intense than "drier" cyclones. This rapid intensification even has a meteorological nickname: "bomb," reserved for midlatitude cyclones whose central pressure drops approximately 1 mb per hour every hour for 24 hours. It's not typical to see the pressure drop a whole millibar in just one hour, let alone keep up that pace for a whole day! Such explosive growth sounds more like a hurricane than an extratropical cyclone, and there are some similarities.

An extreme example is the maritime "polar low" that develops over the seas north of Europe. The water there is cold, but the air is even colder. The water transfers warmth and energy to the air at the surface; this is called *sensible heat.* Latent heat is also released into the air when the seawater evaporates and then later condenses into cloud droplets. These two water-based energy sources give the polar low its energy, just as in a hurricane. In fact, the polar low is called an "arctic hurricane" by some meteorologists because of its hurricane-like appearance in satellite pictures (see figure above, an example over the Barents Sea northeast of Scandinavia).

Tromso Satellite Station

In the middle latitudes, cyclones live and die with fronts and jets. But even here, water plays a key role and makes forecasting very difficult. Midlatitude cyclones born over wa-

ter are especially responsive to water-borne energy. One study of a February 1979 nor'easter that became a "bomb" off the U.S. East Coast revealed that the presence of water helped drop the cyclone's pressure by 32 mb (nearly one full inch of mercury). That's the difference between a weak cyclone and a record-setter! The forecasters were taken completely by surprise, which is called a "bust," short for a "busted" or broken forecast. (See Figure 10.21 and Chapter 13 for more details on this "bust.")

The Great Lakes are also big enough to affect the strength of extratropical cyclones that pass over them. Below is a satellite picture of the September 14, 1996 "Hurricane Huron." Notice how it seems to have an "eye" and spiral rainbands, a little like a hurricane. It's *not* a hurricane, but the presence of warm water underneath the cyclone fueled its fires. This extratropical cyclone stalled over Lakes Michigan and Huron for four days and deepened 19 mb over that time—not a "bomb," but impressive nonetheless.

Did the Great Lakes play a role in the cyclone that helped sink the *Edmund Fitzgerald?* Yes, most likely on the back side of the storm where cold air passed over warmer water. There, just as in a polar low, sensible and latent heat would be released into the air. This energy release helped cause the lake-effect snow squalls that enshrouded the *Fitzgerald* in its final minutes.

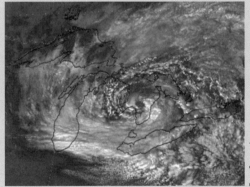

Peter Soussounis

"Hurricane Huron," a midlatitude cyclone influenced by the waters of the Great Lakes.

Hurricane West Wind

Meanwhile, a severe, localized windstorm is developing directly in front of the *Fitzgerald.* Figure 10.15 shows the hourly barometric pressure readings for four weather stations in or near the path of the cyclone. At most of these locations, the pressure rises rapidly after the storm passes. Mariners recognize this as an ominous

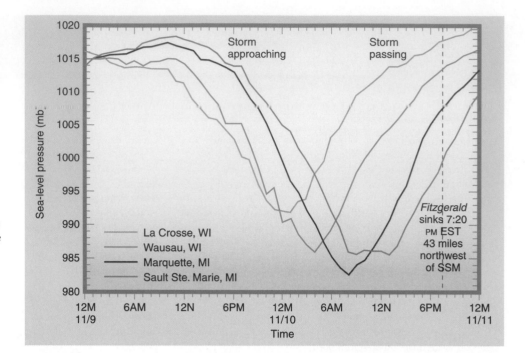

■ **Figure 10.15**

Hourly sea-level pressures at four stations in the general path of the Fitzgerald *cyclone. Notice the pattern of falling and then rapidly rising pressure at each city as the storm approaches and passes. The* Fitzgerald *sank in rising pressure northwest of Sault Ste. Marie, Michigan (SSM).*

sign. According to a saying at least as old as 19th-century British meteorologist Admiral FitzRoy (the captain of *The Beagle* during the voyage on which Charles Darwin conceived of the idea of evolution):

*"Fast rise after low
Foretells stronger blow."*

This folklore forecast is based partly on the force–balance ideas in Chapter 6. Rapid pressure changes behind a low imply that there is a strong horizontal pressure gradient in the area. As we saw in Chapter 6, geostrophic and gradient wind balances both require strong horizontal winds in such cases.

Figure 10.16 proves the wisdom of FitzRoy's rhyme. In the figure, hourly wind observations at Marquette and Sault Ste. Marie, Michigan, tell a tale of increasing, and increasingly gusty, winds after the passage of the cyclone. Marquette is hit first, shortly after noon on the 10th; Figure 10.17 is a rare photograph showing the tempest at Marquette that very afternoon. The pressure rises move east, and Sault Ste. Marie is hammered with gusts of 62 knots (71 mph) after nightfall, knocking out electricity over much of the region.

By the time high winds blow through Marquette and Sault Ste. Marie, the cyclone is well into Canada. These damaging west and northwest winds are far removed from any thunderstorms or tornadoes, nor are they concentrated in an eye wall around the lowest pressure as in a hurricane. These winds stand in the path of the *Fitzgerald,* which is now in a race to safe harbor before the winds and waves sink it.

■ One of the Worst . . .

As daylight dims on the 10th, the *Anderson* and *Fitzgerald* are laboring to make headway in worsening weather. All around the *Anderson,* the winds and seas rise in fury: 43 knots (49 mph) and 3.7- to 4.9-meter (12- to 16-foot) waves at 3:20 PM, 58 knots (67 mph) and 3.7- to 5.5-meter (12- to 18-foot) waves at 4:52 PM. It is profoundly bad timing. At this point in the voyage, treacherous shoals near Caribou Island demand precise navigation as the boats thread the needle between Michipicoten and Caribou (see Figures 10.1 and 10.4).

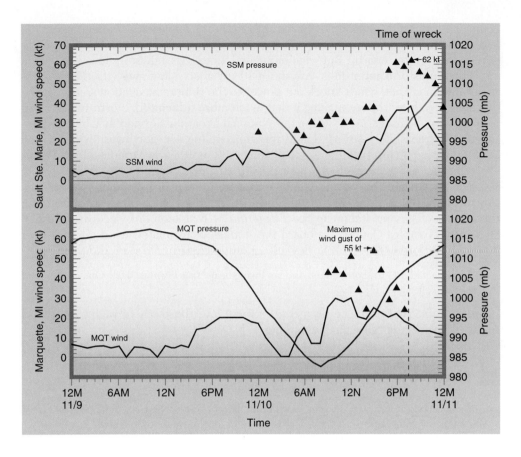

Frederick Stonehouse

■ **Figure 10.16**

Hourly wind speeds (average and gusts) at Sault Ste. Marie and Marquette, Michigan, during the approach and passage of the Fitzgerald *cyclone. Hourly barometric pressure readings are indicated with blue and red lines.*

In the middle of the howling winds, snow, sea, and spray, something goes wrong. Around 3:20 PM, McSorley calls Cooper on the radio:

> "*Anderson,* this is the *Fitzgerald.* I have sustained some topside damage. I have a fence rail laid down, two vents lost or damaged, and a list. I'm checking [slowing] down . . . Will you stay by me 'til I get to Whitefish [Bay]?"
>
> "Charlie on that *Fitzgerald.* Do you have your pumps going?"
>
> "Yes, both of them."

Something has damaged the *Fitzgerald,* inflicting wounds worse than anything that has happened to the boat in seventeen full years on the Great Lakes. It is tilting to one side, or "listing," presumably due to a leak that is letting lake water pour into the boat. The leak is bad enough that even the *Fitzgerald's* large pumps, capable of

■ **Figure 10.17**

The Wreck of the Edmund Fitzgerald author Frederick Stonehouse's photograph of heavy surf at Marquette, Michigan, on November 10, 1975.

removing thousands of gallons of water every minute, apparently do not correct the list for the remainder of its journey. The captain is concerned enough to ask his companion boat to stay nearby. But what happened, exactly, is a mystery forever.

The windstorm intensifies. Around 4:10 PM gusts blow away the *Fitzgerald's* radar antenna. Then winds knock out power to the remote navigation station at the entrance to Whitefish Bay, making it that much more difficult for storm-tossed ships to gain their bearings. At 4:39 PM the National Weather Service in Chicago fine-tunes its forecast for eastern Lake Superior, calling for northwest winds 38 to 52 knots (44 to 60 mph) with gusts to 60 knots (69 mph) early Monday night. Waves are still expected to be 2.4 to 4.9 meters (8 to 16 feet).

But once again the storm exceeds expectations; see Box 10.3 for reasons why this might be so. At 5:00 PM the lighthouse at Stannard Rock north of Marquette, the closest observing station to the *Fitzgerald* at that moment, records a gust of 66 knots (77 mph). Cooper, on board the *Anderson,* estimated wind gusts of over 100 mph. Before 6:00 PM, McSorley tells another ship captain via radio that

> "I have a bad list, lost both radars. And am taking heavy seas over the deck. One of the worst seas I've ever been in."

Blind and ruptured, the *Fitzgerald* steams into the teeth of raging wind and snow, the *Anderson* ten miles behind her.

◼ "Nosedive"

> "Sometime before 7 PM . . . we took two of the largest seas of the trip . . . The second large sea put water on our bridge deck! This is about 35 feet above the waterline!"
> —Bernie Cooper, captain of the *Anderson*

At 7:00 PM, 50-knot (58-mph) winds and 4.9-meter (16-foot) waves are still buffeting the *Anderson's* crew. They are in touch with the *Fitzgerald* by radio and by following the Fitz's blip on the radar. The *Fitzgerald* has struggled to within 24 kilometers (15 miles) of shore and about 32 kilometers (20 miles) of Whitefish Bay.

But even the bay is not safe harbor. Six Native American commercial fishermen are tending their nets in the western part of Whitefish Bay when, in the words of one of the fishermen, "all of a sudden, the lake began to boil and churn enormously. It was like nothing I'd seen before." Another fisherman barely survives his boat's capsizing due to "a giant wall of water coming at us." There is little shelter from this storm.

At 7:10 PM the *Anderson* gives navigation instructions to the radarless *Fitzgerald* up ahead of it. As an afterthought, the *Anderson's* first mate asks, "Oh, and by the way, how are you making out with your problems?" The *Fitzgerald* replies, "We are holding our own."

Immediately after this conversation, another severe snow squall enshrouds the two boats. The *Fitzgerald* is hidden from the radar beam of the *Anderson,* lost in a chaos of snow and sea. Just as suddenly, around 7:30 PM, the snow ends. For the first time in many hours, visibility is excellent. Lights from ships coming north from Sault Ste. Marie shine through clearly on the horizon. A TV/radio tower blinks across the water in Ontario. But the *Fitzgerald* is nowhere to be found: no lights, no radar blip, no radio contact.

Captain Cooper searches for the *Fitzgerald.* Did the Fitz duck into a harbor during the snow squall? How can you lose a 222-meter (729-foot) ore freighter in the middle of a lake? Cooper calls the U.S. Coast Guard in Sault Ste. Marie with his worst fears:

> "This is the *Anderson.* I am very concerned with the welfare of the steamer *Edmund Fitzgerald* . . . I can see no lights as before, and I don't have him on radar. I just hope he didn't take a nosedive."

The *Fitzgerald* is gone, without a mayday, an eyewitness, or a survivor.

BOX 10.3

Cyclone Winds in 3D: Belts, Slots, and Squalls

The November 10, 1975 cyclone surprised everyone by generating the 100-mph wind gusts the *Anderson* reportedly experienced that afternoon. Why did it happen? No one knows, although a variety of pet theories abound in books and on the Internet. Here, in collaboration with Stino Iacopelli, a former student of the co-author, we present our own possible explanation. It's the first proposed explanation to use current three-dimensional conceptual models of winds in the extratropical cyclone.

The Bergen School saw the cyclone as a face in the clouds and used military metaphors (e.g., fronts) to describe it. Today, meteorologists visualize the cyclone the way a cartoonist would, using just a few deft penstrokes to capture the essence of that face. Along with this new way of seeing the cyclone comes a new metaphor borrowed from industry, not war: "conveyor belts" of air that carry air parcels through an ever-moving, ever-evolving cyclone.

The figure below illustrates the approximate locations of these conveyor belts, superimposed on a rare weather satellite picture of our cyclone at 5:00 PM Eastern time on November 10. The "warm conveyor belt" (WCB) is essentially the warm air that rises up and over the warm front. It then peels off toward the east as it is pushed by the strong jet stream above the surface cyclone. The WCB is the main cloud- and precipitation-making air flow in the cyclone.

The "cold conveyor belt" (CCB) is a flow of air ahead of the warm front that is wrapped into the center of the cyclone, rising as it converges and causing clouds. These clouds are the distinctive "comma head" of the extratropical cyclone.

The "dry conveyor belt" (DCB) or "dry slot" is the key to understanding at least some cyclone-related windstorms. The figure below, hand-drawn by meteorologist Ed Danielsen over thirty years ago, gives you a sense of the three-dimensional nature of the DCB. It is a tongue of air dragged down by the jet stream from high aloft, even as high as the lowest part of the stratosphere! This air is very dry; as a result, it appears as a cloudless region in a visible satellite picture—the "dry slot" in Figure 10.8. The strong winds above the cyclone descend in the dry slot and make the surface beneath it a blustery place.

The conveyor belts of an extratropical cyclone, overlaid on top of the weather satellite image for 5:00 PM Eastern time November 10, 1975. The red arrow indicates the WCB, the blue arrow the CCB, and the yellow arrow the DCB. The red X underneath the blue arrow indicates the position of the Edmund Fitzgerald *at the time the satellite picture was taken. (Source: Wilson, E.E., "Great Lakes Navigation Season," 1975, Mariners Weather Log, 20, pp. 139–149.)*

Mariners Weather Log/NOAA

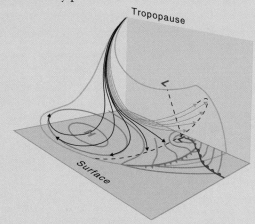

The 3D flow of air above a cyclone. The dry conveyor belt is the cluster of yellow lines that extends downward and then spreads out over the surface low.

To positively identify a dry slot, meteorologists can try at least two different approaches. One is to look for narrow bands of dark (i.e., dry) air leading into a cyclone on water vapor satellite images. Another way is to look for signs of stratospheric air being dragged down into the troposphere.

Water vapor imagery wasn't an option to meteorologists in 1975 because the technology hadn't been invented yet. Fortunately, Nature provided a near-perfect twin of the 1975 cyclone exactly 23 years later, on November 10, 1998. This

BOX 10.3
(continued)

Colorized water vapor image of an intense cyclone over western Wisconsin at 3:15 PM Eastern time on November 10, 1998 (the 23rd anniversary of the sinking of the Edmund Fitzgerald). The colors indicate the temperature of the clouds (dark green = cold thunderstorm clouds). The dry conveyor belt is denoted by the dark region across Illinois and Michigan and the hook-shaped area over extreme southeastern Minnesota and western Wisconsin.

time the satellites were ready and waiting. The figure at the left shows a water vapor image from this 1998 cyclone. The hook-shaped dark region near the center of the cyclone is a spiral arm of the dry slot that has wrapped around the center of the low. Underneath this hook, surface winds gusted to 81 knots (93 mph) at La Crosse, Wisconsin. (The photograph on the front cover of this textbook shows the effects of this windy cyclone on Lake Michigan at the South Haven, Michigan, pier and lighthouse.)

A good indicator of stratospheric air is the presence of larger-than-normal values of ozone. Today, satellites can help us detect ozone from space. In the figure below, the color-coded values depict the amount of ozone over a particular location during the November 10, 1998 cyclone—the higher the number, the more ozone. The narrow line

Values of ozone over the United States during an intense cyclone at 2:00 PM Eastern time November 10, 1998. The higher the number, the greater the ozone in the troposphere over that location. Notice the band of high ozone extending across Lake Michigan. This is the dry conveyor belt (compare to the previous figures).

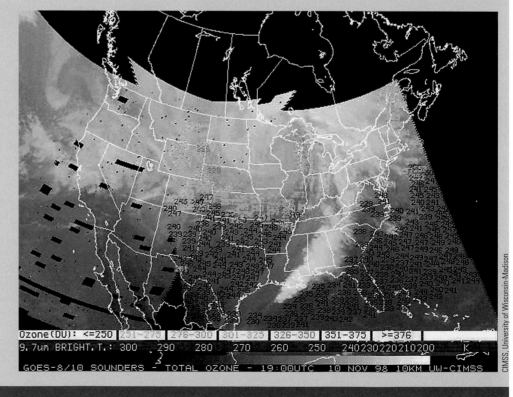

BOX 10.3
(continued)

of high ozone leading to the cyclone (green and yellow numbers) is the DCB.

Back in 1975, satellites couldn't "see" ozone. Instead, surface-based observations, often made at high schools, were used to keep track of daytime ozone pollution. These observations can also be used to look for signs of stratospheric air descending in the dry slot of the cyclone. On the night before the *Fitzgerald* sank, surface ozone values were rising all over the Midwest, particularly in a band stretching through Chicago. The figure below, using data gleaned from several Chicago high schools, tells the story. The ozone values peaked around midnight, right around the time that thunderstorms hit the area (compare to the Madison, Wisconsin, observations in Figure 10.9). Meanwhile, ozone values were also above normal back in Iowa, behind the low. At this same time, winds gusted to 70 mph in the Chicago area, and a man was killed in north-

east Iowa when winds behind the cyclone overturned his airplane. It is possible the *Fitzgerald* cyclone had a well-developed dry slot, just as Figure 10.8 suggested, and high-speed, high-ozone air plunged toward the surface.

Is this why the *Fitzgerald* and the *Anderson* were ravaged by high winds on Lake Superior the next day? No surface ozone data exists in the vicinity of the wreck. But we do know that the *Fitzgerald's* last hours were spent behind the low in heavy lake-effect snow squalls. These squalls, like the Chicago thunderstorms the night before, would help drag any high-speed air above them down to the surface.

Is our modern understanding of the "dry slot" part of the solution to the mystery of the *Fitzgerald* shipwreck? Perhaps. This question won't be settled until the evidence is presented as a formal journal article and the article's results are discussed and debated by today's leading extratropical cyclone experts.

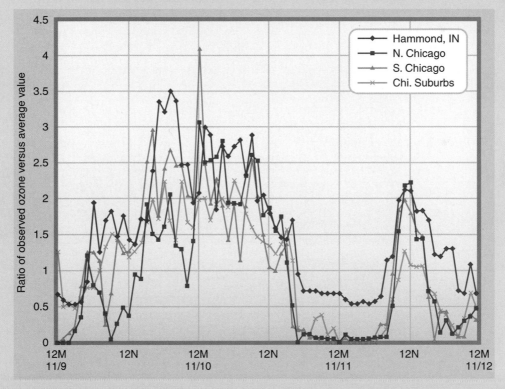

Ozone levels near the surface in and near Chicago on November 9–11, 1975. The values are ratios of the observed ozone concentration versus the November 1975 average for that time. Values greater than 1 indicate greater-than-usual ozone concentrations. Notice the unusually high values of ozone around midnight on the 10th as the extratropical cyclone passed through the area, the unusually low values the next evening in the clean anticyclone air behind the storm, and the human-induced peak in ozone due to pollution around noon on the 11th.

Day Four (and Beyond): Death

The Cyclone

The surface weather map for 7:00 AM Eastern time on Tuesday, November 11, 1975 (Figure 10.18, *A*) betrays little hint of the maelstrom of the past 24 hours. On Monday, cyclone winds blew Lake Michigan waters over a pier and drowned two people in Grand Haven, Michigan. That same day, the cyclone's southwest winds blew across the entire length of Lake Erie from Toledo, Ohio to Buffalo, New York, piling up Erie's waters a full nine feet above normal in Buffalo and killing a Buffalo woman by blowing her off a second-story porch. Showers and thunderstorms spawned by the storm brought rain to every state east of the Mississippi (Figure

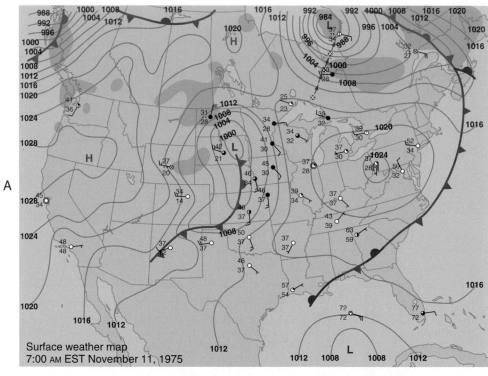

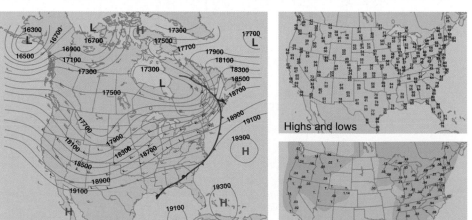

Figure 10.18

*Surface weather (**A**) and 500-mb conditions (**B**) at 7:00 AM (EST) on Tuesday, November 11, 1975. The past day's high and low temperatures and precipitation are indicated in the lower-right panel (**C**).*

10.18, *C*). But on Tuesday, light and variable winds are now firmly in place over the entire Great Lakes.

Bergen School member Sverre Petterssen once said, "Extratropical cyclones are born in a variety of ways, but their appearance at death is remarkably similar." And so it is for the storm that helped wreck the *Edmund Fitzgerald.* Figure 10.18, *A*, shows an occluded cyclone skirting the eastern shore of Hudson Bay—it is the *Fitzgerald* storm. The nearest strong temperature gradients are now hundreds of miles away over the Atlantic. It will die a slow death in the coming days as it crosses northern Canada, its fronts dissolving, the face becoming unrecognizable.

Another low, a youngster, is speeding across the Great Plains toward the Great Lakes. Like its predecessor, it has a potent upper-level trough associated with it (Figure 10.18, *B*). But unlike the *Fitzgerald* storm, it has no warm, moist air to feed upon; dew points ahead of it hover around freezing. It has also reached the occluded stage early in its lifetime (see also Figure 10.14, *A*). Lacking the energy sources of temperature gradients and moisture, it will not grow into a killer.

Yet in death the *Fitzgerald* storm, like all extratropical cyclones, accomplishes tasks needed to keep weather and climate in balance. For example, on this day at 7:00 AM, Great Whale, Quebec, reports drizzle with a temperature of 2.7° C (37° F), up from –8.8° C (16° F) the previous morning. Meanwhile, cooler and drier air has invaded the Gulf Coast of the United States. This poleward transport of heat and moisture is vital to Earth's overall energy balance, as we discussed in Chapter 7. Viewed from this perspective, the cyclone is not a meaningless killer—but instead is part of Nature's broader design for a stable, habitable world.

The *Fitzgerald*

Early on the 11th, the crew of the *Anderson* sights the first debris from the wreck of the *Edmund Fitzgerald:* part of an unused life jacket. Other "flotsam" washes up near shore: more life jackets, a severely damaged lifeboat (Figure 10.19), a stool, a plastic spray bottle. But the *Fitzgerald* itself is not officially located at the bottom of Lake Superior until the next spring, with the help of underwater camera equipment. The end of the *Fitzgerald* was swift and violent, probably a nosedive straight to the bottom. The freighter was torn into two main pieces as it plummeted, the front half landing rightside up, the back half upside down. It lies in 162 meters (530 feet) of water just 27 kilometers (17 miles) from the entrance to Whitefish Bay (Figure 10.20).

Figure 10.19

One of the Fitzgerald's *two mangled lifeboats, recovered by the* Arthur M. Anderson *during the futile three-day search for survivors.*

Fred Plofahan

■ **Figure 10.20**

Artist's rendering of the Edmund Fitzgerald *in its final resting place at the bottom of Lake Superior. (Source: David Conklin Great Lakes Shipwreck Historical Society.)*

Overview of the *Edmund Fitzgerald* Storm

Why did the *Fitzgerald* sink? There is no one simple answer. U.S. government investigators blame non-watertight hatch closures where the iron ore was loaded into the belly of the boat. A dissenting opinion by one of the government's own investigation board members suggests the possibility that the *Fitzgerald* ran aground on the shoals near Caribou Island. Others implicate the unusually high waves that pounded the *Anderson* and others early on the evening of the 10th. Meteorology is not to blame; the NWS forecasts are deemed "excellent." The storm was anticipated more than a day in advance, even if its exact intensity and path were not forecast precisely. Theories and controversy abound; probably no one factor is solely to blame. Sadly, the wreck of the *Edmund Fitzgerald* has become the maritime equivalent of the assassination of President John F. Kennedy, complete with disbelieved government explanations and endless intrigue among its latter-day investigators.

What is rarely, if ever, pointed out is that regardless of the proposed cause—leaky hatch covers, shallow shoals, high waves—the *Edmund Fitzgerald* never would have sunk on a calm, sunny day. An overriding reason for this shipwreck is actually quite simple: a powerful extratropical cyclone and a vulnerable boat crossing paths at the worst of all possible times.

■ The Sailors

The bodies of the 29 sailors on board the *Edmund Fitzgerald* have never been recovered or buried. Bruce Hudson's grieving parents preserve his room—books, photos, musical instruments—exactly as Bruce left them in November of 1975. He never met his daughter Heather, born seven months after the *Fitzgerald* was lost.

There will always be another extratropical cyclone. The Sun and the tilt of the Earth see to this by the endless re-creation of temperature gradients. And as long as there is cargo to be hauled, there will be Great Lakes freighters. But people are unique and irreplaceable and their loss is felt forever. Honor the memory of the *Fitzgerald* and her crew* by never losing your life to the many weapons—wind, rain, snow, and wave—of an extratropical cyclone.

---

*To honor the Fitzgerald's crew and thank their families for the use of the story of their loved ones, a portion of the money you pay for this textbook will be donated to the Great Lakes Mariners Memorial project of the Great Lakes Shipwreck Historical Society, to be located at Whitefish Point, Michigan.

LOST AT SEA

Ernest M. McSorley, 63, Captain, Toledo, Ohio

John H. McCarthy, 62, first mate, Bay Village, Ohio

James A. Pratt, 44, second mate, Lakewood, Ohio

Michael E. Armagost, 37, third mate, Iron River, Wisconsin

George J. Holl, 60, chief engineer, Cabot, Pennsylvania

Edward F. Bindon, 47, first assistant engineer, Fairport Harbor, Ohio

Thomas E. Edwards, 50, second assistant engineer, Oregon, Ohio

Russell G. Haskell, 40, second assistant engineer, Millbury, Ohio

Oliver J. Champeau, 41, third assistant engineer, Milwaukee, Wisconsin

Frederick J. Beetcher, 56, porter, Superior, Wisconsin

Thomas Bentsen, 23, oiler, St. Joseph, Michigan

Thomas D. Borgeson, 41, able-bodied maintenance man, Duluth, Minnesota

Nolan F. Church, 55, porter, Silver Bay, Minnesota

Ransom E. Cundy, 53, watchman, Superior, Wisconsin

Bruce L. Hudson, 22, deckhand, North Olmsted, Ohio

Allen G. Kalmon, 43, second cook, Washburn, Wisconsin

Gordon F. MacLellan, 30, wiper, Clearwater, Florida

Joseph W. Mazes, 59, special maintenance man, Ashland, Wisconsin

Eugene W. O'Brien, 50, wheelsman, Perrysburg Township, Ohio

Karl A. Peckol, 20, watchman, Ashtabula, Ohio

John J. Poviach, 59, wheelsman, Bradenton, Florida

Robert C. Rafferty, 62, temporary steward (first cook), Toledo, Ohio

Paul M. Riippa, 22, deckhand, Ashtabula, Ohio

John D. Simmons, 60, wheelsman, Ashland, Wisconsin

William J. Spengler, 59, watchman, Toledo, Ohio

Mark A. Thomas, 21, deckhand, Richmond Heights, Ohio

Ralph G. Walton, 58, oiler, Fremont, Ohio

David E. Weiss, 22, cadet (deck), Agoura, California

Blaine H. Wilhelm, 52, oiler, Moquah, Wisconsin

The Extratropical Anticyclone

Did you notice the large weather system that moved in immediately behind the *Fitzgerald* cyclone? Probably not. It was an **anticyclone,** or high-pressure system. To complete our understanding of large-scale extratropical weather, we need to learn how to recognize the seemingly boring "face" of the anticyclone.

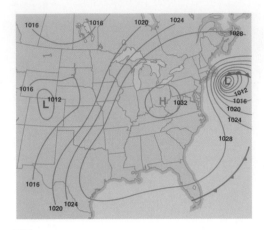

▇ Figure 10.21

Surface weather map for 7:00 PM Eastern Standard Time on February 20, 1979. Notice the stark difference in appearance of anticyclones versus cyclones. The bull's-eye–like low off the U.S. East Coast has fronts and a tight pressure gradient near its center. The blobby high over Ohio covers the entire eastern half of the United States, yet pressures over this vast region differ by only 4 millibars. (Read Box 10.2 and Chapter 13 for more information on this famous storm.) (Source: Kocin, P. and Uccellini, L., Snowstorms Along the Northeastern Coast of the United States: 1955 to 1985, American Meteorological Society.)

The anticyclone is the natural complement to the cyclone, the yang to the cyclone's yin:

- Lows form and grow near fronts, the boundaries of air masses where contrasts of temperature and moisture are strongest. Highs are air masses, with temperature and moisture varying little across hundreds of miles.
- Lows are usually cloudy, wet, and stormy. Highs are usually clear, relatively dry, and calm.
- The pressure pattern of a low looks like a bull's-eye on a weather map, with tightly packed, concentric isobars. A high looks like a large blob with weak pressure gradients near its center (Figure 10.21).
- Lows live for a few tumultuous days. Highs loiter, sometimes for several languorous weeks in summer.

We can explain these features of an anticyclone the same way we did for the cyclone. For example, in Chapter 6 we learned that air in a high diverges at the surface. Diverging air weakens temperature and humidity gradients by spreading out the lines of constant temperature and moisture. Weak gradients mean no fronts in a high; fronts exist instead at the periphery of highs.

Diverging air at the surface in a high requires air to sink toward the surface from above. Sinking air warms adiabatically (Chapter 2) and dries out in a relative sense (Chapter 7). A temperature inversion forms above the ground as a result of the compressional warming of this sinking air. This makes highs stable (Chapter 9) and often cloudless.

Lows can develop strong pressure gradients (bull's-eyes) near their centers because, as we saw in Chapter 6, the pressure gradient force in a low can be counterbalanced by both the Coriolis and centrifugal forces. In a high, the pressure gradient force teams up with the centrifugal force, and the Coriolis force must counterbalance both of them. The stronger the pressure gradient, the harder it is

January anticyclones, 1950–1977   July anticyclones, 1950–1977

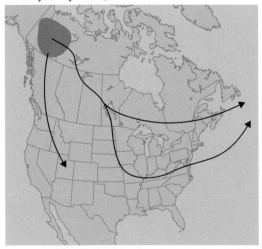
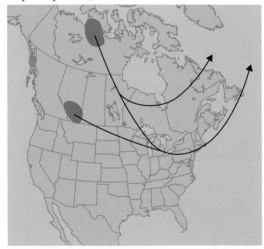

▇ Figure 10.22

Typical regions of anticyclogenesis (shaded) and anticyclone paths (arrows) that affect North America in the months of January and July. Compare this figure to the corresponding data for cyclones in Figure 10.6. (From Synoptic-Dynamic Meteorology in Midlatitudes, Volume II: Observations and Theory of Weather Systems *by Howard Bluestein, copyright © 1992 by Oxford University Press, Inc. Used by permission of Oxford University Press, Inc.)*

for the Coriolis force to do this. The end result is that nature prefers highs with weak pressure gradients, which implies a spreading out of isobars over large distances. Therefore, highs are large and blobby compared to lows; by the geostrophic wind approximation (Chapter 6) this makes the winds almost calm near a high's center.

Highs, being air masses, form where air can sit around for a period of time before being steered elsewhere by jet-stream winds. Figure 10.22 shows these regions of **anticyclogenesis.** Many wintertime highs in the United States originate in northern Canada—they are the cP air masses discussed in Chapter 9. Cold air outbreaks such as the one depicted in Figure 9.8 occur when strong anticyclogenesis over Canada is combined with jet-stream winds from the northwest that guide the high into the United States.

Summertime highs deserve special attention. They share something in common with wintertime lows—both can become cut off from the main jet-stream winds and the surface temperature gradients beneath the jet. Since an extratropical cyclone thrives on temperature contrasts, a cut-off cyclone weakens and dies. But a summer anticyclone, which is an air mass without significant horizontal temperature gradients, can thrive and intensify in this situation. A cut-off high is a form of "blocking," which we discussed in Chapter 7. A blocking high over land in summer can trap and recirculate hot air around and around its center for weeks. This may lead to a heat wave, a drought, and/or an episode of air pollution (see Chapter 15).

We've recognized and explained the identifying features of the anticyclone. Underneath its calm exterior, however, lurks a potential killer. Did you sense any danger? Neither did John F. Kennedy Jr. We close this chapter with his story.

Anticyclones and Another Fitzgerald Tragedy

July 16, 1999 is a hot, hectic Friday in New York City. Manhattanites rush to escape the city's heat and spend the weekend on the coast or in the mountains. Included in this crowd are John Fitzgerald Kennedy, Jr., his wife Carolyn Bessette Kennedy, and her sister Lauren Bessette. He is "JFK Jr.," the son and namesake of the 35th President of the United States. On July 16 they are traveling to a family wedding in Hyannis Port, Massachusetts. But first they will stop to drop off Lauren at Martha's Vineyard, a resort island frequented by the Kennedys south of the Massachusetts coast (see the map in Figure 10.23).

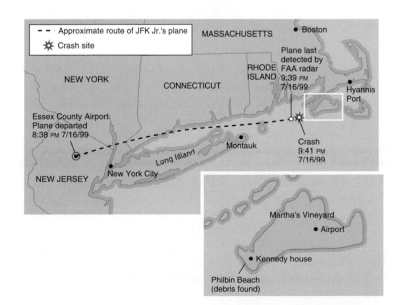

Figure 10.23

The flight path of the Piper Saratoga airplane carrying the Kennedys and Lauren Bessette. (From Newsweek. © 1999 Newsweek, Inc. All rights reserved. Reprinted by permission.)

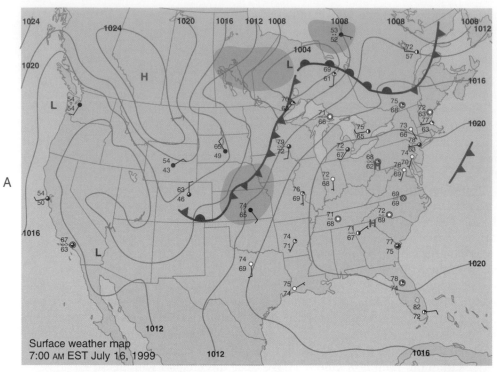

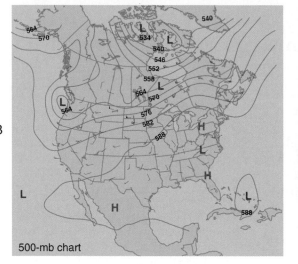

■ **Figure 10.24**

Surface weather (A) and 500-mb conditions (B) at 7:00 AM (EST) on July 16, 1999, showing the anticyclone at the surface and the ridge at the upper levels over the eastern United States. In C, a high-resolution visible satellite picture of the eastern United States is shown for July 16, 1999 at 5:20 PM (EST). The milky, gray region along the coast east and northeast of New York City (NYC) is polluted air.

D. Westphal/Naval Research Laboratory, Monterey, CA

They will fly to their destinations. JFK Jr. is qualified to fly under "visual flight rules"—in other words, when the sky is clear enough to see where you're going. He doesn't yet qualify to use a plane's instrument panel to navigate when visibility is low, however. This shouldn't be a problem on July 16th; there is a high-pressure system sprawling over the eastern United States and the weather reports and forecasts indicate calm and mostly clear skies (Figure 10.24, *A*). The light winds associated with the ridge parked above the high (Figure 10.24, *B*) mean that the surface anticyclone will be stationary for the foreseeable future.

Stuck in weekend traffic, the Kennedys don't reach the Essex County Airport in New Jersey until after 8:00 PM. JFK Jr. prepares his six-seat Piper Saratoga single-engine plane for flight, and the trio takes off at 8:38 PM Eastern time, twelve minutes after sunset. The trip to Martha's Vineyard, less than 320 kilometers (200 miles), should take a little over an hour; the Piper Saratoga's top speed is 307 km/hr (191 mph).

6/16/99 at 7 PM

7/16/99 at 6 PM

Center for Air Pollution Impact and Trend Analysis

Figure 10.25

Photographs of the skyline of Boston, Massachusetts, one month apart, on the evenings of June 16 and July 16, 1999. Notice the poor visibility on July 16, just hours before John F. Kennedy, Jr.'s plane departed for the coastline of southern Massachusetts.

The flight continues uneventfully over Long Island Sound until around the 1-hour mark. Fifty-five kilometers (34 miles) from the Martha's Vineyard airport, Kennedy begins a slow descent from a cruising altitude of 1.7 kilometers (5600 feet). Five minutes later, at 700 meters (2300 feet), the Piper Saratoga turns to the right and climbs to 790 meters (2600 feet). One minute afterward, the plane descends and turns back to the left. Only thirty seconds later, the plane turns to the right and dives, spiraling downward at the astonishing rate of 1430 meters (4700 feet) per minute. At 9:41 PM the Piper Saratoga smashes into the Atlantic Ocean just 12 kilometers (7.5 miles) from the southwestern shore of Martha's Vineyard, killing the Kennedys and Bessette instantly.

High Pressure, Low Visibility

What happened? Investigators quickly ruled out a problem with Kennedy's plane. Instead, pilots and meteorologists alike pointed to the surprisingly low visibility in the "clear" anticyclone over the Northeast that weekend.

The cloudless weather in anticyclones is usually welcome. However, the sinking air and lack of strong winds in a high traps air in the lower troposphere. This air then takes on the characteristics of the region it sits over—the process of "air mass modification" we talked about in the previous chapter. If a high remains over a heavily populated region for more than a couple of days, it can trap and concentrate pollutants from cars and industry.

July 16 was one of the most polluted days of 1999 in the northeastern United States. A week's worth of pollution from the Midwest and the East Coast found its way into the high's circulation. The pollution then traveled on the high's anticyclonic winds and collected over New England and its coastline.

Pollution affects aviators because polluted air is much harder to see through than unpolluted air, even in a cloudless sky. Satellite pictures (Figure 10.24, C) and photographs (Figure 10.25) from July 16 show that the polluted air was much less transparent than cleaner air at other times and places. Other analyses indicate that the temperature inversion caused by the sinking motion in the anticyclone created a layer of extremely polluted air about 1 kilometer (3000 feet) above the ground over the Northeast.

Ozone Pollution on the Day of the JFK Jr. Crash

Figure 10.26

A schematic explaining how the polluted anticyclone contributed to the crash of John F. Kennedy, Jr.'s airplane. Temperature inversions within the anticyclone trapped and concentrated the pollutants in a narrow layer corresponding to the altitude at which Kennedy's plane went out of control, probably because of pilot disorientation. (From Doug Westphal, Naval Research Laboratory, Monterey, CA.)

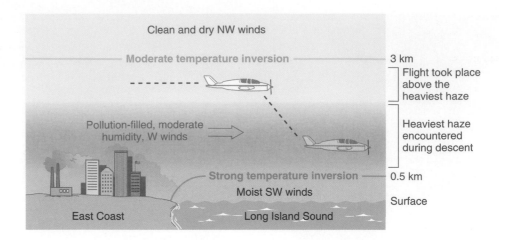

How does this explain JFK Jr.'s plane crash? Prof. Joseph Prospero of the University of Miami suggests the following scenario (see also Figure 10.26): While the plane was cruising at 1.7 kilometers (5600 feet) over Long Island Sound, the flight proceeded normally because the pollutants were trapped below this altitude by the anticyclone. Trouble ensued when the plane descended into the murkiest layer of the pollution, at or a little below 1 kilometer (3000 feet). Visibility would have decreased immediately and drastically. Furthermore, the plane was over water, at night. Any lights or landmarks would have been blotted out by the haze.

At this point an experienced pilot would have navigated the plane using the instrument panel on the aircraft. Kennedy was not yet qualified to do this. Instead, he first tried to steer and climb out of the low-visibility pollution layer—the turn to the right and the ascent to 790 meters (2600 feet). This strategy didn't improve visibility. Nearing land but unable to see any lights due to the pollution, Kennedy turned back to his original course. Then he became disoriented and inadvertently guided his plane into a fatal dive.

John F. Kennedy Jr. once told a friend that he looked forward to a pollution-free future: "Someday we'll be able to see Los Angeles," he joked. Ironically, his inability to see the lights of Martha's Vineyard on July 16, 1999 took his life and the lives of his wife and sister-in-law. An anticyclone, and the pollution trapped within it, played a key role in their deaths.

PUTTING IT ALL TOGETHER

SUMMARY

Extratropical cyclones are low-pressure systems that cause wet and often windy weather, but they are very different than tropical cyclones. Norwegian meteorologists discovered that extratropical cyclones are associated with fronts and that they have a definite life cycle, growing from birth as a frontal wave to maturity as an occluded cyclone to death as a cut-off cyclone over the course of several days.

Extratropical cyclones are born on the downwind side of tall mountain ranges, near warm ocean currents, and beneath strong jet-stream winds. Further growth of these storms may occur if the upper-tropospheric air is diverging above the cyclone. The age and strength of extratropical cyclones can be estimated by looking at satellite pictures and weather maps. Strong cyclones

often have comma-cloud shapes with dry slots and well-defined cold and warm fronts; the presence of an occluded front indicates a mature-but-dying cyclone.

We learned these facts in the context of the storm that helped wreck the *Edmund Fitzgerald* on Lake Superior in 1975. This storm demonstrates that the Norwegian cyclone model, though simplified, can help explain most of the characteristics of a real-life, deadly extratropical cyclone.

The extratropical anticyclone is in many ways the complement to the cyclone. An extratropical cyclone lives at the clashes of different air masses. In contrast, an anticyclone is one big, slow, fairly calm and stable air mass. Even so, a high can become a killer when combined with air pollution, as we saw in the case of John F. Kennedy, Jr.'s fatal airplane crash in 1999.

KEY TERMS

You should understand all of the following terms. Use the glossary and this chapter to improve your understanding of these terms.

Alberta Clippers
Anticyclogenesis
Anticyclone
Baroclinic instability
Comma cloud
Cut-off cyclone
Cyclogenesis
Diffluence
Dry slot
Frontal wave

Nor'easters
Norwegian cyclone model
Occluded cyclone
Occlusion
Open wave
Panhandle Hooks
Pineapple Express
Speed divergence
Warm sector

 # REVIEW QUESTIONS

1. What are the four stages in an extratropical cyclone's life cycle, according to the Bergen School conceptual model used in this chapter? During which stage(s) would you expect to see: An occluded front? Strong warm and cold fronts? A cut-off low? Make a sketch of the low and the fronts at each stage of the life cycle.

2. Why do the birthplaces of cyclones vary in location throughout the year? Would this be true if the Earth was not tilted on its axis?

3. Can a cyclone develop out of a clear blue sky? If so, how?

4. Intense cyclones are more common east of the Appalachian Mountains than east of the much higher Rocky Mountains. Why? (See Boxes 10.1 and 10.2.)

5. Which one of the cyclone types discussed in this chapter is not named for the region in which it is "born"? It didn't get its name from the region it affects, either. So how did it get its name? (Hint: think about how wind flows around a cyclone.)

6. In Figure 10.9, you can see that Madison, Wisconsin's temperature rose but its dew point fell on the day after the cold front passed. Using air mass and radiation concepts from this textbook, explain why this happened.

7. Using the real-life cross-sections in Figure 10.10 as a guide, explain how temperature, dew point, pressure, cloud altitude and type, and wind direction change as a warm front approaches and passes a location. Do the same for the cold front. Compare your answers to the idealized discussion in Chapter 9.

8. You are watching a weather forecast on TV and the meteorologist says, "The low over Colorado is moving east, but it's also digging." In which region will the low be 12 hours later: The northern Rockies? The Great Lakes? The Texas/Oklahoma Panhandles? Hawaii?

9. It's 8:00 AM on Monday, your first day of work at the National Weather Service office in Chicago, Illinois. You notice a developing low pressure system centered over Wichita, Kansas, about 1000 kilometers (600 miles) to the southwest. The winds at 500 mb above the cyclone are blowing toward the northeast at 100 km/hr (60 mph). If the storm maintains its current intensity, when do you forecast that the center of the cyclone will pass near the Chicago area? If the cyclone and the upper-level trough strengthen, do you think the low will pass to the north or the south of Chicago?

10. What are two causes of upper-level divergence? Why does upper-level divergence help intensify a cyclone?

11. According to Box 10.2, was the *Fitzgerald* cyclone a "bomb"? What key fuel source did this cyclone have in limited supply compared to a cyclone just east of Cape Hatteras, North Carolina?

12. Name five different ways that weather associated with an extratropical cyclone can injure or kill people.

13. Assume that you have the power to remove extratropical cyclones from the Earth's weather. How might this affect the wintertime weather in eastern Canada? Could this cause a long-term trend in temperatures in the tropics?

14. Using the "Weather Forecasting/Cyclogenesis" section of the *Blue Skies* CD-ROM, step through the development of two extratropical cyclones while examining the frontal and jet-stream patterns associated with them. Why does the cyclone over the East Coast *not* intensify? Also, compare the appearance of the cyclones in the infrared satellite base maps to the idealized model in Table 10.1. Why does the second cyclone move southeastward as it crosses the Rocky Mountains into the Great Plains? (Hint: see Box 10.1.)

15. Why is it that wintertime Canadian high-pressure systems are colder when they are in Canada than when they reach the Gulf of Mexico several days later?

16. Based on the wind circulation around a high, describe the changes in temperature, dew point, and wind direction as an anticyclone moves across the central United States from west to east.

 # WEB ACTIVITIES

Choose Chapter 10 on the textbook's Web site:
http://info.brookscole.com/ackerman
and select from the following resources:
- Interactive Modules that illustrate and extend your understanding of key topics in this chapter
- Tutorial Quizzes to test your mastery of terms and concepts

 For additional readings, go to the InfoTrac College Edition, your online library, at:
http://www.infotrac-college.com

Thunderstorms and Tornadoes

After completing this chapter, you should be able to:

■ Describe the different types of thunderstorms and the weather conditions that lead to their formation

■ Explain why scientists think tornadoes occur and how the winds in a tornado are estimated using the Fujita scale

■ Characterize the different stages of the life cycles of thunderstorms and tornadoes

■ Identify likely areas for the occurrence of various types of severe thunderstorm weather

Introduction

Prom Night 2001 was a night to remember for all the wrong reasons at Hoisington High School in central Kansas. Will Rubio, a high-school senior at Hoisington at the time, told *The New York Times* what happened on the stormy evening of April 21:

> We were all joking around and dancing like a bunch of idiots . . . The power went out . . . came back on . . . went back off, and they told us to go downstairs. I wasn't really worried, because that's a normal thing around here. We get quite a few storms.
>
> We were down there 30, 40 minutes. Then the principal came down and said, "We're not continuing the prom" . . . My mom had just come in the door . . . She told me that a tornado had hit our house. What tornado? I didn't see it . . . How could there be a tornado without us knowing?
>
> They finally let us go to our house two days later. It was the most unbelievable thing I had ever seen. I'd say a quarter of Hoisington was destroyed. One person died, two came close. . . The destruction was about four blocks, the width of it, and about a mile long. . . My house looked like a little hut. Everything was gone but two walls in the middle of the house . . . After the tornado, we couldn't find my mom's car for three, four days. Finally we found it a block and a half away, crushed. [My senior picture] was lying face down in the debris, and I picked it up. It was all crinkled. And I just started crying and realizing, this did happen.

Will Rubio's story raises many questions. What is a tornado? What causes tornadoes? Why are they so destructive? Why are storms common in Kansas? Are they common at night, in April? How do we know if tornadoes are coming? Why are downstairs areas the safest places to be when they hit?

In this chapter we examine tornadoes and other severe weather in the context of the weather feature that creates them: the thunderstorm. Like the extratropical cyclones we studied in the last chapter, thunderstorms and tornadoes are associated with recognizable large-scale weather patterns that allow meteorologists to predict when and where they may occur. Both thunderstorms and tornadoes also exhibit identifiable life cycles that help us understand how they work, just like cyclones. Thunderstorm phenomena such as tornadoes and lightning are among the most awe-inspiring sights in all of nature, but they can also kill as they did in Will Rubio's hometown. We close the chapter by discussing the precautions you need to take to avoid harm from thunderstorms.

What is a Thunderstorm?

A **thunderstorm** is, as the name implies, a cloud or cluster of clouds that produces thunder, lightning, heavy rain, and sometimes hail and tornadoes. This dangerous mix of weather requires great amounts of energy from the atmosphere. This energy is released when saturated air rises rapidly and high into the atmosphere. For this reason, thunderstorms are associated with tall cumulonimbus clouds that form when air rises or is lifted from the surface. The air in thunderstorms rises so rapidly that their tops may briefly overshoot the tropopause and penetrate the stratosphere (Figure 11.1). These **overshooting tops** can be seen on satellite images either by the visible shadow that they cast on the lower "anvil" cloud top (Figure 11.2, *A*) or by their especially cold temperatures, which are "seen" in infrared satellite pictures (Figures 11.2, *B*).

Not all thunderstorms are the same. Some thunderstorms are brief rainmakers, while others last for hours and form windy lines or clusters. Still others are determined by forecasters to be especially likely to

create severe weather such as extremely high winds, hail, and tornadoes. In the following section, we learn where thunderstorms form and the factors that can lead to different types of thunderstorms.

■ **Figure 11.1**

A photograph of a thunderstorm from the ground, clearly showing the overshooting tops (brightest white clouds at the top) where the thunderstorm's updraft penetrates the tropopause.

John Mecikalski

Identifying Overshooting Tops

Thunderstorm Distribution

Thunderstorms require warm, moist air that rises. For this reason, thunderstorms are most prevalent in regions with mT air masses, and in regions where mountains and frontal cyclones help lift air vertically. Figure 11.3 shows the long-term annual climatology of thunderstorms in the lower 48 United States. Florida leads the nation in thunderstorm frequency, because warm, moist air is present year-round and the sea breeze (Chapter 6) helps create local regions of convergence and lifting even in the absence of large-scale fronts. The far West, in contrast, is too dry, and, near the coast, too cool for thunderstorms to be common there. Eastern Colorado and New Mexico, although typically dry, rival Florida in thunderstorm frequency. This is because warm, moist air rises as it climbs westward along the sloping Great Plains, and because the extratropical cyclone track (Chapter 10) causes rising motion in this region as cyclones "spin up" in the lee of the Rockies. These factors also determine the location of the 40,000 thunderstorms that are seen across the globe at any one time, with most thunderstorms occurring in the tropics and in the midlatitude storm-track regions.

Factors Affecting Thunderstorm Growth and Development

As we learned in earlier chapters, a parcel of air will not rise unless it is *unstable* (i.e. forced upward from the surface and/or warmer than its surrounding environment). More generally, the condition of the environment around warm, moist air makes all the difference between a clear sky, a garden-variety thunderstorm, and a severe thunderstorm with tornadoes. The crucial factors are the environment's temperature, moisture, and wind speed and direction from the ground all the way up to the tropopause.

The first factor we have already discussed in Chapter 9: the *dryline.* A dryline is essentially a moisture front between mT and cT air masses. It provides a focus for convergence and lifting across the southern Great Plains which, in combination with instability aloft, can lead to thunderstorms and severe weather.

Next we turn to stability. To interpret how the environment affects thunderstorm potential and severity, meteorologists have invented several *stability indices* that characterize the atmosphere in a single number.

Perhaps the most common stability index is the **lifted index.** This index follows the same approach we used in Chapters 2 and 7 to discuss how temperature changes

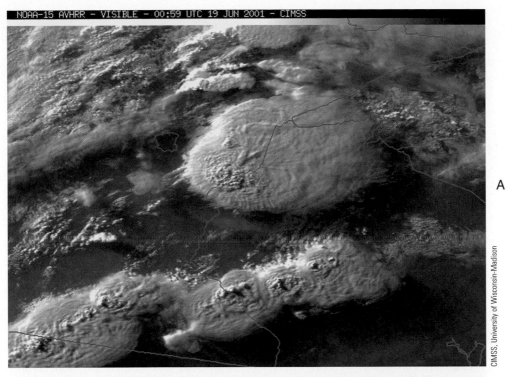

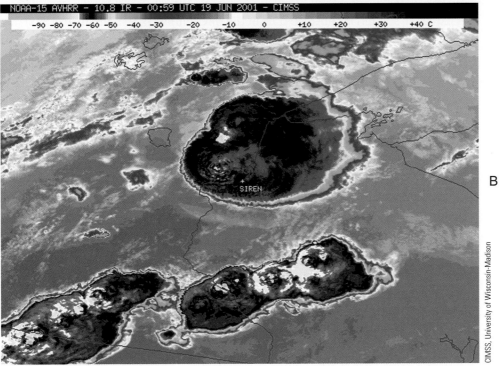

Figure 11.2

*Two low-Earth-orbiting satellite views of thunderstorms, at visible (**A**) and infrared (**B**) wavelengths. Overshooting tops are the bubbly features protruding from the broad, white anvils in **A** and which show up as the very coldest cloud-top temperatures in **B**. These thunderstorms occurred during the afternoon and evening of June 19, 2001 over the upper Midwest United States. The large supercell thunderstorm in the top center of the image is crossing the border from Minnesota into northwest Wisconsin. A tornado spawned by this supercell 7 minutes after these satellite images were made was responsible for three deaths and sixteen injuries along a 0.8-kilometer (0.5-mile) wide path through the small town of Siren, Wisconsin. The line of thunderstorms to the south of the Siren storm produced hail 11.5 centimeters (4.5 inches) in diameter.*

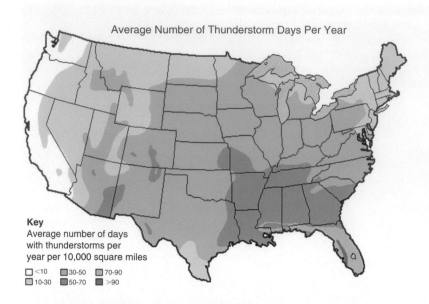

Average Number of Thunderstorm Days Per Year

Key
Average number of days
with thunderstorms per
year per 10,000 square miles

☐ <10   ▨ 30-50   ▨ 70-90
☐ 10-30 ▨ 50-70   ■ >90

■ Figure 11.3
The average number of thunderstorm days per year across the lower 48 United States. The Southeast United States leads the nation with more than 50 thunderstorm days each year. Note the secondary maximum in eastern New Mexico and Colorado. (Source: http://k12ocs. ou.edu/teachers/graphic/ TstormFreq.gif.)

www

Severe Weather Indices

in a rising air parcel. Simplified, the index starts with an air parcel from the surface, lifting and cooling it dry adiabatically to saturation, and then lifting and cooling it moist adiabatically to 500 mb. To compute the lifted index, the temperature of the parcel at 500 mb is subtracted from the environment's temperature at 500 mb, as measured by a radiosonde or satellite. If the observed 500-mb temperature is colder than the lifted air parcel, then the parcel is unstable and will be able to keep on rising and form a cumulonimbus cloud. The lifted index is negative in these cases.

For this reason, negative values of the lifted index can be related to thunderstorm severity. A lifted index of between 0 and −3 (degrees Celsius) indicates that the air is marginally unstable and unlikely to lead to severe thunderstorms. Values between −3 and −6 indicate moderately unstable conditions. Values between −6 and −9 are found in very unstable regions. Lifted index values less than −9 reflect extreme instability. The chances of a severe thunderstorm are best when the lifted index is less than or equal to −6. This is because air rising in these situations is much warmer than its surroundings and can accelerate rapidly and create tall, violent thunderstorms. (However, this index does *not* tell the probability of occurrence of a thunderstorm; other indices predict this.)

Figure 11.4 shows very negative values of the lifted index (yellow- and red-shaded areas) derived for regions ahead of a long line of severe thunderstorms along a cold front extending from Michigan to Texas in late October of 2001. These regions were placed under severe weather watches (Chapter 1) because of the extreme instability of the atmosphere and the lifting of air near the surface as the cold front and squall line approached.

A third factor affecting thunderstorm type is the change of environmental wind speed and direction from the ground up, known as *vertical wind shear*. Small amounts of shear lead to upright and majestic but (as we will soon see) shorter-lived thunderstorms. Large amounts of shear, such as are found in and near extratropical cyclones (Figure 11.5), cause thunderstorm clouds to tilt. If the wind changes vertically in direction and increases in speed, the effect on the thunderstorm is the same as "putting English" on a cue ball in billiards, and the result is the same: the thunderstorm itself spins. Rotating thunderstorms generally cause the worst severe weather. Although indices exist to estimate the impact of vertical wind shear on thunderstorms, they are beyond the scope of this text. The simple rule-of-thumb is, "the more vertical wind shear, the more severe the thunderstorm."

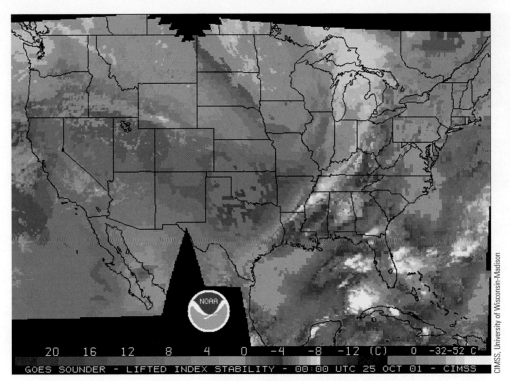

Figure 11.4

Satellite-derived values of the lifted index across the United States on the evening of October 24, 2001. The gray and white regions extending southward from Michigan to Louisiana are a squall line, in which the satellite cannot estimate the index. The yellow and red regions just to the east of the squall line represent areas with lifted-index values less than −4, indicating the possibility of severe weather. Most of the rest of the United States is shaded in beige, indicating regions with positive lifted-index values where severe weather is unlikely. Black regions are gaps in the satellite data coverage.

Two other factors affecting thunderstorm growth and development are shown in Figure 11.5. A low-level jet stream of air often develops over the Great Plains in the spring and summer. This jet occurs primarily at night and so is also called a **nocturnal low-level jet.** This low-level jet is an important ingredient in the formation of severe weather in the United States, because it supplies moisture and energy to Great Plains thunderstorms after the Sun goes down.

In addition, a **capping inversion** can develop in which hot dry air at about 700 mb overlies warm, moist air near the surface. As we saw in Chapter 7, inversions usually prevent surface air from rising. During some hot summer days, however, the hot, dry air acts like a lid on a boiling pot, concentrating energy below it until the near-surface air is warmed enough by the Sun to "blow the cap" and rush upward violently. This can lead to rapidly developing severe weather.

Now that we have connected thunderstorms with our understanding of the atmosphere from previous chapters, we can categorize the different types of thunderstorms, linking their characteristics to the factors described above.

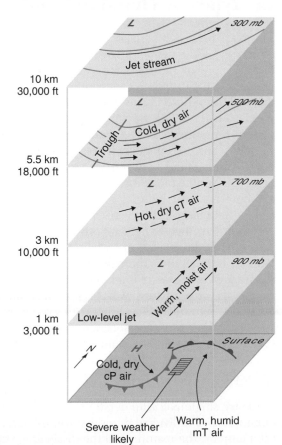

Figure 11.5

A schematic profile through a severe weather pattern, illustrating the vertical changes in temperature, moisture, and winds that contribute to severe thunderstorms and tornadoes. (Source: Adapted from Ahrens, C., Meteorology Today, 6th ed., Brooks-Cole, 2000, p. 408.)

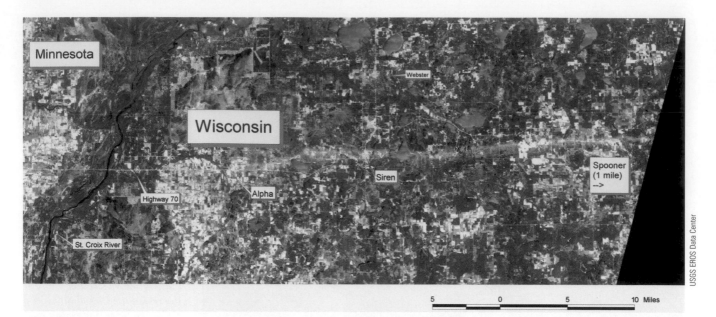

 Figure 11.6

A LANDSAT photograph of the damage path of the tornado spawned by the supercell thunderstorm in Figure 11.2, taken the morning after the tornado hit Siren, Wisconsin. Vegetation shows up as red in this satellite image, lakes and rivers are blue, and roads and towns are generally aqua or white. The fuzzy whitish blur extending left-to-right from the small town of Alpha through Siren and toward Spooner is damage to the landscape caused by the tornado.

Types of Thunderstorms

Thunderstorms can be classified by either their severity or their structure. For example, the NWS defines a **severe thunderstorm** as a thunderstorm that produces one or more of the following: wind gusts of 50 knots (58 mph or 93 km/hr) or greater, hail 0.75 inches (1.9 centimeters) in diameter or larger, or a tornado. The damage from these storms can be so devastating that it can be seen from space. Figure 11.6 is a satellite view of the region in the path of the tornadic thunderstorm shown in Figure 11.2. The path of damage appears as a whitish blur across northwest Wisconsin, as if a pencil eraser had been dragged across the landscape.

To understand how thunderstorms, severe and nonsevere alike, are *created*, however, we must look instead at the structure of these storms. Thunderstorms are composed of basic building blocks referred to as cells. A **cell** is a compact region of a cloud that has a strong vertical updraft.

There are two basic categories of thunderstorm cells, **ordinary cells** and **supercells.** Ordinary cells are a few kilometers or miles in diameter and exist for less than an hour, while supercells are larger and can last for several hours. Multicell storms are composed of lines or clusters of thunderstorm cells, either ordinary cells, supercells, or a mixture of both. Ordinary single-cell thunderstorms are the most common, but multicell and supercell storms are responsible for the vast majority of severe weather reports associated with thunderstorms. Next we examine the structure of ordinary single-cell, multicell, and supercell thunderstorms.

Ordinary Single-Cell Thunderstorm

Ordinary single-cell thunderstorms are short-lived and localized single-cell thunderstorms. An individual thunderstorm cell has a life cycle with three distinct stages: the cumulus, the mature, and the dissipating stages (Figure 11.7).

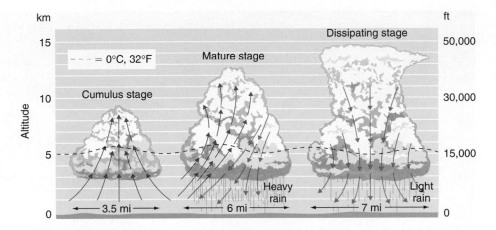

Figure 11.7

The life cycle of an ordinary thunderstorm cell contains three stages: cumulus, mature, and dissipating. Arrows at bottom indicate approximate widths of the base of the storm at each stage.

Growing a Thunderstorm

The **cumulus stage** is the initial stage of a thunderstorm. Warm air near the ground rises and cools initially at the dry adiabatic lapse rate. The rising air parcel approaches saturation as the relative humidity increases, until condensation occurs. The altitude to which a parcel of air needs to be lifted before saturation is reached is called the *lifting condensation level,* or *LCL.* The LCL marks the bottom of the cloud, its *base.* The cloud grows as moist air continues to be lifted and the growing cell expands both vertically and horizontally.

As the cloud is growing, dry air from its surroundings mixes into the cloud at the cloud edges. This is called **entrainment.** This mixing of drier air temporarily lowers the relative humidity and can cause droplets to reduce in size because of evaporation. The evaporation quickly saturates and cools the air within the cloud. Entrainment helps to produce different sizes of drops within the cloud; this is necessary for cloud particle growth by collision and coalescence (Chapter 4).

The moist air that flows upward through the cloud forms the updraft. As the cloud grows beyond the freezing level (the altitude where the air temperature is below freezing) it is composed of both water drops and ice particles, which supports further growth of the cloud particles (Chapter 4). Many clouds never develop beyond the cumulus stage. However, when conditions are right, the cumulus cloud builds into a cumulonimbus as precipitation forms and the mature stage in the life cycle of the thunderstorm cell begins.

The **mature stage** of the ordinary-cell thunderstorm begins when precipitation starts to fall from the cloud. During the mature stage the thunderstorm produces the most lightning, rain, and even small hail. The updrafts in the cumulonimbus become organized and strong, providing the vertical motion needed for cloud-droplet growth as discussed in Chapter 4. Eventually the particles get too large for the updraft to support them in the air, and the cloud particles fall and form the downdraft. As the particles fall out of the cloud base, where the air is unsaturated, they begin to evaporate. This evaporation causes a cooling of the surrounding air, making the air denser and thereby enhancing the downdraft. When the downdraft reaches the ground it spreads out and interacts with the updraft.

The **dissipating stage** of a thunderstorm occurs when the updraft, which provides the required moisture for cloud development, begins to weaken and collapse. During this stage of the thunderstorm life cycle, the downdraft dominates the updraft and the cumulonimbus begins to disappear. Without an updraft the precipitation ends and the cloud begins to evaporate as dry environmental air is entrained into the cloud. In this way, the ordinary-cell thunderstorm snuffs itself out, eliminating the upward supply of high-humidity air needed for thunderstorm formation and maintenance.

The **air-mass thunderstorm** is an example of a single-cell thunderstorm that does not produce severe weather. Air-mass thunderstorms develop in warm humid

Figure 11.8

A photograph of an ordinary thunderstorm.

mT (maritime tropical) air masses and are very common in the southeastern United States, particularly in Florida. They are most prevalent in the summer and complete their life cycle in about one hour (Figure 11.8).

Air-mass thunderstorms cannot produce severe weather because the precipitation falls into the updraft. The drag produced by the falling raindrops eventually extinguishes the updraft before severe weather can occur.

Multicell Thunderstorm

Many thunderstorms are **multicell** storms. Multicell storms are composed of several individual single-cell storms, each one at a different stage of development: cumulus, mature, and dissipating (Figure 11.9). With some cells in the dissipating stage, and others in the cumulus stage, a multicell thunderstorm can last for several hours.

Two basic types of multicell storms are the squall line and the mesoscale convective complex.

Figure 11.9

A schematic of a multicell thunderstorm with cells in each stage of development: cumulus, mature, and dissipating. The red arrows represent the warm updraft and the blue arrows are the cool downdraft. The storm is moving from left to right. (Source: Adapted from Lester, P., Aviation Weather, 2nd ed., Jeppesen, 2001, pp. 9–14.)

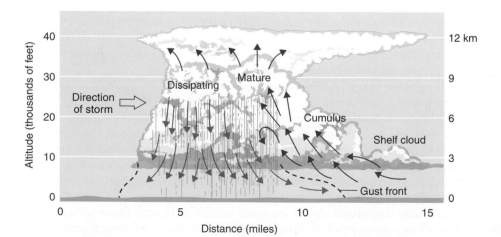

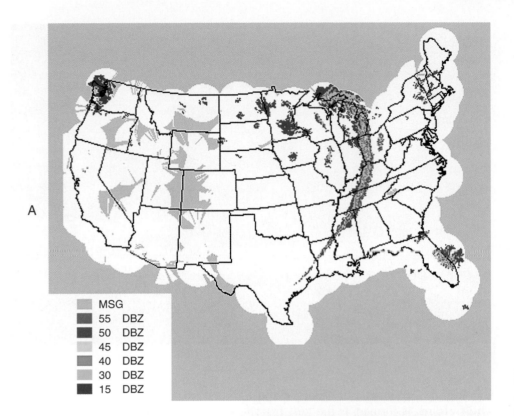

MSG
55 DBZ
50 DBZ
45 DBZ
40 DBZ
30 DBZ
15 DBZ

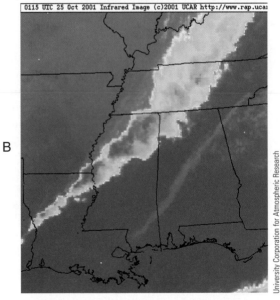

0115 UTC 25 Oct 2001 Infrared Image (c)2001 UCAR http://www.rap.ucar

University Corporation for Atmospheric Research

■ **Figure 11.10**

A, *A composite national radar image of a squall line at 8:15 PM Central Daylight Time on October 24, 2001 (compare to Figure 11.4). The most intense thunderstorms in the squall line are in yellow and red and extend from Detroit, Michigan, to just northwest of Houston, Texas. Gray regions are those without radar coverage, usually because of reflection by high mountains.* **B,** *A close-up colorized infrared satellite image of the same squall line at the same time as in* **A.** *Here, the squall line is bearing down on the Southeast United States. The coldest cloud tops associated with the highest thunderstorms are in blue, and the warmer surface is in red.*

Squall Line

A **squall line** is composed of individual intense thunderstorm cells arranged in a line, or band, as seen in radar and satellite imagery (Figure 11.10, A and B). They occur along a boundary of unstable air, which gives them a linear appearance. Squall lines often have life spans of approximately 6 to 12 hours and extend across several states simultaneously.

Strong environmental wind shear causes the updraft to be tilted and separated from the downdraft in squall-line thunderstorms, as seen in Figure 11.9. This prevents the precipitation from falling into the updraft and quenching it, as happens in the single-cell thunderstorm. The dense, cold air of the downdraft forms the **gust front,** which helps to lift the warm, moist air flowing toward the storm.

http://info.brookscole.com/ackerman

The Photographer's Window

Figure 11.11

The shelf cloud extending out and beneath the main thunderstorm cloud signifies the arrival of the gust front.

Clouds are sometimes observed above the gust front. One such cloud, called a **shelf cloud,** is formed as the gust front forces air near the surface to rise, and is caused by the cool downdraft air lifting warm surface air to its condensation level. While a shelf cloud is very ominous (Figure 11.11), it does not produce damaging weather, although it can precede severe weather by a few minutes.

The weather conditions favorable for the formation of a squall line are divergence aloft and a broad, low-level inflow of moist air. Squall lines are often observed ahead of a cold front, as was the case in Figures 11.4 and 11.10. The low-level winds ahead of the front supply the moist air required to develop and maintain thunderstorms. Divergence aloft induces lifting near the surface, causing the storms to develop and grow.

Mesoscale Convective Complex

The **mesoscale convective complex,** or MCC, is another severe storm composed of multiple single-cell storms. An MCC is a complex of individual storms that covers a large area (100,000 square kilometers or about 40,000 square miles) in an infrared satellite image, roughly the size of the entire state of Iowa. MCCs, like squall lines, are long lived and last for more than 6 hours. MCCs often begin forming in the late afternoon and evening, and reach mature stages during the night and toward dawn. In satellite images MCCs appear as a cluster of thunderstorms that give the appearance of a large circular storm with cold cloud-top temperatures less than $-40°$ C. Figure 11.12 shows an infrared satellite image of a MCC that formed late in the day on July 7, 1997 over Nebraska. Lifted indices for the region were less than -8, indicating that any thunderstorms that formed would be severe.

MCCs often form underneath a ridge of high pressure, as indicated in Figure 11.12. This is because, as we learned in Chapter 6, upper-level divergence can occur in a ridge. This promotes rising motion underneath the ridge, leading to thunderstorm growth if conditions nearer the surface are unstable. The MCC on July 7, 1997 produced several reports of heavy rainfall amounts of 10–15 centimeters (4–6 inches) across parts of Kansas, hail up to 4.5 centimeters (1.75 inches) in diameter, and damaging wind gusts of up to 95 km/hr (60 mph).

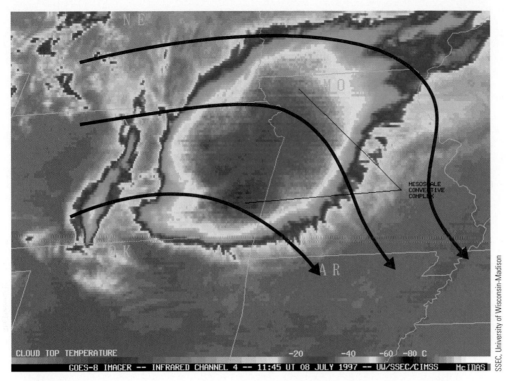

Figure 11.12

Infrared satellite image of a mesoscale convective complex (MCC) over Kansas and Missouri early in the morning on July 8, 1997. The image is color enhanced to highlight the cold cloud-top temperatures. Note the circular shape of the coldest areas, which define the MCC. Overlaid are the wind directions at the 250-mb level, indicating that the MCC is occurring in the vicinity of an upper-level ridge.

Unlike the squall line, MCCs do not require large amounts of vertical wind shear to survive. The MCC is a multicell storm comprised of convective cells in different stages of their life cycles. For MCCs to exist the individual thunderstorms that comprise the system must support the formation of other convective cells. The downdraft of individual cells of the MCC form and enhance the updraft of neighboring cells, as shown in Figure 11.13.

For an MCC to last a long time also requires a good supply of moisture from low levels of the atmosphere. As we noted earlier, this is accomplished over the Great Plains by the nocturnal low-level jet. MCCs are unique in that they are maintained by this low-level jet. The MCCs often move eastward as the southerly low-level jet turns eastward overnight under the influence of the Coriolis force. The low-level jet is lifted over the downdrafts of mature cells, and this maintains thunderstorm development over a long period of time. As the low-level jet weakens

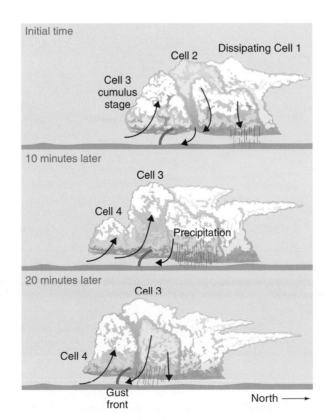

Figure 11.13

A time-lapse schematic showing how the downdraft of thunderstorm cells in an MCC assist in the development of updrafts, leading to new thunderstorms.

at sunrise, so does the thunderstorm complex. MCCs usually reach peak intensity in the early morning hours (midnight to 3:00 AM), causing an otherwise inexplicable *nighttime* peak in thunderstorm frequency and intensity over the Plains and upper Midwest.

Despite their severity, MCCs also do good. The heavy rains of MCCs are an important source of water for the corn and wheat belts of the United States.

Supercell Thunderstorm

The supercell thunderstorm is a single-cell storm that almost always produces dangerous weather. Supercell thunderstorms produce one or more of the following weather conditions: strong wind gusts, large hail, dangerous lightning, and tornadoes. The severity of these storms is primarily due to the structure of the environment in which the storms form.

The development of a supercell requires a very unstable atmosphere and strong vertical wind shear (speed *and* direction). Often in supercell environments, wind direction at the surface is southerly, while the winds aloft are from the west. Wind speed may be 24 km/hr (15 mph) at the surface and over 160 km/hr (100 mph) at the 500-mb level. Under the influence of this strong vertical wind shear, the entire thunderstorm cell rotates!

The structure of a rotating supercell is illustrated in Figure 11.14. The turning of the wind direction causes the updrafts and downdrafts to wrap around one

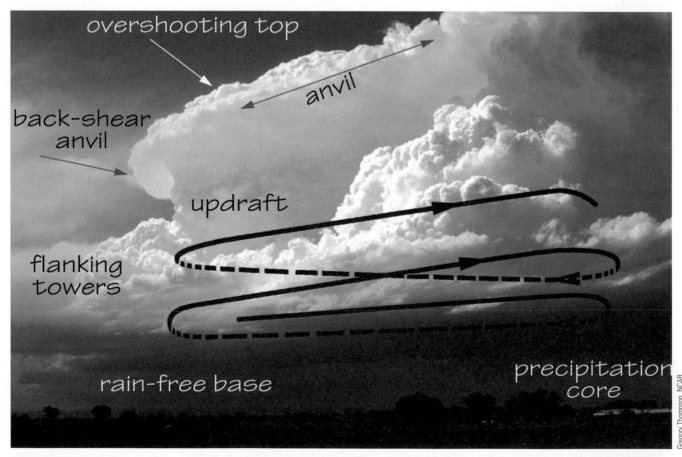

■ **Figure 11.14**

A photograph of a supercell thunderstorm overlaid with a schematic of its windflow and structure, as explained in the text. Photograph and annotations courtesy Gregory Thompson, National Center for Atmospheric Research..

another. The updraft enters the supercell near the rain-free base and slants upward toward the back of the thunderstorm. Overshooting tops indicate where the intense updraft penetrates the stratosphere at the top of the storm. Strong winds aloft blow the updraft downwind where it exits through the anvil, helping to maintain one strong and local updraft.

The main downdraft location of the supercell is located in the precipitation core below the cloud base. To the left of the downdraft are the rain-free cloud base and the flanking line of cumulus towers, which parallels the gust front. Research suggests that as the downdraft behind the flanking line intersects the updraft, rotation results and causes tornadoes to form.

The surface weather conditions associated with a supercell are depicted in Figure 11.15. The region of heaviest rain and hail forms a hook around the updraft, which is near the junction of the gust fronts. At this junction a tornado may form, a possibility we explore shortly. Once initiated, supercell storms sustain themselves for several hours because the updrafts and downdrafts do not interfere with each other.

The favored region for the formation of supercells is the southern Great Plains of the United States in spring. This is because of the extreme instability and the combination of low-level and upper-level wind conditions that exist during this time in these locations. However, a remarkable outbreak of supercells occurred over Illinois on April 19, 1996 (Figure 11.16), causing more than a year's worth of tornadoes in Illinois in just a few hours!

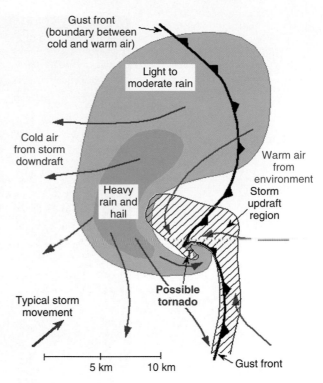

Figure 11.15

A schematic diagram of the surface conditions associated with a typical supercell thunderstorm. A tornado is most likely to be found at the intersection of the downdrafts whose leading edges are depicted as cold (gust) fronts. Heavy rain and hail form a hook-shaped region on radar, usually wrapping around and to the west of the intersection of the downdrafts. (Source: http://k12ocs.ou.edu/teachers/graphic/ SupercellSlice.gif.)

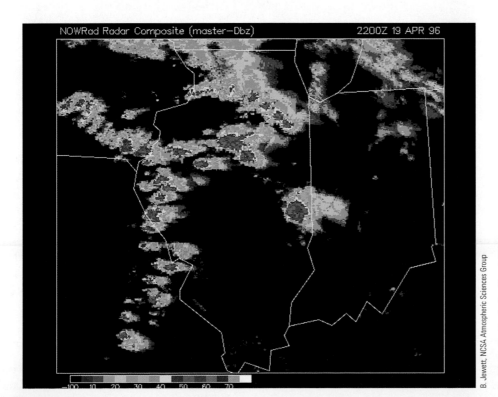

Figure 11.16

Weather radar image from 5:00 PM Central Standard Time on April 19, 1996, depicting a widespread outbreak of supercell thunderstorms over Illinois and Missouri. On this day 36 tornadoes were reported in Illinois, a single-day record for the state.

Merilee Thomas

Figure 11.17

Why chase tornadoes when they can come to you? Audra Thomas poses in front of a majestic tornado passing across the Thomas farm near Beaver City, Nebraska, on April 23, 1989.

The Tornado

Tornadoes are rapidly rotating columns or "funnels" of high wind that spiral around very narrow regions of low pressure beneath a thunderstorm (Figure 11.17). The funnel itself is visible because of moisture that condenses in the rapidly rising and cooling air in the funnel, and also because of dust and debris that are sucked into the vortex. Tornadoes nearly always rotate cyclonically and often move toward the northeast along with the parent thunderstorm. The name "tornado" may stem from the Latin word *tornare,* meaning "to turn" and from the Spanish word for thunderstorm, *tronada.* In the United States, they are also called *twisters* and, as in *The Wizard of Oz, cyclones.* A tornado whose circulation does not extend to the ground is called a **funnel cloud.**

Tornado Formation and Life Cycle

Tornadoes are usually less than 1.6 kilometers (1 mile) wide and are short lived, rarely lasting more than a half hour. Because of their small size, brief existence, and violent nature, tornadoes are particularly difficult to understand. Therefore,

BOX 11.1

Storm Chasers

Don Lloyd

Severe thunderstorms produce damaging, and sometimes deadly, weather events. Because of the danger these storms pose, they are studied at a distance, and unlike hurricanes, they are not probed by aircraft. Because they are still not fully understood, eyewitness observations by "storm chasers" have been instrumental in allowing us to learn more about severe thunderstorms and tornadoes. Storm chasers are trained individuals who seek to gather data, such as video or radar, of severe thunderstorms to further understand their formation and movement. In the photo at right, storm chaser Don Lloyd captured a F5 tornado raking eastern Wisconsin on July 18, 1996.

Storm chasers are exposed to the hazards that often accompany a tornado, such as large hail, lightning, and heavy rains that make driving difficult and dangerous. Storm chasers must have excellent knowledge of thunderstorms and geography. They know that it is better to approach a storm from the southeast to west quadrants so that they can see the tornado without being in its path. They also know it is better to chase in the flat Plains, where tornadoes are visible far away, than in the hills or in the moist Gulf Coast region where cloud bases are low, obscuring the tornado funnel. They chase during the day but not at night, when tornadoes are hard to see. They avoid approaching a severe thunderstorm from the north or east as it exposes them to the region of heaviest rain and hail, a region known to storm chasers as the *bear's cage*.

Although it may seem like thrillseeking, the best chasers practice sound science: They make their own forecasts, use advanced onboard equipment when in the vicinity of severe weather, and share the data they gather with researchers. The chances of actually seeing a tornado are small, perhaps once every ten chases even for experienced chasers. The low odds of success, the long hours of driving, and the cost of travel all make storm chasing a hobby for relatively few severe-weather fanatics. Even so, observations from storm chasers, Doppler radar, satellites, and other instruments have enabled us to better understand the conditions needed for tornado development.

emphasis has been placed on up-close observations of tornadoes, including those made by "storm chasers" (Box 11.1).

No one yet knows exactly how tornadoes form, but close observations of them have revealed many clues about their genesis. Most tornadoes, and the vast majority of damaging tornadoes, drop down out of supercell thunderstorms. Supercells are, as we have seen, rotating thunderstorms. However, their rotation is too slow to explain the fast spinning of tornadoes.

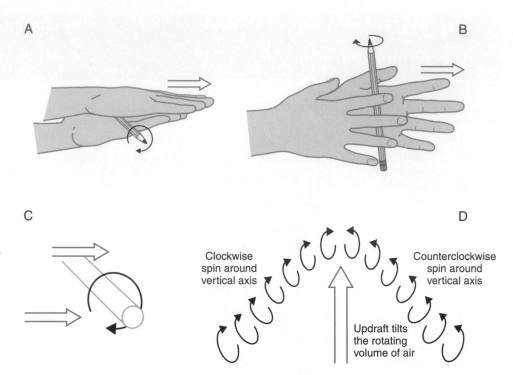

■ **Figure 11.18**

*By moving your hands back and forth, you can make a pencil between them rotate around a horizontal axis (**A**) or, by turning your wrists, around a vertical axis (**B**). Similarly, winds in the vicinity of a thunderstorm can cause a volume of air to spin horizontally (**C**). When pushed upward by the thunderstorm's updraft, it can become two vertically rotating columns of air (**D**). The column on the right in **D** is called the mesocyclone.*

It appears that vertical wind shear near the Earth's surface can increase the spin in some (but not all) supercells. Figure 11.18 illustrates how this can occur. First, to picture how this process works, put a pencil between the palms of your hands, with the pen lying across the lower palm from thumb to pinky and both hands parallel to the ground. Slide your top hand back and forth over the other and notice that the pencil rotates around a horizontal axis (Figure 11.18, *A*). Now, keep sliding your hands while you turn them so that your thumbs are on top. Notice that the pencil is now rotating around a *vertical* axis (Figure 11.18, *B*).

A similar process is at work beneath and inside a tornadic supercell thunderstorm. The changes in wind speed and direction in the vicinity of the gust front are like your hands in Figure 11.18, *A:* they cause spin around a horizontal axis (Figure 11.18, *C*). In the case of the thunderstorm, it is the air, not a pencil, that is spinning. The vertical updraft inside the thunderstorm then tilts this spinning air so that it spins in the vertical, just as you did with the pencil.

Air is not rigid like a pencil, however, and the result is two columns of vertically spinning and rising air, one rotating clockwise (on the left in Figure 11.18, *D*) and the other rotating counterclockwise (on the right). In some cases the thunderstorm will then split into two parts, with the clockwise portion moving to the left of the original storm path and the counterclockwise cyclonic part moving to the right. The cyclonic part is called a **right-mover** and is usually the more intense of the two parts. This is because the right-mover is the more southerly of the two halves of the thunderstorm and has better access to the warm, moist air flowing into the region.

This spinning air in the right-moving supercell is like a miniature extratropical cyclone: it has a lower pressure than its surroundings because of the divergence aloft and the latent heating caused by condensation that warms the air in it. It is therefore called a **mesocyclone,** "meso" meaning "middle" (i.e., mid-sized) in Greek. This mesocyclone is 5 to 20 kilometers (3 to 12.5 miles) wide and can extend well up toward the top of the thunderstorm. As it stretches vertically, the mesocyclone becomes narrower and rotates more quickly, a consequence of the Conservation of Angular Momentum we discussed in Chapters 9 and 10. However, the mesocyclone is *not* a tornado, and many tornadoes form without mesocyclones high above them.

For reasons that are still not fully understood, a tornado sometimes forms in the following way: The cloud base underneath the updraft on the rear side of the

Figure 11.19

The ominous approach of a rotating wall cloud is a sign that a tornado may develop at any moment.

thunderstorm may lower, forming an ominous-looking, rotating **wall cloud** (Figure 11.19). As air continues to flow into the wall cloud, a rapidly rotating column of air much smaller than the mesocyclone may protrude below the wall cloud. As water vapor condenses in the air rushing up into this column, a funnel cloud may form and reach the ground, becoming a tornado.

Once formed, tornadoes exhibit a fairly regular four-stage life cycle, as shown in Figure 11.20 for a carefully observed tornado in Oklahoma in 1973. The first stage is the *organizing stage,* during which a funnel cloud picks up debris as it reaches the surface and widens. The *mature stage* follows when the tornado is often at its peak intensity and width. The tornado reaches the *shrinking stage* when its funnel

Figure 11.20

The life cycle of the Union City, Oklahoma, tornado on May 24, 1973, one of the best-studied tornadoes in history. The boxed line diagrams show what the tornado looked like at various times (shown in 24-hour military time on the graph) during its lifetime, including any dust or debris witnessed near the ground (stippled areas in boxes). The path and width of the tornado are denoted by the two lines stretching across and down the right-hand side of the figure; the tornado is widest just to the west (left) of Union City. The four stages of the life cycle are indicated beneath the tornado path. (From American Geophysical Union, The Tornado, *1993.)*

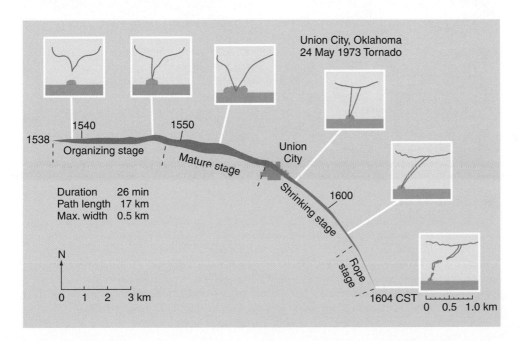

narrows, and ends with a *decaying* or *"rope" stage*. At this point the funnel thins out to a very narrow, rope-like column, after which it eventually dissipates.

Radar Observations of Tornadoes

Visual observations help explain many outward features of tornadoes, but weather radar lets us see into the heart of tornadic thunderstorms. As we noted earlier with reference to Figure 11.15, the pattern of heavy rain inside a supercell forms a kind of hook around the region most likely to produce a tornado. Weather radar beams reflect off this region, which appears as a **hook echo** and usually prompts a tornado warning (Chapter 1) when detected by the NWS.

In addition, modern Doppler radar is also able to detect the rapid change in direction of wind around a spinning vortex. This appears as a couplet of red and green colors on Doppler radar images of the velocity of particles in the funnel itself. This couplet signifies winds next to each other that are moving away from and toward the radar beam, respectively. This couplet is called a **tornado vortex signature,** or **TVS,** and is a reliable indicator that a tornado is forming. Figure 11.21 shows both the hook echo and the TVS from a tornadic supercell that occurred during the April 19, 1996 severe weather outbreak in Illinois (see also Figure 11.16).

Even in the absence of hook echoes and tornado vortex signatures, it is possible to identify severe thunderstorms using radar. As noted above, rotating supercells often are right-movers. On a time-lapse radar loop they stand out because their paths are to the right of the paths of other nonsevere thunderstorms and rain areas. Watch a radar loop on a television weather program during severe weather, look for the right-movers, and make your own amateur severe weather forecasts.

Tornado Winds

Tornado winds are the stuff of legend; some early estimates gauged them to be faster than the speed of sound! Modern-day estimates based on damage and a few

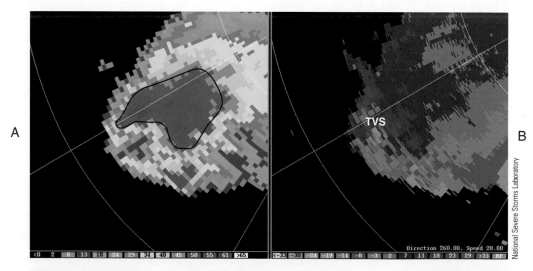

Figure 11.21

Doppler radar reflectivity (A) and velocity (B) images for a tornadic supercell in Illinois during the April 19, 1996 severe weather outbreak (see Figure 11.18). On the reflectivity image, the hook echo is the hook-shaped region of the highest (reddest) reflectivity caused by the heavy precipitation around the tornado itself (compare to the schematic in Figure 11.15). The velocity image at right reveals a small region where bright green and bright red colors are close together, signifying high winds that are blowing toward and away from the radar beam, respectively. This occurs in the spinning column of a tornado and is known as a tornado vortex signature.

TABLE 11.1 The Fujita Scale for Tornadoes

Category		Damage	Wind speed range	
			km/hr	mph
F0	Light	Tree branches broken; heavy damage to crops; chimneys damaged.	64–118	40–73
F1	Moderate	Trees uprooted, some snapped; mobile homes overturned; moving cars pushed off road.	119–181	74–112
F2	Considerable	Large trees uprooted and snapped; mobile homes destroyed; roofs torn off houses; railroad boxcars pushed off track.	182–253	113–157
F3	Severe	Most trees in a forest uprooted or snapped; walls torn off well-constructed frame houses; trains overturned; autos lifted off ground and moved.	254–332	158–206
F4	Devastating	Trees debarked by flying debris; well-constructed frame houses leveled; autos thrown some distance.	333–419	207–260
F5	Incredible	Trees completely debarked; strong frame houses lifted off foundations and demolished over some distance; steel-reinforced concrete structures badly damaged; autos become missiles and fly distances of 100 meters.	420–513	261–318

detailed Doppler radar measurements place an upper limit on tornado winds of about 480 km/hr (300 mph) near ground level. Since the force of wind is proportional to the square of the wind speed, this makes the strongest tornadoes four times as damaging as the winds in a Category 5 hurricane on the Saffir-Simpson scale (Chapter 8).

On May 11, 1970, a destructive tornado in Lubbock, Texas, hit downtown and even the local National Weather Service office. How fast were the tornado's winds? To answer this question, meteorologist Ted Fujita created the **Fujita scale** to estimate the winds of a tornado after the fact based on the damage caused by them (Table 11.1). As with the Saffir–Simpson scale, the higher the "Fujita number," the more severe the storm, but on a scale from 0 to 5. Based on painstaking surveys of tornado damage performed during the past three decades, it has been found that only about 1% of all U.S. tornadoes attain the violent F4 and F5 categories. However, these rare tornadoes account for almost 70% of all deaths by tornadoes.

Figure 11.22 depicts the path width, length, and Fujita rating for the April 26, 1991 Wichita/Andover, Kansas, tornado, which killed 17 people, obliterated a mobile home park, and just missed destroying B-1 military airplanes at McConnell Air Force Base that may have been loaded with nuclear weapons. Comparing this figure to the life cycle in Figure 11.20, notice that the highest winds (highest Fujita number) and greatest width occur during the middle or mature stage of the tornado. This is common, but is not always the case; tornadoes can also be very destructive during the "rope" stage as the funnel tightens and, conserving angular momentum, spins even faster. Five minutes after the Wichita/Andover tornado dissipated, another weaker tornado formed from the same thunderstorm and chased two television cameramen into an overpass across the Kansas Turnpike, videotapes of which were shown all over the world.

Balance in a Tornado

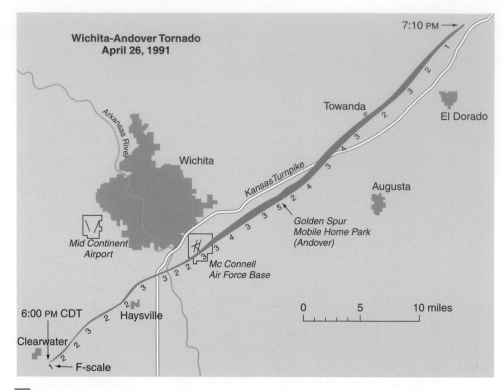

Figure 11.22

The track, width, and Fujita rating of the Wichita/Andover, Kansas, tornado on April 26, 1991. This tornado reached its peak intensity of F5, implying winds up to 512 km/hr (318 mph), just as it reached the Golden Spur Mobile Home Park. Its long path length (46 miles) and lifespan (1 hour and 13 minutes) reflect the fact that weather conditions on that day were ideal for severe weather and violent tornadoes. Lifted index values in Norman, Oklahoma, south of Wichita, were −6 on the morning of this outbreak. (Source: Grazulis, T., The Tornado: Nature's Ultimate Windstorm, Oklahoma University Press, 2000. p. 110.)

How a tornado achieves such high winds is still not completely understood. Some of the worst tornadoes ever observed, such as the F5 that destroyed large parts of Xenia, Ohio, on April 3, 1974, appear to be tornadoes-within-tornadoes (Figure 11.23). This phenomenon is called a **multiple-vortex tornado.** As in a hurricane (Chapter 8), winds are strongest where the forward speed of the storm adds to the tornado's winds, just to the right of the storm path. Where the winds of a suction vortex are combined with this effect (see right-hand side of Figure 11.23), the total winds can exceed 400 km/hr (250 mph) and cause incredible damage. A few houses or a block away, however, the winds may be less than half as fast. This explains why tornado damage can be so hit-and-miss even in the path of a giant twister.

The intense horizontal and vertical winds in a tornado can cause unheard-of damage. Objects weighing a kilogram or two (up to a pound) can be sucked up into the parent thunderstorm and travel over 100 miles downwind from an F5 tornado. In this respect, the images of airborne cows and houses in the movies *Twister* and *The Wizard of Oz* are not as unrealistic as they might initially seem.

Tornadoes are not the only damaging winds caused by thunderstorms. In the next chapter we will explore some non-tornadic winds associated with thunderstorms.

Tornado Distribution

Tornadoes form in regions of the atmosphere that have extremely unstable air (large negative values of the lifted index), large amounts of vertical wind shear, and weather systems such as fronts that force air upward. The United States provides

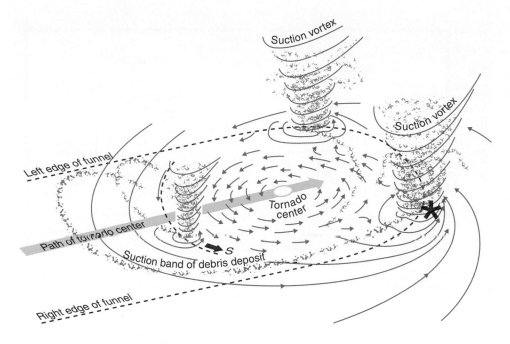

Figure 11.23

Dr. Ted Fujita's schematic of the multiple-vortex tornado, showing how the whirls-within-whirls of the suction vortices can lead to intensified winds in certain locations of the main tornado vortex. On the far right-hand side the effects of the main vortex, the suction vortex, and the forward speed of the tornado itself all combine for maximum winds at the location of the asterisk at far right. (Source: Grazulis, T., The Tornado: Nature's Ultimate Windstorm, Oklahoma University Press, 2000. p. 111.)

these three ingredients in abundance, and so it is not surprising that the majority of the world's reported tornadoes are "born in the USA."

Within the United States, tornadoes can occur in nearly every state and in every month of the year. Figure 11.24 shows the average number of tornadoes per year and reveals a **tornado alley** of highest frequency in the Great Plains centered on Texas, Oklahoma, and Kansas. However, the distribution of violent F4 and F5 tornadoes is a bit different, with additional "alleys" in the Southeast and the Midwest. Killer tornadoes are concentrated in the southern United States from Arkansas to Alabama.

Tornado season is based on when the ingredients for severe weather come together in a particular place. Since vertical wind shear is closely related to the presence of a jet stream, tornado season moves north and south during the year with the jet. Figure 11.25 shows tornado occurrence in Minnesota and Mississippi for each month of the year, compared to the national average. Minnesota's tornado season peaks in June when the jet stream and midlatitude cyclones move over the state. Mississippi, in contrast, has two seasons: early spring and late fall, both of which are times when the jet stream and vigorous lows move through the Southeast.

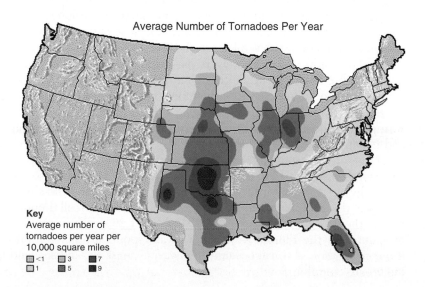

Average Number of Tornadoes Per Year

Key
Average number of
tornadoes per year per
10,000 square miles

<1 3 7
1 5 9

Figure 11.24

Average yearly number of tornadoes per 10,000 square miles across the lower 48 United States. The traditional "tornado alley" from Texas to Kansas stands out as the region of highest tornado occurrence. (Source: http://k12ocs.ou.edu/teachers/graphic/TornadoFreq.gif.)

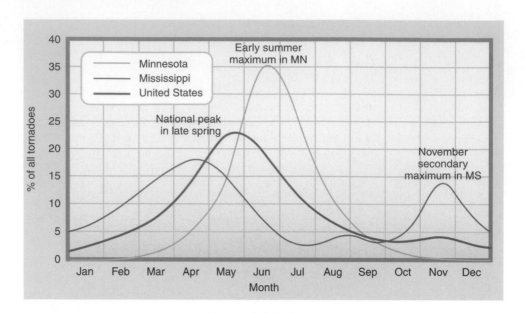

Figure 11.25

Tornado occurrence by month in Minnesota, Mississippi, and the United States as a whole. Note that tornadoes are most common in Minnesota in June, but there is a double peak during April and again during November in Mississippi. (From Tom Grazulis, The Tornado, *University of Oklahoma Press, 2001, p. 267.)*

Nationally, tornado season lasts from approximately March through June, but every month sees some tornadoes in some part of the United States.

Tornadoes can also happen at any time of day or night. However, they thrive on solar heating and in some cases the ability of warm, moist air at the surface to penetrate the capping inversion. Therefore, the most likely times for tornadoes are late afternoon or early evening. Over half of all U.S. tornadoes occur during the hours of 3:00 PM to 7:00 PM local time.

It is sometimes said that tornadoes must be attracted to mobile homes because killer tornadoes, such as the 1991 Wichita/Andover storm, often hit mobile home parks. However, the reverse is actually true: for economic reasons, the number of mobile homes in the violent tornado alley of the Southeast United States has quadrupled in just twenty years. Since mobile homes are especially vulnerable in high winds (see Box 11.2), this may partly explain why killer tornadoes are concentrated in the South while overall tornado occurrence is highest in the Great Plains.

■ The Waterspout

The night-and-morning showers caused by the sea/land breezes along the coast (Chapter 6) appear benign. However, it's not uncommon along the Gulf and Atlantic coasts to hear of "special marine warnings" because of **waterspouts** sighted just offshore in connection with sea-breeze showers.

Like tornadoes, waterspouts are narrow spinning funnels of rising air that form underneath clouds. But while tornadoes generally develop in association with immense cumulonimbus clouds, waterspouts usually form underneath shorter cumulus clouds. As in a tornado, low pressure at the center of a waterspout sets up a pressure gradient that drives air inward. This air rises and cools, causing condensation and making the funnel visible.

Even at their most intense, waterspouts are only as strong as weak tornadoes, with winds less than 160 km/hr (100 mph). This is partly because the wind patterns that help the waterspout spin are limited to the area near the surface. In contrast, we learned earlier that many tornadoes form inside supercell thunderstorms that rotate all the way up, from base to anvil. Figure 11.26 shows an airborne view of a waterspout near the Florida Keys, where waterspouts are fairly common because of the combination of warm ocean water, warm, moist air, and sea/land breezes helping create cumulus clouds.

J. Golden, NOAA/American Red Cross

Figure 11.26

A waterspout in the Florida Keys, as photographed by meteorologist Joe Golden.

BOX 11.2

Severe Weather Safety

The most important lesson you can learn from this book is how to protect yourself from injury or death during dangerous weather. Here we discuss some safety rules for the three biggest severe weather killers in the United States today: tornadoes, lightning, and flash floods.

First of all, you need access to reliable weather information. Commercial television and radio are often good sources of current weather updates. Keep in mind that the watches and warnings they report are issued by government meteorologists.

The national system of severe weather watches and warnings has saved untold numbers of lives in the United States since the 1950s. NOAA's improved Weather Radio network puts nearly everyone in the nation within range of government weather broadcasts. These radios can sound an alarm whenever severe weather warnings are issued for your area. You should respond quickly when these warnings are issued. For example, today's tornado warnings from the National Weather Service provide, on average, about 10 minutes of lead time before the tornado strikes. What should you do? Below are some safety guidelines. More information is also available at the nearest NWS forecast office.

Tornado

In general, try to get as low as you can. This is because tornado winds decrease close to the ground because of friction. Go into a tornado shelter, or the basement, or into a small interior room on the lowest floor of a building, such as a bathroom or closet. Protect yourself from flying debris and stay away from windows. Don't bother opening or closing them, as you put yourself at risk of being hit by flying debris. Also, opening windows may increase tornado damage. In particular, protect your head. Wear a bicycle or motorcycle helmet if one is available.

If you are away from home during a tornado, you are at greater risk. As always, you need access to current weather information. In 1994 twenty people died in an Alabama church when a tornado blew in an 18-foot-high, 60-foot-wide wall of the church during Palm Sunday services. The death toll was heavy because news of severe weather watches and warnings for the area never reached the church.

If you are away from a sturdy home, you must seek adequate shelter immediately in the event of a tornado. At school or in a dorm, follow the severe weather safety plan in place for that building. Avoid auditoriums, gyms and eating areas; their large, high roofs can blow off and the walls can collapse. If you are in a mobile home or car, leave it and go to a strong building. Many people are killed when cars and mobile homes are overturned in high winds. If there are no shelters nearby, get into the nearest ditch or depression and protect yourself from flying debris.

Television videos have wrongly popularized the notion that it is safe to drive at high speeds away from a tornado and, if necessary, to hide under a highway overpass as the tornado passes overhead. Don't do it. It is far safer to take shelter in a sturdy building instead. A highway overpass creates a "wind tunnel" effect underneath it and can increase the amount of damage due to a tornado.

Lightning

Because of the vast differences in the speed of light and the speed of sound, the flash of lightning precedes the rumble of thunder. It takes sound waves 5 seconds to travel 1 mile (or 3 seconds per kilometer), whereas the flash of lightning travels the same distance in less than 1/100,000th of a second. For example, if 15 seconds elapse between a lightning bolt and the arrival of its thunder, the bolt was 5 km (3 miles) away.

However, just because a lightning bolt is a few miles away, this doesn't mean you are safe! The NWS advocates the 30/30 rule: take cover if you hear thunder within 30 seconds of the lightning, and then wait at least 30 minutes after the last lightning flash before resuming outside activities.

Follow these safety rules in the event of lightning. If you are outside, get into an enclosed building. Avoid being in or near high places or in open fields. If outside avoid all metal objects, such as flagpoles, metal fences, golf carts, baseball dugouts, and farm equipment. If you are in a forest, seek shelter in a low area under a thick growth of bushes or small trees. In open areas, go to a low place and crouch down on the balls of your feet. Don't lie down; if lightning strikes nearby, you minimize your risk of burns by crouching instead of lying. Stay away from open water. On the water you are usually taller than your surroundings and vulnerable to a direct lightning strike. In the water you can be shocked, since water can conduct electricity from a lightning strike over long distances. Fully enclosed, all-metal vehicles with the windows rolled up usually provide good shelter from lightning—but avoid contact with any metal.

Flash floods

Stay away from streambeds, drainage ditches, and culverts during periods of heavy rain. Move to high ground when threatened by flooding. Stay out of flooded areas. Never drive your car across a flooded road, even if you think the water is shallow. Most flash-flood-related deaths occur when people drive into floodwaters. Never underestimate the power of moving water!

Keep safety in mind even after the storm passes. Stay clear of downed power lines. Do not touch them. If you smell gas or suspect a gas leak, turn off the main valve, get everyone out of the structure quickly, and open the windows. If there is a power outage, use flashlights instead of candles, which have open flames that might start a fire. After high winds use caution when walking around trees, as trees and tree limbs may be weakened and could fall unexpectedly. Deal with immediate problems, such as helping injured people, until professional help arrives.

Other Thunderstorm Severe Weather

Other types of severe weather produced by thunderstorms include lightning, flooding, hail, and high winds. These weather phenomena are some of the most spectacular sights in nature. A single thunderstorm can generate all of these perils, or only one. We explore some of these below.

Lightning

Lightning is a huge electrical discharge that results from the rising and sinking air motions that occur in mature thunderstorms (Figure 11.27). Each year in the United States lightning kills, on average, 85 people and injures approximately 300 people. Lightning also causes several hundred million dollars of property damage each year. While your chances of being struck by lightning are small (about 1 in 600,000) it is important to understand how nature's fireworks operate.

Lightning can travel from cloud to cloud, within the same cloud, or from cloud to ground. In-cloud lightning discharges are far more common than cloud-to-ground discharges and are not as hazardous. The processes that lead up to this electric discharge, or lightning flash, are the same for these three types.

Though lightning appears to be a continuous almost instantaneous flash of light to human eyes, high-speed photography shows that a "lightning bolt" is actually a series of flashes. Let's break down cloud-to-ground lightning into a sequence of split-second events. To explain the sequence, let's consider lightning striking a tall building (Figure 11.28).

• *Charge separation in the cloud.* Meteorologists do not fully understand how a cloud gets electrified. Current research indicates that electric charges get distributed throughout the cloud by the collision of ice particles with ice-covered snowflakes or "graupel" (see Chapter 4) at different temperatures. When the collisions occur at temperatures below −15° C (5° F), the ice crystals become positively charged while the graupel acquires a negative charge. At temperatures warmer than −15° C, the collision induces a positive charge on the graupel. The updrafts that maintain the cloud storm carry the particles to different regions of the cloud. The ice crystals are moved upward to the top of the storm while the graupel collects lower in the cloud. In-cloud lightning is the surge of electric current that passes between the negatively and positively charged regions of the cloud.

Harald Edens

Figure 11.27

Lightning brilliantly illuminates the night sky.

■ **Figure 11.28**

*A sequence of events that leads to a cloud-to-ground lightning strike. **A,** Charges collect in the base of the cloud. **B,** As negative charges build up near the base of the cloud, the ground repels negative charges and changes from its usual negative to a positive charge. Lightning formation has begun with the pilot leader. **C,** The stepped leader connects the cloud to the ground. **D,** The return stroke surges upward.*

- *Ground becomes positively charged.* Opposite charges attract, and like charges repel. As the cloud base becomes negatively charged, the objects on the ground becomes positively charged. The atmosphere is resistant to the flow of electricity, which allows the development of a very large difference between the charges on the cloud and ground and establishes conditions for lightning strike. The voltage begins to build as the negative charges continue to collect near the base of the cloud. Air is a good insulator; it can separate voltages as great as 9,000 volts per meter (3,000 volts per foot). Lightning results when the voltages climb above this value.

- *Lightning formation begins.* Once the charge difference becomes so large that the atmosphere can no longer insulate the two regions, negative charges near the cloud base begin to move toward the ground. This is initiated when a small pocket of positive charges collects at the ground below the cloud. The initial discharge of negative charges near the cloud base is called the **pilot leader.** Once this first step occurs, electrons flow downward toward the ground into the pilot leader, and continue to surge down in a sequence of events toward the ground. This flow of charge creates **stepped leaders,** which attempt to establish a conductive channel from the cloud to the ground for electrons to flow through and neutralize the charge difference between the cloud and ground. Stepped leaders

Blue Skies

Severe Weather/Lightning

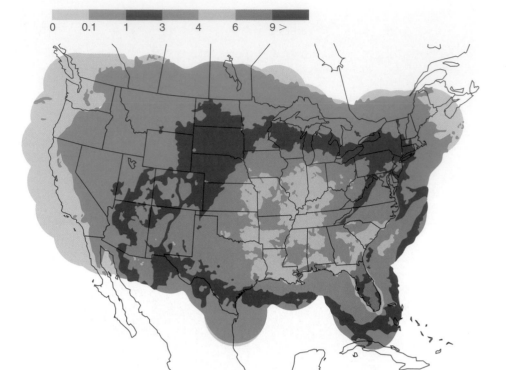

■ **Figure 11.29**

The average number of cloud-to-ground lightning flashes per kilometer across the lower 48 United States, as measured electronically by the National Lightning Detection Network for the years 1989 through 1998. Over 216 million flashes were recorded during this period. (Source: Orville, R. and Huffines, G., "Cloud-to-Ground Lightning in the United States: NLDN Results in the First Decade, 1989–1999," Monthly Weather Review, May 2001, Figure 3.)

Heat Lightning

propagate toward the ground in distinct steps that look like branches. The stepped leaders are very faint and are about 50 meters (160 feet) long and about 2.5 centimeters (1 inch) wide. By the time a stepped leader surges downward, objects near the ground (particularly tall, sharp, metal objects) have become positively charged. As the channel nears the ground, a spark occurs to complete a narrow channel that serves as a conduit for electrons to flow through.

- *Brilliant flash is observed.* The instant a channel is established from the cloud to the ground, the current flows upward through this charged channel, generating a brilliant flash known as the **return stroke.** After the initial return stroke, negative charges from higher in the cloud move toward the ground; these are called **dart leaders.** The dart leaders may generate additional return strokes if they reach the ground. What appears to the human eye as a single lightning stroke is actually a series of return strokes that occurs too fast for the eye to distinguish

During the past two decades, meteorologists have devised methods for recording the location and other characteristics of cloud-to-ground lightning bolts striking the United States. Figure 11.29 shows a recently published decade-long climatology of 216 *million* lightning strikes. Notice that lightning is generally most prevalent where thunderstorms are most common (refer to Figure 11.3). Floridians are at the most risk from lightning, not only because of the frequency of lightning but also because of the outdoor lifestyle of Floridians that often makes them the highest objects on the water or on the golf course—natural targets for lightning. Box 11.2 explains how to protect yourself from lightning.

■ Flash Floods and Flooding

A **flood** is a substantial rise in water that covers areas not usually submerged. A flood occurs when water flows into a region faster than it can be absorbed (i.e., soaked into the soil), stored (in a lake, river, or reservoir), or removed (in runoff or a waterway) into a drainage basin. Common causes for floods are high-intensity rainfall, prolonged rainfall, or both.

Matthew C. Larsen, U.S. Geological Survey, 2000

■ **Figure 11.30**

Torrential rains triggered by a stalled cold front caused over 1.2 meters (nearly 50 inches) of rain in mountainous coastal Venezuela in December 1999. Most of this rain fell in only 3 days. The ensuing flash floods roared through the third story of the apartment building in this photo, causing a partial collapse. Approximately 30,000 people were killed by these flash floods.

Floods pose the greatest weather-related threat to human life, killing over 130 people each year in the United States (see Table 11.2). Not all floods are associated with thunderstorms, although most flood-related deaths in the United States result from floods caused by slow-moving thunderstorms or series of thunderstorms that move over the same region. As we have seen, hurricanes (Chapter 8) and extra-tropical cyclones (Chapter 10) can also produce flooding. Thunderstorms are of particular concern because they can produce a very dangerous condition known as a *flash flood*.

A **flash flood** is a sudden, local flood that has a great volume of water and a short duration (Figure 11.30). Flash floods occur within minutes or hours of heavy rainfall, or because of a sudden release of water from the break-up of an ice dam or constructed dam. The worst flash flood in the United States occurred in Johnstown, Pennsylvania, on May 31, 1889. A dam broke as a result of heavy rain and structural problems, resulting in a 12-meter (36- to 40-foot) wall of water that swept through the town and killed 2200 people.

Rainfall intensity and duration are two key elements of a flash flood. Topography, soil conditions, and ground cover also play important roles. Steep terrain can cause rain water to flow toward and collect in low-lying areas, causing water levels to rise rapidly. If the soil is saturated with water, it cannot absorb more and so the excess water runs off the land quickly. On the other hand, extremely dry soil conditions can also be favorable for flooding. Dry

TABLE 11.2   Weather-Related Deaths in the United States

Weather Event	Average Deaths per Year
Flood	136
Lightning	85
Tornado	73
Hurricane	25
Hail	1

soil often can develop a hard crust over which water will initially flow as if the ground were concrete.

Flash floods occur within about six hours of a rain. A flood is a longer-term event and can last weeks or months. Record flooding along the upper Mississippi River in the summer of 1993 resulted from prolonged rains in the upper Midwest of the United States and caused up to $10 billion in damage. These kinds of floods occur during blocking events (Chapter 7) in which thunderstorms develop and move over the same region repeatedly. The worst such flood in U.S. history occurred in 1927, when nearly all the tributaries of the Mississippi breached their banks because of thunderstorm rains. Seventy thousand square kilometers (27,000 square miles) of land along the lower Mississippi was flooded to depths of 30 feet, 700,000 people were left homeless, and the $1 billion in damage was nearly a third of the U.S. government's annual budget at the time!

Floods are natural phenomena and do have benefits. Large seasonal floods have resulted in productive farmland, such as in central North America and along the Nile river, by bringing nutrient-rich, fine soil to the flooded region. Floodwaters also refill wetlands and replenish groundwater. While flooding is a natural event, humans increase the likelihood of flooding by changing the character of the land through such actions as paving with asphalt and removing vegetation on hillsides. This promotes rapid runoff and flooding.

■ Hail

Hail is precipitation in the form of large balls or lumps of ice (Figure 11.31). Hailstones begin as small ice particles that grow primarily by accretion and therefore require abundant supercooled water droplets. Hailstones can be as large as oranges and grapefruits. How do they get so large?

Remember from Chapter 5 that precipitation particle size depends on how long it stays in the cloud and the amount of water available for growth. When a hailstone is cut in half (see Figure 11.31), rings of ice are observed. Some rings are milky white; others are clear. This ringed structure suggests a cycling of the hailstone through the storm. To explain this ringed structure let's consider how hail grows in a supercell thunderstorm.

Hail embryos are small particles of ice that grow into hailstones. Embryos form along the edge of the storm's main updraft and circle around the updraft (Point 1 in Figure 11.32). As the embryos rise in the storm they grow, reaching sizes of approximately one millimeter. Circulations within the storm move the embryos toward the front of the storm where the updraft is weaker. The ice particle then falls because of gravity along the front (or forward) edge of the main updraft (Points 2 and 3 in Figure 11.32). Because the updraft is tilted, a growing hailstone can fall into the leading edge of the strong updraft (where temperatures are near freezing and water accumulates on the ice) and be carried back up repeatedly into the storm to be refrozen (Points 4, 5, and 6 in Figure 11.32). As the hailstone vertically cycles through the storm, it collides with supercooled water droplets (Chapter 4) of different sizes, growing larger with each collision. Eventually the hailstone falls through the cloud to the ground (Point 7 in Figure 11.32).

Small supercooled droplets freeze quickly when they collide with a hailstone. The rapid freezing traps air bubbles that cause the ice to appear white because of multiple scattering (see Chapter 5). Larger supercooled droplets freeze slowly, spreading over the hailstone and allowing air bubbles to escape. Accretion of large supercooled droplets forms the clear layers of a hailstone. This explains the ringed structure of the inside of a hailstone.

The production of large hail requires a strong updraft that is tilted and an abundant supply of supercooled water. Since strong updrafts are required to generate large hailstones, it is not surprising to observe that hail is not randomly distributed

NCAR/UCAR/NSF

■ Figure 11.31

Hailstones grow inside of thunderstorms, with clear and white rings developing as the stone cycles through the tall, moist cumulonimbus cloud. Hailstones fall out of the cloud when the updraft no longer can support the stone's weight. This hailstone is baseball-sized. However, the largest hailstone on record was found in Coffeyville, Kansas, in September 1970. It weighed over 0.7 kilograms (1.5 pounds) and had a diameter of 14 centimeters (5.5 inches).

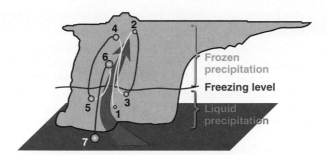

■ **Figure 11.32**

The life cycle and path of a hailstone in a supercell thunderstorm. (Source: http://k12ocs.ou.edu/ teachers/graphic/ HailstoneLifecycle.gif.)

in a thunderstorm but occurs in regions near the strong updraft. The curtain of hailstones that falls below the cloud base is called the **hailshaft.** These regions are often said to appear green to observers on the ground, although recent research suggests that heavy rain as well as hail can create this optical phenomenon. As the storm moves, it generates a **hailswath,** a section of ground covered with hail.

Hailstorms can severely damage crops, automobiles, and roofs. As seen in Figure 11.33, in the United States large hail is most common in southeastern Wyoming and eastern Colorado, not in Florida where thunderstorms are most common. This is because the dry air of the high Plains allows falling hailstones to preserve themselves by self-cooling through melting and evaporation. In contrast, the moister air of the Gulf Coast does not allow as much evaporation, and the rapidly melting hailstones simply turn into large raindrops.

For this reason, hail is of more concern to residents of Denver, Colorado, and Cheyenne, Wyoming, than to those in stormier Tampa, Florida. A hailstorm in and near Denver on July 11, 1990 caused $625 million in property damage. The storm caused a power outage, leaving people stranded on a Ferris wheel and exposed to the storm's fury—forty-seven of whom where injured by the storm's softball-sized hail. However, most hail damage is to crops. Hailstorms, which occur worldwide and frequently during the growing season, destroy approximately 1% of the world's annual agricultural production.

The great loss of property attributable to hailstorms has generated efforts to suppress or prevent hail. Unfortunately such efforts have not been fruitful and many farmers, particularly in the midwestern United States, purchase crop-hail insurance for economic protection.

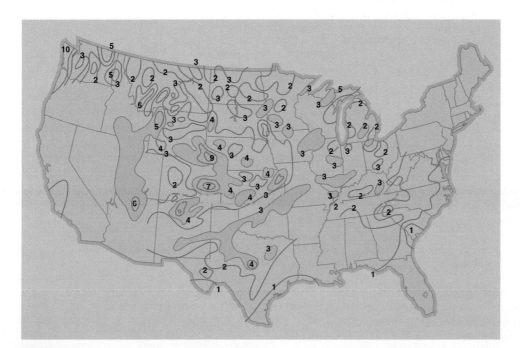

■ **Figure 11.33**

The number of days each year with hail 1.9 centimeters (0.75 inch) in diameter or larger. How does the frequency of large hail compare and contrast to the distribution of thunderstorms and lightning in Figures 11.3 and 11.29? (Source: Illinois State Water Survey.)

PUTTING IT ALL TOGETHER

 SUMMARY

Thunderstorms produce lightning, thunder, tornadoes, floods, hail, and other severe weather. All thunderstorms form in unstable air masses. Indices such as the lifted index allow meteorologists to quickly assess the stability of the atmosphere by using observations of the atmosphere above a particular location. Severe thunderstorms grow in an unstable environment that also has vertical wind shear, a change in wind speed and/or direction with increasing altitude. Environmental vertical wind shear is important for severe thunderstorms as it helps to separate the updraft from the downdraft. This enables the storm to last longer and grow more severe.

The basic building block of a thunderstorm is the cell. A thunderstorm can be composed of a single cell or multiple cells. Air-mass thunderstorms are single, ordinary cells that are not associated with severe weather. Single-cell thunderstorms follow a predictable life cycle from cumulus and mature stages to a dissipating stage in an hour or less. Squall lines and circular mesoscale convective complexes are examples of multicell thunderstorms, both of which can last for hours and produce severe weather. Supercells are single-cell thunderstorms, but they last longer than ordinary thunderstorms and often produce severe weather.

Supercell thunderstorms develop when the environmental wind changes in direction with altitude, causing the supercell to rotate. This rotation promotes the development of tornadoes. Tornadoes and large hail are produced by thunderstorms that grow in unstable atmospheres with vertical wind shear.

Tornadoes are violently rotating vertical columns of air that stretch from the cloud base to the ground in the updraft region of a storm. A sign of tornadic conditions is a rotating wall cloud or a funnel cloud extending downward from the base of the thunderstorm.

Tornadoes appear on radar as hook echo reflections and/or mesocyclones with particles rushing toward and away from the radar beam, forming a tornado vortex signature. Once formed, the tornado usually grows in size and strength, and then narrows to a thin "rope" before dissipating. Tornadoes are usually less than 1.6 kilometers (1 mile) wide and last for less than half of an hour.

Tornado winds can be as high as 480 km/hr (300 mph) and are estimated using the Fujita scale. The Fujita scale is damage based and runs from 0 (light damage) to 5 (incredible damage). Only about 1% of U.S. tornadoes are F4 or F5 tornadoes, but they cause almost 70% of tornado deaths. Many U.S. tornadoes occur in the Great Plains in spring, but they can occur over nearly the entire country in every month of the year.

Lightning is a huge electrical discharge that results from rapid rising and sinking air motions within the thunderstorm that cause collisions of ice particles with graupel. A typical lightning flash is a composite flash composed of several lightning strokes. The stepped leader with subsequent dart leaders leads each stroke. The return stroke is the part of the lightning flash that we typically observe.

The most life-threatening thunderstorm weather is flooding, particularly flash flooding. A flash flood occurs suddenly and floods a region with a great volume of water in a short time span. Rainfall intensity and duration are two key elements of a flash flood. Topography, soil conditions, and ground cover also play important roles. Severe long-term flooding can occur when thunderstorms repeatedly drench a region.

Hail occurs when the strong updrafts and downdrafts in a thunderstorm cause repeated freezing of supercooled water on small ice particles. The curtain of hail that falls near the updraft, the hail shaft, can cause severe damage to cars and agriculture as the thunderstorm passes over urban and rural areas. Although thunderstorms and lightning are most common in the Southeast United States, the Front Range of the Rocky Mountains is a preferred area for hailstorms.

Thunderstorms and their attendant severe weather can threaten your life. While there are many different dangers associated with severe thunderstorms, we can summarize who is most at risk from severe weather. People who are outdoors in the open, under trees, or on the water are most at risk from lightning. People in mobile homes and automobiles are at higher risk when tornadoes are nearby. People in automobiles are also at risk in flash-flooding conditions, as are people in low-lying areas and canyons.

Thunderstorms are complex weather systems and no one chapter can cover all of their aspects. In the next chapter, we look at small-scale wind patterns, some of which are created in the vicinity of thunderstorms.

 KEY TERMS

You should understand all of the following terms. Use the glossary and this chapter to improve your understanding of these terms.

Air-mass thunderstorm	Multicell
Capping inversion	Multiple-vortex tornado
Cell	Nocturnal low-level jet
Cumulus stage	Ordinary cell
Dart leader	Overshooting top
Dissipating stage	Pilot leader
Entrainment	Return stroke
Flash flood	Right-mover
Flood	Severe thunderstorm
Fujita scale	Shelf cloud
Funnel cloud	Squall line
Gust front	Stepped leader
Hail	Supercell
Hailshaft	Thunderstorm
Hailswath	Tornado
Hook echo	Tornado alley
Lifted index	Tornado vortex signature
Lightning	(TVS)
Mature stage	Wall cloud
Mesocyclone	Waterspout
Mesoscale convective complex (MCC)	

 REVIEW QUESTIONS

1. Explain the differences between a severe thunderstorm and an air-mass thunderstorm.
2. What causes lightning? What causes thunder?
3. What should you do when a tornado warning is issued for your location?

4. Describe how a hailstone can grow to the size of a grapefruit.

5. Are thunderstorms likely if air is converging in the upper troposphere?

6. If the surface temperature is 20° C, the surface dew point is 20° C, the moist adiabatic lapse rate is 5° C per kilometer, and the 500 mb (5.5 kilometer altitude) temperature is −1.5° C, what is the lifted index equal to? If thunderstorms occur, are they likely to be severe?

7. Draw a vertical profile of temperature, dewpoint temperature, and wind speed that would be favorable for the formation of a severe thunderstorm.

8. What are the three stages of an ordinary thunderstorm life cycle? When would you expect lightning to occur and why?

9. What should you do if you were caught by surprise by a thunderstorm while you are in a large, open area?

10. Discuss the differences and similarities between a cold front and a gust front.

11. Explain why a tilted updraft is necessary for long-lived thunderstorms and the formation of hail.

12. Why are thunderstorms most common in Florida? Why are tornadoes more common in Oklahoma than in Florida?

13. You are standing outside and you see a shelf cloud. What type of weather phenomenon is nearby? Later you see a wall cloud. What type of weather phenomenon may soon occur?

14. A television meteorologist can tell viewers that a "Cat 5" hurricane is approaching the coast, but an "F5" tornado can only be identified after the fact. With reference to the Saffir–Simpson and Fujita scale definitions in Chapter 8 and this chapter, explain why there is a difference between the two scales.

15. What is the "rope" stage of a tornado?

16. "Heat lightning" is the term popularly applied to lightning that is seen on summer nights and is not accompanied by thunder. Do you think that lightning can occur without creating thunder? See the text's Web site for an explanation of this phenomenon.

17. Weather lore states that in thunderstorms you should
    Beware the oak,
    It draws the stroke;
    Avoid the ash,
    It draws the flash;
    But under the thorn,
    You'll come to no harm.
    Relate this folklore to the safety precautions discussed in Box 11.2.

18. Using the *Blue Skies* CD-ROM exercise "Severe Weather/Lightning," examine the life cycle of a lightning stroke. Which portion of the stroke is the brightest: the stepped leader, the return stroke, or the dart leader? If you saw the stepped leader forming, would you have time to take cover?

■ WEB ACTIVITIES

Choose Chapter 11 on the textbook's Web site:
http://info.brookscole.com/ackerman
and select from the following resources:

• Interactive Modules that illustrate and extend your understanding of key topics in this chapter

• Tutorial Quizzes to test your mastery of terms and concepts

 For additional readings, go to the InfoTrac College Edition, your online library, at: http://www.infotrac-college.com

Small-Scale Winds

After completing this chapter, you should be able to:

■ Relate the concept of turbulence to "friction" in the atmosphere

■ Explain what the dominant force(s) are in most small-scale winds, and why

■ Name and locate on a map likely locations for the occurrence of various small-scale winds

Introduction

The phone rings in the office of a government meteorologist. The caller wants to know the wind conditions at a spot along a road at a certain time on a day many months ago. The reason: The caller was moving his new dishwasher in the back of his pickup truck when it blew off the truck and was destroyed. He wants specific weather information proving to the insurance company that a gust of wind did in his dishwasher.

The meteorologist is stumped. She knows that even in today's high-tech world we simply do not have weather information on such small scales of time and space. Furthermore, we also don't have as good an understanding of small-scale winds as we do of large-scale winds. Ironically, meteorologists comprehend more about the winds in an extratropical cyclone or a hurricane than they do about the local winds that blow in our faces every day! The meteorologist tells the caller, "I'm sorry, the best we can do is an hourly observation of winds 30 miles away from the scene of the disaster." It's more likely that a gust of wind from a passing car, not a weather system, caused the dishwasher's demise. But she can't quite know for sure.

In this chapter we tackle the topic of small-scale winds. For our purposes, "small-scale" usually means winds that occur in a timespan you can perceive (usually minutes or hours) and across distances you can see (often a few tens of miles or so). This is the great frontier of meteorology, because so little is known about weather on these dimensions. It's not for lack of trying; as we'll see, it is simply a fact of the atmosphere that when meteorologists have to "sweat the small details," they end up perspiring a whole lot. Such is the maddening difficulty of small-scale meteorology, the gentle breezes and sudden windstorms that defy explanation.

Two unifying principles guide our study of these bedeviling winds. These principles are (1) the balance of forces we learned about in Chapter 6, and (2) the geographic features of the local landscape. The Coriolis force, which is so crucial for explaining the large-scale features of the atmosphere addressed in Chapter 7, is usually negligible for small, quick wind patterns. In the absence of the Coriolis force, the pressure gradient and frictional forces dominate. Knowing this helps us understand small-scale winds. For all of their complexity, small-scale winds generally boil down to the interaction between a relatively strong pressure gradient and whatever is in its way—a mountain, a valley, a lake, a dusty plain, or even an airplane.

Because geography is so crucial in the way that small-scale winds develop, we will study them by taking a tour of local winds across the United States. The diversity of America's landscapes and meteorology creates a wide assortment of winds, spanning most of the types observed worldwide. After a short introduction to the messiness of turbulence, we will follow the Sun and take an east-to-west journey through America, its small-scale winds, and the small-scale winds of the world.

Friction in the Air: Turbulent "Eddies"

Friction is a familiar concept to us: driving a car, sanding furniture and striking a match all involve one rough surface coming into contact with another. Air, however, doesn't have any rough surfaces. How, then, is there any friction? In Chapter 6 we ascribed friction to the contact between the air and the Earth's surface, but now we will explore this concept in more detail.

The friction in a fluid, such as air, is called **viscosity.** You hear this word on TV in relation to motor oil in cars. In the atmosphere, viscosity means the same as for motor oil: the higher the viscosity, the more friction there is.

Viscosity comes in two scale-dependent varieties. There is friction at the smallest scales when molecules bump into each other. This happens in particular near boundaries, such as the ground (which is a rough surface). This is called *molecular viscosity.* But if molecular viscosity were the only kind of friction, then the atmosphere from just above the ground on up would never feel the effects of friction.

The real "friction" in the atmosphere arises from the jostling of the wind with human-sized swirls of air, not tiny molecules. These swirls are called **eddies,** the same name given to swirls of water in a stream or in the ocean. They arise in the atmosphere when the wind blows over or around obstacles such as trees or buildings. Daytime heating by the Sun also leads to eddies; in addition, the atmosphere naturally develops eddy motions, especially near the Earth's surface. At the smallest scales, the eddies themselves lose their energy to molecular viscosity.

These invisible eddies impede the smooth flow of wind by causing slower-moving air to mix with higher-speed air. It's similar to traffic merging onto a crowded highway: the right-lane traffic jams up as slower cars from the on-ramp mix into the main flow of traffic. In the same way, eddies mix air from the surface, where winds are slow, with faster-moving air higher up. As a result, the overall wind slows down.

For this reason, the analogy is made between friction and the effect of eddies on the wind. Meteorologists call this slowing-down of wind the *eddy viscosity.* The jostling of air with the swirls, as well as the ever-changing motions within the swirls themselves, leads to very irregular fluctuations in the wind. This irregular, almost random pattern of wind is called **turbulence,** and the eddies are called *turbulent eddies.* The fluctuations we call *gusts.* Figure 12.1 schematically illustrates the relationship between eddies, turbulence, and wind gusts.

What is turbulence, *really?* If you know, please tell the world's greatest scientists right away, because they don't know yet. The atmosphere is enormously complicated at small scales. A famous physicist once told the British Association for the Advancement of Science:

> "I am an old man now, and when I die and go to heaven there are two matters on which I hope for enlightenment. One is quantum [physics], and the other is the turbulent motion of fluids. And about the *former* I am rather optimistic."

Therefore, you can be content with a definition of turbulence that comes from experience with aircraft flights: "bumpiness due to small-scale changes in the wind." Box 12.1 explores the fascinating and unsolved problem of clear-air turbulence. To summarize what we've covered so far in the context of our understanding of the atmosphere: the atmosphere contains wind patterns at all different scales. At the smaller scales, winds are slowed down and made irregular—turbulent—by the effect of eddies. This friction-like process is a "brake" on the natural tendency of the pressure gradient force (PGF) to push air from high to low pressure at all scales. And at the tiniest scales, true friction—the rubbing-together of molecules—does

Kelvin–Helmholtz Instability

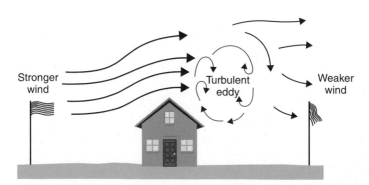

■ **Figure 12.1**

The relationships among eddies, turbulence, and wind gusts.

BOX 12.1

Clear-Air Turbulence

Airplanes and turbulence don't mix. Commercial and especially private aircraft are roughly the same size as large turbulent eddies high up in the atmosphere. This means that planes travel from one bump to another very quickly. The result can be in-flight chaos. One airline pilot in the 1970s recalled from a particularly severe encounter with turbulence:

> "I felt as . . . one might expect to encounter sitting on the end of a huge tuning fork that had been struck violently. Not an instrument on any panel was readable to their full scale but appeared as white blurs. . . . Briefcases, manuals, ashtrays, suitcases, pencils, cigarettes, flashlights flying about like unguided missiles."°

For this reason, pilots avoid regions of turbulence. They know to avoid the parallel lines of clouds near mountains, lenticular clouds, and the rainy or dusty swirls of microbursts.

However, nature doesn't always provide a visual indicator for turbulence. Sometimes, in nearly clear skies high up at cruising altitude, planes will suddenly encounter the same sort of jarring bumpiness. This is called *clear-air turbulence,* abbreviated "CAT" for short. It is one of a pilot's worst nightmares.

What is clear-air turbulence? One of the main theories is that wind shear—the change of wind as you go up, which we discussed in Chapters 8 and 11—self-develops its own gravity waves. These waves then rapidly break, like ocean waves on a beach. Waves breaking on a beach generate a lot of foam; the "foam" of a breaking atmospheric gravity wave is turbulence, and planes flying through it will encounter bumps and jolts.

What does CAT look like? By definition, you can't see it. Under the right conditions, however, clouds can form and reveal the process outlined above. A beautiful example is shown below.

°(Source: Lester, P., *Turbulence,* Jeppesen, 1994, vi–vii.)

In the photograph, you can see the different crests of the waves in different stages of breaking. On the left, the wave crest is just beginning to turn over. At right, it's a foamy mess. This process is happening all the time in the atmosphere; only rarely do clouds or a high mountain warn of its presence.

An airplane's encounter with CAT doesn't just make for a good story; it is a destroyer and even a killer on occasion. On December 28, 1997, United Airlines Flight 826 carrying 393 people to Honolulu from near Tokyo hit heavy turbulence over the Pacific Ocean. Passengers who happened to be wearing their seat belts at the time described floating "like we were in an elevator falling down," according to the Associated Press. Those not wearing seat belts left dents where their heads crashed into the cabin ceiling. One woman was killed due to severe head trauma and at least 102 people were injured, some of them seriously. Fortunately, CAT this severe doesn't happen very often. Even when it does happen, CAT persists for only a few minutes in most cases.

A phenomenon as silent, invisible, small, and fleeting as CAT is a major challenge for weather forecasters. Tried-and-true rules of thumb exist to steer airplanes around likely areas of wind shear, such as jet streams. This isn't enough. Even today airplanes fly into unforecast CAT on a daily basis. Progress toward understanding this special brand of turbulence has been slow—just as slow as for every other type of turbulence.

Now you know why the airlines tell you to keep your seat belt fastened tightly at *all* times!

Kay Ekwall

take place and robs the eddies of the energy they steal from the larger-scale wind. Meteorologist L.F. Richardson, the hero of our next chapter on weather forecasting, expressed this complicated chain of events in a memorable little rhyme:*

> Big whirls have little whirls that feed on their velocity
> And little whirls have lesser whirls and so on to viscosity—
> in the molecular sense.

A Tour of Small-Scale Winds

In Chapter 7 we studied a few circulation systems that together spanned the globe. In this chapter, we will discover a wide variety of small-scale winds that occur locally in parts of the United States and across the globe. Time and again we will find that a small-scale wind can be explained by the interaction between a PGF and the topography of the region in which the wind occurs. We will also find, not surprisingly, that the change of seasons plays a governing role in the exact nature and role of the wind.

To explore these winds, let's take a tour of small-scale winds in the lower 48 United States (Figure 12.2), relating them as we go to these winds' overseas cousins. Our tour follows the Sun from east to west and examines the winds in each region.

The East and South

Coastal Fronts and Cold-Air Damming

We begin at the northeastern extreme of the United States, in blustery New England (Box 12.2). Cold air formed at these higher latitudes is often trapped between the warmer coast and the high Appalachian Mountains. At the small end of the synoptic scale, this may lead to a *back-door cold front* slipping southward down the Atlantic coast. The interaction between this air and the mountains also has consequences at smaller scales. For example, wintertime extratropical cyclones along the East Coast can draw warmer air above the Gulf Stream onshore, where it clashes with the colder air inland. The boundary between these two air masses is a miniature version of a stationary front and is called a **coastal front.** Coastal fronts add

*Source: Richardson, L.F.: Weather Prediction by Numerical Process.

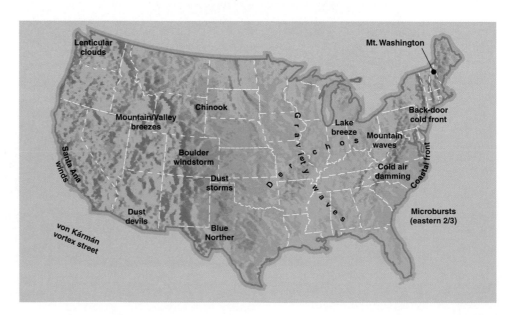

Figure 12.2

Geographic summary of small-scale winds across the contiguous (lower 48) United States. (Source: Richardson, L., Weather Prediction by Numerical Process, Cambridge University Press, 1922.)

BOX 12.2

The Windiest Place on Earth

Mountain ranges on Earth extend upward several kilometers into the atmosphere. Rarely are mountain peaks isolated enough that strong winds blow over rather than around the peak. Mount Washington in northern New Hampshire, at 1.9 kilometers (6,288 feet, to be exact) above sea level, rises well above the surrounding landscape and is perfectly situated geographically to penetrate the swift winds associated with East Coast extratropical cyclones. As a consequence, Mount Washington experiences what is called "the world's worst weather," a combination of persistent clouds, heavy snow, cold temperatures, and record-setting high winds.

The world record for the highest wind speed ever measured on land or sea is not from a hurricane or a tornado, but from the Mount Washington Observatory high up in the clouds (see photograph). On April 12, 1934, a wind gust of 372 km/hr (231 mph) was observed there! Never again has an official wind-recording device observed near-surface winds of this magnitude. A report of 380 km/hr (236 mph) due to a typhoon that hit Guam in 1997 turned out to be a major over-

estimate; it's difficult to measure winds accurately at such high speeds. During the May 3, 1999 Moore, Oklahoma, tornado discussed in Chapter 11, a portable Doppler radar estimated its winds at 512 km/hr (318 mph). However, these tornado winds were measured at several hundred feet above the surface and probably were much slower near the surface. Therefore, the Mount Washington record still stands.

In February 2001, the average wind on Mount Washington was 82.9 km/hr (51.5 mph), with hurricane-force winds on 24 of 28 days, and a peak gust of 233 km/hr (145 mph) was recorded on February 10. Combined with 1.4 meters (56 inches) of snow that month and a monthly average temperature of only –15.2° C (4.7° F), worse weather is hard to imagine. The *average* wind speed on the mountaintop is 56.8 km/hr (35.3 mph), and winds exceed hurricane force on an average of 104 days each year. As a result, wind chill temperatures can regularly drop below –73° C (–100° F). For the meteorologists at the Mount Washington Observatory, home is the windiest place on Earth.

Mount Washington Observatory Collection

Mike Smith, WeatherData, Inc.

Figure 12.3

A downward-plunging microburst with rain (right) "splashes" against the ground near Wichita, Kansas, on July 1, 1978. This sequence of photographs, taken at intervals of 10 to 60 seconds, helped confirm that microbursts develop vortex-like spins on their leading edges after hitting the ground (upward curl of rain near center of last two photographs).

one more layer of complication to the chore of forecasting the impact of extratropical cyclones along the East Coast, because a coastal front often separates cold air and heavy snow from warmer air and rain across a distance of only a few miles.

The stubborn entrenchment of cold air that is pinned against high mountains is called **cold-air damming.** Cold-air damming can cause transportation nightmares in winter. As we learned in Chapter 4, freezing rain is likely when warm air overruns a shallow layer of below-freezing air. Cold-air damming is the classic case of shallow cold air. When warm, moist air from the Gulf of Mexico or the Atlantic overruns a case of cold-air damming, a damaging ice storm in the Carolinas is a definite possibility. (This may explain why the title of an American Meteorological Society research journal article was once misprinted as "cold-air *damning.*")

In other parts of the world, the **harmattan** of western Africa and the **southerly buster** in southeastern Australia resemble the winds we have just discussed. The harmattan develops when cool air from the Sahara Desert in winter moves south and west and displaces warmer, more humid coastal air. The southerly buster is a mesoscale cold front that causes quick temperature drops (10–15 °C) and high winds (up to 100 km/hr) during spring and summer. The cold ocean region south of Australia supplies the cold air, and the mountains of eastern Australia help funnel and intensify the winds of the "buster."

Microbursts

One particularly dangerous small-scale wind develops when rain falling from a thunderstorm evaporates underneath the cloud, cooling the air beneath it. This cold, heavy air plunges to the surface and "splashes" against the ground like a bucket of cold water. The air then rushes sideways and swirls upward due to the pressure gradient between the cold air and the warm surroundings. This wind is a **microburst.** It is also sometimes called a "downburst," or a "macroburst" if its path of destruction exceeds 4 kilometers (2.5 miles). Figure 12.3 illustrates the microburst lifecycle in a series of famous photographs.

The winds from a microburst can cause as much damage as a small tornado, flattening trees and power lines. Microbursts that occur near airports are particularly dangerous. Strong winds from above, below, and sideways buffet aircraft in just a few seconds. Planes that are landing or taking off are pushed into the ground, causing deadly crashes (Figure 12.4). Microbursts have led to major air disasters in New York City, Charlotte, New Orleans, and Dallas, killing many hundreds of people. (Microbursts also occur frequently near Denver. The higher frequency of thunderstorms in the East and South, combined with heavy airline travel in these regions, focuses attention on microbursts in those regions.)

The scariest microburst-related aviation event was a disaster that *didn't* happen. On August 1, 1983, Air Force One was ferrying President Ronald Reagan back to Andrews Air Force Base near Washington, DC. The President and his plane landed on the dry runway, uneventfully, at 2:04 PM Eastern time. An approaching thunderstorm then generated a massive microburst that, *less than seven minutes later,* caused winds above 150 mph

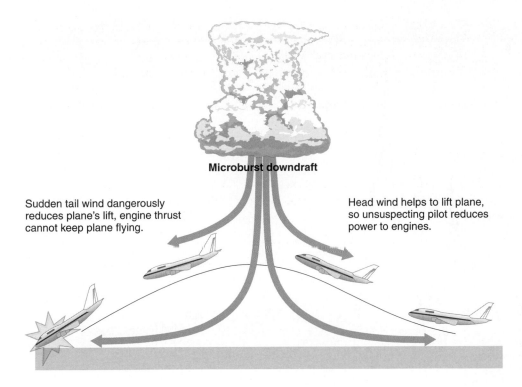

Microburst downdraft

Sudden tail wind dangerously
reduces plane's lift, engine thrust
cannot keep plane flying.

Head wind helps to lift plane,
so unsuspecting pilot reduces
power to engines.

Figure 12.4

Schematic illustrating the effects of a microburst on a plane flying underneath a thunderstorm during takeoff or landing.

to blow across that same runway (Figure 12.5)! The microburst even had a nearly calm "eye" and a second round of high winds, a little like a mini-hurricane. If President Reagan's plane had approached the airport just a few minutes later, a crash would have been unavoidable.

Spurred by disasters and close calls such as the President's, the U.S. government has spent millions of dollars on microburst detection equipment at airports. Few microburst-related disasters occur today, thanks to this technology and to extensive pilot training. Crashes related to microbursts today are usually attributed to poor decisions by pilots.

Blue Skies

Severe Weather/Microbursts

Gravity Waves

Straight lines are few and far between in nature. However, sometimes long straight lines of clouds will appear in the sky, only to disappear a few minutes later. These clouds occur when the air is jostled. This jostling can be caused by wind blowing over a mountain, or by a growing thunderstorm that blocks the wind's path like a mountain, or by complicated changes in winds at the jet-stream level. No matter the cause, the result of this jostling is very similar to throwing a rock into a pond: waves develop.

These atmospheric waves are known as **gravity waves,** because their alternating pattern of high and low pressure is maintained with the help of gravity. When made visible by clouds, gravity waves in the atmosphere look a lot like ocean waves and they are very similar to them in most ways. Air goes up in the crests of the waves, cools, becomes saturated, and forms clouds. Air in the troughs of the waves sinks and dries out. This is why the clouds caused by gravity waves form parallel straight lines.

Mountain-generated gravity waves are visible, even by satellite, many times each year over the central Appalachian Mountains when near-surface winds blow perpendicular to the mountain ridges. The mountainous regions west of Washington, DC, often produce these waves, which are called "lee waves" because they are downwind of the mountains. Figure 12.6 shows a classic example of the clouds produced by these waves; the wind in the figure is blowing toward the southeast. Pilots

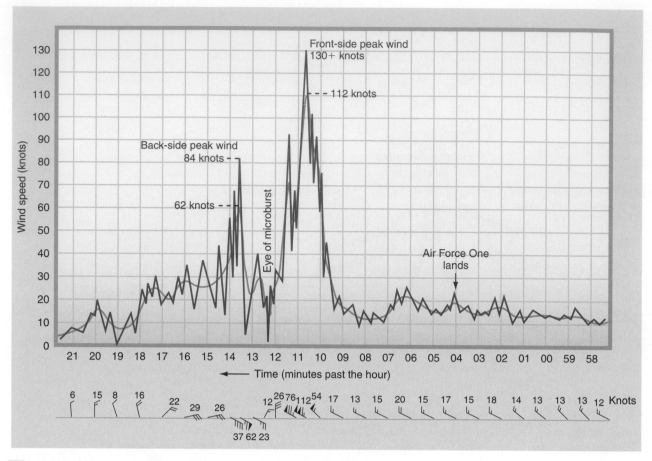

Figure 12.5

The chronology of a microburst at Andrews Air Force Base near Washington, DC, shortly after President Reagan landed there on Air Force One on August 1, 1983. Time goes from right to left in this figure and observed winds are shown in the blue line. (Source: Fujita, T., The Downburst, Author, 1985, p. 108.)

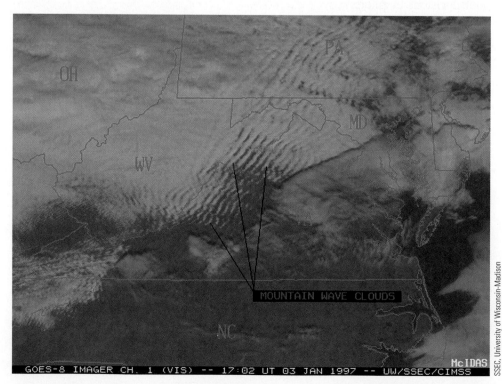

Figure 12.6

Lines of clouds caused by gravity waves in the lee of the Appalachian Mountains, as seen by a weather satellite on January 3, 1997.

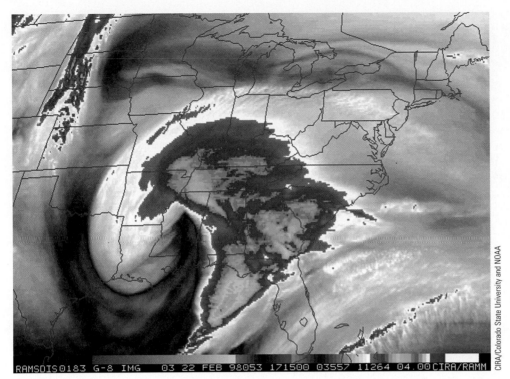

RAMSDIS@183 G-8 IMG   03 22 FEB 98053 171500 03557 11264 04.00CIRA/RAMM

CIRA/Colorado State University and NOAA

■ **Figure 12.7**

Colorized water vapor image of the upper-level cyclone that helped trigger the Birmingham, Alabama, gravity-wave windstorm at 11:15 AM on February 22, 1998. The reddish hook in the image is a region of dry air from the stratosphere that is wrapping around the cyclone. (Source: www.srh.noaa.gov/ bmx/february_22_1998/ february_22_1998.html.)

know to avoid these "wave trains" of parallel lines of clouds, for they are likely to harbor clear-air turbulence (refer to Box 12.1).

Other gravity waves form because of wind changes in the jet stream that send out "ripples" of waves. One particularly impressive case occurred in Alabama on the morning of February 22, 1998. A powerful upper-level cyclone (Figure 12.7) unexpectedly triggered gravity-wave ripples that moved northward across the entire state of Alabama, a distance of over 400 kilometers (250 miles), in only 3 hours. In downtown Birmingham the surface pressure dropped 10 millibars in just 17 minutes (Figure 12.8). The corresponding tight pressure gradient caused winds over 22 m/s (51 mph). Houses and trees exposed on the sides of small mountains in the Birmingham

Exploring Gravity Waves

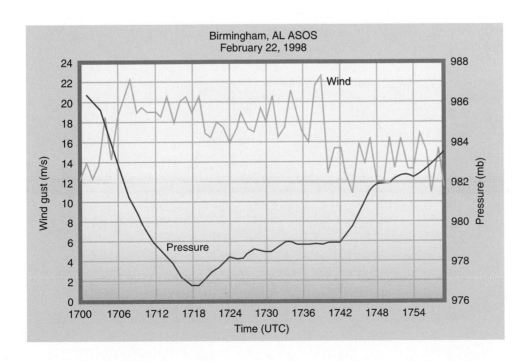

■ **Figure 12.8**

Automated observations of wind and pressure at Birmingham, Alabama, during the February 22, 1998 gravity-wave–induced windstorm. Notice the rapid changes in both variables. (Source: www. srh.noaa.gov/bmx/february_22_ 1998/february_22_1998.html.)

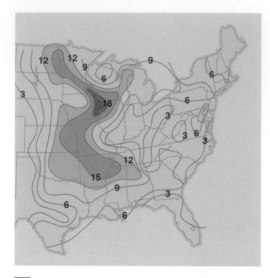

Figure 12.9

A 25-year climatology of gravity waves across the lower 48 United States, based on surface observations of hourly pressure drops of at least 4.25 mb. (Source: Adapted from Kippel, L., Bosart, L., and Keyser, D., "A 25-year Climatology of Large-Amplitude Hourly Surface Pressure Changes over the Conterminous United States," Monthly Weather Review, *January 2000, p. 58.)*

Poetry in Motion: Wake Lows

area, unsheltered by the effects of friction, experienced even higher winds; roof and tree damage was extensive. Yet no thunderstorm was involved; the winds were all part of the waves generated by the sloshing of the jet stream high above. The local weather forecasters were taken by surprise. We learn in the next chapter that some of these same forecasters made a brave, and accurate, forecast of a record snowfall in Alabama just a few years before. This story illustrates that forecasting small-scale winds can be extremely difficult, even harder than the most complicated large-scale weather event.

Because of the wide variety of ways in which they are triggered, gravity waves can be observed in many parts of the United States and the world. For example, the **morning glory** of northern Australia is a spectacular linear cloud up to 1000 kilometers (621 miles) in length that forms on the leading edge of a gravity wave. Closer to home, Figure 12.9 shows the observed distribution of non–mountain-related gravity waves across the United States. They are most common in the southern United States and also in our next stop: the Midwest.

The Midwest

West of the Appalachians, we encounter the Midwest. In Chapter 9 we explored lake-effect snow and in Chapter 10 we learned about the localized windstorm that helped sink the *Edmund Fitzgerald.* Now we investigate other small-scale Midwestern winds, some of which are also related to the presence of the immense Great Lakes.

Lake Breezes

During warm summertime days, the Great Lakes are usually colder than their surrounding coastlines. These lakes are so large that local wind circulations develop because of the resulting pressure gradient, just as they do along the world's ocean coastlines during the daytime. We called the daytime circulations along ocean coastlines the *sea breeze;* by analogy, winds that blow onshore during the day around the Great Lakes are called **lake breezes.**

A classic example of a lake breeze is shown in Figure 12.10. The cloudless region ringing Lake Michigan is cool lake air; the region encircling it that is dotted with cumulus clouds is the warmer land air. The lake air at this time is 6–8° C (10–15° F) cooler than the air over inland areas.

Figure 12.10

A Lake Michigan lake breeze on the afternoon of July 13, 2000, as viewed from satellite. Notice the absence of white dots (cumulus clouds) near the lake, especially on the east side over western Michigan. This is the region of the lake breeze.

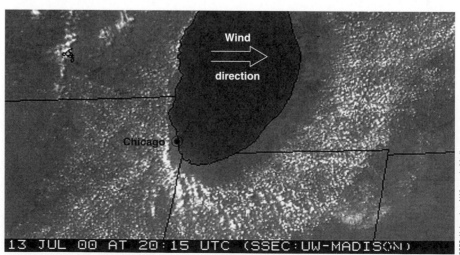

Because the Great Lakes are right in the middle of extratropical cyclone storm paths in summertime, it's common to see lake breezes in combination with larger-scale wind patterns. Notice that the ring of cloudless air in Figure 12.10 is not centered over Lake Michigan, but instead is shifted to the east. This is because the prevailing winds were from the west at the time of the satellite picture. You can also see that south of Chicago (lowest center part of the satellite image) the cumulus clouds are growing larger. The boundary between the lake breeze and land air can sometimes be a focal point for thunderstorm development, just like a small-scale front.

Derechos

At the opposite extreme from the gentle lake breeze is the **derecho.** A derecho (pronounced *deh-RAY-cho,* the Spanish word for "straight" or "right") is an hours-long windstorm associated with a line of severe thunderstorms. It is due to *straight-line winds,* not the rotary winds of a tornado; hence its name.

The extreme winds of a derecho—up to 240 km/hr (150 mph) in the worst cases—come about in the following way. Derechos are often associated with a quasi-stationary front in mid-summer. The front isn't moving, but it is the boundary between different air masses with different temperatures. Above these fronts, a relatively strong summertime jet stream may exist (see Chapters 6, 7 and 9). If the atmosphere just north of the front is very unstable, with lifted indices (Chapter 11) of less than –6, the front may trigger rapidly developing thunderstorms. A line of thunderstorms that forms in the vicinity of the stationary front can, via its cold downdrafts, drag down the high-speed air from above. This can cause the high winds of a derecho.

At the same time, the jet stream aloft pushes the line of thunderstorms outward, causing it to bend or "bow" outward. This is called a *bow echo* when it is seen on weather radar. Because the thunderstorms are pushed rapidly by the jet stream, they cover lots of territory—up to 1000 kilometers (over 600 miles). Derechos leave significant property damage, and even entire forests flattened, in their wake.

Figure 12.11 shows a radar image of a severe derecho that caused three deaths and 70 injuries and left 600,000 people without power in Michigan on May 31, 1998.

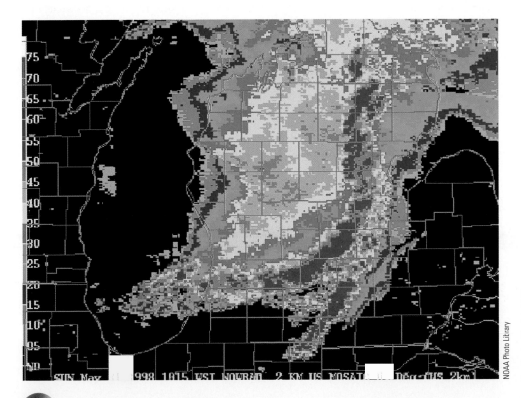

Figure 12.11

Radar image of a derecho moving through lower Michigan on May 31, 1998. Notice the curved, bowed-out nature of the red area of strongest storms.

NOAA Photo Library

http://info.brookscole.com/ackerman

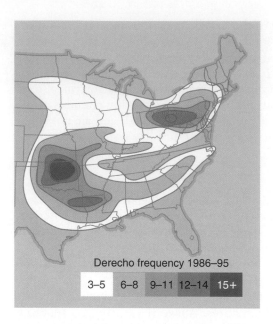

■ **Figure 12.12**

The number of derecho windstorms occurring from 1986 to 1995 across the United States. (Source: Adapted from Bentley, M., and Mote, T., "A climatology of Derecho-Producing Mesoscale Convective systems in the Central and Eastern United States, 1986–1995. Part I: Temporal and Spatial Distribution," Bull Amer Met Soc, 79, 2527–2540.)

Notice the characteristic bowed-out pattern of the squall line on radar. This line of storms moved at a forward speed of 113 km/hr (70 mph), carving a path of non-tornadic destruction all the way from South Dakota to New York state.

Derechos are confined to the eastern two-thirds of the United States. The ten-year climatology of derechos shown in Figure 12.12 reveals that they are most common west of the Appalachians in Ohio and western Pennsylvania, and also in our next stop: the southern Great Plains states.

■ The Great Plains

The Great Plains are nearly flat, they tilt upward gradually from east to west, and at their western edge they are close to high mountains, especially the Front Range of the Rocky Mountains. These facts help explain the strong small-scale winds of this region, which farmers try to slow down through the use of windbreaks (Box 12.3).

■ Blue Northers

We learned in Chapter 11 that the topography of the Great Plains creates a warm, moist, low-level jet that can blow across the Plains from south to north. In winter, the Plains also provide a clear path for cold air pushing south out of Canada. The Rocky Mountains help channel this cP air rapidly southward. As a result, cold fronts can sometimes zoom across the 2400 kilometers (1500 miles) of the western Plains, from the Dakotas to Texas, in only a couple of days.

As the cold front enters Texas, the temperature clash at the fronts can be extraordinary. Temperatures can drop tens of degrees Fahrenheit in only a few hours. However, the air on both sides of the front is extremely dry, having originated in the dry high hills of the Southwest to the south, and bone-dry Canada to the north. Therefore, the front is accompanied by clear blue skies. The only sign of changing weather is that the mercury in the thermometer drops so fast you can see it move! This is why the fierce north wind behind a fast-racing cold front in west Texas is known as a **blue norther.** In other parts of the world, blue-norther–type winds go by the names of **norte** (Mexico) and **buran** (Russia).

BOX 12.3

Using Turbulence to Advantage: Snow Fences and Windbreaks

On the Great Plains and in the nation's snow belts, wind can be an enemy. Snow or dust picked up by small-scale winds can reduce visibility, cover roads, and shut down normal life for days on end. Unfortunately, there is no "off" switch on the winds. However, there is a "slow motion button" that can be used: the effect of turbulent eddies. This is the concept behind two human creations, the snow fence and the windbreak.

The presence of obstacles such as fences and trees slows down the wind by causing turbulent eddies to develop. The wind is broken up into a swirl of eddies and its overall speed is reduced. This is why a wooded, fenced-in suburb is generally much less windy than a shoreline.

Snow fences use this concept to keep snow from blowing across land and roadways. Snow flying on high winds past a snow fence will, instead of blowing straight downwind, get caught up in the turbulent eddies the fence creates. Some of the snow will slow down just past the fence and drop to the ground. As more and more snow collects behind the snow fence, it presents an even larger obstacle to the wind. By the time the wind subsides, a large pile of snow accumulates behind the snow fence. If the fence is located upwind of a road or a farm, this is snow that *didn't* drift over the asphalt or bury the barn. In addition, in cold, arid regions such as the Dakotas, the snow acts as a blanket in winter and a moisture source in spring, keeping the soil warmer and wetter than it would be without snow cover.

The same idea holds for windbreaks (see photograph) on the dusty Great Plains. Blowing dust in spring can be a great

Steve Ackerman

hazard to transportation; it is also precious topsoil leaving the area. To prevent this, lines or belts of trees called "windbreaks" or "shelterbelts" have been planted from place to place across the Plains. The wind blowing past these windbreaks is chopped up periodically into slower turbulent eddies. Windbreaks thus function like the "speed bumps" used in shopping-center parking lots to keep cars from going too fast. They keep the winds from roaring across the hundreds of miles of flat land and lifting freshly plowed soil into the air.

For an example of what can happen when there are no trees, all the soil is plowed, and the winds blow hard, read Box 12.4 on the "Dust Bowl."

Blue northers can be killers. Farmers or hunters unprepared for their arrival can freeze to death, particularly if the initial clear-sky front is followed by snow. These high winds can, in times of drought, lead to **dust storms.** Box 12.4 examines dust storms and the infamous "Dust Bowl" of the 1930s in more detail.

■ Chinooks

On the western edge of the Great Plains, mountains meet flatlands. When air moves down these mountain slopes, it goes down in a hurry. The peaks of the Rockies reach well over 3 kilometers (10,000 feet) in altitude; in contrast, the Colorado plains are only 1.5 kilometers high. Similarly, the highest Black Hills of western South Dakota are over 2 kilometers (7000 feet) tall, but the nearby towns and cities are less than 1 kilometer (about 3000 feet) in elevation. We learned in previous chapters that when air is brought downward in the atmosphere, adiabatic compression causes a warming and drying-out of the air. A dry, warm wind is thus created whenever the large-scale pressure gradient forces air down the slopes of the Rockies, or the Black

BOX 12.4

Dust Storms and the "Dust Bowl"

Dust storms are to dust devils what a mesoscale convective complex (Chapter 11) is to a thunderstorm: a larger, longer-lived version that has its own unique look. Below is a photograph of an approaching dust storm.

Meteorologist David Smalley took this picture from the roof of the Business Administration building on the campus of Texas Tech University in Lubbock, Texas, on October 2, 1983. The 300-meter (1000-foot)-tall dust storm, powered by winds gusting out of a thunderstorm, reached his location just *one minute* after the picture was taken. The pressure jumped 4.4 mb in the next half hour—proof that this wind, like all the others we study in this chapter, is driven by strong pressure gradients.

Dust storms lasting several days are not uncommon in the far western Great Plains of the Texas and Oklahoma Panhandles. As we learned in Chapter 10, extratropical cyclones are often born in the Panhandles, causing windy conditions. Dryline thunderstorms can kick off dust storms with microbursts. And

D. Smalley

Hills of western South Dakota, onto the flatlands. Plains residents call this dry, warm wind a **chinook.**

The chinook's nickname of "snow eater" arises because a chinook quickly warms a location above freezing while dropping its relative humidity into the single digits. Snow is able to melt and evaporate rapidly under these windy conditions. Not far from Rapid City, South Dakota in 1943, a chinook rocketed the temperature up from −20° C (−4° F) to 7° C (45° F) in only *two minutes,* a world record! In other parts of the world, this type of wind is called a **foehn** (European Alps), a **puelche** (west slopes of the Andes Mountains in South America) or a **zonda** (east slopes of the Andes in Argentina).

On the Great Plains, these extreme changes are mostly a curiosity, significant only for farmers and the comparatively few residents of the area. We will see that when the same type of wind occurs on the heavily populated West Coast, it is as big a threat to life and property as El Niño or earthquakes.

BOX 12.4
(continued)

blue northers can usher in dust on their leading edges as they race southward from Canada. A pressure gradient, plus dry ground, is all that is needed. That, plus overly aggressive farming practices, can even lead to a "Dust Bowl."

In the early decades of the 20th century, American farmers moved into the Great Plains as part of a drive to convert the region into an agricultural paradise. The farmers came from the East and the South. However, those regions typically receive double, triple, or quadruple the annual rainfall of the Panhandles. Furthermore, typical wind speeds in the East and South are lighter than in the high Plains.

Every available piece of soil in the Panhandles was broken, first by horse and plow and later by tractor. No trees, no windbreaks (Box 12.3), no natural grasslands—nothing but wheat as far as the eye could see.

In the summer of 1931 a drought began, not unlike other droughts in the Panhandles in other times. The difference was, this time all the soil was plowed and ready to go airborne. When the winds came, dust storms developed early and often: 14 dust storms in 1932, 38 in 1933. The sky turned black, thousands of years of topsoil literally gone with the wind. Dust blew and drifted like snow—except that dust doesn't melt and instead had to be swept out of homes and shoveled off of all exposed areas. Crops died, year after year. Animals died of starvation. Children died from pneumonia triggered by dust inhalation. It was a meteorological and ecological disaster that one journalist dubbed the "Dust Bowl" (see photograph at right).

The drought lasted almost a decade, until late 1939. By that time, 25% of the population of the Panhandles had fled, many to California.

This mass exodus became the inspiration for John Steinbeck's classic novel *The Grapes of Wrath.*

Improved soil conservation efforts, including the planting of windbreaks, helped keep the dust down during the later years of the Dust Bowl. Today, some of those same safety measures are being ignored; for example, many windbreaks have since been cut down. Will history repeat itself during the next severe Great Plains drought?

NOAA

The West

The Western United States enjoys the richest array of small-scale winds because its geography is the most varied, from tall peaks to flat deserts. Many of the local wind patterns we have already studied, such as microbursts, mountain waves and cold-air damming, are frequently observed in the mountain West as well. Other winds are unique to the complex geography of the West. Once again, though, we can explain their existence as the combination of pressure gradients interacting with geography and topography.

Mountain/Valley Breezes and Windstorms

We've already discussed how temperature differences between ocean and coastline, or a Great Lake and its shores, can lead to pressure gradients that drive local wind

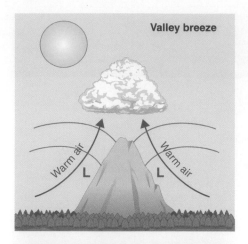

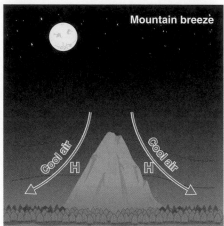

Figure 12.13

Schematic explaining the mountain and valley breezes. Purple lines are isobars. (Source: Adapted from Ahrens, C., Meteorology Today, *6th ed., Brooks/Cole, 2000, p. 261.)*

circulations. The same thing can happen along the slopes of high mountains such as the Rockies. These small-scale winds are called **mountain and valley breezes** and closely resemble the ocean and land breezes of Chapter 6.

Figure 12.13 is a schematic of how these breezes develop. During the day, the thin air above the high mountainsides warms quickly. The warm air rises and creates local low pressure along the slopes. Air from the lower valleys moves in to replace it, creating an upslope breeze that becomes strongest around noon. This is the valley breeze.

At night, the high mountain slopes cool very quickly. This cold, dense air forms a local high-pressure area. The pressure gradient drives a gentle breeze down the slope into the valley that is strongest just before sunrise. This is the mountain breeze.

Mountain and valley breezes are usually gentle, just a few kilometers or miles per hour. Even so, they can have a profound impact on local weather and climate. In some bowl-shaped valleys in the western United States, bitter-cold air sinks overnight during winter, creating a reverse treeline *below* which it is too cold for trees to survive! Mountain and valley breezes are also found throughout the world. For example, the **hira-oroshi** is a breeze that occurs along the mountainous shoreline of Lake Biwa west of Tokyo, Japan.

In mountainous regions with steep-sided, snow-covered large plateaus, the cool mountain breeze is anything but gentle. In some cases wind gusts can exceed 160 km/hr (100 mph)! These more violent relatives of mountain breezes are called **katabatic winds.** They occur all over the world and are called the **bora** (Adriatic coast of Europe), the **mistral** (French Riviera), and the **fall wind** in Greenland and Antarctica. In the United States, Colorado and the Columbia River valley experience katabatic winds. Their violence is due to the strong pressure gradient built up by the chilling of air over the high snowy plateau, and the steep slopes that allow the wind to rush downhill quickly.

Still other mountain winds affect the mountainous regions of the world. Any time there is a strong pressure gradient across the mountains, the wind tries to find a place to break through the barrier of rock. Low points in the mountains, called "gaps" or "passes," become wind tunnels in these cases. The **gap winds** that develop can easily exceed 160 km/hr (100 mph) in the strongest cases. The **squamish** of British Columbia, the **levanter** of the Strait of Gibraltar between Spain and Africa, and the **tehuantepecer** on the Pacific Coast of Central America are all examples of gap winds.

In Boulder, Colorado, just northwest of Denver, downslope winds associated with mountain gravity waves can cause particularly ferocious **Boulder windstorms.** Winds well over 160 km/hr (100 mph) have been observed in some cases! Figure 12.14 depicts the winds from the Groundhog Day Boulder windstorm of 1999. These windstorms cause considerable property damage; in areas near Boulder, building codes have been changed to reduce the frequency and amount of damage.

Dust Devils

A much more benign wind blows along the sands of the desert Southwest. There, intense daytime heating helps spin up thin, rotating columns of air called **dust devils** (Figure 12.15). Unlike its cousins the tornado and the waterspout, the dust devil appears to be a creature created solely by solar heating. The Sun above Arizona bakes the ground until the surface air becomes unstable and rises, creating a local low-pressure center usually only a few meters (yards) across and 100 meters (330 feet) tall. As the little vortex begins to spin, dust and sand are drawn into the circulation and a dust devil is born. It usually lasts only a few minutes.

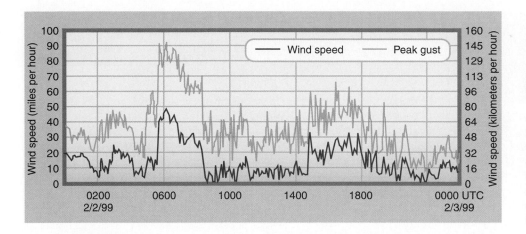

■ **Figure 12.14**

Winds in the Boulder, Colorado, windstorm of February 2, 1999. Notice how quickly the winds increase and decrease, and how consistently strong they are for hours. (Source: http://www.ucar.edu/ communications/staffnotes/ 9902/wind.html.)

Unlike waterspouts and tornadoes, dust devils do not require any clouds, showers, or thunderstorms above them; they form under clear skies in a hot Sun. Like waterspouts, the winds in a dust devil are much weaker than in tornadoes, and they rarely cause damage. Close relatives of the dust devil are the **willy-willy** of Australia and the **simoom** whirlwind of the African and Arabian deserts.

■ Lenticular Clouds

As our tour approaches the mountains of the Coastal Range, we encounter another small-scale wind and cloud feature that combines the beauty of the dust devil with the beast of a Boulder windstorm. It is called a **lenticular cloud** (Figure 12.16).

The name *lenticular* means "lens-shaped." These clouds hang over mountainous regions for hours, moving little if at all. To modern eyes, lenticular clouds look like the hovering motherships in space alien movies. In fact, the first modern "sighting" of a UFO occurred near Mount Rainier, Washington, and was probably a type of lenticular cloud. They are also seen in Alaska and Hawaii, and some of the most impressive lenticular clouds occur near the isolated Norwegian volcanic island of Jan Mayen northeast of Iceland.

N. Renno

■ **Figure 12.15**

Two dust devils, large and small, as seen by glider flying over a field in Arizona.

C. Johnson

Figure 12.16

A lenticular cloud at sunset on January 3, 1996 near Tehachapi, California, as photographed by college student Cynthia Johnson. This smooth, sculpted cloud is a sign of severe mountain-related turbulence nearby.

What does a lenticular cloud have to do with small-scale winds? Everything. When winds blow across high mountain ranges in certain circumstances, vigorous gravity waves develop downwind of the mountains. Air rising on the crest of the wave just past the mountain becomes saturated, forming the lenticular cloud. Because the wind and the mountain anchor the wave crest in the same place, the cloud is stationary—just as a river eddy downstream from a boulder remains in the same spot. The swirling pattern of winds around and over the mountain sculpts the lenticular cloud into unique and ever-changing shapes.

The beauty of the lenticular cloud sounds a warning alarm to pilots. The wind circulations beneath lenticular clouds are extremely turbulent (Figure 12.17), despite the seemingly calm, smooth appearance of the cloud. Aviators know to avoid these situations if at all possible.

Santa Ana Winds

As our tour begins its westward descent over the coastal mountains of California, we discover yet another downslope wind. This wind, the **Santa Ana wind,** combines the characteristics of its close kin the chinook with the damage potential of a Boulder windstorm.

The Santa Ana wind occurs when the pressure gradient caused by an anticyclone over the Rockies, in combination with friction, forces already-dry air from the mountainous West down the Coast Range in northern California, or down the San Gabriel Mountains in southern California. This happens most often in the autumn months when anticyclones are common in the Rockies. As with the chinook, a Santa Ana causes the temperature to increase and the relative humidity to plummet.

There are a few key differences between chinooks and Santa Anas, however.

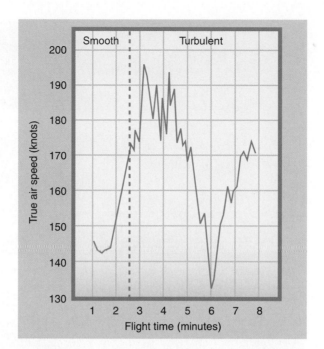

Figure 12.17

The airspeed of a plane flying downwind below the region of lenticular clouds. Notice the rapid transition from smooth flying to turbulent conditions with large and abrupt fluctuations in airspeed. (Source: Adapted from Lester, P., Turbulence, Jeppesen, 1994, p. 5-34.)

There is no snow for the Santa Ana to "eat" in fall in California. The trees and grasses of California are naturally dry in autumn. Also, over 30 million people live in California, as opposed to less than one million in all of South Dakota. As a result, the Santa Ana wind turns some of America's largest metropolitan areas into bone-dry, roasting tinderboxes.

One spark, cigarette butt, or lightning strike later, regions affected by the Santa Ana wind become huge roaring fires. On October 20, 1991, a Santa Ana wind with speeds above 32 km/hr (20 mph) dropped the relative humidity in the Oakland, California, vicinity to less than 10% for an entire day. The remains of a brush fire flared up, and the inferno was on. When the fire was finally extinguished, 25 people were dead, over 2000 buildings were destroyed, and damage was assessed in the billions of dollars.

Four years after the Oakland conflagration, a similar Santa Ana wind–related fire burned Point Reyes National Seashore just north of San Francisco. Near the coast, temperatures were above 80° F, but dew points were only in the teens. The smoke from the raging fire spread 1600 kilometers (1000 miles) out to sea as revealed by satellite photographs (Figure 12.18).

Numerous winds around the world bring hot, dry conditions to regions downwind of mountains and deserts. The **berg wind** of South Africa, the **leveche** of Spain, the **khamsin** of Egypt, the **leste** of the Canary Islands and the **sirocco** of the Mediterranean all combine aspects of the Santa Ana wind with other types of circulations we have studied in this text, such as the dryline (Chapter 9).

von Kármán Vortex Street

Before we conclude our American tour of winds, we peek out the airplane window at an amazing sight: a long interlocking chain of eddies rippling in the clouds downwind of an island (Figure 12.19). This chain of eddies is called a *von Kármán vortex street*, named for CalTech engineering professor Theodore von Kármán, who pioneered modern research on this and many other topics in turbulence and aerodynamics.

These vortices are caused when a mountain interrupts the smooth flow of wind past it and causes the airflow to be deflected mainly laterally, instead of vertically as in the case of mountain waves or lenticular clouds. The wind closest to the

Figure 12.18

Satellite image of the smoke plume from the Point Reyes, California, fire, caused by a Santa Ana wind in October 1995. San Francisco Bay is easily visible just to the right of the origin of the smoke.

mountain feels its frictional effect because of turbulence and slows down, but the deflected wind on either side of the mountain proceeds downwind quickly. This creates horizontal wind shear that causes the deflected winds to roll up into interlocking pairs of vortices, one cyclonic and the other anticyclonic. Even though the eddies may extend for hundreds of kilometers downwind, the von Kármán vortex street is a fundamentally small-scale process. It does not require the Coriolis force for its existence.

These vortices show up best in moist marine atmospheres with just enough low-level cloud cover to make them visible from the air or from satellite. Some of the most beautiful examples of von Kármán vortices are found along the coast of Baja California, as in Figure 12.19, and downwind of the Canary Islands west of Africa. They are not dangerous, but instead are stunning visual reminders of the fascinating pageant of small-scale winds that blow across our planet.

The Big Picture

Let's step back and survey small-scale winds. Geography is destiny for these winds. The Hoosiers of flat Indiana need never fear a Santa Ana, just as residents of Minot, North Dakota, will never experience a coastal front. But each region of the country has small-scale winds. Their visible signatures are all around us, occurring at any time and place where the combination of pressure gradient and geography is right.

Small-scale winds act locally, but they are found globally. This has profound consequences for weather. Worldwide, these winds connect the synoptic weather of

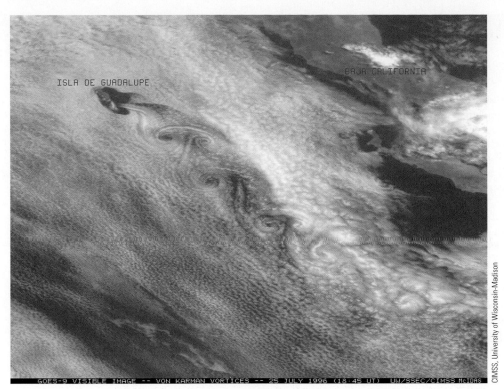

ISLA DE GUADALUPE

BAJA CALIFORNIA

GOES-9 VISIBLE IMAGE -- VON KARMAN VORTICES -- 25 JULY 1996 (18:45 UT) UW/SSEC/CIMSS McIDAS

CIMSS, University of Wisconsin-Madison

Figure 12.19

A visible satellite picture of a von Kármán vortex street downwind of Guadalupe Island just offshore Baja California on July 25, 1996.

extratropical cyclones and anticyclones (Chapter 10) and the mesoscale weather of thunderstorms (Chapter 11) with the motions of the atmosphere at the smallest scales, and thus with the surface of the Earth itself. Figure 12.20 summarizes the global extent of small-scale winds, relating each overseas example with its closest American cousin.

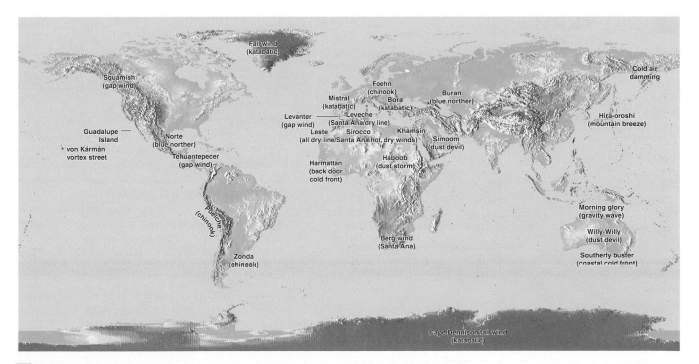

Figure 12.20

Selected small-scale winds from around the world. Their American counterparts are listed in parentheses below the names.

http://info.brookscole.com/ackerman

PUTTING IT ALL TOGETHER

 ### SUMMARY

Winds occur on every imaginable scale. Small-scale winds are driven primarily by the pressure gradient force and are slowed down by the effects of "friction." The main type of friction in a fluid such as the atmosphere is actually turbulence caused by swirling eddies of different sizes. These eddies are stirred up by the Sun, by wind blowing around obstacles, and by the atmosphere itself. This causes the characteristic gustiness of winds that is frequently observed.

An amazing variety of small-scale winds exists in the United States. Some winds are gentle, such as those that slide mountain breezes down the sides of the Rockies. Other winds are violent, such as derechos and microbursts dropping out of thunderstorms, or Boulder windstorms blowing out of the Front Range of the Colorado Rockies. A few are deceptive, such as the smooth lenticular cloud that hides a commotion of turbulence underneath its polished sides. There are dry winds that eat snow and cause fires. And there are rippling gravity waves that "draw" lines in clouds and, on rare occasions, spark windstorms. Small-scale winds are everywhere.

These kinds of winds are not limited to the United States. A dazzling array of small-scale winds exists worldwide. For all of their diversity, however, small-scale winds on our planet can generally be attributed to the interaction between a pressure gradient and geographic features such as mountains, plains, lakes, or coastlines. The colorful, unique names and rich histories of these local winds disguise the fact that they are all "brothers under the skin."

 ### KEY TERMS

You should understand all of the following terms. Use the glossary and this chapter to improve your understanding of these terms.

Berg wind	Lenticular cloud
Blue norther	Leste
Bora	Levanter
Boulder windstorms	Leveche
Buran	Leveche
Chinook	Microburst
Cold-air damming	Mistral
Derecho	Morning glory
Dust devils	Mountain breezes
Dust storms	Norte
Eddies	Puelche
Fall wind	Santa Ana wind
Foehn	Simoom
Gap winds	Sirocco
Gravity waves	Southerly buster
Haboob	Squamish
Harmattan	Tehuantepecer
Hira-oroshi	Turbulence
Katabatic winds	Valley breezes
Khamsin	Viscosity
Lake breezes	Willy-willy
	Zonda

REVIEW QUESTIONS

1. A tiny swirl in the bathtub can look a lot like an extratropical cyclone. Using facts from this chapter, explain how they differ.

2. Why do meteorologists link turbulence with friction, when in fact there aren't any rough surfaces rubbing together in the atmosphere?

3. Go to your kitchen or bathroom sink and make three different sizes of eddies. What causes them, how big are they, and how long do they last?

4. On a U.S. map that includes elevations, find the highest mountain peaks in the entire eastern United States. Does their location surprise you? How does their location affect cold-air damming events?

5. Would mountain or valley breezes be more intense on the *northern* or the *southern* slopes of a mountain? Why would there be a difference?

6. Based on your understanding of gap winds, explain why the streets of skyscraper-filled Manhattan can be so windy on a day when the winds in the suburbs around New York are light to moderate.

7. A real estate agent tries to sell you a house in Colorado that's near the bottom of a valley, ringed by tall mountains. The view is gorgeous. What meteorological advantages and disadvantages are there to purchasing a house in this location? How would your heating and cooling costs be different if the house were up near the mountain peaks?

8. It's a hot, sunny day in Dallas, Texas. A low is approaching from the west, however. What different kinds of small-scale winds could lead to turbulence for a plane taking off and ascending, or descending and landing, near Dallas?

9. Why are windmills and wind turbines placed on tall towers?

10. An old folklore saying goes, "The winds of the daytime wrestle and fight/Longer and stronger than those of the night." Based on what you've learned about turbulent eddies, explain why this is usually true.

11. The planet Mars has very high mountains and flat plains. It's very dry (few clouds, no thunderstorms) and dusty. Its axis tilts about the same as Earth's, so that Mars's tropics receive a lot of sunshine all year long. Based solely on this information and facts found in this chapter, what kinds of small-scale winds would you expect to find on Mars?

12. It's the week before your wedding. You're in your seventh-story apartment watching television with your fiancé(e). Suddenly, the local TV weatherman comes on, shows the radar, and points out a "bow echo" approaching quickly. What kind of small-scale wind is coming? What should you do immediately? (This is a true story!)

13. Using the "Severe Weather/Microbursts" section of the *Blue Skies* CD-ROM, experience the challenge of flying an airplane through a microburst.

14. A neighbor says, "I don't care about all those other watches and warnings, the only thing that scares me is a tornado." Tell the neighbor what other kinds of small-scale winds can cause damage almost as severe as a tornado, and why.

WEB ACTIVITIES

Choose Chapter 12 on the textbook's Web site:
http://info.brookscole.com/ackerman
and select from the following resources:
- Interactive Modules that illustrate and extend your understanding of key topics in this chapter

- Tutorial Quizzes to test your mastery of terms and concepts

 For additional readings, go to the InfoTrac College Edition, your online library, at:
http://www.infotrac-college.com

CHAPTER

13

Weather Forecasting

Introduction

"What will you do after college?" This question inspires excitement, but also dread, in many students. Your future is wonderful to ponder, but difficult to predict. So many different variables can affect your decisions. So many twists and turns may lie ahead!

Similarly, the question "What will the weather be tomorrow?" has captivated, but also vexed, meteorologists for decades. Weather forecasting is one of the most fascinating aspects of meteorology. However, an accurate forecast requires a thorough knowledge of all the variables of the atmosphere, and also an understanding of how the atmosphere "twists and turns" and changes over time. Until the last century, scientists often assumed that precise weather forecasting was impossible.

It's not surprising, then, that the founder of modern weather forecasting was someone who could not easily answer the question, "What will you do after college?" Lewis Fry Richardson (below) graduated from Cambridge University in England in 1903, but he avoided majoring in any one branch of science. After college, he changed from one job to another and from one science to another. When he was 26, Richardson's mother noted that he was having a "rather vacant year with disappointments"—an early version of a slacker! A few years later, Richardson became interested in numerical approaches to weather forecasting. The rest is history: Richardson created, from scratch, the methods of modern weather forecasting that are now used worldwide. His remarkable story will guide us through the intricacies of computer-based forecasting.

Before we get to Richardson, however, we will survey the various ways that weather forecasting can be done *without* a computer. From the dawn of civilization until about 1950, humans had to rely on a variety of very imperfect forecasting techniques. This was true during World War II, when a handful of meteorologists had to make the most important forecast in human history: whether or not to launch the "D-Day" invasion of Europe (below). Democracy or dictatorship? For a brief moment in history, the answer depended on an accurate weather forecast. The meteorologists in 1944 used methods that you can use today, and we'll learn about them here too.

Hedvig Bjerknes

NARA

Methods of Forecasting by People

Folklore

Long before computers and The Weather Channel, humans needed weather forecasts. Farmers and sailors in particular needed to know if storms were approaching. Over time, various **folklore forecasts,** often in the form of short rhymes, were devised and passed down through the generations. While memorable, the folklore forecasts are of uneven quality—some good, others laughably bad.

Some of the earliest folklore forecasts in Western culture, such as the following, come from ancient mariners:

Red sky at night, sailor's delight
Red sky at morning, sailor take warning.

This saying is fairly accurate. A clear western sky at sunset allows the Sun to shine through the atmosphere, its light reddening due to Rayleigh scattering (Chapter 5) and then reflecting off of clouds in the eastern sky. Clouds to the east usually move away; recall from Chapter 7 that storms in the middle latitudes generally travel to the east under the influence of jet-stream winds. The reverse is true in the morning, when the red sunlight shines on storm clouds approaching from the west. However, this folklore doesn't work at all in overcast conditions, or at tropical latitudes (see Chapter 8) where weather often moves from east to west.

The ballad "Sir Patrick Spens," a Scottish poem from the Middle Ages, contains another famous mariner's folklore forecast:

Late late yester'en I saw the new moon
Wi' the auld moon in her arm,
And I fear, I fear, my dear master,
That we will come to harm.

The poet's image of an old moon inside of a new moon refers to an optical effect caused by high ice-crystal clouds; probably a halo (Chapter 5). High clouds often occur well in advance of a midlatitude cyclone's warm front and are the first sign of approaching bad weather that could sink a boat (Chapters 9 and 10).

This folklore forecast in "Sir Patrick Spens" is remarkably accurate. Notice in Chapter 10 that halo-making cirrostratus clouds were reported in the vicinity of the *Edmund Fitzgerald* wreck site some 36 hours before the shipwreck. However, this forecast requires the Moon (or Sun) to be up and visible through thin high clouds, conditions that are somewhat hit-or-miss. Also, not all warm fronts produce extensive regions of high clouds, and not all high clouds are the result of a warm front. Nevertheless, this folklore forecast seems to work: a dedicated amateur meteorologist in Britain in the 1800s observed 150 halos in six years, and noted that nearly all of the halos preceded rain by 3 days or fewer. (However, it often rains every few days in Britain, so this statistic is less impressive than it initially seems.)

One famous forecast that *doesn't* work is Groundhog Day, known as Candlemas Day in medieval Europe:

If Candlemas Day is bright and clear
There'll be two winters in that year;
But if Candlemas Day is mild or brings rain,
Winter is gone and will not come again.

What the weather is at one location in Europe or Pennsylvania on February 2 tells us very little about the weather for the rest of the winter season. The popularity of

Groundhog Day in the United States has much more to do with clever marketing than it does with forecast accuracy.

Even though the Groundhog Day festivities don't help with weather forecasting, some accurate folklore forecasts do involve animals and plants that are weather-sensitive. One example is the saying:

*When spiders' webs in air do fly
The spell will soon be very dry.*

This forecast seems to be well grounded. Here is one possible explanation for why it works: Spiders want to catch insects in their webs, and insects will often breed after a rain. However, spiders don't want to have their webs covered with water; a sopping-wet spider web is easily detected, making it a lousy trap for insects. So spiders spin webs at the end of a period of rain as the air dries out—for example, after a cold frontal passage in spring. As you know from Chapter 10, high-pressure systems follow cold fronts and often lead to a spell of dry weather. The spider doesn't know meteorology, but its actions do fit into a predictable pattern of weather changes that humans can anticipate with some success.

As meteorology developed into a scientific field in the 19th century, folklore forecasts were gradually replaced with rules-of-thumb based on observations of clouds, barometric pressure, and winds. We have already encountered one such example in Chapter 10: FitzRoy's rhyme about fast pressure rises leading to strong winds behind a low-pressure system. Figure 13.1 shows a modern example, known as a "decision tree," that uses concepts such as pressure gradients and the lifted index from Chapters 6 and 11 to predict the probability of high winds over west Texas. Box 13.1 includes several much simpler rules-of-thumb that *you* can use to make your own weather forecast when you are away from the TV, radio, or computer.

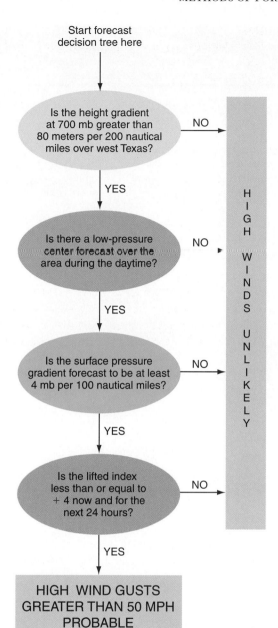

■ **Figure 13.1**

A modern rule-of-thumb "decision tree" for forecasting the occurrence of high winds in west Texas, adapted from work by National Weather Service meteorologists Greg Murdoch and Jim Deberry. The logic behind this decision tree is as follows: Strong pressure (or height) gradients cause strong winds. Meanwhile, low pressure aloft and a relatively low lifted index ensure that strong winds above the surface can be easily dragged down to the ground during the day, when solar heating stirs up the atmosphere into turbulent eddies. When all of these factors are present, high wind gusts are likely. (Source: Adapted from http://www.srh.noaa.gov/maf/html/wind.htm, Figure 7)

■ Persistence and Climatology

Other forecasting methods rely on the continuity of weather from one day to the next, or from one year to the next. A **persistence forecast** is simple: the weather you are having now will be the weather you have later. The accuracy of this forecast is very dependent on where you are, the type of upper-troposphere winds that exist over your location, and how long a forecast you want.

BOX 13.1

Personal Weather Forecasting

Would you like to be a weather forecaster? Below is a table that can help you forecast the weather based only on what you can see: no weather instruments necessary! Both "typical" and "possible" forecasts are included. Many of them are actu- ally forecasts based on the Norwegian cyclone model in Chapter 10. For more precise forecasts for specific regions, you need the guidance of numerical models of the atmosphere, which are discussed later in this chapter.

Clouds/Precipitation	Clouds heading toward which direction	Surface winds from which direction	Typical Forecast	Possible Forecast
Cirrus shield covering the sky	E or NE	NE	Cloudy with chance of precipitation within two days	Continued fair weather; or, major storm coming
Wispy cirrus	SE	NW	Fair weather and unseasonably cool	Bitter cold at night in winter if clouds go away
Cirrocumulus in bands	E or NE	E or NE	Changing weather soon	Major intensifying storm coming
Nimbostratus and stratus, rain and fog	NE	E or SE	Turning partly cloudy and warmer, rain ending	May warm up even at night
Nimbostratus and stratus, rain and fog	NE	NE	Windy with cold rain	Rain turning to snow if surface air cold and dry enough
Towering cumulus to the west	E or NE	S or SW	Thunderstorms soon; then clearing and turning colder	Severe weather possible soon
A few puffy cumulus	Stationary	Light and variable	Partly cloudy; cold in winter, hot in summer	Scattered thunderstorms if very clear (cold, dry) above clouds
Stratocumulus with flat bases	SE	NW and gusty	Winds dying down at sundown and cool	Increasing high clouds by morning if next storm approaches
Hazy and humid	Stationary	Calm	Hot; unseasonably warm at night	More of the same for a week or longer
Clear with new snow on the ground	—	Light and northerly	Rapidly dropping temperatures after sunset	Record low temperatures by morning
Cumulonimbus with continuous lightning in the distance	Anvil top pointed just to your right	Toward the cloud and gusty	Thunderstorm with heavy rain soon	Severe weather imminent with hail and/or a tornado
Saucer-shaped lenticular cloud near mountains	Stationary	Across the mountains	Partly cloudy; high winds downslope of the mountains	UFO reports!

For example, a 24-hour persistence forecast works pretty well in places like Hawaii or southern Florida. These locations are rarely under the direct influence of a strong jet stream. However, persistence can be a terrible forecast where the jet stream is strong. In these regions, growing low-pressure systems cause rapidly changing weather. Britain and the northeastern United States are locations where persistence forecasts for more than a few hours in the future are not very good. The reverse is true in situations where the upper-level winds are "stuck" in a "blocking"

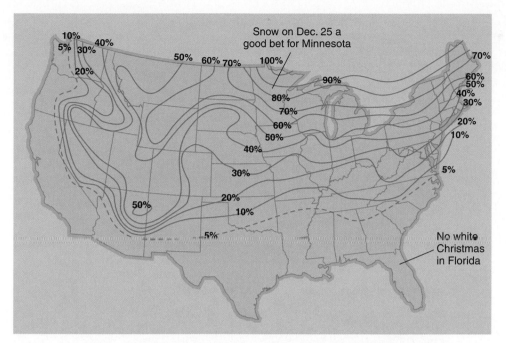

■ **Figure 13.2**

Climatological forecast probability of a "White Christmas" across the contiguous United States. (Source: Adapted from Ahrens, C., Meteorology Today, *6th ed., Brooks/Cole, 2000, p. 361.)*

pattern for several weeks. In these cases—for example, during a drought or the Flood of 1993 in the midwestern United States—a persistence forecast may be fairly accurate for days or even weeks at a time.

A **climatology forecast** relies on the observation that weather for a particular day at a location doesn't change much from one *year* to the next. As a result, a long-term average of weather on a certain day or month should be a good guess as to the weather for that day or month. The most obvious climatology forecast for the Northern Hemisphere middle latitudes is, "Cold in December, warm in July." You don't need to be a meteorologist to make that forecast! Its success derives from the fact that weather, although changeable, is strongly determined by the tilt of the Earth and the global energy budget discussed in Chapter 2.

Climatology forecasts can be quite specific. A favorite is the "White Christmas" forecast: What is the percent chance that at least one inch of snow will be on the ground at a particular location on December 25 (Figure 13.2)? This forecast depends on the past thirty years of weather observations on Christmas Day across the United States. It would change somewhat if more, fewer, or different years were used in the climatology.

Today's numerical forecast methods still use climatological statistics as a "reality check." They make sure that the computer models aren't going off the deep end, climatologically speaking.

■ Trend and Analog

We know that persistence forecasts are doomed to failure, ultimately, because the weather *does* change. An approaching cyclone brings precipitation and falling barometric pressure. Thunderstorms sprout ahead of a cold front. A **trend forecast** acknowledges that weather does change, but assumes that the weather-causing patterns, such as an extratropical cyclone, are themselves unchanging ("steady-state") in speed, size, intensity, and direction of movement.

Benjamin Franklin pioneered the concept of trend forecasting in 1743. Clouds ruined an eclipse for Ben in Philadelphia, but his brother in Boston saw the eclipse clearly. The clouds did not reach Boston until many hours later. Franklin discovered that "nor'easter" cyclones frequently moved up the East Coast, a trend that could be used to forecast the weather (Figure 13.3).

http://info.brookscole.com/ackerman

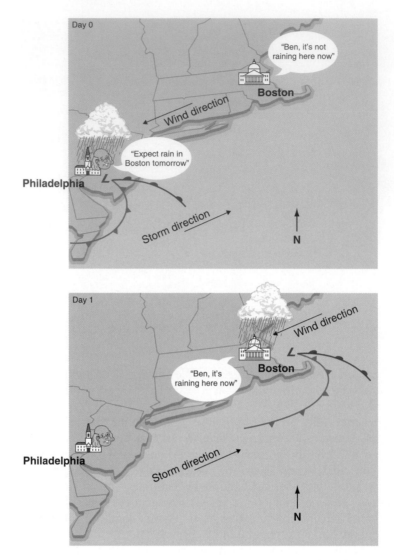

Figure 13.3

A trend forecast based on the assumption that a midlatitude cyclone moves, unchanging, up the East Coast. (Source: Adapted from Gedzelman, S.D., The Science and Wonders of the Atmosphere, John Wiley and Sons, 1980, p. 14.)

Trend forecasts work quite well for a period of several hours. Forecasting for such a brief period is called **nowcasting.** This is because large-scale weather systems such as cyclones don't change very much over a short time period. However, in Chapter 10 we learned that cyclones have definite life cycles and can change remarkably in size, speed, and intensity from one day to the next. Therefore, the accuracy of trend forecasts declines quickly when they are made for longer than a few hours ahead. Modern nowcasting combines trends with high-resolution numerical forecast model information.

The **analog forecast** also acknowledges that weather changes, but unlike the trend method, it assumes that weather patterns can evolve with time. The key—and flawed—assumption for the analog forecast is that *history repeats itself,* meteorologically speaking. The analog forecaster's task is to locate the date in history when the weather is a perfect match, or analog, to today's weather. Then the forecast for tomorrow is simple: whatever happened in the day after the analog will be the weather for tomorrow. The forecast for the day after tomorrow is whatever happened in the second day after the analog, and so forth.

Analog forecasting therefore requires many years of weather maps and an efficient way to compare one map to another. One approach, borrowed from personality testing in psychology, is to categorize the weather into a small number of **weather types.** If today's weather has a "Type A personality" (Figure 13.4), then you can create a multi-day forecast based on how the weather evolved in Type A cases. If today's

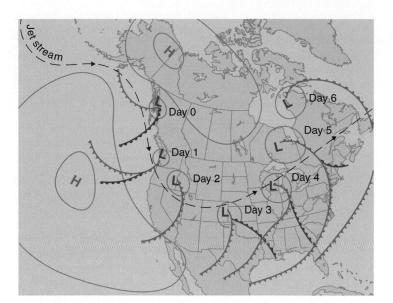

Figure 13.4

North American Weather "Type A" used in the heyday of analog forecasting in the 1940s. The goal of analog forecasting is to identify a weather type that matches today's weather and to use the left-to-right sequence of weather in that type as a forecast for the weather for the next several days. Compare this weather type with the life cycle of the Edmund Fitzgerald cyclone in Chapter 10. (Source: Adapted from Krick, I., and Fleming, R., Sun, Sea and Sky: Weather in Our World and in Our Lives, Lippincott, 1954, p. 97.)

weather isn't Type A, then you try to match it to one of the other types (six main types were identified in the 1940s and 1950s) and hope for a perfect match.

But does meteorological history repeat itself? Very rarely, although we saw in Chapter 10 that strong Great Lakes cyclones do recur around November 10, and Type A does resemble the evolution of the *Fitzgerald* storm of 1975. For how long could the historical analog help you forecast the weather—a day? Possibly. A week? Unlikely. A month? Impossible. Even so, meteorologists used analog forecasting by types widely in the United States from 1935 until about 1950. Overblown claims for its long-term accuracy then led to a strong backlash among scientists.

Recently, researchers have returned to weather typing via sophisticated statistical approaches. Today we know that certain weather patterns in widely separated locations—for example, Alaska and the U.S. East Coast—vary in relation to one another in a predictable way during global phenomena such as El Niño–Southern Oscillation (Chapter 8). These long-distance relationships are called *teleconnections* and are used by forecasters today to make general forecasts months into the future. In short-term forecasting, an analog method called "pattern recognition" is still used by weather forecasters to supplement today's computerized methods. But in the end the complexities of weather, like human personalities, defy simple categorization.

A Real Life-or-Death Forecast: D-Day, June 1944

It was the largest military invasion of all time. Nearly three million Allied soldiers and support personnel, 4000 watercraft, and 600 warships assembled along the southern coast of England during World War II in the spring of 1944. Their goal: to cross the narrow, often stormy English Channel into Nazi-occupied France. The decision of when to launch the invasion was largely in the hands of meteorologists.

This forecast could not rely on folklore; it was one-of-a-kind. Ships, boats, aircraft, and paratroopers all required special weather conditions. The Allied commanders determined that suitable weather for the invasion required all of the following for the coast of southern England and the Normandy region of France:

1. Initial invasion around sunrise
2. Initial invasion at low tide

3. Nearly clear skies
4. At least 3 miles of visibility
5. Close to a full Moon
6. Relatively light winds
7. Nonstormy seas
8. Good conditions persisting for at least 36 hours, preferably for 4 days

And, to make the task of forecasting ever more difficult:

9. At least 2 days' advance forecast of these conditions.

To create this forecast, the Allied supreme commander General (later U.S. President) Dwight "Ike" Eisenhower assembled three teams of the world's best meteorologists:

- Cal Tech meteorology professor (and onetime classical pianist) Irving Krick and his protégés formed the American team. Krick strongly advocated analog forecasting.
- An "odd couple" led the British Royal Air Force team. Sverre Petterssen of the Bergen School was an expert in the science of air masses, cyclones and upper air patterns. C.K.M. Douglas made his forecasts by looking at weather maps and guessing what would happen next by pattern recognition and his intuition.
- Cambridge engineer Geoffrey Wolfe and Rhodes scholar mathematician George Hogben headed the third team, from the British Royal Navy. Their expertise was the forecasting of waves on the English Channel.

The three forecast teams devised separate forecasts using their diverse methods. Over the phone, they hashed out their differences with their leader, British meteorologist James Stagg. Stagg molded the three forecasts into one consensus forecast and presented it to Eisenhower.

The immense Allied invasion force came together in England in late spring of 1944. Eisenhower asked his meteorologists a crucial initial question: What were the odds, month-by-month, that the weather required for the invasion would actually occur based on past experience? The meteorologists made a fateful climatology forecast. The odds were 24-to-1 in May, 13-to-1 in June, and 33-to-1 in July. The reasons for the long odds: low tide with a full Moon is a fairly rare event. Also, persistent calm weather because of "blocking highs" is more common in western Europe in late spring than in summer.

Based on these odds and the need for more time to assemble the invasion force, Eisenhower settled on early June, probably June 5, as the "D-Day" of invasion. Because of the requirement of a low tide around dawn, the only alternative would be two weeks later. After that, the climatological odds for success would drop quickly.

However, just as May turned to June, placid weather turned stormy over Europe. A series of strong midlatitude cyclones developed and deepened (Figure 13.5, A to C), a winterlike pattern not seen in the Atlantic in June in the past forty years! Fronts crossed the English Channel every day. Climatology, persistence, and even analog forecasts did not promise much insight into this weather pattern.

What was the all-important forecast for June 5? The three teams of forecasters debated this point for a week. Krick's team consistently found historical analogs that called for acceptable weather that day. Petterssen's theories and Douglas's intuition just as consistently indicated deteriorating weather. By June 4, the Royal Navy forecasters also foresaw weather unfit for the invasion. Stagg presented Eisenhower with a bleak, windy forecast.

Early on June 4, Eisenhower canceled the invasion for the next day—even though the weather overhead was clear and calm! But the forecasters were right; the 5th was a windy, rainy mess in the Channel, a potential disaster for the invaders. Meanwhile, the largest invasion force in history waited anxiously. Further delays meant lost time and lives.

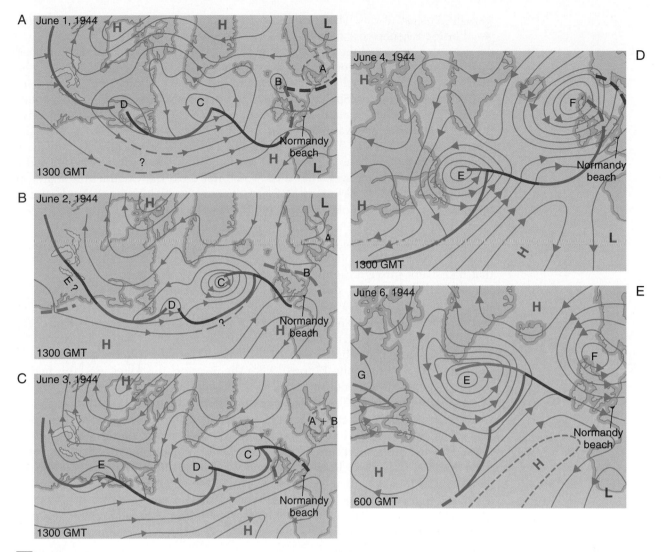

■ **Figure 13.5**

*Allied surface weather maps for the North Atlantic at 1300 Greenwich time for June 1 (**A**), June 2 (**B**), June 3 (**C**), June 4 (**D**), and at 600 Greenwich time on June 6 (**E**), the dawn of D-Day. Letters **A** through **G** on the maps indicate midlatitude cyclones in alphabetical order of their time of origin. Question marks indicate where the Allies did not have adequate weather observations. (*From Some Meteorological Aspects of the D-Day Invasion of Europe, pp. 165–166. Roger H. Shaw and William Innes, editors, American Meteorological Society © 1984. Reprinted with permission.*

As if on cue, the weather suddenly changed for the better. Just a few hours after the decision to cancel the June 5th invasion, one of Stagg's deputies telephoned him. He told Stagg, "A cold front has turned up from somewhere and is already through the Irish west coast" (Figure 13.5, *D*). Defying the Norwegian model and trend and analog forecasts, several lows had merged into a giant over northern Britain. Their combined winds were pushing cold air and better weather rapidly southeastward toward the English Channel. Meanwhile, a large low over the Atlantic was slowing down.

Would there be a gap between the two storm systems? The teams hastily drew up new forecasts. The analog forecasters said the cold front would rush through the Channel rapidly. But why believe them, when their consistent forecast for good weather on June 5th was a failure? However, both the Royal Air Force and Royal Navy teams concurred. For once, the different forecast approaches all converged on a single forecast. It was: There *might* be *just enough* time of *marginally acceptable* weather for Eisenhower's force to cross the Channel successfully.

Eisenhower seized the initiative. Before dawn on June 5, he gave the go-ahead for the Normandy invasion. Ironically, rain and gales battered his location as he made the fateful decision! Twenty-four hours earlier, he had canceled the same invasion under clear skies. Such was his faith in his meteorologists' forecasting ability.

Meanwhile, German meteorologists also noticed the cold front that would give the Allies a chance at invasion on June 6th. But their warnings did not reach the German military leaders. The Nazis had spent a tense May expecting the invasion. Then rainy weather at the beginning of June and the gloomy forecasts signaled a chance for a breather. Surely the Allies would not cross the Channel until later in the month! Field Marshal Rommel left the front to be with his wife on her birthday. Other high-ranking Nazi officers went hunting and to a Paris nightclub. The revised German forecast went unheeded.

As a result, the Allies completely surprised the German army. June 6 was, as forecast, marginally acceptable invasion weather. Northwest winds behind the front that had swept through the previous day (Figure 13.5, *E*) churned up rough seas. However, the element of surprise gave the Allies the edge against one of history's most formidable armies. Partly as a result, World War II in Europe ended earlier and more definitively than it would have otherwise. Sverre Petterssen hailed the forecast for the gutsy Allied invasion of Normandy as "meteorology's finest hour."

The forecasters' celebration was short-lived. The other window of opportunity in June for D-Day was June 17 to 20. After the invasion, the Allies used this period to resupply their troops, now engaged in fierce fighting along the coast of France. What was the forecast? The Allied meteorologists all agreed this time: excellent weather. Instead, the English Channel experienced its worst windstorm in forty years. The climatological forecast was for a less than 1% chance of gale-force winds in the Channel in June. Instead, gales blew along the entire length of the Channel for more than three days. It is said that this windstorm did more damage to the Allied forces in four days than the Germans had in the two weeks since D-Day!

A weather forecast had helped win World War II. Even so, it was obvious that there had to be a better way to make forecasts. Folklore, climatology, persistence, trends, analogs, and intuition were not enough. Even the Norwegian cyclone model couldn't explain everything.

There *was* a better way. It had been discovered and tried out over a quarter-century earlier, in wartime, in France. And then forgotten. This method, numerical weather prediction, would revolutionize meteorology and make weather forecasting a true science. Its prophetic discoverer: Lewis Fry Richardson. We now turn to his story to understand how modern weather forecasts are made.

L.F. Richardson and the Dawn of Numerical Weather Forecasting

Richardson's inspiration came from the father of modern meteorology, Vilhelm Bjerknes. In the early 1900s Bjerknes proclaimed "the problem of accurate [forecasting] that was solved for astronomy centuries ago must now be attacked in all earnest for meteorology." Astronomers can forecast eclipses years in advance. Why can't we forecast tomorrow's weather? Bjerknes' answer: meteorologists don't yet possess the two basic ingredients for an accurate forecast. They are:

1. The current conditions of the atmosphere over a wide area, known as **initial conditions;** and
2. Mathematical calculation of the atmosphere's future conditions, using the physical laws that govern the atmosphere's changes. In a nearly exact form, they are called the **primitive equations** because they are so complicated that they cannot be solved with algebra and pencil and paper.

BOX 13.2

Modeling the Equations of the Air

The "primitive equations" of the atmosphere are actually very easy to say in words. They are the conservation principles of momentum, mass, energy, and moisture, combined with the Ideal Gas Law from chemistry. What is hard about these equations is solving them for the variables meteorologists care about—the wind or the temperature.

For example, the equation governing the change in the west to east wind u can be written in word form as:

Change of u at a point over time = Advection of u by the wind at that point
+ West–east pressure gradient force at that point
+ Coriolis force turning the north–south wind at that point
+ Friction at that point

This equation is all tangled up. Here's why: what we want to solve for, u, isn't all by itself in one term like in a simple high-school algebra problem. The west–east wind is on the left-hand side. It's also hiding in the advection term. It affects the north–south wind in the Coriolis force term. And it's embedded in the poorly understood effects of turbulent friction. This messy situation, where what you're solving for is also part of the answer for it, is called "nonlinearity." There's no simple solution to this equation; it's the mathematical equivalent of trying to run while standing on your own shoelaces. This is one of many reasons why computers, not pencil and paper, are used to make forecasts.

But computers don't do algebra easily. Everything is 0's and 1's to them—numbers. A way must be found to translate that word equation into pure numbers. One method for doing so is called "finite-difference approximations." This is the technique that L.F. Richardson used in his first-ever numerical weather forecast.

A finite-difference approximation works a lot like a strobe light at a dance hall or disco. A strobe light flashes on and off rapidly; a dancer sees snapshots of his or her surroundings in the intermittent light. Similarly, a finite-difference approximation "sees" what's going on for an instant . . . and then sees nothing . . . and then sees things for another instant . . . and then nothing . . . and so on, as the model marches forward in time.

For example, the left-hand side of our word equation can be written as this finite-difference approximation:

Change of u at a point over time is approximately equal to ($u_{now} - u_{earlier}$) ÷ the time elapsed between now and earlier

Computers can subtract and divide at lightning speeds, so this kind of approximation suits them perfectly. The approximation lies in the fact that a "finite" or measurable amount of time may have elapsed between *now* and *earlier* time. This isn't the same as the instantaneous change in u at a point, which is what the original primitive equations require. But it's close enough in many instances.

Similarly, the advection term involves changes in wind over distance, and the finite-difference approach replaces the change with

($u_{here} - u_{there}$) ÷ distance from here to there

In this case, the "strobe effect" is in space, not time; the model "sees" and calculates weather variables at some places, but not others.

The great advantage of finite-differencing is that nasty math is reduced to lots and lots of arithmetic calculations. Computers can do these effortlessly, even though the drudgery would drive a human being to distraction.

The big drawback is that a forecast based on the finite-difference approximation is *not* the same thing, exactly, as the original problem. It is a model of the real thing, just as a robot can be a model of a human being. Robots can short-circuit or blow up; humans can't. In other words, models can have special problems that the real-world situation doesn't.

In the case of modeling the atmosphere, the big difference between the model and reality is that the atmosphere is a fluid that is always everywhere in time and space. In the model, the finite-difference calculations are only done *here* and *there* and *now* and *earlier*. This strobe effect is the Achilles' heel of modeling; we explore it further in Box 13.4.

Bjerknes admitted that this was a "problem of huge dimensions . . . the calculations must require a preposterously long time." He turned his attention to air mass and cyclone studies instead, revolutionizing meteorology. But his grand plan for a science of weather forecasting was left undone.

Into this void stepped Richardson. During his "slacker" period, Richardson gained on-the-job experience reducing difficult equations to lots and lots of simple arithmetic formulas that could be solved with calculators. The technique of approximating tough real-world problems with numbers is called **numerical modeling.** The numerical formulas used are called a **model,** just as a realistic approximation of a train is called a "*model* railroad." (See Box 13.2 for a more in-depth explanation of this topic.)

The Forecast Factory Fantasy in the College Classroom

Then in 1911 a vision came to Richardson. In this "fantasy" he saw a vast circular room like a theater. In this room, people at desks representing every part of the globe were solving the primitive equations numerically. It was a kind of meteorological orchestra, with numbers replacing notes. The combined effort was not a symphony but instead a forecast.

This visual metaphor enthralled Richardson. He would approximate the difficult equations of the atmosphere with a numerical model. All Richardson needed was some data to use for initial conditions, the primitive equations, and "a preposterously long time" to run his numerical model and make a forecast. Or, because the forecast would take a very long time to do by hand, he planned to do a "post"-cast of a weather situation that had already happened, as a test case for his approach.

World War I gave Richardson his opportunity. He was a member of the Quaker religion and therefore a pacifist, morally opposed to war. Married and in his mid-thirties, Richardson could have easily dodged the draft in England. Instead, in 1916 he registered as a conscientious objector to the war. For the next two years, he *voluntarily* drove an ambulance in war-torn France, caring for the wounded while bombs exploded nearby.

During lulls in the fierce fighting, Richardson would fiddle with his equations and his methods of solving them. On a "heap of hay" he attempted the first-ever mathematical weather forecast as Bjerknes had envisioned it. The task was an immense undertaking for one person, so he limited his "forecast" to a prediction of wind and barometric pressure at a point in Germany for the morning of May 20, 1910, using data gathered across Europe earlier that same morning. Even with such a limited "forecast," it took Richardson six weeks to compute it by hand!

What did Richardson get for all his troubles? A *really* bad forecast. According to his calculations, the pressure at one spot in Germany (Figure 13.6) was supposed to change 145 mb in just six hours. That's the difference between a major hurricane and an intense high-pressure system! Since his "forecast" was for an actual day many years earlier, Richardson could go back and compare his result to reality. May 20, 1910 was tranquil, and the barometric pressure in Germany had changed a couple of millibars at most, not 145. His forecast fantasy had flopped; his approach was apparently a failure.

Richardson put a good face on his failed test-case forecast and called it "a fairly correct deduction from . . . somewhat unnatural initial [conditions]." Meteorologists

■ **Figure 13.6**

An actual map from Richardson's famous forecast. The circled X's are stations at which upper-level weather observations were available on May 20, 1910. The M's and P's are the model's gridpoints at which Richardson calculated momentum (i.e., wind) and pressure changes, respectively. The P near München (Munich, Germany) is the location for which Richardson obtained his erroneous forecast of a pressure change of 145 mb in 6 hours. The lines on the map divide Europe into a grid that facilitates the numerical forecast. (From Richardson's Weather Prediction by Numerical Process, *Cambridge University Press, 1922, p. 184. Copyright © 1922 Cambridge University Press. Reprinted with permission.)*

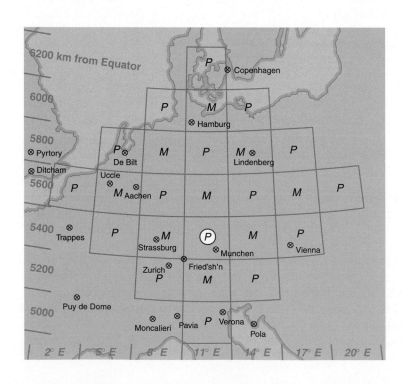

L. F. Richardson, Pioneer and Prophet

The hero of this chapter, L.F. Richardson, left an eternal mark on science by doggedly pursuing simple questions with difficult solutions. Early in his career, Richardson learned numerical solution methods by studying "unsexy" problems. A prime example: the flow of heat through peat bogs (swamps). A Cambridge University friend was shocked: "*Dried peat?*" Richardson replied, "Just peat." Silly as it sounded, it was the ideal training ground for Richardson's trailblazing work on numerical weather prediction a few years later.

Around the time of his now-famous forecast, Richardson also became interested in turbulence (Chapter 12)—another phenomenon that is easy to see but difficult to explain. Today, the onset of atmospheric turbulence is predicted using a special parameter known as the "Richardson number," in honor of his pioneering efforts on this subject.

Richardson's work on turbulence garnered interest among those wanting to study the movement of poison gas in the atmosphere. His pacifist beliefs offended, Richardson promptly quit meteorology. He went back to school and, at the age of 48, earned a second bachelor's degree in, of all things, psychology. He studied human perception, gradually becoming interested in the psychology of war.

In the years during and after World War II, Richardson combined his mathematical prowess, his understanding of complex phenomena from meteorology, and his Quaker pacifism. He tried to quantify and predict (and thus avoid) war in much the same way that he had forecast the weather, with data and equations. The political scientists of his era were completely befuddled at this elderly meteorologist with his equations of war (right, a photograph from a scientific meeting in 1949; note the sleeping woman and the exasperated man in the row in front of Richardson!). Richardson encountered great difficulty getting any attention, or even a publisher, for his laborious calculations.

Yet Richardson was, as usual, on the right track. Decades after his death in 1953, scholars realized the worth of his approach. One short paper of his was rediscovered, titled "Could an arms-race end without fighting?" In the nuclear Cold War days of 1951, Richardson's equations said, "yes, without a shot being fired," if one side outspent the other on armaments and the weaker nation bankrupted itself. Almost forty years later, the Berlin Wall came down as an outspent

Soviet Union relinquished power quietly in the face of American military buildups. Coincidence—or prophecy?

In the 1960s a mathematician named Benoit Mandelbrot stumbled across an obscure paper of Richardson's. Its simple subject: how long is a coastline? What a simple question! That is, until you realize that the shorter a ruler you use, the longer a coastline is. Mandelbrot extended Richardson's work and discovered that the length of the coastline is intimately related to "fractal geometry," an aspect of chaos theory.

Meanwhile, in Boston MIT meteorologist Ed Lorenz, using a numerical model based on Richardson's ideas, discovered "sensitive dependence on initial conditions." This, too, is a hallmark of chaos, and is discussed at the end of this chapter. Together, Lorenz and Mandelbrot, intellectual heirs of Richardson, brought chaos theory to the forefront of 20th century science (not to mention the movie "Jurassic Park").

L.F. Richardson was so far ahead of his time that he won no major scientific awards or prizes while he was alive. He was the antithesis of a scientific prodigy, doing his best work after the age of thirty. For much of his life he taught in obscure universities. He was not an inspiring lecturer (as the photo suggests). He did his research "on the side." He was a loner and a dreamer. Today, however, Richardson's name adorns meteorological prizes, peace-studies foundations, and fundamentals of mathematics and meteorology. This pioneer of modern meteorology was also one of the most creative and visionary scientists of the entire 20th century.

O. Ashford

generally applauded his Herculean effort, but no one followed in his footsteps. As we saw earlier, World War II forecasts employed every technique *except* Richardson's. By that time, Richardson himself had quit meteorology on moral grounds because the science was being used for military purposes.

But Richardson's forecast was not a total failure. Richardson, as usual (see Box 13.3), was simply too many decades ahead of his time. As we'll see, a combination of meteorological and computer advances were needed to turn his failure into success. The invention of the electronic computer during World War II revived interest in numerical weather forecasting. The modern computer could shave the "preposterously long time" of forecast-making from weeks to hours. In 1950, a simplified computer next-day forecast was made in 24 hours, not 6 weeks. This meant that forecasts could be made for real-life situations, not just test cases. The era of real-time forecasting by numerical modeling had begun! By 1966, advances in computers and meteorology theory finally allowed meteorologists to use Richardson's approach on an everyday basis.

Today, weather forecasts worldwide follow a numerical process that Richardson would instantly recognize as his own—with a few key improvements.[1] In the following section, we compare today's methods versus Richardson's in order to understand the process better. Along the way, we discover why Richardson's forecast failed!

The Numerical Weather Prediction Process, Then and Now

Step One: Weather Observations

Richardson made a "forecast" for May 20, 1910 because it was one of the few days in history up to that time when there had been a coordinated set of upper-air observations. Even so, as Figure 13.6 reveals, he wasn't working with much data—just a few weather balloon reports.

A numerical forecast is only as accurate as the observations that go into the forecast at the beginning of its run, the "initial conditions." Because weather moves from one place to another rapidly, tomorrow's weather is influenced by today's weather far upstream, and next week's weather can be affected by today's weather a continent away. For this reason, forecasters must have lots of data worldwide. Today we have global sources of data of many different types, to give the forecast the best possible start. Figure 13.7 shows the types and the spatial distribution of weather data for a typical six-hour period in the modern era of numerical weather prediction. Notice how much more data is available now versus 1910! Surface observations (Fig. 13-7, *A*), radiosondes (Fig. 13.7, *B*), and satellite measurements (Fig. 13.7, *C*) supply a majority of the data used for model initial conditions. The newest wide-bodied jet airliners are now providing forecasters with even more upper-tropospheric data along major flight paths (Figure 13.8).

This vast and continuous data-collection process is overseen by the World Meteorological Organization (WMO) and relayed worldwide. In the United States, the National Centers for Environmental Prediction (NCEP) receive the data for use in their forecast models.

Deriving Winds from Satellite Images

Step Two: Data Assimilation

In order to do their work, most numerical models look at the atmosphere as a series of boxes. In the middle of each box is a point for which the model actually calculates weather variables and makes forecasts. The result of this three-dimensional boxing-up of the atmosphere is known as the **grid;** the point in the middle is the **gridpoint;** and the distance between one point and another is called the **grid spacing.** Figure 13.9 illustrates this "gridding" of the atmosphere in a model. These models are called, appropriately enough, **gridpoint models.**

Gridpoint models of the atmosphere can get fussy when the data in the initial conditions isn't obtained at exactly the location of the grid points. Notice how

---

[1] The very latest numerical models actually compress and combine the steps that follow. To understand the overall approach, however, we keep each step distinct—as was the case for all models until recently.

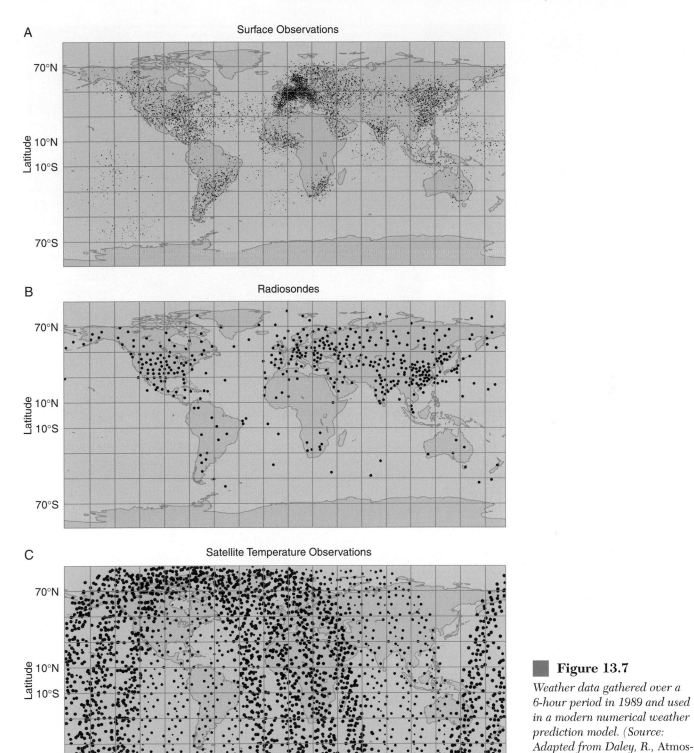

Figure 13.7

Weather data gathered over a 6-hour period in 1989 and used in a modern numerical weather prediction model. (Source: Adapted from Daley, R., Atmospheric Data Analysis, Cambridge University Press, 1991, pp. 14–15.)

irregularly spaced and sparse the observations were in Richardson's test case (see Figure 13.6). He needed data at the exact locations of the P's and M's in Figure 13.6, but instead his observations were scattered randomly. The process of creating an evenly spaced data set from irregularly spaced observations is called **interpolation.** It was a key part of Richardson's forecast, and remains so today for some models.

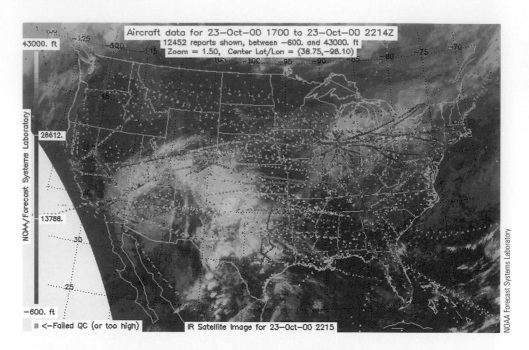

Figure 13.8

Graphic showing the distribution of aircraft meteorological data gathered by wide-bodied jets during a 5-hour period on October 23, 2000 over the continental United States. Blue colors indicate flight observations in the upper troposphere or lower stratosphere; red colors indicate observations nearer to the ground as the plane was taking off or landing.

Richardson then inserted the interpolated data into his numerical model. Observed data can be "bumpy" and out of balance (Chapter 6). Observations are likely to include little gusts and swirls that do not reflect the large-scale weather. Putting this bumpy data into a forecast model is a lot like driving a sport–utility vehicle (SUV) with no shock absorbers over a bumpy road. The variables predicted by the forecast model, just like the SUV, can bounce around out of control. As a result, the forecast can, like an SUV, "crash" and become useless. No one knew this in Richardson's day, however, and so he inserted out-of-balance data into his model.

In addition, Richardson unknowingly put a big pothole in his data. Figure 13.10 shows the sea-level pressure from Richardson's original calculations. The little bull's-eye low-pressure center over Germany didn't exist in real life. The forecast's winds did not agree with this pressure pattern; the winds and pressure gradients were completely out of balance. Instead, Richardson had made a math error that caused the low to show up, erroneously, in his interpolated pressure data.

Today, meteorologists know that a crucial step in forecasting is to take the unbalanced "shocks" out of the data. This is called **data initialization,** and can be thought of as the mathematical equivalent of shock absorbers. The goal of data initialization is essentially to balance things out and fill in the potholes in noisy real-life data before that data enters the model. An example of this balancing act is shown in Figure 13.11. This figure is based on the exact same data of Richardson's in Figure 13.10, except it's been balanced-out by data initialization. Notice that the pressure patterns in the two figures are almost identical, except that the pothole low-pressure center has been eliminated mathematically.

The multiple jobs of interpolating and balancing the data for use in numerical models are collectively called **data assimilation.** For those who like food more than SUVs, this step can be summarized metaphorically in the following way: meteorologists use data assimilation to "cook" the raw data they feed into the numerical model so that the forecast doesn't get poisoned.

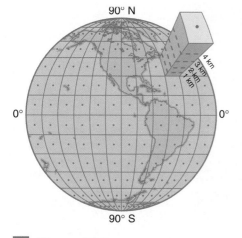

Figure 13.9

Dividing the world up into boxes for use with a gridpoint model. Forecasts and calculations are made only at the gridpoints, which are depicted as dots on this diagram.

■ **Step Three: Forecast Model Integration**

Once the data is "cooked," the model's formulas "ingest" it and the real forecast begins. Millions of arithmetic calculations may be made in order to get forecasts for

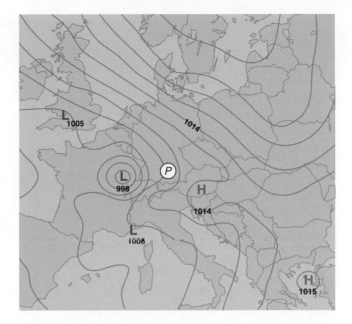

■ Figure 13.10

Garbage in: the unbalanced initial conditions for Richardson's model forecast, including a bull's-eye low that shouldn't be there just to the left of center near the "P" in Germany. (Courtesy Dr. Peter Lynch, Irish Meteorological Service.)

atmospheric variables at a later time. Then these forecasts are used as the new initial conditions for a forecast at a still later time. The whole process piggybacks on itself and marches forward in time. Mathematicians call this *integration*. This step puts together the two key ingredients of the forecast: the observed data and the model's formulas, which are approximations to the actual primitive equations describing the atmosphere.

Richardson solved his formulas by hand and obtained his erroneous result in this step. However, if he had been able to use the smoothed initial conditions in Figure 13.11, his forecast would have been for a pressure change of about 1 mb, not 145 mb! That's much more realistic. So Richardson wasn't mistaken; his forecast was, as he claimed, poisoned by "unnatural" initial conditions. And he was typically ahead of his time: mathematical "shock absorbers" weren't used in meteorology until about forty years later.

Today, supercomputers have replaced Richardson's manual calculator. Forecasts that took hours in 1950 now can be performed in seconds on a personal computer.

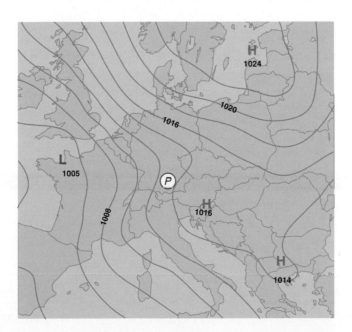

■ Figure 13.11

The same data as in Figure 13.10, but balanced by data initialization. The bull's-eye low near Germany has been removed. (Courtesy Dr. Peter Lynch, Irish Meteorological Service.)

Figure 13.12

Computing power used for making weather forecasts at the European Centre for Medium-Range Weather Forecasts (ECMWF) during the 1980s and 1990s. Note that the values on the vertical axis increase by a factor of 10. Computer power at ECMWF has increased almost a factor of 10,000 in less than 20 years and rivals the world's fastest computers! (Source: Gordon, A., et al., Dynamic Meteorology: A Basic Course, Arnold Press, 1998, p. 208.)

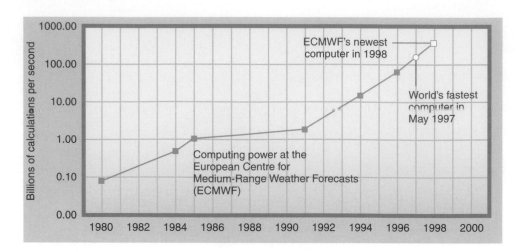

Simple Numerical Forecast Model

However, today's models are also much more detailed, requiring even more computing power than the latest PC. For example, a 24-hour global forecast for just five weather variables can take over *one trillion* calculations. The end result is that numerical weather forecast models always use the fastest supercomputers in the world, "pushing the envelope" of computing as much as any other science (Figure 13.12).

In this step, the physical distance between the interpolated data points becomes very important. The smaller the grid spacing, the easier it is for the model to "see," or, resolve, small-scale phenomena. This is called good or fine **resolution.** Wide spacing between gridpoints is called poor or coarse resolution. Figure 13.13 shows examples of both coarse and fine resolution; notice that fine resolution simply implies more gridpoints in a given area than coarse resolution.

The downside is that fine resolution in a gridpoint model means more points at which the model must make forecasts. So, a forecast model with very fine resolution takes a very long time to compute, even on a supercomputer. Even worse, a coarse-resolution forecast with grid spacing that is too wide can "blow up" and crash even if the original data were silky-smooth. (Why? See Box 13.4.) Richardson didn't know this. His forecast would have gone haywire even if he had owned meteorological "shock absorbers" and he had spent six more weeks to calculate the pressure in Germany for a still later time. Today, this fine-resolution requirement on gridpoint

Figure 13.13

The various grid spacings of the Nested Grid Model (NGM). Each dot represents a gridpoint at which the model makes calculations. Notice that coarse grid spacing means fewer gridpoints per unit area than fine resolution.

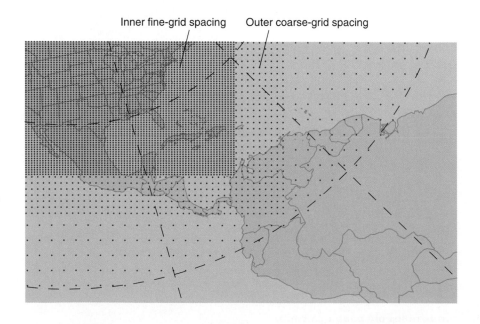

"Blowing Up" a Forecast Model

BOX 13.4

In Box 13.2, we identified the Achilles' heel of gridpoint models of the atmosphere. It is related to the distance between adjacent gridpoints, known as the "grid spacing," and the time elapsed between one forecast calculation and the next, known as the "time step."

The problem itself is the "strobe effect." If a strobe light at a dance flashes too slowly, dancers can't see where everyone is in between flashes. Dancers are moving too fast for the strobe, and they'll bump into each other and cause a big mess.

Similarly, if weather simulated in a forecast model moves too fast for the time step of the model, then the forecast will be a big mess. When this happens, meteorologists say the model "blows up" because the mess consists of larger and larger, totally unrealistic numbers for the weather variables the model is calculating.

Richardson didn't know this, because the field of numerical modeling was in its infancy in the early 1920s. This problem hadn't been discovered yet! In 1928, three mathematicians named *Courant*, *Friedrichs*, and *Lewy* created a criterion that, if violated, would lead to the "blowing up" of a finite-difference model. This "CFL" criterion is:

Speed of fastest winds in model ≤ Grid spacing ÷ time step

The upshot of the CFL criterion is that a modeler cannot arbitrarily choose a horizontal grid spacing without also taking into account the time step of the model. If you want fine horizontal resolution to see small-scale weather, you *must* have fine time resolution too. Otherwise, the model "blows up" and the forecast crashes. For a horizontal resolution of 50 kilometers, a model typically needs a time step on the order of just 10 minutes.

What does it look like when a model "blows up"? Try it yourself. A simplified "barotropic" model is located on the text's Web site. Nice troughs and ridges turn into total messes when the horizontal grid spacing and time step don't satisfy the CFL criterion. The forecast is ruined. The model is useless.

In short, the CFL criterion is the "there's no such thing as a free lunch" rule translated into numerical weather forecasting. Meteorologists want fine resolution in both time and space, just like we want a free lunch. But both inevitably come at a price. For meteorologists, the price is high: their best models are required to make forecasts every few minutes. These minute-by-minute forecasts are not released to the public; imagine the information overload! These miniforecasts are used by the model itself, and only by the model, to satisfy the CFL criterion. This wastes valuable computer time and vastly increases the computational requirements of numerical weather prediction. But it is a necessary chore. Without it, numerical weather forecasts would "blow up" in meteorologists' faces.

---

models is a major reason why forecast models take hours to run, even on the most powerful supercomputers.

■ Step Four: Forecast Tweaking and Broadcasting

Richardson made a "forecast" for one place in Germany. He translated his results into a sea-level pressure prediction. Then he broadcast his result in the form of a book. But the time elapsed from forecast to broadcast was over four years!

Today, forecast centers follow an approach similar to, but much faster than, Richardson's. Computers create forecasts of variables that forecasters care about: temperature, dew point, winds, and precipitation. Forecasters also look at other more complex variables. Atmospheric spin, also known as *vorticity* (see Chapter 10) turns out to be especially useful in forecasting. Some computer forecasts are for as short as 6 hours into the future, others as much as 15 days ahead. Then the forecasts are transmitted worldwide in both pictures and words. Using this information, everyone from the NCEP's colleagues at local National Weather Service offices to your local TV weather personality adds his or her own "spin" to the information, creating a specific forecast tailored to the consumer's needs. (See Box 13.5 on the special tasks of private-sector weather forecasting.)

In this last step, some of the other forecasting methods—especially climatology and modern-day analog approaches—sneak in the "back door" of the modern forecasting process. They are used to make small improvements on the numerical

A Day in the Life of Private-Sector Meteorology

Historically, government-sector meteorology has been the best-known branch of (and employer in) meteorology. As a consequence of Federal government downsizing during the 1980s and 1990s, this is no longer the case.

Today, private-sector meteorology satisfies the demand for customized meteorological information that is beyond the ability and mission of the National Weather Service to provide. This information can be too small-scale, or too specialized, or too risky for the government to deal with. For these reasons, private-sector meteorology is rapidly growing.

The most visible branch of private-sector meteorology is TV weather, from local stations to The Weather Channel. These on-air personalities often have meteorology degrees but usually rely on the National Weather Service for most of their forecasts. However, their specialized skills in graphics and in communications exceed what the government can provide.

Private-sector meteorology is much more than TV weather, however. One of the largest private-sector meteorology firms is WeatherData, Inc., in Wichita, Kansas. A typical workday for WeatherData meteorologists begins with looking at and even creating their own weather maps, often for an hour or more. They consult satellite and radar loops to get a three-dimensional view of the weather. This way, the forecasters dig deep into what's actually happening, a necessary prerequisite to making forecasts that push the envelope of current techniques.

Two or three hours into the work shift, it's time for weather discussion among the various WeatherData meteorologists. What are the key aspects of the weather to focus on today? Where are the trouble spots that matter to WeatherData's clients? These questions dominate the discussion. Then forecasts for specific clients are fine-tuned, based on the consensus of the weather discussion.

Who are WeatherData's clients? They range from major airlines to railroads to newspapers. New WeatherData forecasters often hone their forecasting skills by making and then

sending out forecasts to prominent newspapers such as the *Los Angeles Times* and the *Dallas Morning News*. The hours of studying the maps and forecast models is condensed into numbers for individual cities' high and low temperatures and a graphic symbol for the kind of weather expected that day. According to WeatherData Vice President of Customer Operations Todd Buckley, newspapers get more detail, graphics, and accountability from WeatherData than from official government sources. The newspapers pay for this service.

During this particular shift, fog and low visibilities are keeping planes over the United States from landing at large airports. It's not severe weather, but nevertheless the costs are large: as much as $10,000 per hour per aircraft flying in a holding pattern. Airlines incur costs of tens of millions of dollars per year due to these delays. To stave off these costs, Southwest Airlines, among others, pays WeatherData to provide up-to-the-minute and site-specific forecasts of visibility and other aviation-related weather variables. When planes crash, it's often WeatherData CEO Mike Smith (in the center in the photo above) who is called on to provide expert meteorological testimony in court cases.

forecast. But these days, unlike in Richardson's time or in World War II, forecasters are most likely to depend on what "the *model* says." Bjerknes' plan and Richardson's method, once a failure, are now the foundations of modern meteorology.

Modern Numerical Weather Prediction Models

The weather forecasts you've grown up hearing are based on a small number of numerical models of the atmosphere. These models coordinate the boring but essential work of performing all of the arithmetic calculations Richardson did by hand.

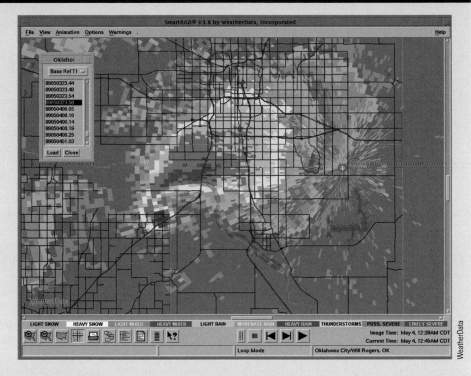

Another consumer of WeatherData information is the railroad industry. Trains carrying freight in high-profile double-stack cars often go through mountain passes where winds can be very strong. If the winds are above a certain speed, the double-stack of freight can blow off, leading to a derailment. A railroad can spend $1,000,000 to $5,000,000 in damages and delays due to this type of derailment. Burlington Northern Santa Fe Railway contracts with WeatherData to provide wind forecasts at specific mountain passes in the Western United States—a level of forecast detail unavailable from the National Weather Service.

Trains can also be blown over by tornadoes. However, it's a hit-or-miss proposition whether or not a tornado will actually collide with a train. If the two collide, the cost is in the millions. If not, the cost of stopping a train for a tornado that doesn't occur, or doesn't threaten that section of track, is about $200 per train per minute of waiting—not chicken feed. So, railroads and other users need pinpoint-precise tornado warnings instead of the county-by-county warning system provided by the National Weather Service.

To meet this need, WeatherData has developed its own software that combines government Doppler radar graphs (Chapter 5) with very specific maps showing major transportation arteries such as roads and railway lines. This was critical during the May 3, 1999 tornado that ravaged Oklahoma City (see figure above). This tornado crossed the BNSF tracks, destroying communications lines alongside the tracks and blowing debris onto the tracks. Thanks to WeatherData's warnings, however, no trains were hit, saving lives as well as millions of dollars.

Heat waves aren't as exciting as tornadoes, but they can be just as devastating to energy companies. An electric utility must pay top dollar to obtain energy immediately in order to meet the spike in demand caused by air conditioners running full blast; otherwise, a power blackout could occur. Sometimes the unexpected costs can run into the tens of millions. For this reason, utilities such as Southern California Edison hire WeatherData to give them precise, agonized-over temperature forecasts.

Once the forecasts are sent to the clients, the work shift is over (unless severe weather threatens, in which case everyone pitches in). When the employees exit through the glass doors of WeatherData, there is a sense of job satisfaction. Jeff House, lead meteorologist at WeatherData, explains, "The work is rewarding, as one sees how many lives are saved or how much money is saved by clients."

They are converted from arithmetic formulas into complicated computer programs comprised of tens of thousands of lines of computer code. Hidden in all the details is the stepwise scientific process we examined in the last section, particularly data assimilation and model integration.

Like other complicated pieces of equipment—a car, for example—these computer programs are given names and have their own unique characteristics. Professional forecasters swear they even have identifiable quirks and "personalities." Now that we've talked about the numerical weather prediction process in general, we can look at the individual models that actually help make the forecasts for tomorrow and beyond.

Short-Range Forecast Models

The first truly modern numerical forecast model was called the "Limited Area Fine Mesh Model," or **LFM** for short. Meteorologists at NCEP's forerunner, the National Meteorological Center (NMC), developed it in the early 1970s. It closely followed Richardson's methods, but for the United States, not Europe.

The LFM was, as its name advertised, "limited." It was better than flipping a coin or reading the Farmer's Almanac. However, it made forecasts only for North America. Its longest forecast was for 48 hours from the present. And its grid spacing was a very coarse 160 kilometers (100 miles). The LFM, like a nearsighted person, could only "see" larger details and not the finer details of weather, even features several hundred miles across. As a result, it's not a surprise that the LFM didn't forecast the 1975 *Edmund Fitzgerald* cyclone perfectly (see Chapter 10). The LFM is no longer used by NCEP, but it was the forerunner of the later forecast models.

In the early 1980s the Nested Grid Model (**NGM**) was implemented at NMC as an improvement to the LFM. Its name refers to the very fine inner or "nested" grid over the United States (see Figure 13.13), the region the NMC forecasters cared about most. The NGM also focused more attention on the jet-stream winds; the LFM's ability to "see" in the vertical was as fuzzy as in the horizontal. More and smoother data were incorporated into the model calculations.

In the 1990s, U.S. government scientists began using two new models, the **Eta** and the Rapid Update Cycle (**RUC**). These models incorporate continued improvements in resolution in all three dimensions. The RUC model is run every three hours and the Eta model is run four times every day. These are vast improvements upon Richardson's pace of one forecast computed every 6 weeks! However, these models are limited to short-term forecasts, a few days at most.

Meanwhile, the British developed the United Kingdom Meteorological Office (UKMO) model, known as **UKMET.** It is similar to the NGM, but is run for somewhat longer forecasts, which are called "medium-range forecasts."

Medium-Range Forecast Models

For forecasts a week or so into the future, NCEP has relied on a slightly different forecast technique. So far, all of the models we've looked at have been gridpoint models that chop the atmosphere into boxes. It's as if the atmosphere were a football stadium and the gridpoint model "saw" the sellout crowd only in terms of one person per row or aisle. However, as we learned in Chapter 10, the atmosphere, like crowds in football stadiums, can do "The Wave." In fact, the atmosphere can be described completely in terms of different kinds of waves, such as Rossby waves (Chapter 10) and gravity waves (Chapter 12).

Spectral models use this concept as the basis for a very different forecast approach. Instead of calculating variables on points here and there, the spectral approach takes in the whole atmosphere at once and interprets its motions as the wigglings of waves. The primary advantage is that computers can compute waves more efficiently than data at lots of points on a map.

Spectral models, because they can run on a computer faster than gridpoint models, have been used to make longer-range and more global forecasts than gridpoint models. The "Aviation" (**AVN**) or "Spectral" model was developed at NMC in 1980. Its forecasts were especially useful for transcontinental airline flights—hence the nickname "Aviation." Today, the AVN makes 3-day forecasts for the entire globe. A special version of the AVN, generically called the *Medium Range Forecast* (**MRF**), uses more data and makes global forecasts for up to 15 days in advance.

For decades, United States numerical weather prediction models were the best in the world. Today, the best model is probably the one designed at the European

TABLE 13.1. Summary of numerical weather prediction models discussed in the text
(Based on information from Dr. Stephen Jascourt, UCAR/COMET, valid January 2002.)

Model	First used	Horizontal resolution (KM)	Number of vertical grid points	Makes forecast for	Forecast issued at (A.D. or EST)	Longest forecast	Best used for
Richardson's	1916–1918	200	5	One spot in Germany	1922 for a day in 1910	3 hours	History lessons!
LFM	1971	160	7	North America	Twice a day	48 hours	Not currently used
NGM	1982	80–160	18	Hemisphere, best for North America	7:00 PM, 7:00 AM	48 hours	Large-scale weather in U.S.
Eta	1993	12	60	North America	7:00 PM, 10:00 PM, 6:00 AM, noon	84 hours	Wide variety of U.S. forecasts
RUC	1994	40 (20 in spring 2002)	40 (50 in spring 2002)	Lower 48 states	Every 3 hours	12 hours (24 hours in spring 2002)	Short-term forecasts over central U.S.
UKMET	1972	60	30	Globe	Twice a day	6 days	Longer-range forecasts
AVN	1980	About 80	42	Globe	Every 6 hours	3 days	Overseas and U.S. forecasts
MRF	1985	About 80	42	Globe	7:00 PM	15 days	Longer-range forecasts
ECMWF	1979	About 40	31	Globe	7:00 AM	10 days	Wide variety of forecasts

Centre for Medium-Range Weather Forecasts, or **ECMWF.** The ECMWF model is a type of spectral model, but with more waves (i.e., better resolution) than the AVN and special attention to data assimilation. It is used to make forecasts as long as 10 days in advance. Instead of creating a different medium-range model for each nation, the European countries have collaborated on one very good model that runs on the world's fastest computers (as shown in Figure 13.12).

However, no single numerical weather prediction model is perfect. Each one is a little different. These small differences can lead to large differences in forecasts. Forecasters today tend to look at what *all* the models expect for the future. Then, based on their experience with the models, pattern recognition, and even good old-fashioned hunches, the forecasters develop their own forecasts.

You can try this yourself. At our text's Web site, we provide links to current forecasts from most of these models. View them as often as you like and for the weather variables and regions that you care about. Like professional forecasters, you can compare different models' forecasts to what actually happens and develop your own forecasting savvy! To help you, Table 13.1 summarizes the characteristics of these models, and Figure 13.14 shows an example of visualized model output.

In a nutshell, today's forecasts are based on science and very complicated computer programs. However, human meteorologists are still a very important part of the forecast process. They use the models' predictions as "guidance" for the forecast that gets issued to the public. We see this interplay of humans and forecast models in the next section: the successful forecast of the biggest storm of the 20th century in the eastern United States.

Current Model Weather Forecasts

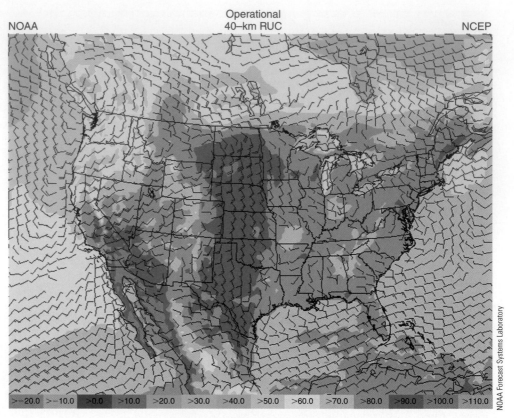

Operational
40–km RUC

NOAA   NCEP

> −20.0 > −10.0 > 0.0 > 10.0 > 20.0 > 30.0 > 40.0 > 50.0 > 60.0 > 70.0 > 80.0 > 90.0 > 100.0 > 110.0

Surface temperature/winds (°F/Knots)
12–hr fcst valid 27–Jun–01 21:00Z

NOAA Forecast Systems Laboratory

 Figure 13.14

An example of a 12-hour forecast of surface temperature and winds from the "RUC" model, which was made on the morning of June 27, 2001 and shows the conditions forecast for 12 hours later, the afternoon of the 27th. Temperatures are in degrees Fahrenheit and winds are in knots. Notice the hot (red and magenta) regions in the Desert Southwest and in the region of warm southerly winds in the Great Plains, as well as the cool conditions over the Great Lakes.

A Real Life-or-Death Forecast: The "Storm of the Century," March 1993

The Medium-Range Forecast

The eastern United States basked in unseasonable warmth in early March of 1993. Cherry trees were on the verge of blossoming in Washington, D.C. Thousands of college students headed south by car or plane toward the Gulf Coast beaches for spring break.

There was just one problem. By late Monday, March 8, medium-range forecasts at the National Meteorological Center (NMC; renamed NCEP in 1995) began calling for an epic winter storm that weekend. Snow, wind, and bitter cold were on the horizon? Unbelievable!

Overseeing the forecast process at NMC in Washington was meteorologist Louis Uccellini. He knew only too well how unreliable the official forecasts could be. In 1979, the worst snowstorm in over fifty years had surprised the nation's capital (see Chapter 10, Figure 10.21). It dropped up to 20 inches of unforecast snow on Washington. The nation's snowbound leaders, who vote every year on the National Weather Service's budget, took note.

Uccellini studied that storm and pioneered efforts to understand why the forecast had "busted." In March of 1993, he was the one in charge of the nation's forecast, and by his own admission "sweating bullets." Before the end of the week, President Clinton and two Cabinet secretaries would be getting personal briefings on the forecasts of an impending blizzard.

But on March 8, the only sign that the early spring weather would come crashing to a halt was in the output of numerical weather prediction models at NMC and overseas. The MRF forecast 5 days into the future called for an extratropical cyclone to develop very far to the south, over the Gulf of Mexico, and then curve up the East Coast and strengthen. The MRF suggested that by the morning of the 14th (Figure 13.15, *A*) the storm would be a potent 983-mb low just east of Washington.

However, during the previous month the MRF had made one wrong forecast after another. Strong cyclones popped up on the forecast maps but were no-shows in real life. Could the forecasters trust the MRF now? The NMC meteorologists skeptically accepted the MRF's guidance. The words "chance of snow" entered the extended forecast for the nation's capital on Tuesday.

In the days ahead, Uccellini's group compared the results from the world's best medium-range forecast models as they continually ingested new data and calculated new predictions. On Tuesday night (Figure 13.15, *B*), the MRF continued to predict a Gulf low that would intensify and hug the East Coast. However, the ECMWF model painted a much less dire picture. It envisioned a weak 993-mb low far out to sea over the Atlantic. And the UKMET forecast was for a monster 966-mb low over Lake Ontario! Which model was right? The NMC forecasters stuck with the MRF guidance, but played it safe and kept the low at 984 mb, the same as the last forecast.

By the night of Wednesday the 10th, the forecast picture flip-flopped (Figure 13.15, *C*). Now the European Centre model called for an intense low over Erie, Pennsylvania, not the weak Atlantic low it had predicted only a day earlier. This time it was the British model that called for a weaker low out to sea! The American MRF, in contrast, more or less stayed the course laid out in its earlier model runs. The NMC forecasters trusted the MRF's more consistent forecast. An intense winter storm *would* hit the East Coast that weekend—even if the computer models were disagreeing as vehemently as the human forecasters before D-Day!

By the next day, Uccellini's group at NMC had gained enough confidence to do something unprecedented: warn the nation several days in advance of a killer storm. The vague "Chance of snow" gave way to official statements warning of "a storm of historic proportions forecast over the mid-Atlantic." This was

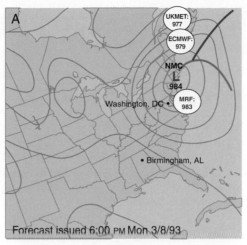

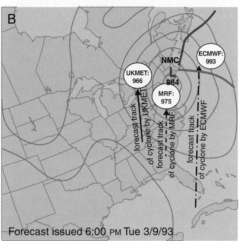

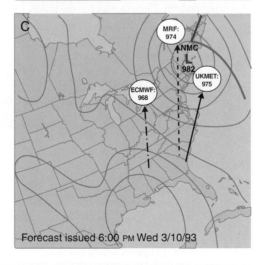

■ **Figure 13.15**
*Medium-range forecasts of the "Storm of the Century" as of 6:00 AM EST on March 14, 1993 from three numerical models: the MRF, ECMWF, and the UKMET models. The top panel (**A**) shows the forecasts for the 14th issued on March 8; the middle panel (**B**) shows forecasts for the 14th issued on March 9; and the bottom panel (**C**) shows forecasts for the 14th issued on March 10. In each panel, numbers indicate the lowest predicted central pressure of the cyclone at that time. The L and fronts on each panel represent the NMC meteorologists' own forecasts based on a synthesis of the model projections. (From* The Life Cycles of Extratropical Cyclones, *Shapiro and Gronas, eds., American Meteorological Society, 1999. Reprinted with permission.)*

a big risk, both professionally to the meteorologists and financially to the country. Millions of dollars in lost business would result if the weekend turned out to be nice and sunny.

The Short-Range Forecast for Washington, D.C.

The time for medium-range forecasts had ended. On March 11th, the first short-term storm forecasts from the higher-resolution models came out. Using them, NMC and local NWS meteorologists could make specific snowfall forecasts.

The LFM, NGM, Eta, and AVN models all predicted that a strong low (or lows) would move northeastward out of the Gulf of Mexico into Georgia. Then, as a classic "nor'easter" (Chapter 10) it would deepen as it moved north to Chesapeake Bay.

The NMC meteorologists held their collective breath and predicted a huge 963 mb low—strong as a Category 3 hurricane—would pass just east of Washington on Saturday night (March 13). As Friday the 12th dawned, they stuck to that forecast. The cyclone would pummel the nation's capital initially, but the track of the cyclone would be close enough to the city for the low's warm air to invade the region and change the snow to freezing rain or sleet. This would keep snow accumulations down near the coast (Figure 13.16, A). The cyclone would be much stronger than the surprise snowstorm back in 1979. However, because of a slightly different storm path, the overall snow totals this time would be about half as much as in 1979, perhaps as much as 30 cm (1 foot), the sum of the forecasts in Figure 13.16.

But was all this really going to happen? A local TV weatherman told his viewers, "When those NWS guys start using terms like 'historic proportions' . . . you know this one will come through." Meanwhile, Uccellini, aware of the models' disagreements and past forecast failures in D.C., sweated a few more bullets. Washington and the entire U.S. East Coast braced for a record storm that would bury the Appalachian Mountains with feet of snow and dump a wintry mix of snow, sleet, and freezing rain on the large coastal cities.

Figure 13.16

NMC's 12-hour snow forecasts for the "Storm of the Century" (left) and what actually happened (right) for two overlapping time periods on March 12–13, 1993. The region of 16 inches of snow in the lower left figure was the most snow ever forecast by NMC in a 12-hour period. (Source: Adapted from The Life Cycles of Extra-Tropical Cyclones, *Shapiro and Gronas, eds., American Meteorological Society, 1999. Reprinted with permission.)*

A   12-HOUR SNOW FORECAST

ISSUED: 2:00 PM EST March 12

ACTUAL

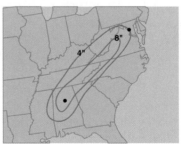

Snowfall 6:00 AM EST March 12 to 6:00 PM March 13

B   12-HOUR SNOW FORECAST

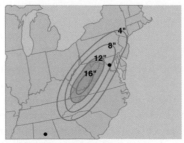

ISSUED: 7:30 PM EST March 12

ACTUAL

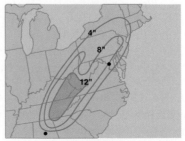

Snowfall 12:00 AM EST March 13 to 12:00 PM March 13

◼ The Short-Range Forecast for Birmingham, Alabama

Tough forecasts weren't confined to the East Coast that day. An intensifying Gulf low with bitter-cold air also means a chance of snow for the Gulf Coast states—even at spring break. In Birmingham, Alabama, a metropolitan area of one million nestled in the foothills of the southern Appalachians 400 kilometers (250 miles) from the Gulf, March snows aren't unheard of. But when National Weather Service meteorologists in Birmingham started calling for six inches or more of snow on Friday, March 12, afternoon and evening, eyebrows went up and panicked residents cleaned out supermarket shelves.

Six inches of snow? This was even more than the NMC models suggested (see Figure 13.16, *A*). The Birmingham forecasters relied on their local experience and knowledge of the city's hilly topography, too small-scale for the forecast models to "see." They had a hunch that, despite the blooming trees and flowers and warmth, the conditions were just right for a climatology-defying snowstorm in Birmingham.

Even so, there was no proof that the storm would come together. Proof would come shortly.

◼ The Storm of the Century Appears

On the morning of Friday the 12th, meteorologists' eyes turned to the Texas Gulf Coast. There, all the ingredients for cyclone growth discussed in Chapter 10—upper-level divergence, temperature gradients, even warm, moist water—came together in a once-in-a-century mix. The storm exploded, Uccellini said later, "like an atom bomb." The cyclone engulfed the Gulf. By afternoon, oil rigs off the coast of Louisiana were reporting hurricane-force winds. In its first 24 hours of life, the cyclone's central pressure would drop almost 30 mb. By midnight, the storm had the characteristics of a major cyclone: strong fronts, a massive "comma cloud," and a powerful squall line (Figure 13.17).

◼ **Figure 13.17**

An infrared satellite picture of the "Storm of the Century" making landfall over the Florida Panhandle at 1:00 AM Eastern time on March 13, 1993. Estimated surface pressures are shown with yellow lines; yellow- and red-shaded regions indicate high clouds such as thunderstorms.

The incredible pressure gradient caused by the dropping pressure led to the first widespread Southern blizzard on record—snow combined with cold and howling winds. The atmosphere became so unstable that thunderstorms developed in the cold air, dumping several inches of snow every hour. In Birmingham, a radio transmitter on top of 300-meter (1000-foot) Red Mountain was struck twelve times by eerie green lightning during "thundersnow."

Nearly 50 University of Wisconsin students heading for Panama City, Florida, during spring break were stranded in Birmingham when their bus skidded off a mountainous road. Boasted one, "Don't they have sand or salt around here? Back home this is nothing." The thundersnow raged on.

A Perfect Forecast

The numerical models hadn't foreseen the cyclone's explosive development over the Gulf (although the out-of-date LFM had been giving hints). NMC forecasters now revised the models' predicted central pressures downward. The models were often accused of going overboard with their doomsday forecasts. But on the 13th, nature itself was "over the top!"

What did this mean for the East Coast? The Eta and NGM models now indicated that the cyclone would move inland over Virginia with a central pressure well below 960 mb. Importantly, this storm track would turn all snow to rain over the East Coast cities. The AVN, however, insisted on the same overall scenario that had been based on recent guidance from the very similar MRF model: a 960-mb cyclone over Chesapeake Bay. The forecasters knew that the AVN had a reputation for forecasting East Coast cyclone tracks accurately. NMC thus ignored the "Armageddon" predictions of the Eta and NGM and based their forecast on guidance from the AVN. They were right. At 7:00 PM on the 13th, the storm of the century was centered over Chesapeake Bay, where barometers measured its central pressure at a record-low 960 mb, exactly as NMC had predicted. *It was a perfect forecast of a once-in-a-lifetime event.*

Nowcasting in D.C.

National Weather Service forecasters in Washington, using the NMC forecasts and their local knowledge, had given 1 to 2 days' notice of impending winter storm and blizzard conditions. On the morning and afternoon of the 13th, bands of thunderstorms of snow and sleet rotated around the cyclone into the Washington vicinity. The computer models couldn't "see" these narrow bands, and neither could older weather radars, but Doppler radar (see Chapter 5) could. The local forecasters used the radar to make very short-term forecasts of precipitation type, intensity, and winds. Instead of a vague "heavy snow today" forecast, Washington residents received hour-by-hour nowcasts of the tiniest details of a cyclone the size of half a continent.

The Aftermath

The "Storm of the Century" buried the eastern United States in enough snow that, if melted, would cover New York State with 30 centimeters (one foot) of water. The amount and extent of the snowfall was unprecedented in American history (Figure 13.18).

In Birmingham, the official total of 33 centimeters (13 inches) achieved the rare "hat trick" of snowfall records. It was the most snow ever recorded in the city in one day, one month, and even one whole year! In the hills around Birmingham, a retired National Weather Service forecaster carefully measured 43 centimeters (17 inches) of snow. The forecast had been an underestimate, but it gave residents the warning they needed that a major storm was on the way.

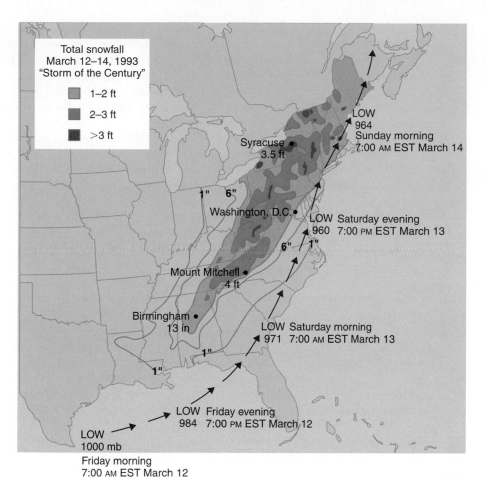

Total snowfall
March 12–14, 1993
"Storm of the Century"

- 1–2 ft
- 2–3 ft
- >3 ft

LOW
964
Sunday morning
7:00 AM EST March 14

Syracuse
3.5 ft

1"   6"

Washington, D.C.

LOW   Saturday evening
960   7:00 PM EST March 13

6"   1"

Mount Mitchell
4 ft

Birmingham
13 in

LOW   Saturday morning
971   7:00 AM EST March 13

1"

1"

1"

LOW   Friday evening
984   7:00 PM EST March 12

LOW
1000 mb

Friday morning
7:00 AM EST March 12

■ **Figure 13.18**

The path of the "Storm of the Century" on March 12–14, 1993, along with the total amount of snowfall reported across the southern and eastern United States. (Source: Adapted from The Life Cycles of Extra-Tropical Cyclones, *Shapiro and Gronas, eds., American Meteorological Society, 1999. Reprinted with permission.)*

The deep white blanket of snow and an intense anticyclone behind the low caused Birmingham temperatures to drop to –17° C, or 2° F—an all-time March record (Figure 13.19). Meanwhile, half the state of Alabama was without power and therefore heat in most cases. "Prairie Home Companion" storyteller Garrison Keillor, by chance in Birmingham to do his live public-radio variety show, sang: "Oh Susannah, now don't you cry for me/I'm going to Alabama for a couple of days to *ski.*"

Washington (Figure 13.20) and other East Coast cities such as New York City fared just as predicted. They generally received about a foot of snow, which turned to ice and slush as the low and its warmer air moved up the coastline.

■ **Figure 13.19**

The "Storm of the Century" in the hills of Birmingham, Alabama. Wet, wind-driven snow is defying gravity and hanging off the roof of the co-author's boyhood home. The deep blue sky at top left is a sign of the rapidly sinking and drying air of the intense anticyclone that followed the storm. The record snow led to massive power outages that cut off heat for days to tens of thousands of homes in Alabama, including this one.

H. Knox

http://info.brookscole.com/ackerman

J. McGuire

Figure 13.20

The March 1993 "Storm of the Century" buried East Coast cities with heavy, but not record-breaking, snowfall.

Some East Coast residents grumbled that this wasn't the century's worst storm; they'd seen worse. They didn't realize that much worse weather had indeed occurred, as forecast, just a few miles inland. Three million people lost electric power nationally. Storm winds gusting up to 232 km/hr (144 mph) shut down airports across the East. This caused the worst aviation delays in *world* history up to 1993. Many spring breaks were spent by students sleeping on airport carpets.

Other regions were dealt a cruel blow by the unexpected ferocity of the cyclone's winds and thunderstorms. Forty-seven people died in Florida as a result of tornadoes and storm–surge-like flooding. Cuba suffered $1 billion in coastal flood damages. Severe weather battered Mexico's Yucatan peninsula. Forty-eight people died at sea in this truly "perfect storm."

However, the $2 billion in damages and up to 270 deaths attributed to the storm in the United States were a fraction of what might have been. What if the year had been 1939 instead of 1993? Folklore, persistence, climatology, trends, and analogs—the methods of forecasters in 1939—could not have forecast such an extreme event as the Storm of the Century. Today's numerical models of the atmosphere, Richardson's fantasy come true, could and did make the right forecast—but only when coupled with the wisdom of the human meteorologists.

At NMC, Louis Uccellini could stop sweating bullets. Numerical weather prediction had saved the day. Afterward, he observed: "We [meteorologists] attained a level of credibility that we never had before and we haven't lost since. People took action based on our forecasts and they take action today . . . that's the future."

Why Forecasts Still Go Wrong Today

Even with forecasting successes like the Storm of the Century, meteorologists are not satisfied. "We always want to push the limits," Louis Uccellini says. The reason is that numerical weather prediction isn't perfect. As we saw earlier, even in the best of forecasts there are times and regions where the models are wrong. Even when tomorrow's forecast is right, it's fair to ask: why couldn't we have known it weeks, not days, in advance?

The limits of prediction today have their roots in Richardson's original forecast. Richardson's model, as we've seen, had several limitations:

1. He didn't have much data to work with.
2. The forecast was for a small area of the globe.
3. The forecast was for much less than one day.
4. The complicated nature of the equations forced him to make approximations.
5. His surface data were "bumpy" and made a mess of the forecast.
6. His model's wide grid spacing would have made a mess of his forecast if "bumpy" data hadn't done it already.
7. The forecast itself inspired little confidence because of its unrealistic results.

These limitations directly relate to today's numerical forecast models. We examine a few of them here.

Imperfect Data

The data "diet" of today's numerical models still includes a large helping of radiosonde observations. However, the number of radiosonde sites in the United States has actually declined over the past few decades. It is easier to ask the U.S. Congress for money for exciting weather satellites than for boring weather balloons. Satellite data are global in coverage, but researchers in data assimilation are still trying to figure out how this data can be "digested" properly by the models. In fact, it took nearly 20 years of research and testing before satellite data made a large positive impact on the quality of NMC/NCEP model forecasts in the Northern Hemisphere. In addition, important meteorological features still evade detection, especially over the oceans. The model results are only as good as the data in its initial conditions. This was a reason for Richardson's forecast failure, and explains some forecast "busts" today, including the surprise blizzard that shut down the federal government (once again) in Washington, D.C. on January 25, 2000.

■ Faulty "Vision" and "Fudges"

Today's forecasts also involve an inevitable tradeoff (see Box 13.4) between horizontal resolution and the length of the forecast. This is because fine resolution means lots of points at which to make calculations. This requires a lot of computer time. A forecast well into the future also requires lots of computer time because each day the model looks further into the future requires millions or billions more calculations. If fine resolution is combined with a long-range forecast, the task would choke the fastest supercomputers today. You wouldn't get your forecast for weeks! Future improvements in computing will help speed things up.

In the meantime, however, the models will still not be able to "see" small-scale phenomena such as those discussed in Chapter 12, not to mention clouds, raindrops, and snowflakes. To compensate for this fuzzy "vision" of models, the computer code includes crude approximations of what's not being seen. These are called **parameterizations** (Figure 13.21). Even though much science goes into them, these approximations are nowhere close to capturing the complicated reality of the phenomena. This is because, as we learned in Chapter 12, the smallest-scale phenomena are often the most daunting to understand. Therefore, it is not an insult to meteorologists' abilities to say that parameterizations are "fudges" of the actual phenomena.

Forecast models have to parameterize or "fudge" all the small-scale phenomena listed above. Worse yet, the atmosphere's interactions with other spheres, such as the ocean or the land, also have to be approximated—usually very poorly. This seemingly trivial part of modeling turns out to be a critical area for forecast improvement today.

■ **Figure 13.21**

About twenty different meteorological processes in this figure require parameterization in forecast models because they are too small and/or too complicated for the models to compute precisely. A few parameterized processes are labeled in the illustration. (Source: http://deved. meted.ucar.edu/nwp/pcu1/ic1/ 2_2.htm)

Would Richardson have believed it? Of course. His 1922 forecast book includes entire exhaustive sections devoted to the proper handling of the effects of plants, soil moisture, and turbulence on his forecast. Today's best models devote the most computer time to their parameterizations of tiny phenomena. Their successful approach can be summarized simply: "sweat the small stuff."

 ## Chaos

Why settle for a 15-day forecast? Assume you own the world's fastest supercomputer in the year 2020. If the trend in Figure 13.12 holds, by the year 2020 your computer could do *quadrillions* of calculations each second! With that much computing power, why not make a forecast for each day out to one month in advance? Surprisingly, you would *not* get a better forecast. Brute-force numerical weather forecasting, with extremely fine resolution, has its limits.

The reason for these limits is a curious property of complex, evolving systems like the atmosphere. It is called "sensitive dependence on initial conditions," and is a hallmark of what's popularly known as **chaos theory.** Chaos in the atmosphere does not mean that everything is a mess. Instead, it means that the atmosphere—both in real life and in a computer model—may react *very* differently to initial conditions that are only *slightly* different.

Because we don't know the atmospheric conditions perfectly at any time, chaos means that the resemblance between a model's forecast and reality will be less and less with each passing day. (The same is true for different runs of the same model with slightly different initial conditions.) Meteorologists believe that a 2-week forecast is the eternal limit for a forecast done Richardson's way. No amount of computer improvements, parameterization advances, or complaining will change this limit.

Forecasts of Forecast Accuracy: Ensemble Forecasting

Is chaos then the eternal roadblock for numerical weather prediction? Not quite. In the past decade, meteorologists have figured out how to use chaos theory to give us something Richardson sorely needed—confidence in the accuracy of a forecast. This method is called **ensemble forecasting** and it works for both short- and long-range forecasts.

Here's a recipe for ensemble forecasting. Make a numerical forecast for a certain day. However, there is no guarantee that the data used by the model for the initial conditions were measured precisely and accurately. For example, radiosonde measurements of temperature have an error range of about one degree Fahrenheit. So make a different forecast using slightly different initial conditions that are within the error range of the observing instruments. Then make yet another forecast with yet another set of slightly different initial conditions. Repeat for many different sets of initial conditions. Then compare all the different forecasts, which are called the "ensemble." If most or all of the different forecasts agree, then there is a high degree of confidence that their prediction will become reality. If the different forecasts give wildly different results, then there is lower confidence in whatever forecast is eventually chosen. In the latter case, forecasters also get a "feel" for what the possibilities are for a rare weather event, such as a severe cyclone.

A recent overseas example illustrates the usefulness of ensemble forecasting. On the day after Christmas Day 1999, a severe extratropical cyclone crossed the English Channel with a forward speed of nearly 110 km/hr (70 mph) and a central pressure near 960 mb. Extreme wind gusts close to 180 km/hr (110 mph) swept through Paris and across much of France, Germany, and Switzerland, causing damage in the billions of Euros (see the introduction to Chapter 6).

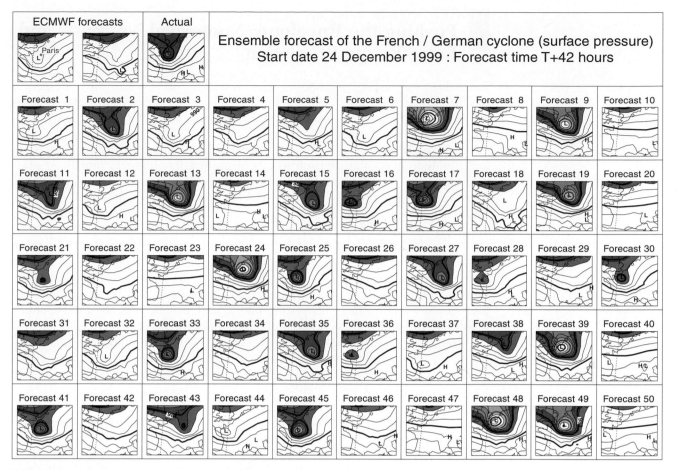

■ **Figure 13.22**

Top row: Lower-resolution (left) and high-resolution (center) ECMWF model forecasts of surface pressure versus the observed analysis (right) during a severe extratropical cyclone on December 26, 1999 over western Europe. Bottom five rows: Ensemble forecasts of the same storm using 50 ECMWF model runs. Compare the "Actual" conditions to each panel forecast. Notice that many of the ensemble model runs closely resemble what actually happened. This indicates that the chance of a severe storm was unusually high. (Source: http://www.ecmwf.int/pressroom/ newsletter/pdf/88_2.pdf)

Could any numerical model anticipate such a swift and severe storm? The top row in Figure 13.22 suggests the answer is "no." The official forecasts of the state-of-the-art ECMWF model reveal little hint of the deep cyclone that was observed, the shaded region in the panel labeled "Actual."

However, the logic behind ensemble forecasting says that a single model run is not enough to base a forecast on. The 50 forecasts shown in the bottom five rows of Figure 13.22 are all different simulations of this same storm using the same ECMWF model at the same resolution as the forecast in the first image in the top row. The other difference is that each of the 50 model runs were begun with slightly different initial data that account for the imperfections in our weather observing systems. Of these 50 model runs, almost half show signs that very low pressure would develop over western Europe. With this knowledge, forecasters can immediately recognize that the risk for an epic storm was unusually high, even if one or two model runs showed no hint of it.

Ensemble forecasting requires a sophisticated understanding of the atmosphere and computer models. Also, it takes exceptional amounts of computing power to perform 50 runs of the ECMWF model. Nevertheless, this technique will soon become familiar to the public as a routine part of media coverage of the weather.

Pushing the Envelope: Numerical Nowcasts and Long-Range Prediction

Ensemble forecasting exemplifies the versatility of Richardson's numerical weather forecasting approach. Emboldened by such successes, meteorologists would like to forecast the weather at two opposite ends of the time scale: minutes and months.

We have already mentioned nowcasting in the context of using observations and trends to make short-term forecasts of up to a few hours in length. At present, the use of very high resolution models to help with these forecasts of thunderstorms has not proved very beneficial. In short, meteorologists still don't understand enough about the atmosphere at the smallest scales to make this approach work.

At somewhat larger scales, the news is better. Until very recently, short-range hurricane forecasts were based largely on satellite and aircraft observations and climatology, not on computer models using Richardson's approach. However, hurricane modeling is making some strides in the early years of the 21st century, as complex models that include both atmospheric and oceanic effects are becoming feasible to run on today's lightning-fast supercomputers.

Long-range forecasting runs up against the 2-week limit imposed by chaos theory. However, ensemble forecasting can extend the range of reliable forecasts all the way to this 2-week limit. Recently, media weather discussions have begun showing forecasts for the next week and beyond, a visible sign of the rising confidence in longer-range forecasting since the early 1990s.

Beyond the 2-week limit, forecasters mostly use statistical approaches based largely on past observations. For example, the known effects of El Niño (see Chapter 8) on weather and climate in certain regions can be used to create forecasts of temperature and precipitation versus average conditions from a few months to a year in advance. And, as we saw in Chapter 8, hurricane seasons can be forecast statistically with good results many months in advance. Unfortunately, in-between the two-week limit and multi-month statistics lies the much-desired one-month forecast. A combination of approaches may eventually lead to useful monthly forecasts—but don't hold your breath!

The Proper Perspective

Why is it that the task of weather forecasting seems to be so complex and error-prone compared to predicting, say, a solar eclipse? L. F. Richardson explained it first and best: "the [forecasting] scheme is complicated because the atmosphere is complicated." But far from being a failure and a waste of time, modern weather forecasting is a triumph.

In just a half-century, numerical weather prediction has turned weather forecasting into a true science. Figure 13.23 charts this progress in terms of decreasing forecast errors. In the top part of this figure, errors in NMC/NCEP forecasts for North American weather at 500 mb are shown to have plummeted from the days of human forecasters in the 1950s to nearly "perfect" computerized forecasts by the late 1990s. A forecast of jet-stream winds today for three days from now is as accurate as a similar forecast made in 1980 for just a day and a half later.

These improvements are due to better science, better computers, and new generations of enhanced computer models of the atmosphere, as illustrated in the bottom part of Figure 13.23 for forecasts of sea-level pressure over North America. Precipitation forecasts are harder because of the spotty nature of rain and snow, but

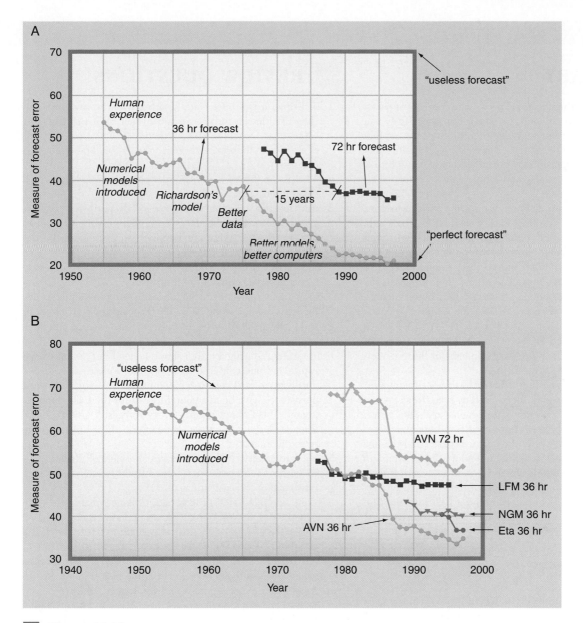

■ **Figure 13.23**

*Errors in NMC/NCEP model North American forecasts for 36-hour forecasts at 500 mb (**A**) and mean sea-level pressure forecasts (**B**). Notice the steady decline in errors since the introduction of numerical models in the late 1950s. In **B**, the accuracy of NCEP models relative to each other is depicted. (Source: Adapted from Kalnay, E., Lord, S., and McPherson, R., "Maturity of Operational Numerical Weather Prediction: Medium Range," Bulletin of the American Meteorological Society, vol. 79, no. 12, pp. 2753–2892.)*

even so today's forecast of 2.5 centimeters (1 inch) of rain 2 days from now is as accurate as a 1-day forecast of 1 inch of rain made 20 years ago. Similar improvements have been made for medium-range forecasts as well. As a consequence, today's media place much more emphasis on and confidence in multi-day weather forecasts than they did just a few years ago.

These improvements in weather forecasting are steady but generally unspectacular, except in the case of a Storm of the Century. In your lifetime, you will grow accustomed to—and spoiled by—better and longer-range weather forecasts. As you do, please try to remember that weather forecasting was a confusion of competing unreliable methods as recently as World War II!

http://info.brookscole.com/ackerman

PUTTING IT ALL TOGETHER

 SUMMARY

People can forecast the weather in a wide variety of ways. They can recall folklore rhymes, or watch the skies. They can assume that today's weather will persist into tomorrow, or that the trend of weather will continue. They can even assume that the weather will be typical from a climatological perspective. These methods are hit-or-miss and inadequate for most modern needs.

Numerical weather prediction is the solving of the equations of the atmosphere—a "model"—on a computer. Lewis F. Richardson created the first numerical forecast model. Even though his "forecast" turned out to be wrong, his approach was sound. Modern forecast models ingest observational data, take out the "potholes" via data assimilation, and then integrate the models forward in time to obtain a forecast. With these models, meteorologists can make precise and accurate weather forecasts for the first time in history. The LFM, NGM, Eta, RUC, UKMET, AVN, MRF, and ECMWF are nicknames of modern numerical forecast models. Forecasts from most of these models are available on the World Wide Web.

Modern weather forecasting today fuses advanced computer modeling with human insight. Together, they save lives and fortunes through increasingly accurate predictions, as in the "Storm of the Century" in March 1993. Limits exist on how good forecasts can become, however. Imperfect data, imperfect knowledge of how the atmosphere works, limits on computing power, and chaos make forecasts go wrong. Even so, modern weather forecasting is one of the great achievements of modern meteorology and all of science and will continue to improve during your lifetime through new techniques such as ensemble forecasting, which provides forecasts of the accuracy of a forecast.

KEY TERMS

You should understand all of the following terms. Use the glossary and this chapter to improve your understanding of these terms.

Analog forecast	LFM
AVN	Model
Chaos theory	MRF
Climatology forecast	NGM
Data assimilation	Nowcast
Data initialization	Numerical modeling
ECMWF	Parameterizations
Ensemble forecasting	Persistence forecast
Eta	Primitive equations
Folklore forecasts	Resolution
Grid	RUC
Gridpoint models	Spectral models
Grid spacing	Trend forecast
Initial conditions	UKMET
Interpolation	Weather types

REVIEW QUESTIONS

1. In Britain it was once said, "If it rains on St. Swithin's Day (July 15) it means forty more days of rain." What kind of forecast is this? Do you trust it? Why or why not? Can you think of a region in the world where it actually could rain forty days in a row?

2. What is the difference between a persistence forecast and a trend forecast?

3. You travel to a region where the north–south temperature gradient is very strong. Do you think a persistence forecast will be reliable? (Hint: review the concept of the thermal wind relationship in Chapter 7.)

4. How might global warming, if real, affect the usefulness of climatology forecasts in the future?

5. Some meteorologists say that chaos theory proves that analog forecasts cannot work. Why do they say this? (Hint: think about the small differences between one weather map and another.)

6. In Chapter 10, Figure 10.8 shows a small but potent extratropical cyclone centered over Wichita, Kansas, and headed northeast toward the lower peninsula of Michigan. Would persistence have been a good forecast for the Lake Superior region? What about a climatology forecast? Would a trend forecast have accurately anticipated the storm's future strength and path?

7. Is a wrong answer always a failure in science? Explain how Richardson's forecast woes helped lead to much more accurate weather forecasts today.

8. If the November 1975 "Fitzgerald" cyclone (Chapter 10) happened today, do you think it would be forecast better than it was by the LFM model? Answer with reference to the details in Table 13.1 and Figure 13.23.

9. You are an aviation forecaster and you need a good forecast for an airplane flying over Japan in the next 6 hours. Which American forecast model would you look at, and why?

10. NCEP is having a picnic for its forecasters and their families in Washington, D.C., in a week. Which one of their own models would the NCEP forecasters look at to decide whether or not the picnic will be rained out? Which overseas model would they look at first? Why does this overseas model make the most accurate forecasts?

11. A radiosonde launching site costs roughly $100,000 a year to operate. Why don't we lobby the government to shut down all the radiosonde sites and spend the money instead on multimillion-dollar supercomputers?

12. Which type of data-gathering method is the backbone of modern numerical weather prediction? Why has satellite data *not* been extensively used until recently?

13. Why is it that different models can give widely varying predictions for the same time and the same region, as shown in Figure 13.15?

14. Why can't today's numerical models make forecasts for individual thunderstorms five days ahead of time?

15. A grandparent says, "Weather forecasts always were terrible, always will be terrible." After reading this chapter, do you agree? How could you use the examples of D-Day and the

"Storm of the Century" to describe the advances of modern weather forecasting?

WEB ACTIVITIES

Choose Chapter 13 on the textbook's Web site:
http://info.brookscole.com/ackerman
and select from the following resources:

• Interactive Modules that illustrate and extend your understanding of key topics in this chapter

• Tutorial Quizzes to test your mastery of terms and concepts

For additional readings, go to the InfoTrac College Edition, your online library, at:
http://www.infotrac-college.com

Past and Present Climates

After completing this chapter, you should be able to:

■ Identify the main climate zones around the world

■ Explain how scientists study past climates

■ List the natural processes that can affect climate and describe how each one can lead to climate change

Introduction

Los Angeles residents do not expect to have snow at Christmas. No polar bears are found roaming the deserts of Arizona, and diamondback rattlesnakes do not inhabit arctic ice fields. Tropical rain forests teem with life, but life is much sparser in the great deserts such as the Sahara. Summer is the off-season for hotels and restaurants in Florida, but the busiest time of year for the same businesses in Maine and on Cape Cod. *Life adapts to climate.* Climate, the overall weather that prevails from year to year in an area, profoundly affects the distribution and abundance of life forms and the activities of people.

Today, Antarctica has a frigid climate where humans can survive only because of our ability to build shelter from the weather elements. A recent analysis of temperatures by Dr. Susan Solomon (below), a senior scientist at the National Oceanic and Atmospheric Administration, indicates that an unusually cold Antarctic autumn contributed to the death of Captain Robert F. Scott and his four comrades on their 1500-kilometer (900-mile) trek back from the South Pole in March 1912. Temperatures were 10° to 20° colder than expected during the race to the South Pole. The cold weather cut in half the distance the explorers could travel in a day. A

S. Solomon

blizzard trapped them in a tent, where they froze to death 18 kilometers (eleven miles) from a supply depot.

Scott's party had lost the race to the South Pole to Roald Amundsen, who reached the pole a month before. However, the Scott expedition revealed that Antarctica once basked in warmth. Among the 16 kilograms (35 pounds) of rocks the expedition collected were fossils of *Glossopteris,* a seed fern. This fossil is scientific evidence that the current ice-covered continent was once fertile.

Climates usually remain the same during a person's lifetime, but climates have also changed remarkably during the Earth's lifetime. This chapter will define climate and describe how and why climate changes over time.

Defining Climate

Climate is to weather what a friend's *personality* is to his or her mood. In other words, weather captures the atmosphere's short-term ups and downs, whereas climate sums up its long-term behavior. More precisely, climate can be defined as the collective state of the atmosphere for a given place over a specified interval of time. There are three parts to this definition:

1. *Location,* because climate can be defined for a globe, a continent, a region, or a city. In this chapter we concentrate on regional and global-scale climates.
2. *Time,* because climate must be defined over a specified interval. In Chapter 3 we used 30-year averages, whereas in studying Earth's history we may use averages of a century or longer.
3. *Averages and extremes of variables* such as temperature, precipitation, pressure, and winds. In Chapter 3 we focused on temperature, but in this chapter we look at both temperature and precipitation.

We have already made use of climate to understand the atmosphere. For example, in Chapter 3 we compared the 30-year averages of monthly temperature of different regions to understand the controls of the seasonal temperature cycle. In Chapters 11 and 12 we used climate data to study the preferred locations and frequency of thunderstorms and windstorms. In this chapter we learn how to classify various climates and explain climates of the distant past.

Climate Controls

The five basic factors that affect climate are very similar to the temperature controls we studied in Chapter 3. They are latitude, elevation, topography, proximity to large bodies of water, and prevailing atmospheric circulation. Latitude determines solar energy input. Elevation influences air temperature and whether precipitation falls as snow or rain. Mountain barriers upwind and downwind can affect precipitation of a region as well as temperature. Topography also affects the distribution of cloud patterns and thus solar energy reaching the surface. The thermal properties of water moderate the temperature of regions downwind of the region. Atmospheric circulation is somewhat less regular than other factors. But consistent large-scale circulation patterns, such as the position of the subtropical high-pressure belts or the Intertropical Convergence Zone (ITCZ), exert a systematic impact on the climate of a region. These surface and atmospheric features produce variations in temperature and precipitation that create different climate patterns.

Classifying Today's Climate Zones

Are you a "Type A" personality? Have you ever taken a personality test that categorized you as an "ENFP"? Pop psychology tries to categorize the complexities of human personalities into a few types, which are usually represented by one or more letters. Scientists in all disciplines use classification for similar reasons: to make sense of complex natural systems.

Climatologists use this approach with climate. There are many different regional climates across the world. To make sense of this variability climatologists use classification schemes that resemble the personality categories of psychology. The difference with climate classification is that the categories identify regions, not people, that have similar characteristics.

A challenge in designing a climate classification scheme is that climates, like personalities, do not have clear dividing lines. We often classify the climate of a location using descriptive terms for temperature and precipitation: for example, hot and dry, warm and wet, or cold and dry. We choose temperature and precipitation because these two weather parameters are extremely important to life. Temperature and precipitation together determine the environmental conditions under which certain plants and animals flourish while others perish.

Between 1918 and 1936 Vladimir Köppen (pronounced KEPP-in) devised the climate classification scheme that is most widely used today. He used vegetation and temperature as natural indications of the climate of a region. Improvements have been made to the original **Köppen scheme,** particularly by Glenn Trewartha and Lyle Horn of the University of Wisconsin.

The current Köppen classification scheme has six main groups, each designated with a letter: Tropical Moist (**A**), Dry (**B**), Moist with Mild Winters (**C**), Moist with Severe Winters (**D**), Polar (**E**), and Highland (**H**). Some groups are described by two- and three-letter designations that are reminiscent of personality classifications. The second letter typically refers to whether and when a dry season occurs, and the third letter denotes differences in temperatures. Figure 14.1 organizes and summarizes these climate zones by their temperature and precipitation characteristics.

Tropical Humid (A)	Dry (B)		Moist Subtropical and Midlatitude (C)		Severe Midlatitude (D)	Polar (E)	Highland (H)
Af, tropical wet, no dry season	**BWh**, Subtropical desert located in low-latitudes and dry	**BWk**, Midlatitude dry desert	**Cfa**, humid subtropical, hot summer, no dry season	**Cwa**, humid subtropical, hot summer, and brief winter dry season	*Humid Continental D climates have a severe winter* **Dfa**, no dry season and hot summer **Dfb**, no dry season, warm summer **Dwa**, winter dry season and hot summer **Dwb**, winter dry season and warm summer	**EF**, Perennial ice	**H**
Am, tropical monsoonal, short dry season	**BSh**, Subtropical steppe, a low-latitude semi-dry climate	**BSk**, Midlatitude steppe, semi-dry midlatitude desert	**Csa**, Mediterranean, dry hot summer	**Csb**, Mediterranean with dry and warm summer	*Subarctic D climates all have cool summers* **Dfc**, Severe winter with no dry season **Dfd**, Extremely severe winter with no dry season **Dwc**, Severe winter and winter dry season **Dwd**, Extremely severe winter and winter dry season	**ET**, No summer	
Aw, tropical wet and dry, dry season in winter			**Cfb**, Marine west coast, no dry season and warm summer	**Cfc**, Marine west coast, no dry season and cool summer			

Figure 14.1

Overview of the main climatic groups with respect to temperature and precipitation. In general, temperature decreases from left to right and precipitation from bottom to top of the figure. (Source: Adapted from Aguado, E. and Burt, J., Understanding Weather and Climate, 2nd ed., Prentice-Hall, 2001, p. 414.)

Understandably, the Köppen climate zones are closely related to both geography and the global circulation of the atmosphere. Figure 14.2 shows the global distribution of the Köppen climate zones, and Figure 14.3 depicts this distribution in relation to the large-scale wind patterns we studied in Chapter 7, such as the ITCZ. Next, we discuss the characteristics of each zone. We use plots of climate data called **climographs** to depict the characteristic monthly mean temperature and precipitation of each of these climates.

Discovering Climate Types

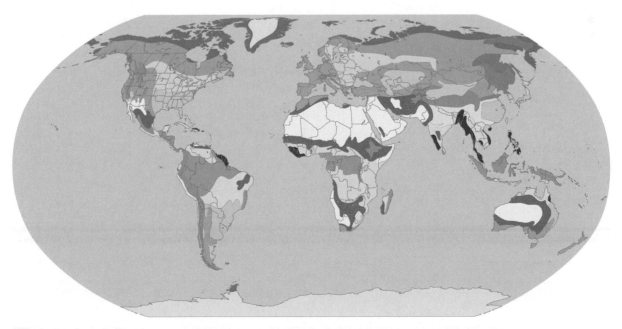

Figure 14.2

A world map of the Köppen climate classification scheme. The colors correspond to the climate types defined in Figure 14.1.

Figure 14.3

A cross-section of pressure, wind, and cloud patterns in latitude, relating the Köppen climate classes to the global scale atmospheric circulation patterns discussed in Chapter 7. (Source: Adapted from Trewartha, G., and Horn, L., An Introduction to Climate, 5th Edition, McGraw-Hill, 1980, p. 209.)

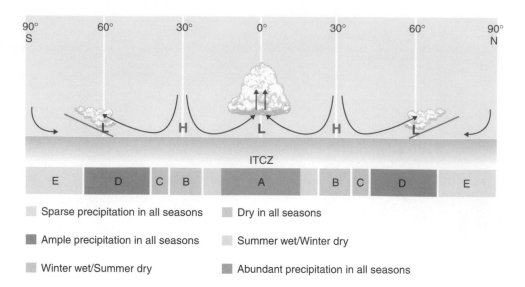

Tropical Humid Climates

The mean monthly temperature of tropical humid (**A**) climates is high, at least 18.3° C (65° F). The range of the annual temperature is small, typically less than 10° C (18° F). As a result, killing frosts are absent in **A** type climate regions. The diurnal variation in temperature in **A** climate regions is often larger than the annual variation.

Although **A** climates have abundant rainfall, typically more than 100 centimeters (39 inches) per year, they can have different precipitation patterns. **A**-type climate zones are therefore subdivided into three subtypes: tropical wet climates (**Af**), tropical wet-and-dry climates (**Aw**), and tropical monsoon (**Am**) climates. Examples of these climates are Iquitos, Peru (**Af**), Pirenopolis, Brazil (**Aw**), and Rochambeau, French Guiana (**Am**). Figure 14.4 shows the annual temperature and precipitation patterns of these cities.

The tropical wet climates (**Af**) have temperatures that are distributed fairly uniformly throughout the year. Precipitation each month averages between 17.5 and 25.4 centimeters (6.9 and 10 inches). **Af** climates also have a diurnal precipitation

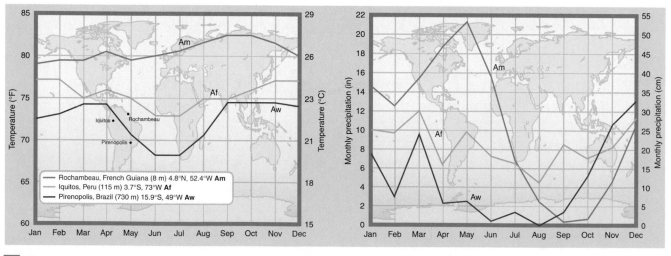

Figure 14.4

*Monthly temperature and precipitation for three tropical climate regimes: Iquitos, Peru (**Af**), Pirenopolis, Brazil (**Aw**), and Rochambeau, French Guiana (**Am**). The altitude, latitude, and longitude of each station are given in the legend.*

pattern, with most thunderstorms occurring in the afternoon, triggered by solar heating of the surface. Vegetation in **Af** climates is very lush, as in the tropical rain forests of Brazil and the Congo. Analysis of Figure 14.2 shows that **Af** regimes often border **Aw** climates.

Tropical wet-and-dry climates (**Aw**) have wet and dry seasons. Summers are wet and winters are dry in **Aw** climates. This seasonal rainfall is linked to the seasonal migration of the ITCZ. There is a cool season in **Aw** climates, which occurs during winter. The vegetation of **Aw** climates is typically savanna or tropical grasslands with scattered deciduous trees, as in the grasslands of Africa.

Tropical monsoon climates (**Am**) are climates that have a short dry season (see Figure 14.4). Monthly average temperatures of **Am** climates are uniform throughout the year. These climates tend to occur in regions that have seasonal onshore winds that supply ample moist air. Orographic lifting can also help to enhance the precipitation of **Am** regimes.

■ Dry Climates

Dry climate zones (**B**) are located in regions where potential evaporation and transpiration exceeds precipitation. Rainfall is highly variable in **B** climate zones. Most of the land regions of the world are designated as **B** climate zones! The descending branch of the Hadley cell or a rain shadow caused by mountain barriers causes the lack of precipitation in many of the **B** climate zones.

There are two subtypes of the **B** climates: steppe or semiarid (**BS**) and arid or desert (**BW**). Inspection of the climate zone map in Figure 14.2 indicates that the **BS** climates are transition zones, situated between humid climates and desert climates.

Dry climates are those for which evaporation exceeds precipitation. Dry climates span the tropics and the poles. **BSh** and **BWh** are warm dry climates, typical of tropical regions. Figure 14.5 illustrates the temperature and precipitation of Dakar, Senegal, and Cairo, Egypt, which are examples of **BSh** and **BWh** climates, respectively. The precipitation peak over Dakar results from the seasonal migration of the ITCZ (see Chapter 7). The annual mean precipitation of Cairo (**BWh**) is much lower than that of Dakar (**BSh**), and its annual temperature range is larger.

Hollywood's portrayal of deserts is usually one of hot sweltering days, an intense Sun, and large sand dunes. The cities of Dakar and Cairo tend to conform to this image. But not all dry climates are hot, tropical deserts. **BSk** and **BWk** climates,

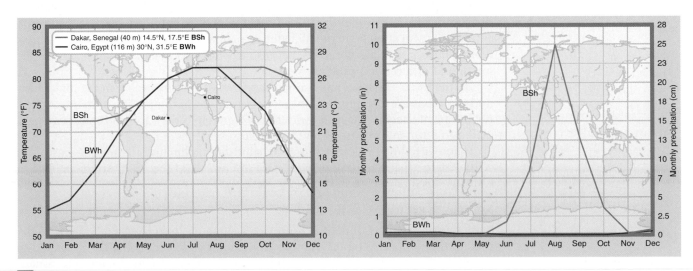

■ Figure 14.5

*Monthly temperature and precipitation for the dry subtropical climates of Dakar, Senegal (**BSh**), and Cairo, Egypt (**BWh**).*

http://info.brookscole.com/ackerman

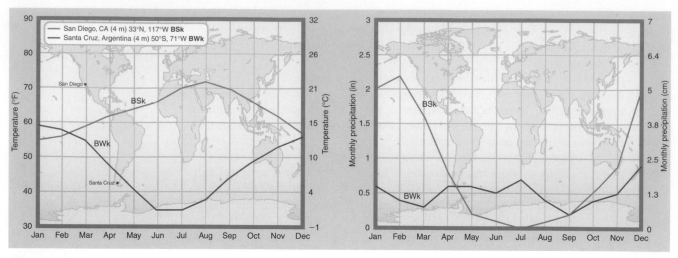

Figure 14.6

*Temperature and precipitation for dry climate regimes of the middle latitudes at San Diego, California (**BSk**), and Santa Cruz, Argentina (**BWk**). December and January are summer months in the Southern Hemisphere.*

such as San Diego, California, and Santa Cruz, Argentina, are cool and dry climates at nontropical latitudes (Figure 14.6). **BWh** and **BSh** climates have a mean annual temperature of 18° C (64° F) or higher while **BWk** and **BSk** have a mean annual temperature below 18° C. **BSk** and **BWk** climates also typically have more precipitation and less evaporation than their warmer counterparts, **BSh** and **BWh.**

The difference between these two climates is in total annual precipitation. **BSk** climates typically have more precipitation than **BWk** climates. **BWk** climates are often located in the rain shadows of large mountain ranges or the interior of continents. **BSk** climates are midlatitude steppe regions and have annual temperatures similar to those in **BWk** regions.

Moist Subtropical and Midlatitude Climates

Moist subtropical and midlatitude (**C**) climates are characterized by humid and mild winters. At least 8 months of the year have temperatures above 10° C (50° F), with the coolest month below 18.3° C (65° F) and above –3° C (27° F). Geographically, the subtropics lie between the tropics and the middle latitudes; however, subtropical *climates* also often lie in the midlatitude regions. This is where the largest annual temperature ranges are observed, because tropical and polar air masses govern the weather at different times of the year.

In the tropics, seasons are distinguished by wet and dry cycles; in the middle latitudes seasons are distinguished by annual variations in temperature. In the tropical regions plants become dormant with a lack of precipitation. In subtropical climates, plants go dormant because of low temperatures.

There are three major subgroups of subtropical climates, the marine west coast (**Cfb** or **Cfc**), the humid subtropical (**Cfa** or **Cwa**), and the Mediterranean (**Csa** or **Csb**). We examine these next.

Marine West Coast (Cfb, Cfc)

Summers and winters of marine West Coast climates are typically mild with no dry season, though precipitation can vary throughout the year. The **Cfb** regime has a warm summer, and the **Cfc** has a cool summer. **Cfb** and **Cfc** climates are usually near the coast. Examples of these climates are those of Bergen, Norway (**Cfb**), and Reykjavik, Iceland (**Cfc**) (Figure 14.7).

The characteristic temperature and precipitation of marine west coast climates are determined by the advection of air over ocean currents to the land. This

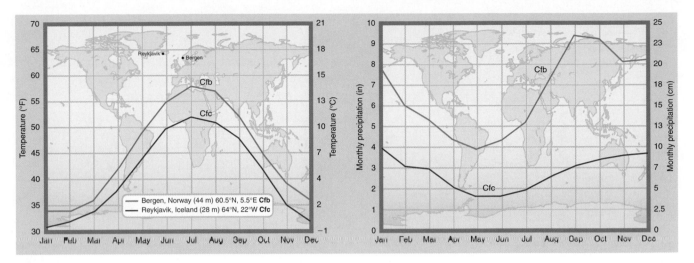

■ Figure 14.7

*Marine west coast climates such as those at Bergen, Norway (**Cfb**), and Reykjavik, Iceland (**Cfc**), have mild temperatures throughout the year with no dry season. **Cfb** have warm summers and **Cfc** have cool summers. This climate regime does not always have to lie on the west coast of a continent, as the name would imply.*

moderates the annual range in temperature. When cool water is upwind, the summer high temperatures are held down. Warmer ocean currents upwind lead to milder winter temperatures. The coldest month of the year has an average temperature above freezing, making snowfall rare.

The name of this climate type suggests that these climates lie along the west coasts of continents. This is because cool ocean currents predominate along west coasts in the subtropics (see Chapter 8). However, these climates are also found along southeastern Australia and southeastern Africa.

Humid subtropical (Cfa, Cwa)

Humid subtropical climate locations, such as New Orleans or Hong Kong, have hot summers (Figure 14.8). Summer daytime high temperatures typical of this regime are in the 27° to 32° C (80° to 90° F) range. The humid conditions keep the low

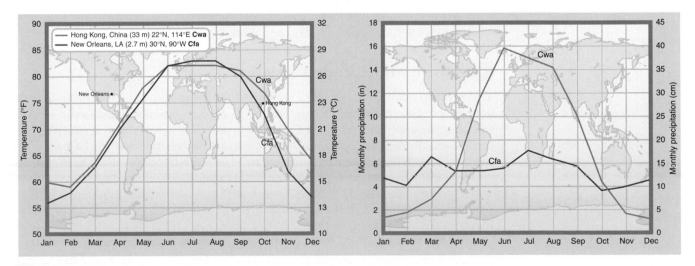

■ Figure 14.8

*Humid subtropical climates such as those at New Orleans, Louisiana (**Cfa**), and Hong Kong, China (**Cwa**), are hot and wet. The **Cwa** regime has a drier winter than summer while the **Cfa** is wet all year.*

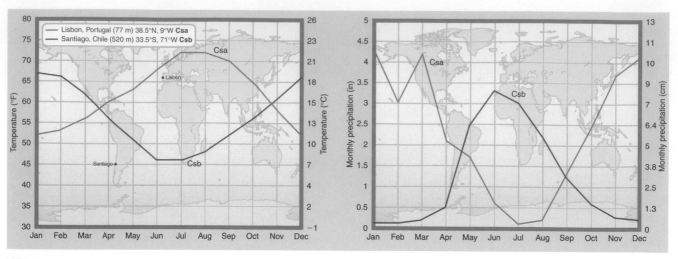

Figure 14.9

*Mediterranean climates at Lisbon, Portugal (**Csa**), and Santiago, Chile (**Csb**) have a distinct dry summer with maximum precipitation in winter. June, July, and August are winter months in Chile.*

temperatures in the evening from getting very cool. Winter temperatures are mild. While mean temperatures may be above freezing in winter, it is not uncommon for daily minimum temperatures to drop below 0° C (32° F). Precipitation in humid subtropical climates is plentiful, 75 to 250 centimeters (30 to 100 inches) per year. Summer precipitation is usually associated with convection; extratropical cyclones (see Chapter 10) provide the winter precipitation. **Cfa** climates are wet year-round while **Cwa** regions have a definite seasonal precipitation cycle, with a brief dry period in the winter.

Mediterranean (Csa, Csb)

A dry summer and a wet winter characterize the "Mediterranean" climate of Lisbon, Portugal, and Santiago, Chile (Figure 14.9). The lack of precipitation in summer is associated with the presence of a semipermanent high-pressure system. Summer temperatures range from hot to mild, and winter temperatures are mild. When Mediterranean climates are located along a coast, winter temperatures are also very mild. Winter temperatures can drop below freezing if the region is far from the moderating influence of a large body of water.

Severe Midlatitude Climates

The severe midlatitude climates (**D**) tend to be located in the eastern regions of continents. The temperature range of the **D** climate regimes is generally greater than that seen in the west coast **C** climate types. The average temperature of the coldest month of a **D**-type climate regime must be less than –3° C (27° F). The warmest month has an average temperature exceeding 10° C (50° F). These climate types typically have snow on the ground for extended periods of time.

There are two basic **D** climate types: humid continental and subarctic. The second letter **f** indicates that the climate has no dry season, while a second letter of **w** indicates a dry season in winter.

For **D** climates, a third letter of **a, b,** or **c** indicates a hot summer, a warm summer, or a cool summer, respectively. A hot summer climate (**Dfa** or **Dwa**) has a warmest month of about 22° C (72° F) with at least 4 months above 10° C (50° F). Finally, a **D** type climate with **d** as a third letter indicates an extremely severe winter with a cool summer. Next we examine some types of **D** climates.

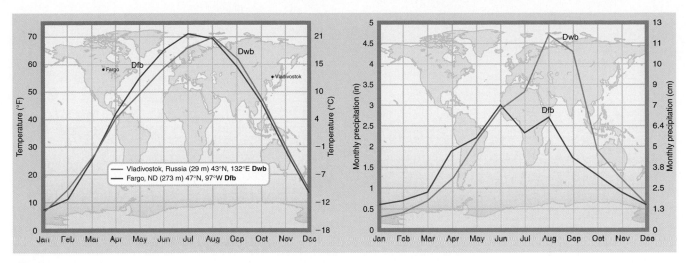

Figure 14.10

*Climographs for Vladivostok, Russia (**Dwb**), and Fargo, North Dakota (**Dfb**). The existence of a winter dry season at Vladivostok separates these two climate types.*

Humid Continental (Dfa, Dfb, Dwa, Dwb)

Humid continental climates have a large range in temperature; each subgroup has severe winters and cool-to-warm summers. The climates in the subgroup denoted by an **f** (e.g. **Dfa, Dfb**) do not have a dry season. Examples of the **Dfa** and **Dwb** climates are Fargo, North Dakota (**Dfb**), and Vladivostok, Russia (**Dwb**) (Figure 14.10). Both cities have a large annual temperature range. Vladivostok has a strong summertime maximum in precipitation, while Fargo's range in monthly average precipitation is smaller.

Subarctic (Dfc, Dfd, Dwc, Dwd)

Subarctic climates have a very large range in annual temperature. Winters are very long and cold, and summers are brief and cool. Fairbanks, Alaska (**Dfc**), and Verkhoyansk, Siberia (**Dfd**), are examples of subarctic climate regimes (Figure 14.11). Both have very cold winters with monthly average temperatures below freezing for

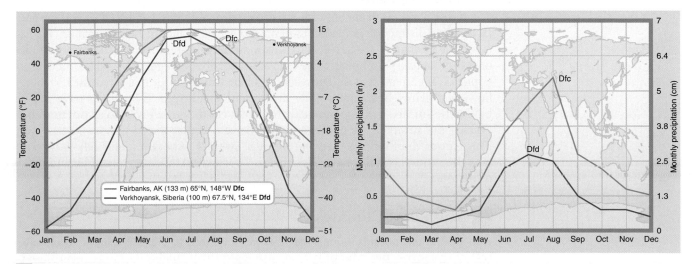

Figure 14.11

*Fairbanks, Alaska (**Dfc**), and Verkhoyansk, Siberia (**Dfd**), are examples of subarctic climates. Both cities have a very large annual range in temperature and small amounts of precipitation.*

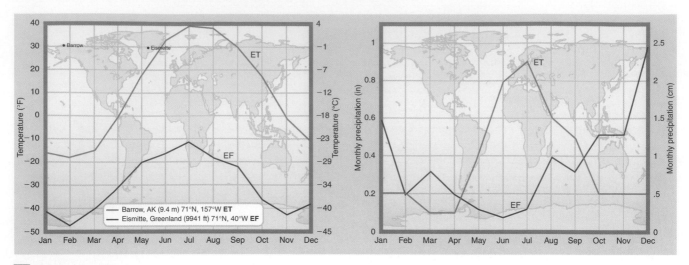

■ **Figure 14.12**

*Polar climates such as those at Barrow, Alaska (**ET**), and Eismitte, Greenland (**EF**), have long, cold winters and are typically poleward of 70°.*

five months. Precipitation is greater in summer than winter for both cities, but is never more than a few inches per month. The poleward displacement of the mid-latitude cyclones leads to this maximum precipitation in summer. These dry, severe conditions are best suited to coniferous forests, known as **boreal forests** in North America and **taiga** in Asia.

Polar Climates

Polar climates (**E**) occur poleward of the Arctic and Antarctic circles. Polar climates are extremely cold and have little precipitation (Figure 14.12). The mean temperatures of polar climates are less than 10° C (50° F) for all months. This cutoff temperature is the minimum temperature for tree growth. Annual precipitation, which is mostly frozen, is less than 25 centimeters (10 inches) of melted water. However, these regions are not considered deserts because precipitation still exceeds evaporation. The climates have a marked seasonal temperature cycle that corresponds directly to the solar input.

A distinction is made between two polar climate types: tundra (**ET**) and ice caps (**EF**). This distinction is made based on the warmest month being warmer (**ET**) or colder (**EF**) than 0° C (32° F). Greenland and the Antarctica Plateau are examples of **EF** climates. **EF** climate zones have essentially no vegetation, while tundra occupies **ET** climate zones. The vegetation of tundra is primarily mosses, lichens, flowering plants, and some woody shrubs and small trees. **ET** regions have a layer below the surface that is perennially frozen, a condition referred to as **permafrost.** During the summer, enough energy is received so that the top layer of soil thaws. This causes the tundra to become wet and swampy. About a meter below the surface the ground is still frozen. This frozen layer may extend downward for hundreds of meters.

Highland Climates

A location's elevation is an important variable related to climate. Highland climates (**H**) characterize the type of regions associated with high elevations regardless of other variables such as latitude. These climate zones are complex and affected by differences in latitude, altitude, and exposure to solar energy. For this reason, a wide variety of climates are exhibited in **H** climate zones.

As noted in Chapter 3, temperature usually decreases with height in the troposphere. For this reason, highland climates exhibit a wide range of temperatures, but are usually cooler the higher up they are. The temperature of a mountain location will also depend heavily on the slope angle and aspect, which influence the amount of solar energy it receives. A common feature of a highland climate is the large diurnal temperature variation. Rapid daytime heating and nighttime cooling occurs because of the thin, dry air.

Mountainous regions have large variations in precipitation. As we have seen throughout this text, mountains cause lifting and abundant precipitation on their windward sides, but lead to sinking and dry rain-shadow conditions on their leeward sides. Therefore, the amount of precipitation in **H** climate zones depends critically on the orientation of the highland, the amount of atmospheric moisture present, and the prevailing wind directions.

Past Climates: The Clues

Now that we have classified today's climates, it is natural to ask: Have these climates always been the same? To answer this question, climatologists study a fascinating variety of clues from the Earth's past. However, the study of past climate does not simply yield knowledge for its own sake. A fuller understanding of past climates enables scientists to better predict future climates and assess the impact of humans on climate, as we will see in Chapter 15.

Past climates can be divided into two main, unequal categories. **Historical climate** refers to the climate of the past several thousand years, during which humans have kept some sort of record of climate conditions. The *instrumental record* covers the period during which scientific instruments have been used to quantify climate, an era that began around 1600. *Historical data* consists of nonquantitative records of weather, such as diaries. **Paleoclimate** ("paleo-" is a Greek root that means "ancient") refers to climate conditions that existed in the billions of years before the dawn of human civilization. *Paleoclimatology* is the study of climates of the distant past.

Describing today's climates is rather straightforward because of the large number of observations we have available for analysis. Determining past climates is more challenging. Paleoclimatologists use environmental records to infer past climate conditions. To determine climates of the distant past, paleoclimatologists seek remains from the period that reflect the climate at the time of their creation. The goal is to discover any past changes of climate and the causes of these changes.

Determining the global and regional climates of the distant past is like solving a mystery. Scientists have to think like detectives, gathering evidence wherever and however it can be found. They look for clues hidden in libraries, trees, ice, land, and oceans that help to show what past climates were like and what caused them. Next they examine some sources of information that paleoclimatologists use to study past climates. We start with the climate clues that are closest to our own era and work backwards in time; Figure 14.13 summarizes these clues, the eras that they tell us about, and how detailed in time their information is.

Historical Data

Historical documents can provide valuable clues about past climates. Farmers' logs, travelers' diaries, newspaper articles, and other written records often include descriptions of weather conditions. Holy men of the Shinto faith have recorded observations of when Lake Suwa in Japan freezes over in winter and when the ice starts breaking apart. These observations provide proxy records of seasonal temperatures for this region for more than 500 years. Similar records have been kept at other lakes across the continents, and these data indicate that lakes and rivers in the

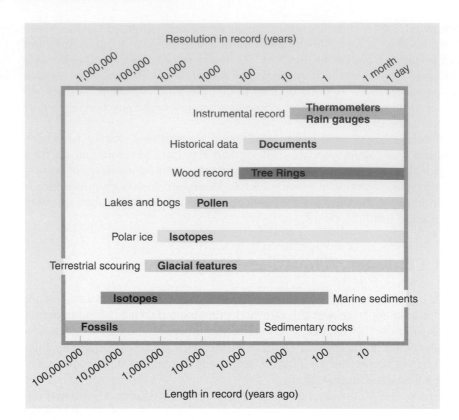

Figure 14.13

Climate clues and the lengths of time for which they provide evidence for past climate conditions. (Source: Adapted from Imbrie and Imbrie, Ice Ages: Solving the Mystery, *MacMillian, 1979.) Adapted from Imbrie and Imbrie,* Ice Ages: Solving the Mystery, *MacMillan, 1979.*

Northern Hemisphere are freezing later and thawing sooner than they did 150 years ago. On average, the freezing now occurs 8.7 days later and the thaw about 9.8 days earlier than they did 150 years ago.

Paintings, drawings, and other works of art can also provide clues to past climates. Cave paintings in the Sahara Desert (Figure 14.14) dated between 5000 and 6000 BC depict hippopotami and crocodiles, suggesting a much different climate than today.

Figure 14.14

Cave drawings of various hoofed animals in the Sahara Desert and Saudi Arabia provide evidence that these regions were at one time very wet.

Tree Rings

Climatic conditions and weather variations influence tree growth. The diameter of a tree trunk increases as the tree grows. In regions with distinct growing seasons, this growth appears as concentric rings (Figure 14.15). Trees generally produce one ring a year, and indirectly record environmental conditions each year.

The width of tree rings can be used to gather information about climates from several thousand years ago up to recent times. The width of each ring indicates how fast the tree grew during a particular time period. It is a function of available water, temperature, and solar radiation. Thick rings indicate favorable growing conditions while thin rings suggest poor growing seasons. Some tree species are more susceptible to temperature variations, and others are more sensitive to variations in water availability. The tree rings of different species are helpful in determining what caused the growth spurt or suppressed it.

The science of studying tree rings to ascertain a climatic condition is known as **dendrochronology.** This method of research relates tree ring width to modern-day precipitation, which turns the width of tree rings into a yardstick for estimating precipitation for periods when no human observations of precipitation are available.

For example, Figure 14.16 shows the amount of precipitation for Iowa as determined indirectly using dendrochronology. This analysis indicates that the 1930s, a period known as the "Dust Bowl," were very dry (see Chapter 12). For these years, dendrochronologists can compare the tree ring widths with historical measurements of precipitation made by meteorologists. This gives them confidence in interpreting the widths of the tree rings in terms of precipitation amounts. According to the tree ring data from Iowa, very dry times also occurred in 1700 and 1820—periods for which we have little in the way of historical precipitation data. But by using the relationships between tree ring width and observed precipitation in modern times, dendrochronologists can estimate with confidence the precipitation for these periods as well. In this way, dendrochronology allows climatologists to infer climate conditions long before the dawn of modern scientific instruments.

Pollen Records

Trees leave climate clues in addition to the width of their growth rings. Trees also produce pollen that can accumulate in a given location, such as a lake. While this might not be good if you have certain allergies, pollen is useful to paleoclimatologists.

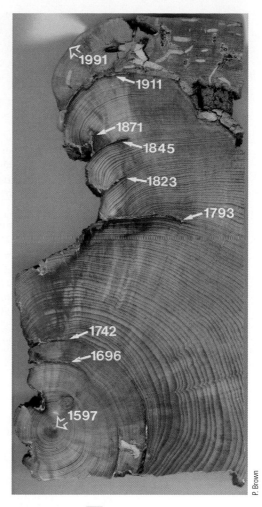

P. Brown

Figure 14.15

The growth of a tree trunk appears as rings in regions with distinct growing seasons. The width of each ring indicates how fast the tree grew during a particular year.

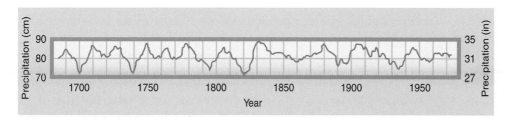

Figure 14.16

Measuring and counting tree rings can provide clues about past climates. The graph shows a plot of annual precipitation in Iowa derived from analysis of tree-rings. Notice the dry periods in the decades around 1700, 1740, 1820, 1890, and 1930. (From Duvick, N.D. and T.J. Blasing, 1981. "A Dendroclimatic Reconstruction of Annual Precipitation Amounts in Iowa Since 1680." Water Resources Research, 17:1183–1189. Reprinted with Permission.)

BOX 14.1

Dating Ancient Climates

How do we know how old rocks and fossils are? Radioactivity is used to date ancient climate. Some elements undergo radioactive decay in predictable ways. We can get an idea of the age of a sample of soil or rock or pollen grains by measuring how much of these elements is present in them.

A few heavy elements will spontaneously disintegrate into simpler elements, losing one or more neutrons in the process. The original element is referred to as the parent isotope, and the simpler element is called the daughter isotope. Atoms of the same element that have different numbers of neutrons in their nuclei are called *isotopes*. For example, carbon (C) has three isotopes: C^{12}, C^{13}, and C^{14}. For a particular element, the rate of decay is constant and is expressed in terms of the *half-life*—the time it takes for a given amount of the parent isotope to decay to half that amount. The age of rocks and minerals can be determined by measuring the amount of daughter isotope relative to the amount of the parent isotope. The ratio of the amount of the daughter isotope

produced to that of the parent isotope remaining provides an accurate dating method.

All living things contain carbon. There is about one C^{14} atom for every 100 billion C^{12} atoms in a living creature. Once a plant or animal dies, the C^{14} begins to decay, with a half-life of 5,760 years. The percentage of C^{14} that remains is determined by how long ago the plant or animal died. Radiocarbon dating, as this method is referred to, is good for samples up to 50,000 years old and has an uncertainty of about 15%.

Uranium is used in dating inanimate objects, such as rocks. Uranium-238 decays into Lead-206 with a half-life of 4.5 billion years. Lead-206 is created only from the decay of uranium. So, the ratio of Uranium-238 to Lead-206 tells us how old an object is (in the absence of any human-produced radioactivity). If there are equal amounts of Uranium-238 and Lead-206, then the object is 4.5 billion years old. This is true for the oldest Earth and Moon rocks, and this is how we know the age of the Earth and the Moon.

This is because pollen degrades slowly and each species of pollen is identifiable under a microscope by its distinctive shape. When buried in lake sediments, pollen can provide a regional climate record similar to that provided by tree rings. While pollen grains do not give the same detailed year-to-year information that tree rings do, they are valuable in extending our understanding of climate backward for tens of thousands of years by using **radiocarbon dating** of the pollen grains (Box 14.1).

A pollen record extracted from a northern Minnesota bog provides evidence of a changing climate over the last 11,000 years (Figure 14.17). The oldest (deepest) layers with pollen indicate that spruce trees, which require cool climates, were the dominant species. The pollen record indicates that the spruce trees were replaced by pine species, which require a much warmer climate. About 8600 years ago the dominant tree changed to oak, which require drier conditions than pine. Oak and birch were the dominant trees between about 3900 and 2700 years ago. As the climate again became warmer and moister, the oak declined and the pine became the dominant species. Therefore, pollen records provide evidence that climate has changed repeatedly in parts of the United States during the past several thousand years.

■ Air Bubbles and Dust in Ice Sheets

Clues to climate can be buried in ice just as in lake sediments. Bubbles of air trapped in ice provide windows to the past for atmospheric chemists and climatologists. Air bubbles get trapped in **glaciers** and ice sheets (Box 14.2) as snow gets compressed. Glaciers that exist today can hold gas bubbles tens or hundreds of thousands of years old. These trapped bubbles provide a record of the concentration of atmospheric gases such as carbon dioxide (CO_2) and methane (CH_4) over the past 200,000 years.

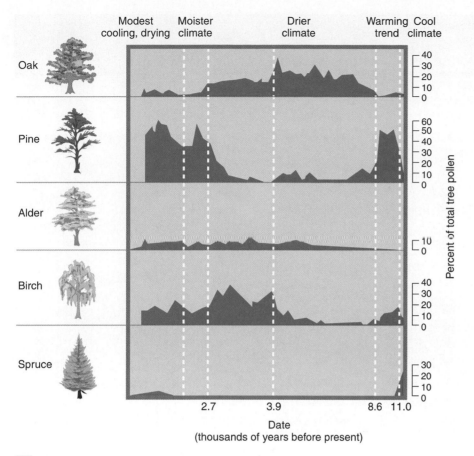

Figure 14.17

Tree pollen layered in lakes and bogs provides evidence of changing climate conditions. Pines prefer warm, moist climates while oak trees do well in drier climates. The percentage of pollen from a given species provides evidence for paleoclimatologists to determine previous climate conditions. The pollen samples used to construct this diagram come from a bog in northwestern Minnesota and provide climate clues for the past 11,000 years. (Source: Adapted from Graedel, T., and Crutzen, P., Atmosphere, Climate and Change, Scientific American Library, 1997, p. 82.)

Figure 14.18 shows the concentration of atmospheric carbon dioxide and methane obtained for a 2083-meter (1.3-mile) long ice core cut from the ice sheet at Vostok, Antarctica. The core is dated by counting the number of layers of ice, much like counting tree rings. A core this long has bubbles that date back to over 150,000 years ago. Also shown on this figure are estimates of temperature changes during this period. The warmer temperatures are apparently related to higher concentrations of carbon dioxide and methane, but cause and effect are very difficult to separate using only paleoclimate data. We explore this relationship between temperature and "greenhouse gases" further in Chapter 15.

Dust in ice sheets can be caused by climate-changing volcanoes (see Chapter 3) or by dry, windy conditions that lead to soil erosion. The soil can be transported by small-scale winds in the form of dust storms (see Chapter 12). For example, dust storms from the Sahara Desert (Figure 14.19, *A*) inject dust into the global circulation. Global-scale winds can carry this dust as far away as the poles, where the dust can then be detected in ice cores. Comparing the ice core data in Figure 14.19, *B*, to Figure 14.18, we find that the colder periods of Earth history (20,000, 60,000, and 160,000 years ago) are usually much dustier. But did the dust block out the Sun and cause the colder temperatures, or did the colder temperatures create drier and dustier conditions? Based on this evidence, we can't say for sure.

BOX 14.2

Glaciers and Icebergs

Glaciers form on land when the accumulation of ice and snow in winter exceeds summertime melting. As the snow accumulates, ice crystals are crushed under the pressure. Trapped air is expelled, forming bubbles. Eventually, this forms larger ice crystals and the glacial ice compacts and has a blue appearance. This blue color arises from the fact that ice weakly absorbs red light, but scatters blue light back to the eye.

When the ice gets to be about 30 meters thick, its weight causes it to flow downhill, even though it is a solid. How fast it flows is a function of how steep the land is and the size of the glacier. The speed can range from a few centimeters to ten meters per day. On land, the front edge of a glacier melts when it reaches a region with above-freezing temperatures. If the glacier ends in the ocean, great blocks of ice can break away before they have a chance to melt and become icebergs. This process is called *calving*.

The accompanying figure is a satellite image of a calving process that occurred in Antarctica during the spring of 2000. This iceberg is approximately 298 kilometers (185 miles) long and 37 kilometers (23 miles) wide—a surface area about twice that of the state of Delaware! It is about 400 meters (a

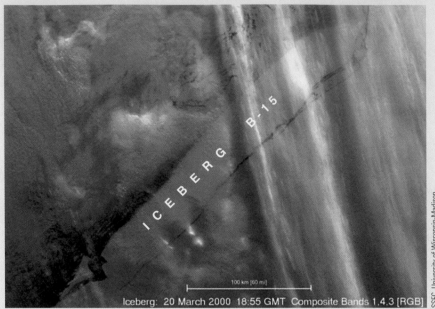

quarter-mile) thick, with over 90% lying below the water. The iceberg, called B-15, contains approximately 12.9 trillion liters (3.4 trillion gallons) of water. The glacier that gave birth to B-15 moves toward the ocean at a rate of about 0.8 kilometer (0.5 mile) per year, and calves an iceberg this size every 50 to 100 years. In fact, B-15 is not even its biggest offspring. In 1956 an iceberg broke off that was about 335 kilometers (208 miles) long by 97 kilometers (60 miles) wide!

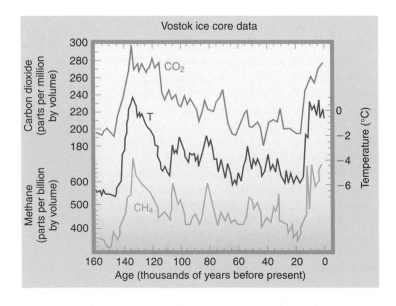

Figure 14.18

Concentration of atmospheric carbon dioxide and methane determined from analysis of the chemistry of air bubbles trapped in an ice core cut from Vostok, Antarctica. Variations in these gases are correlated with changes in temperature.

A

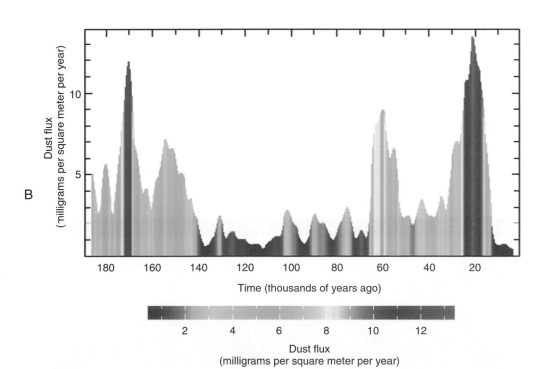

B

Figure 14.19

A, *A dust storm blows westward from the Sahara Desert, as seen from the NASA SeaWiFS satellite. These dust storms can carry Saharan soil across the Atlantic, eventually spreading the dust worldwide.* **B,** *The dust concentration in the Vostok ice core during the past 185,000 years. Compare this figure to Figure 14.18 to see how temperature and dustiness may be related. (Source: http:// ingrid.ldeo.columbia.edu/ SOURCES/.ICE/.CORE/. VOSTOK.)*

In addition, the acid content of the dust in ice sheets can be used to distinguish dust storms from volcanic explosions. Highly acidic dust containing sulfuric acid is a clear indication of volcanic activity, as we will see later in this chapter.

Marine Sediments

Materials have been deposited in layers on the ocean floor for millions of years. The deeper the layer, the older the material. These deposits can include soil from wind erosion or floods, ash from volcanic eruptions, and shells of animals. In ocean sediments, the shells of animals are primarily calcium carbonate ($CaCO_3$), a compound that makes up limestone. The calcium carbonate is very useful for tracking past climates by the relative amounts of different versions of oxygen atoms, the 'O' in $CaCO_3$. Different versions of the same element, such as oxygen, are called *isotopes*, and they have different atomic weights.

Most oxygen atoms have an atomic weight of 16. This isotope oxygen is denoted as ^{16}O. Other isotope oxygen atoms can have an atomic weight of 18 (^{18}O). ^{16}O is much more common than ^{18}O. These two isotopes of oxygen are found in calcium carbonate in ocean sediments. The ratio of ^{18}O to ^{16}O ($^{18}O/^{16}O$) provides a clue to past climates.

Foraminifera are microorganisms that live in the oceans and have hard shells made of calcium-containing compounds, including calcium carbonate. The relative amount of isotopes ^{18}O to ^{16}O in the shells of these marine protozoans is related to the amount of continental ice that was present when they were alive. The proportion of ^{18}O to ^{16}O in ocean water is partly controlled by the volume of water in continental ice sheets. Since ^{16}O is lighter than ^{18}O, it can evaporate from water more easily. So, the precipitation that forms from ocean water and falls on glaciers has relatively more of the lighter isotope in it. The lighter water molecules tend to accumulate in snow and ice that form the glaciers. If more glacial ice accumulates, more of the ^{16}O isotope is retained in the ice sheets.

As a result, during colder climatic periods there is a higher concentration of ^{18}O in ocean water than during warmer periods. During colder climatic periods, as foraminifera construct their shells, they incorporate relatively more ^{18}O than ^{16}O. As the foraminifera die, their shells settle on the ocean floor and provide a record of the isotope ratio. When we pull sediment cores from the ocean floor, we can obtain an indirect record of climate during the past 2 to 3 million years!

Figure 14.20 shows the departures from the average $^{18}O/^{16}O$ ratio over the past 300,000 years. Note that warm periods alternate approximately every 100,000 years. This periodic oscillation in Earth's climate is a major clue as to what mechanisms alter climate, as we will see shortly.

Fossil Records

Fossils also provide useful insight into the distant past. They provide a means to track life through the ages because they are an integral part of the rocks in which

■ Figure 14.20

Variations in average temperature as determined from the ratio of ^{18}O to ^{16}O measured from fossil shells over the last 800,000 years. Periods as warm as today occur infrequently. Notice that warm and cold periods recur approximately every 100,000 years. (Source: Adapted from Imbrie, J. and Imbrie, J.Z. "Modeling the climatic response to orbital variations," Science, vol. 202, pp. 943–953.)

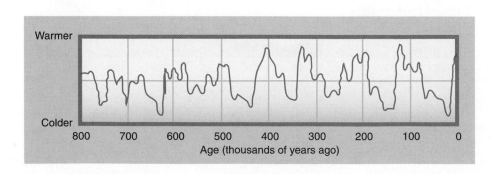

they are found. The age of the rocks can be dated (refer to Box 14.1), providing evidence for climate extending back hundreds of millions of years. Fossil ages can also be deduced from the layer of ground in which they lie, if that layer has conspicuous characteristics that can be definitively dated.

Fossils reveal ancient animal and plant life that can be used to infer climate characteristics of the past. For example, tropical plants often have pointed tips so that the moisture can drip off the leaf. Plant fossils that have pointed leaves indicate a warm and moist climate. Large numbers of a given fossil also indicate favorable climate conditions for these organisms. In this way, climatologists can infer the climate of Earth long before the first humans walked on the planet—a remarkable feat of science!

Past Climates: The Change Mechanisms

As detectives, climatologists are not content to know the "what" of past climate. Paleoclimate research is a "who-dun-it." In other words: What mechanisms have forced climate to change as it has over the entire span of Earth's history (Table 14.1)? How do these mechanisms cause these changes, and over what periods of time do they operate? Below we examine several of the most prominent *naturally occurring* climate change mechanisms, in approximate order from those that act quickest to those that act on timescales unimaginably long to humans. (In Chapter 15 we will take up the subject of climate change caused by humans themselves.)

Volcanic Eruptions

Previously in this text we have discussed how volcanic eruptions have affected local, regional, and even global climate. Locally, the ash may immediately reduce sunlight reaching the surface and cause cooling. Globally, sulfuric acid droplets launched into the stratosphere by the eruption can reflect sunlight and cool much of the globe for a period of several years (see Chapter 3). A series of eruptions therefore has the potential to affect climate on the time scale of a decade or more.

Because volcanoes emit sulfur compounds that turn into sulfuric acid in the atmosphere, sharp increases in surface acidity are an indication that fallout from a volcanic eruption was deposited onto the Earth's surface. This allows paleoclimatologists to detect volcanic eruptions in ice cores (Figure 14.21). By comparing the timeline of acidity with the timing of known volcanic eruptions, many of the "spikes" of high acidity in this figure can be definitively traced to specific eruptions. Some of the sharpest spikes in Figure 14.21 are for relatively minor eruptions that occurred near Greenland, but large eruptions such as Tambora in 1815 can spread significant volcanic debris worldwide.

Volcanic climate change is also relevant to prehistoric climates tens of millions of years ago, when immense eruptions covered swaths of continents with lava thousands of meters deep.

Asteroid Impacts

Once thought to be a figment of science fiction writers' imaginations, it is now an accepted scientific fact that objects from outer space can and do collide with planets. The collision of the Shoemaker-Levy comet with Jupiter in 1994 provided spectacular evidence that, to paraphrase the poet John Donne, no planet is an island and that extraterrestrial objects can affect a planet's atmosphere. The climate effects of an asteroid impact are probably similar, but much more pronounced and devastating, than those of a large volcanic eruption. Depending on its exact location, a major asteroid impact and the debris ejected by the impact can cause extended darkness, global fires, acid rain, ozone loss and even

http://info.brookscole.com/ackerman

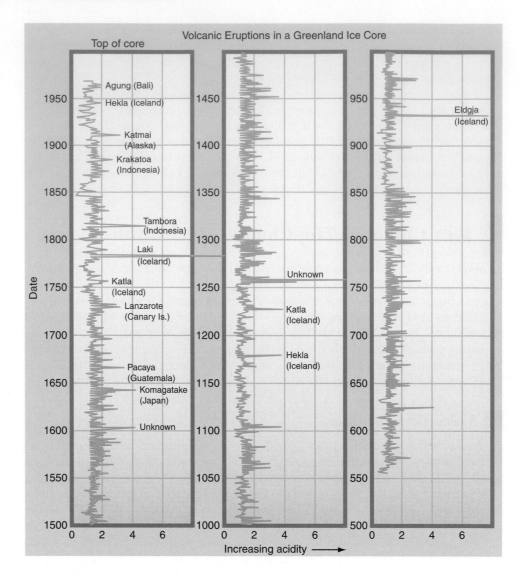

Figure 14.21

The annual acidity of layers of an ice core from central Greenland for the years 553 to 1972. The Icelandic fissure eruptions of Laki in 1783 and Eldgja in 934 stand out, largely because of their proximity to Greenland. In terms of global climate impact, the two most significant eruptions during this period were Krakatoa in 1883 and Tambora in 1815, which led to the "year without a summer" a hemisphere away in North America and Europe. (Source: From Graedel, T., and Crutzen, P., Atmosphere, Climate and Change, Scientific American Library, *1997, p. 94.)*

mile-high tsunamis. The result of these extraterrestrial catastrophes can be a cooling of global climate and widespread extinctions, which in turn can cause long-term changes in climate.

Sixty-five million years ago, the age of the dinosaurs ended abruptly, coinciding with the extinction of approximately 75% of the total number of living species. The father-and-son science team of Luis and Walter Alvarez proposed in the 1980s that this extinction was caused by the indirect effects of an asteroid impact on Earth. In 1990, evidence of just such an impact precisely 65 million years ago was found near the Yucatán Peninsula (Figure 14.22). Named for a local village, the Chicxulub crater is a 189-kilometer (112-mile)–wide impact crater visible in gravity and magnetic field data. The crater size is consistent with a 10- to 20-kilometer (6- to 12-mile) wide asteroid. Remnants of the asteroid have been found in sediments worldwide, confirming its global influence.

The Chicxulub impact was probably not a once-in-history event. There are indications in the fossil record of other extinction events that occur every 26 million years or so (Figure 14.23). Scientists are coming to the realization that on very long time scales the history of Earth, including its climate, may be periodically upset by asteroid and comet impacts.

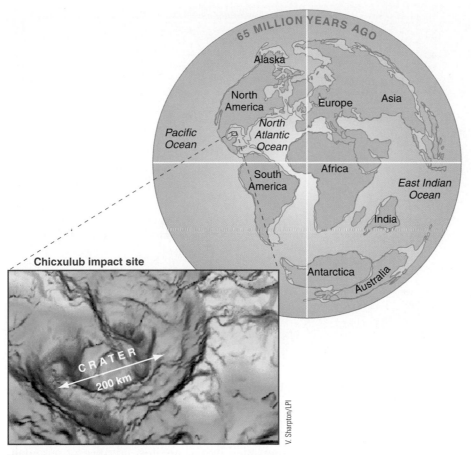

Figure 14.22

A collision with a large meteor is believed to have caused a climate change that resulted in the extinction of 70% to 90% of all species at the end of the Mesozoic Era. The top map indicates the landmasses and the point of contact at the time of the collision. The outlines of the Chicxulub crater are seen today in gravity and magnetic field data of the Yucatán Peninsula region (inset). The blue regions in the inset are low-density rock that was pulverized during the impact. The crater is more than 200 kilometers in diameter.

Solar Variability

Less dramatic than asteroid impacts, but probably just as influential to climate, are variations in the amount of energy the Earth receives from the Sun. Since the invention of the telescope in the 1600s, observers have recorded variations in the numbers of dark spots—"sunspots"—on the Sun's surface. These variations normally follow a regular cycle with peaks 11 years apart (Figure 14.24). This cycle is coincident with an oscillation in solar energy output of a few watts per square meter. The Sun's output is slightly higher during periods with large numbers of sunspots.

It is difficult to prove a direct relationship between the sunspot cycle and climate. However, some atmospheric scientists theorize that the stratosphere acts as an amplifier of the small variations in solar output and causes a discernible 11-year cycle in climate.

As shown in Figure 14.24, between the years 1645 to 1715 the number of sunspots was dramatically lower than observed before or since. This period is known

Figure 14.23

Extinctions in the fossil record of the past 250 million years. The solid line represents the rate of disappearance of species families in the fossil record per million years. The arrows are spaced 26 million years apart. Notice how well the arrows correspond to the major extinction events, such as the demise of the dinosaurs. This may be an indication that asteroid impacts occur cyclically every 26 million years, caused by as-yet-unknown astronomical patterns. (Source: Adapted from Graedel, T., and Crutzen, P., Atmosphere, Climate and Change, Scientific American Library, 1997, p. 75.)

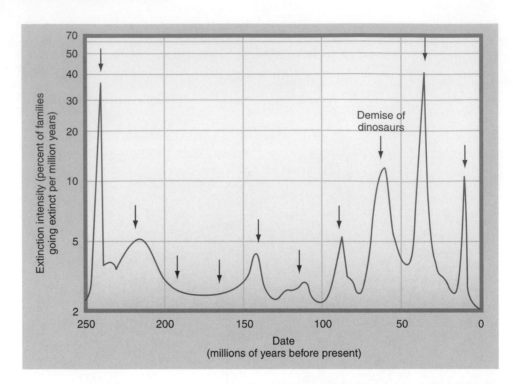

as the **Maunder Minimum.** It is hypothesized that the reduction in solar energy output during this period could have cooled the Earth.

The historical record supplies some evidence supporting this hypothesis. The period between about 1400 and 1850 is called the **Little Ice Age** in Europe. The coldest portion of this period, accompanied by the greatest advance of mountain glaciers, occurred around 1750. In geological terms, the Little Ice Age started and ended very quickly. Around 1570 Europe was 1° to 2° C cooler than it is today. The Thames River in London froze over eleven times in the 17th century, but it has not frozen over in the last 100 years.

Variations in solar output may affect climate on the time scales of decades to centuries. In addition, there is the **weak Sun paradox** that concerns the earliest billion years of the Earth's atmosphere. At the very beginning of Earth's history, the Sun was young and its output should have been only 70% to 80% of its current output, according to the best theories of astrophysicists. If the Earth's climate is sensitive today to the small changes in solar output during the sunspot cycles, then surely the climate of early Earth should have been profoundly cooler with such a faint Sun. However,

Figure 14.24

The yearly average number of sunspots observed since Galileo discovered them in 1610. Notice the regular interval of 11 years between peaks in the numbers of sunspots. Between 1645 and 1715, a period known as the "Maunder Minimum," almost no sunspots were observed.

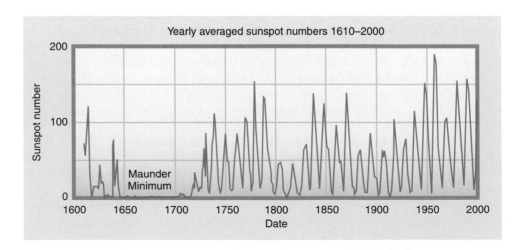

the evidence from paleoclimatology is that the Earth's temperature billions of years ago was almost exactly the same as now. There is currently no satisfactory explanation for this puzzling contradiction, although "greenhouse gases" may have played a role.

Variations of the Earth's Orbit: Milankovitch Cycles

On time scales of sufficient length, the shape of the Earth's orbit around the Sun and the tilt of its axis are not constant. Instead, they oscillate with periods that are tens of thousands of years in length. In the 1860s and 1870s a Scottish janitor named James Croll devised a theory explaining how changes in the tilt and orbit of the Earth could cause climate change. Throughout the early 1900s Serbian engineer Milutin Milankovitch refined and expanded Croll's theory. In the theory's current form, the **Milankovitch cycles** (Figure 14.25) that describe the variation of the Earth's orbit are:

1. **Precession:** The Earth wobbles on its axis once every 27,000 years, similar to a spinning top. This alters the relationship between the solstices and the distance from the Earth to the Sun. For example, 11,000 years ago the Northern Hemisphere summer solstice occurred at perihelion, when the Earth is closest to the Sun. (This is almost the exact opposite is the case today; see Chapter 2.) This "synching up" of summertime with perihelion made the differences between winter and summer more pronounced 11,000 years ago than they are today. More generally, the solstices and equinoxes move slowly forward through the calendar with each passing year, a phenomenon known as *precession*.

2. **Obliquity:** The tilt of the Earth also changes slightly, with a dominant cycle every 41,000 years. The change in angle of inclination is only about one degree from the present tilt, from about 22° to 24.5°. However, as we learned in Chapter 2, the Earth's tilt is a critical factor in climate. For example, these small changes in tilt lead to solar radiation changes of up to 15% in the high latitudes. "Oblique" means neither parallel nor perpendicular, and so changes of the Earth's angle with respect to the Sun often go by the name "obliquity."

3. **Eccentricity:** The shape of the Earth's orbit becomes more or less elliptical on time scales of about 100,000 years. At present the maximum difference between Sun–Earth distances during the year is only 3%, but over the past several hundred thousand years this number has been as small as 1% (a nearly circular orbit) and as large as 11% (a more elliptical orbit). This orbital variation gets its name because the deviation of an ellipse from a perfect circular shape is known in geometry as "eccentricity."

Orbital Cycles and Climate

These three orbital variations take place simultaneously. Like overlapping musical tones, the cycles of orbital variations create overtones and resonances that are not quite the same as the original cycles. The result is that Earth's climate is affected by these Milankovitch cycles on four different periods: 19,000, 23,000, 41,000, and 100,000 years.

The cold periods experienced 20,000, 60,000, and 160,000 years ago as shown in Figure 14.18 probably resulted from the combined effect of Earth's orbital variations. Notice that the cold episodes are separated by lengths of time that closely match the expected periods of climate change due to Milankovitch cycles. However, the match is not quite perfect, and it is difficult to explain how variations of a few percent in solar energy on Earth can lead to the 10° C (18° F) variations in global temperature inferred from paleoclimate data. Research continues on this subject.

Plate Tectonics

On the time scale of millions of years, not even the location of the continents can be considered constant. Continental movements are extremely slow in everyday terms—1 to 10 centimeters (up to a few inches) per year. But over millions of

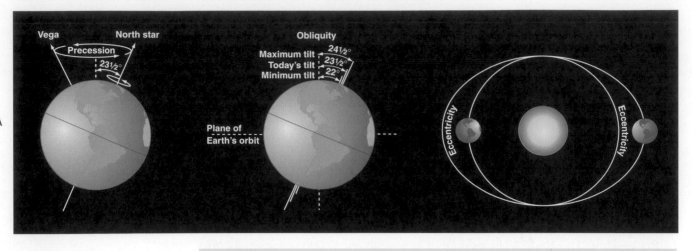

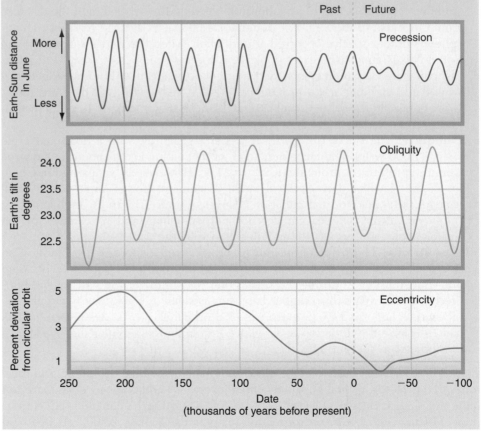

■ **Figure 14.25**

A, *The three orbital variations that lead to climate change: precession, obliquity, and eccentricity (exaggerated in the drawing). (Source: Adapted from Tarbuck and Lutgens,* The Atmosphere, *8th ed., Prentice-Hall, 2001, p. 377.)* **B,** *The cycles of these orbital variations during the past, present, and future. (Source: Adapted from Graedel, T., and Crutzen, P.,* Atmosphere, Climate and Change, *Scientific American Library, 1997, p. 77.)*

years, continents have migrated from the poles to the tropics and the tropics to the poles.

Because the location and latitude of land affects climate, the motions of the continents undoubtedly affect climate on very long time scales. If a continent moves poleward, its climate cools as its solar energy gains decrease. In addition, wherever continents collide mountain ranges are created, which further alters climate.

The theory of continental movement, now known as **plate tectonics,** was pioneered by meteorologist Alfred Wegener, the son-in-law of Vladimir Köppen. Wegener wrote to his future wife in December 1910, "Doesn't the east coast of South America fit exactly against the west coast of Africa, as if they had once been joined (Figure 14.26, A)? This is an idea I'll have to pursue." Wegener's idea was

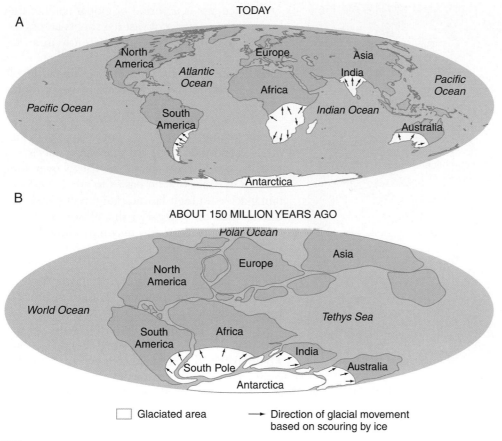

TODAY

A

North America
Europe
Asia
India
Atlantic Ocean
Africa
Pacific Ocean
Pacific Ocean
South America
Indian Ocean
Australia
Antarctica

B

ABOUT 150 MILLION YEARS AGO

Polar Ocean
North America
Europe
Asia
World Ocean
Tethys Sea
South America
Africa
India
South Pole
Australia
Antarctica

☐ Glaciated area → Direction of glacial movement based on scouring by ice

■ **Figure 14.26**

*The map at top (**A**) shows the continents as they appear today. In the distant past (**B**), the conti-nents fit together to form a larger land mass. The shapes of the continents fit together like a jigsaw puzzle. In addition to the jigsaw pieces locking together, the patterns on the surface must match. Similarly, patterns on the Earth's surface must match to support the theory that the continents were truly combined. The figure above illustrates how evidence found on several continents for ice sheets in the distant past tends to confirm the theory of a joined-together supercontinent. (Source: Adapted from Ahrens, C.,* Meteorology Today, *6th ed., Brooks/Cole, 2000, p. 512 and Tarbuck and Lutgens,* The Atmosphere, *8th ed., Prentice-Hall, 2001, p. 373.)*

ridiculed by the scientists of his day, but a half-century later the research of paleo-scientists confirmed its essential truth. Where the continents fit together in the dis-tant past, they also match in rock formation, glacier flow patterns, fossil record, and continuity in mountain ranges.

Approximately 300 million years ago the continents of today were joined in one supercontinent referred to as **Pangaea** (or Pangae) (Figure 14.26, *B*). Approxi-mately 160 to 230 million years ago Pangaea began to drift apart, eventually form-ing Laurasia and Gondwanaland. Laurasia consisted of what are today Asia, Europe and North America. Gondwanaland was comprised of South America, Africa, India, Australia and Antarctica. As the continents continued to drift apart, some land masses collided, forming today's mountain ranges, including the Himalayas and Rocky Mountains.

Notice in Figure 14.26 that Pangaea was essentially a tropical supercontinent, whereas today much of the world's land is located at high latitudes. The trend to-ward mountainous high-latitude land is a "recent" phenomenon, occurring during the past tens of millions of years. This may explain the most notable climate devel-opment of the Earth's recent past: the **ice ages.**

http://info.brookscole.com/ackerman

18,000 YEARS AGO

Greenland
ice sheet

Mountain glaciers

Laurentide
ice sheet

■ **Figure 14.27**

The extent of the ice sheets over North America during the Pleistocene. The arrows indicate the direction of advance of the sheets.

An ice age is a period of global cooling that leads to the creation of vast ice sheets across the Earth's land masses. We are currently in an ice age that began approximately 10 million years ago. Before that, the Earth was comparatively warm for hundreds of millions of years (Table 14.1).

Why are there ice ages? It is possible that the increasing concentration of land at high latitudes, combined with its increasing elevation, have progressively cooled the Earth over long periods of time. It is far easier to create and sustain glaciers in high-latitude mountains and continents than on water. For this reason, scientists think that plate tectonics may have directly contributed to the formation of continental ice sheets.

Just 20,000 years ago, North America was buried under 1.6 kilometers (1 mile) or more of ice, the edge of which extended as far south as what is now St. Louis, Missouri (Figure 14.27). Why has the ice retreated since then? One possibility is that the Milankovitch cycles that led to more pronounced seasonality 10,000 years ago also helped melt the ice sheets during the warm Northern Hemisphere summertime. In this way two different climate change mechanisms can combine to create a complex history of global climate.

■ Changes in Ocean Circulation Patterns

We close with a climate change mechanism that can be both rapid and long-lasting. Today, surface ocean water in certain regions of the North Atlantic and Antarctic oceans sinks because it is cold and salty enough to be denser than the water beneath it. Oceanographers have discovered that the location and strength of this sinking water drives ocean circulations across the globe. However, past changes in the composition of the ocean waters, as well as changes in the locations of the continents, have altered the location of sinking ocean waters. Oceanographers believe that, as a result, ocean circulations in the past have been dramatically different than today's. Because of the influence of ocean circulation on climate, past climates may also have been dramatically different.

One possible example of the effect of ocean circulation on climate occurred at the end of the last advance of ice across the Northern Hemisphere. Probably because of changes in the Earth's orbital patterns, global temperatures warmed about 13,000 years ago. This caused the glaciers covering North America to melt rapidly, raising sea levels at a rate of 1 centimeter (about 0.4 inch) per year. Then temperatures over North America and Europe suddenly cooled for almost 1,000 years. This relatively cold period existed between 11,000 and 10,000 years ago (Figure 14.28) and is known as the **Younger Dryas** (named for the reappearance of a polar wildflower, *Dryas octopetala,* in Europe).

A possible cause for the cooling during the Younger Dryas is the melting of the North American Laurentide ice sheet. The massive floods of fresh water from the melting glaciers may have rushed out of the St. Lawrence River Valley and into the North Atlantic. This could have caused North Atlantic surface ocean water to become much less salty and less dense. If the surface water in the North Atlantic became too light to sink, this could have shut down the "normal" circulation of the North Atlantic for hundreds of years. The change in vertical ocean motions could,

TABLE 14.1 Earth's Geological Periods (Source: Adapted from Bradley, R., *Quaternary Paleoclimatology,* Allen and Unwin, 1985, p. 2.)

Era	Period	Epoch	MYBP (approximate beginning of period)	Features	Climate Features	Mean global temperature Present Cold / Warm	Mean global precipitation Present Dry / Wet
Cenozoic	Quaternary	Holocene	0.01	Age of Mammals	Little Ice Age (1450–1850 AD) Medieval Climatic Optimum (900–1200 AD)		
		Pleistocene	1.6		The Ice Age		
	Tertiary	Pilocene	5.3				
		Miocene	23.8	Earliest hominids			
		Oligocene	36.7				
		Eocene	57.8				
		Paleocene	65		Beginning of a cooling period, leading to glaciers in Antarctica		
Mesozoic	Cretaceous		144	Mass extinction at the end of the Mesozoic era Age of Dinosaurs	Intraglacial climates: no glaciers during the Mesozoic Era		
	Jurassic		208	Break-up of Pangaea			
	Triassic		245	Increasing volcanic activity			
Paleozoic	Permian		286	Final assembly of Pangaea; mass extinction at end of period	Glaciers in southern Africa, South America, and Australia recede, desert regions expand		
	Carboniferous		360	Extensive coal formation	Hot and humid		
	Devonian		408	First amphibians	Warm and dry		
	Silurian		440				
	Ordovician		505	Beginning formation of Pangaea Primitive fish			
	Cambrian		570		Widespread ice age		
Pre-cambrian			4600		Evidence of ice sheets		

in turn, have changed circulation patterns throughout the oceans of the Northern Hemisphere and even the world. In particular, a weakened northward transport of heat by warm ocean currents could have caused the cooling associated with the Younger Dryas in Europe and North America.

http://info.brookscole.com/ackerman

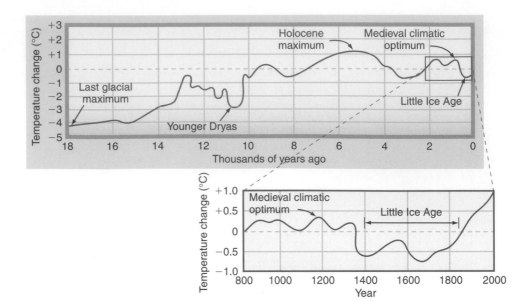

 Figure 14.28

The average global temperature changes over the last 18,000 years. A detailed look at the past 1200 years is shown. (Source: Adapted from Ahrens, C., Meteorology Today, 6th ed., Brooks/Cole, 2000, p. 509.)

On longer time scales, it is possible that changed ocean circulation patterns tens of millions of years ago could have kept Earth's atmosphere warmer and moister than it is today. The key, then as now, is where and how much cold and/or salty water sinks, which drives the rest of the world's ocean circulation.

This concludes our tour of natural processes that affect climate. However, we have not accounted for 100% of the observed changes in Earth's climate. Other factors, including *natural variability,* the ability of climate to oscillate without being forced to, may explain the changes in climate that are not driven by the mechanisms we have discussed here. This is particularly relevant to observations during the past thirty years indicating that Earth's climate may be warming. Separating natural variability from possible human-caused climate change is particularly difficult, as we will see in Chapter 15.

PUTTING IT ALL TOGETHER

SUMMARY

Climate varies from place to place and over time. The five basic climate controls are latitude, elevation, topography, proximity to large bodies of water, and prevailing atmospheric circulation.

Climate classifications are developed to organize the complex and varied climates of the world. The Köppen-based climate classification is the most widely used scheme for describing today's climates. Köppen's scheme has six main groups, each designated with a letter: Tropical Moist (**A**), Dry (**B**), Moist with Mild Winters (**C**), Moist with Severe Winters (**D**), Polar (**E**), and Highland (**H**).

Through a variety of scientific evidence, climatologists have been able to determine the past climates of Earth. This evidence includes historical records, tree rings, pollen deposits, air and dust trapped in ice sheets, sediments found on the ocean floor, and fossils. This evidence indicates that during most of the past 500 million years the Earth enjoyed a warmer and more congenial climate than we experience today. This mild climate has been repeatedly punctuated by cooler periods, including ice ages.

Scientists have proposed several different mechanisms by which Earth's climate can change naturally on a variety of time scales. Emissions from volcanic eruptions can cool the Earth for a few years. Asteroids and comets can crash into the Earth, causing mass extinctions and the dawn of new eras in Earth's biological and climatic histories. The Sun's energy output can vary over short and very long time scales. The Earth's orbit also varies on time scales of tens of thousands of years, altering the distribution and amount of solar energy reaching Earth. The motions of the continents over millions of years have probably triggered eras of worldwide warmth as well as ice ages. Changes in the ocean circulation may also have abruptly changed climate through the millennia.

For the past 20,000 years, records of climate are rather complete. Over this timespan the Earth's climate changed from a period of extreme glaciation to a warmer interglacial period. Alternating episodes of relatively cooler and warmer temperatures have characterized the past several thousand years, including the Younger Dryas about 11,000 years ago and the more recent Medieval Climatic Optimum (900 to 1200 AD) and Little Ice Age

(1400 to 1850 AD) in and near Europe. Changes in ocean circulation and solar output may explain these oscillations in climate. Recent increases in global temperature may be a result of human activity, which we examine in the next chapter.

 KEY TERMS

You should understand all the following terms. Use the glossary and this chapter to improve your understanding of these terms.

Boreal forests
Climate
Climographs
Dendrochronology
Dry climates (**B**)
Eccentricity
Glacier
Highland climates (**H**)
Historical climate
Ice ages
Isotopes
Köppen scheme
Little Ice Age
Maunder Minimum
Milankovitch cycles

Moist subtropical and midlatitude climates (**C**)
Obliquity
Paleoclimate
Pangaea
Permafrost
Plate tectonics
Polar climates (**E**)
Precession
Radiocarbon dating
Severe midlatitude climates (**D**)
Taiga
Tropical humid climates (**A**)
Weak Sun paradox
Younger Dryas

 REVIEW QUESTIONS

1. What is the difference between weather and climate?
2. Describe the climate where you live in terms of the Köppen climate classification scheme. Using Figure 14.2, find a location on a different continent that has the same classification as your hometown.
3. What features separate a tropical climate from the climate of the poles?
4. What type of parameters should be included in a climate classification scheme that would be useful to city planners? To farmers?
5. Develop a climate classification scheme based on precipitation type and amount that would be useful to people who want to avoid regions with lots of snow or ice storms.
6. How would you classify the climate of the dinosaur era in terms of the Köppen classification scheme? Visit the animation of Earth's geologic and climatological history that is linked to on the text's Web site to see if your classification is accurate.
7. Consider how might you design a climate classification scheme that is based on factors other than temperature and precipitation. For example, how could you use the frequency of air mass types to categorize climate?
8. A million years from today, human activities from our time might be studied to determine the climate of our age. Describe how a human activity of today could leave behind a useful signature of today's climate?
9. Explain why the deepest layers of ice on a glacier are the oldest.
10. What was the Little Ice Age? Was it really an ice age? Why or why not?

11. Why are lower sea levels associated with ice ages?
12. Over a century ago, trees cut down from old-growth forests in the upper midwestern United States routinely sank in the cold waters of Lake Superior while being floated downstream to furniture makers. The cold lake waters have preserved these logs, and today furniture makers are raising them from the depths to make custom furniture. Can you think of a way to use these trees for paleoclimate purposes? How could you get around the problem of not knowing exactly when the trees were cut down?
13. What is plate tectonics? Why is it important in climate change studies?
14. What climate change mechanisms might cause a mass extinction in the near future?
15. Discuss how a mile-high ice sheet would affect climate conditions because of its height, distribution across a continent, and temperature.
16. Based on the number of craters on the Moon, we expect to find 400 large craters (50 kilomerers wide) on Earth. However, only about 160 known impact craters, with only about 14 craters larger than 50 kilometers wide, have been found on Earth so far. Why do we find so few craters on Earth?
17. What is the Maunder Minimum, when did it happen, and how might it be connected to climate change in the past? Answer the same questions with regard to the Younger Dryas.
18. Use Figure 14.25 and the discussion in the text to explain how Milankovitch cycles should affect the Earth's climate in the *future*—for example, the next 50,000 years.
19. Following are the recent average temperatures of the Northern Hemisphere and Southern Hemisphere for winter, summer, and the year. The annual range is given as well as the differences between the Hemispheres. Discuss the causes of these observed differences.

	Winter	*Summer*	*Year*	*Annual Range*
Northern Hemisphere	8.1° C (46.6° F)	22.4° C (72.3° F)	15.2° C (59.4° F)	14.3° C (25.7° F)
Southern Hemisphere	9.7° C (49.5° F)	17.0° C (62.6° F)	13.3° C (55.9° F)	7.3° C (13.1° F)
Difference	−1.6° C (−2.9° F)	5.4° C (9.7° F)	1.9° C (3.5° F)	7.0° C (12.6° F)

 WEB ACTIVITIES

Choose Chapter 14 on the textbook's Web site
http://info.brookscole.com/ackerman
and select from the following resources:

- Interactive Modules that illustrate and extend your understanding of key topics in this chapter
- Tutorial Quizzes to test your mastery of terms and concepts

 For additional readings, go to the InfoTrac College Edition, your online library, at:
http://www.infotrac-college.com

CHAPTER

15

Human Influences on Climate

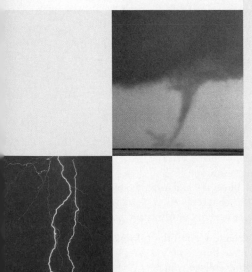

After completing this chapter, you should be able to:

■ Name the major pollutants in our atmosphere and describe their impacts

■ Explain the roles of feedback mechanisms in climate change

■ Relate acid rain and the ozone hole to human-caused changes to the atmosphere

■ Discuss the possibility of global warming and the roles of land-surface changes, water vapor, clouds, and the oceans in climate change

 Introduction

John Ruskin taught what he saw, and by the end of the 19th century he didn't like what he saw. A professor of fine arts at Oxford University in England, Ruskin once said, "I have come to the conclusion that it is not Art that I loved but Nature." But the Industrial Revolution fouled the landscapes Ruskin adored with clouds of smoke and soot.

Ruskin's knowledge of the atmosphere was based on observations of storms and sunsets in diaries and sketchbooks that spanned a period of 50 years. In 1884, certain that his data reflected a fundamental change in the atmosphere, Ruskin gave a lecture entitled "The Storm-Cloud of the Nineteenth Century." In this lecture he called the 1880s a "period which will assuredly be recognized in future meteorological history as one of phenomena hitherto unrecorded in the courses of nature." He described, probably for the first time, the phenomenon later known as acid rain. Ruskin also captured the impact of humans on climate in a clever turn of phrase by charging that "the empire of England, on which formerly the Sun never set, has become one on which he never rises." Later scholars agreed that Ruskin bore witness to an actual change in the atmosphere; he first saw the darkening "plague-cloud" when the burning of sooty coal increased dramatically across Europe.

A century later, atmospheric scientists following Ruskin's lead have testified publicly that humans are responsible for global climate change. In this chapter we explore how scientists have reached these conclusions using complex computer models of climate. Are the scientists brave and accurate prophets in the mold of John Ruskin, or are they falsely "crying wolf"? This chapter is intended to supply you with the knowledge to develop informed opinions concerning the often-contentious topics of pollution, acid rain, the ozone hole, and global warming.

 Observations of Global Warming

Over the past two decades, the global average surface temperature has increased noticeably (Figure 15.1). While a few warm winters and hot summers in one location or region do not mean global warming, the observed warming trend over the last two decades may indicate a significant global change.

Is this temperature change a natural fluctuation in our climate, or is it a result of human activities in the last 150 years or so? This question is a focus of this chapter, and is one of the most important scientific questions in the world today. But natural variations in climate, as discussed in Chapter 14, make it difficult to distinguish long-term trends caused by humans.

This trend in global warming has resulted in a global debate about what measures might be taken to mitigate this warming. Most of these measures involve changing human activities, such as reducing the burning of fossil fuels and deforestation. Because of the potential impact on society, global warming is a frequent topic in news reports. Sometimes these reports suggest that scientists are arguing whether greenhouse gases will change the climate. In most scientific discussions, the issue is not whether greenhouse gases will induce a climate change. The issues are what the effects will be and how these can best be detected.

The debate about global warming has not arisen in isolation. Rather, it is part of two larger stories: how climate changes, according to the ongoing research of atmospheric scientists; and how humans have changed

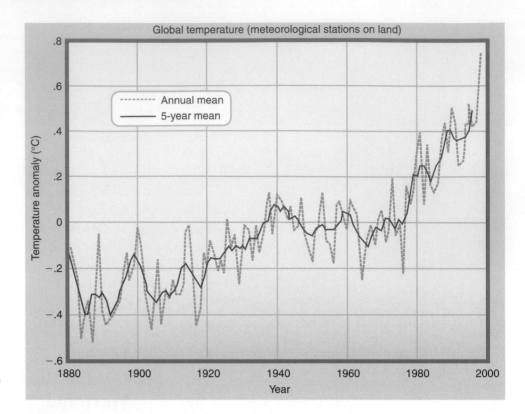

 Figure 15.1

Trend in the departure of surface temperature over the last 120 years. Notice the trend of increasing temperature over the last two decades.

the Earth's atmosphere. To place the global warming debate in its proper context, we will first examine these topics in some detail. We will see that climate change is an exceptionally complex topic, and that there is ample reason to believe that humans can and do make large changes to our atmosphere.

Feedback Mechanisms

As we have seen in previous chapters, climate is the result of an interplay of different mechanisms. To understand climate *change*, we need to understand how and when different mechanisms lead to these changes.

Feedbacks occur when one change leads to some other change, which can act to either reinforce or inhibit the original change. A **positive feedback mechanism** is one that enhances an existing trend of a change in climate. It is through positive feedback mechanisms that small changes can lead to large ones.

You can learn about positive feedback by experimenting with a microphone and an amplifier with a stereo speaker attached. If you stick the microphone into the speaker, here is what happens. First, the sounds that the microphone detects are amplified and emitted by the speaker. Then the microphone picks up these amplified sounds, which are then amplified still more and emitted by the speaker. This cycle repeats continuously. In just a few seconds, a deafening high-pitched sound develops even in a room that was initially quiet. You have probably heard this annoying sound at a concert, large lecture course, or another event that used microphones.

Sound engineers have long called this effect "feedback." This name arises because the sound into the microphone is repeatedly "fed" back into the amplifier, getting louder and louder with each cycle from the microphone to the speaker. The "positive" part of the name means that the amplification of the sound grows with time.

Positive feedbacks can occur in climate and contribute to rapid climate change, just as positive feedback with sound equipment leads to a rapid change in noise level. For example, the amount of water vapor in the atmosphere is a positive feedback mechanism involved in climate change. As the air temperature warms there is increased evaporation from surface waters, resulting in higher atmospheric water content. As the atmosphere warms, more water vapor can exist in the atmosphere. This causes the atmosphere to warm even further, causing more water vapor to evaporate into the atmosphere, enhancing the warming. While human emissions of carbon dioxide are attributed to the current trends in global warming, it is the water vapor feedback that may cause most of the warming.

Another example of a positive feedback mechanism is the **ice-albedo temperature feedback.** Ice sheets affect climate by reflecting more sunlight than other types of surfaces, such as bare ground and areas covered by vegetation. This reduces the amount of the Sun's energy that can warm the planet's surface. Other things being equal, the more ice, the cooler the Earth. Ice-covered areas can expand when the atmosphere along the margins of established ice regions cool. The increased area of ice reflects even more sunlight and further reduces the amount of solar energy absorbed by the surface. This reduction in solar energy gains causes further cooling and results in the formation of more ice, and so on. This is a positive feedback loop: more ice causes a cooling which reduces the temperature so that more ice develops, leading to a further cooling (Figure 15.2).

Also note that a retreat of an ice sheet is also a positive feedback since it would cause a warming, due to the lower albedo, and lead to a warming and a further retreat of the ice sheet. Positive feedbacks can cause "less, less, less" as well as "more, more, more." A classic example of this in human culture is a stock market crash, in which financial worries amplify and cause an accelerating trend of dropping stock prices.

A **negative feedback mechanism,** in contrast, damps out an existing trend of climate change. A simple example of a negative feedback mechanism involves the influence of carbon dioxide (CO_2) concentration on plant photosynthetic rates. In the process of photosynthesis plants use carbon dioxide and water to make sugar. An environment rich in carbon dioxide accelerates the growth of many plant species. This is a negative feedback: increasing carbon dioxide concentrations allow plants to grow faster and thereby increase the overall photosynthetic rate, which then *removes* increased amounts of carbon dioxide from the atmosphere. So, by a negative feedback loop more carbon dioxide results in a decrease of carbon dioxide, and the end result is an equilibrium instead of an accelerating trend as in the case of a positive feedback.

Of course, there are other limiting factors in a plant's ability to respond to an enriched carbon dioxide environment, such as a lack of water and nutrients. Also, insects and other pests might also enjoy the warmer environment associated with high concentrations of carbon dioxide which can reduce the number of plants. Feedback processes can quickly become complicated even in the simplest climate change scenario.

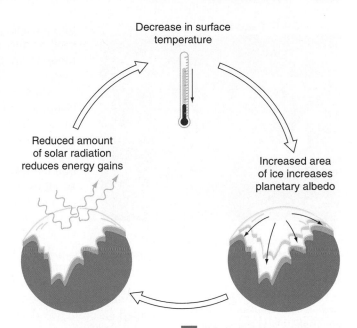

Figure 15.2

Ice-albedo temperature feedback explains how a small change in surface temperture can grow into a larger change.

Air Pollution

Air pollution is composed of airborne solid and liquid aerosols called **particulates** as well as gases that, when in high concentrations, seriously affect the lives of people and animals, harm plants, or threaten ecosystems. Pollutants can arise from

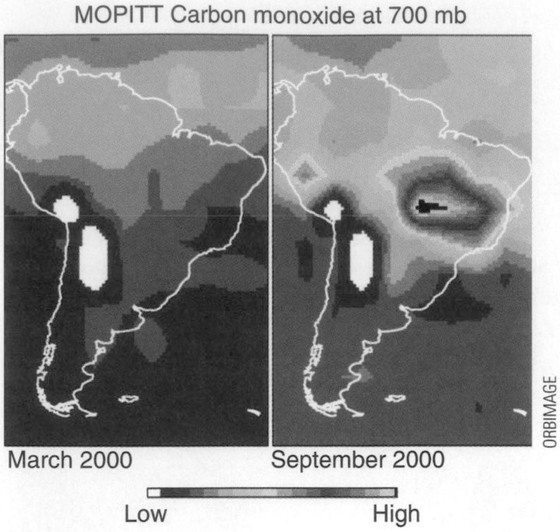

MOPITT Carbon monoxide at 700 mb

March 2000   September 2000

Low   High

Figure 15.3

A research satellite image of carbon monoxide in the lower troposphere over South America in connection with deforestation and burning of forests.

Blue Skies

Atmospheric Chemistry/ Smog

human activities (anthropogenic sources) or from natural sources such as dust storms and volcanic eruptions. In previous chapters we focused on the natural sources, and so here we examine anthropogenic air pollution.

Carbon monoxide (CO), a colorless and odorless gas, is a prime example of a harmful pollutant. Carbon monoxide forms when fossil fuels are not completely burned during combustion, for example in car engines or in forest fires (Figure 15.3). It is very toxic, because it disrupts how red blood cells absorb oxygen. As a result, inhalation of carbon monoxide reduces the body's ability to provide oxygen to the body. In cities, the carbon monoxide levels can approach unsafe levels in confined areas, such as garages and tunnels. Carbon monoxide is also very dangerous in the home where it can be produced by heating devices that are not operating properly.

Lead is a particulate released into the atmosphere when, for example, treated gasoline is burned in car engines. It settles out of the atmosphere and is easily absorbed into the human body. Even small amounts of lead cause brain damage and lower IQs in infants. For this reason, unleaded gasoline was developed and replaced leaded fuels in American automobiles in the 1970s, reducing lead emissions dramatically.

Pollutants can also develop from harmless chemicals that are emitted directly into the atmosphere, but then become noxious gases or particulates after chemically combining with other atmospheric constituents. Human activities produce oxides of sulfur, in particular **sulfur dioxide (SO_2)** and **sulfur trioxide (SO_3)**. These sulfur oxide compounds are released into the atmosphere primarily through the burning of fossil fuels that contain sulfur. Levels of sulfur dioxide and sulfur trioxide can become large when the activities concentrate the compounds over a small region that allows the pollutants to reach high levels, as in urban and industrial areas. Sulfur dioxide is a highly corrosive gas that irritates the human respiratory system. Sulfur trioxide is an important pollutant as it readily combines with water vapor to form droplets of **sulfuric acid (H_2SO_4),** creating acid rain, which we discuss in detail in the next section.

Oxides of nitrogen, in particular **nitric oxide (NO)** and **nitrogen dioxide (NO_2)** are two important air pollutants. Nitric oxide is a byproduct of high temperature combustion, such as in automobile engines and electric power generation. Indeed, atmospheric concentrations of nitrogen dioxide in urban areas are well correlated with the density of vehicular traffic. Nitric oxide is a very reactive gas and quickly forms nitrogen dioxide. Nitrogen dioxide is also a toxic gas that is also emitted by automobile engines. High concentrations of nitrogen dioxide give polluted air its reddish-brown color. In high concentrations, oxides of nitrogen cause serious pulmonary problems.

Hydrocarbons are compounds made of hydrogen and carbon atoms. Examples of hydrocarbons are methane, butane, and propane, which can occur as either a gas or particulate. Hydrocarbons are also called **volatile organic compounds (VOCs).** Hydrocarbons do not appear hazardous in themselves, but during the day, they can combine with nitrogen oxides and oxygen to produce photochemical **smog.**

The main component of photochemical smog is **ozone (O_3).** Ozone is a chemically active molecule and is considered a corrosive gas. In the presence of sunlight, oxides of nitrogen from engine exhaust and hydrocarbons react to form a noxious mixture of aerosols and gases. This mixture includes ozone, formaldehyde, and PAN (peroxyacetyl nitrates). Exposure to high concentrations of ozone irritates the eyes, nose and throat, and causes coughing, chest pain, and shortness of breath. Ozone also aggravates diseases such as asthma and bronchitis.

Table 15.1 summarizes recent trends in emissions of major pollutants for the United States as a whole. While the Clean Air Acts of 1970 and 1990 have led to remarkable improvements in most directly emitted pollutants, tropospheric ozone has

TABLE 15.1	Trends in Air Pollutant Concentrations in the United States from 1991 to 2000					
Pollutant	Lead	Carbon Monoxide	Sulfur Dioxide	Particulates	Nitrogen Dioxide	Smog
% Change	–50%	–41%	–37%	–19%	–11%	–10%

emerged as a growing threat to health. Ozone alerts have become a commonplace occurrence in many urban regions of the United States, and gasolines specially formulated to reduce ozone pollution are sold (at a higher price) in these regions.

These days most smog is photochemical rather than the mixture of smoke and fog that gave "smog" its name. An extreme case of old-fashioned smog occurred on December 5–9, 1953 when an estimated 3500 to 4000 people died in London, England. Over this 5-day period, stagnant moist air combined with the smoke from burning of low-quality coal to produce a lethal mixture.

Air pollution is not just a modern-day problem. Cave dwellers undoubtedly had to address local air quality problems such as smoke from fires. The Hopi Indians had to deal with sulfur dioxide emitted from burning coal to make pottery. England has long had a problem with coal burning. Air quality was so poor that it has been estimated that in 18th-century England half the children died before their second birthday.

Acid Deposition

Acid deposition refers to the falling of acids and acid-forming compounds from the atmosphere to Earth's surface. Air pollution from industrial areas can become acidic and be carried downwind for many miles. When these acids settle on the ground, they can damage plants and aquatic life. This settling may occur as dry particles (dry deposition) or as rain, snow, or fog (wet deposition). When in the form of rain, the acid deposition is referred to as **acid rain.**

The pH scale represents the acidity of a solution. pH levels range from 0 to 14, with 0 being extremely acidic. A pH of 7 is neutral. The scale is logarithmic, so a change in pH from 7 to 6 or from 5 to 4 represents a tenfold increase in acidity.

Normal precipitation has a slightly acidic pH value of approximately 5.5. This is because some atmospheric carbon dioxide dissolves in the water drops as they form and grow. Acid rain forms because of the increased levels of sulfur dioxide and oxides of nitrogen that enter the atmosphere due to burning of high–sulfur-content coal. These gases dissolve in the cloud drops, making the precipitation more acidic, lowering the pH from 5.5 to between 4 and 4.5. This is an increase in acidity of a factor of ten or more.

Industrial sources and petroleum-powered vehicles emit massive quantities of sulfur dioxide and nitrogen oxides into the atmosphere. The United States alone emits approximately 40 million tons per year. These chemicals undergo complex changes when in the atmosphere. Some get dissolved in raindrops, snow, or fog particles and produce weak solutions of sulfuric and nitric acids. As these acids fall to the ground they can accumulate in lakes and can affect ecosystems. For example, there has been a decline in the health of coniferous forests in the Appalachian Mountains from North Carolina to New England that may be related to acid rain.

In North America, acid rain is primarily a concern in the Northeast and Canada, downwind of sources of sulfur dioxide and nitrogen oxides. Figure 15.4 depicts the pH level of precipitation measured over the United States and Canada. Notice the low pH values east of the Mississippi, with the lowest pH values (most acidic) occurring in northeastern United States and Canada. The situation is improving with the 1990 Clean Air Act amendments that mandate reductions in sulfur and nitrogen

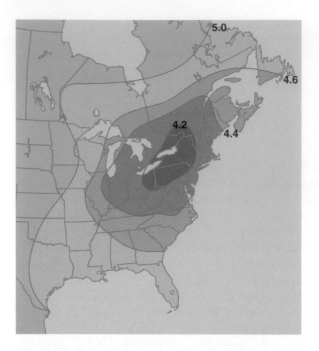

 Figure 15.4

The pH level of rain over North America measured in 1982. Purple shaded regions indicate acid rain regions where the pH is less than 5.0. (Source: Moran and Morgan, Essentials of Weather, *Prentice-Hall, 1995, p. 417.)*

acid-forming compounds. However, the problem is even more severe in Norway and Sweden, where an estimated 6500 lakes are essentially lifeless.

Acid rain is an example of how human activities that emit gases into the atmosphere can impact the environment. An even more dramatic example is the stratospheric ozone hole, which we discuss next.

The Stratospheric Ozone Hole

As we learned in Chapters 1 and 2, ozone is produced in the stratosphere by the combination of three oxygen (O) atoms under the influence of sunlight. The concentration of atmospheric ozone is small, approximately three molecules of ozone for every ten million air molecules. Through absorption of ultraviolet radiation (UV), ozone plays a fundamental role in the radiation budget and the dynamics of life on Earth. Absorption of UV energy causes a heating that produces the increasing temperature with altitude, a characteristic feature of the stratosphere. Ozone absorption of UV also keeps this harmful radiation from reaching the surface. Reduction in amounts of ozone can lead to increased amounts of biologically damaging UV at the surface. Increased amounts of UV can lead to incidents of cataracts and deadly skin cancers known as melanoma.

Observations of the monthly average total column ozone amounts during the past several decades—but not before about 1955—show a minimum amount over the Southern Hemisphere in that hemisphere's spring season (Figure 15.5). During winter, the amounts of ozone over the South Pole region remain fairly constant. A decline in ozone is seen in September and a minimum amount of ozone is observed in October. After October, ozone levels begin to increase. Why does this minimum occur today, and why in October?

In Chapter 2 we explained ozone depletion as the result of the release of chlorofluorocarbons (CFCs) into the atmosphere during the middle decades of the 20th century. This is true, but the reason for the dramatic ozone losses seen over Antarctica is a complex interaction between clouds and atmospheric chemistry, including but not limited to the CFC–ozone interactions depicted in Chapter 2. Here we tell the rest of the story.

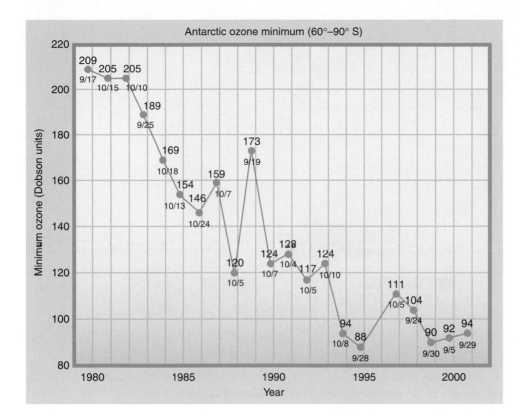

Figure 15.5

Observations of stratospheric ozone loss in the Antarctic, based on satellite observations from 1979 to 2000. The minimum amount of stratospheric ozone and the day of that minimum for each year are indicated on the graph. The total amount of ozone in an atmospheric column at a given location is measured in Dobson units (DU). A typical value of 300 DU translates to a layer of ozone that would be 0.3 centimeters thick at surface pressure, or approximately the thickness of a dime. (Source: http://toms.gsfc.nasa.gov/multi/min_ozone.jpg.)

The winter atmosphere above Antarctica is very cold. These cold temperatures result from the high altitude of the Antarctic continent and the resulting energy losses caused by longwave radiation losses at the surface. Temperatures in the stratosphere can be less than –90° C (–130° F)! In addition, the cold temperatures result in a temperature gradient between the South Pole and the Southern Hemisphere middle latitudes. Recalling the thermal wind law from Chapter 6, these temperature gradients lead to a belt of strong westerly stratospheric winds that encircle the South Pole region. These strong winds, referred to as the *polar vortex,* prevent the transport of warm equatorial air to the polar latitudes. This isolation of the south polar regions from the middle latitudes and tropical regions helps keep the stratospheric air very cold. The isolation from outside air also forces the polar vortex to "stew in its own juices" in terms of its chemical makeup.

These extremely cold temperatures cause water vapor and some nitrogen compounds to condense and form unique types of clouds. These **polar stratospheric clouds (PSCs),** also known as *nacreous clouds,* are composed of ice and frozen nitrogen particles and form in air temperatures colder than approximately –80° C (–112° F). PSCs begin to form during June and dissipate in October, the Antarctic spring.

In the wintertime, chemical reactions on the surface of the particles composing PSCs result in chemical reactions that remove the chlorine from the atmospheric compounds. During the spring, when sunlight again shines on the Antarctic stratosphere, the PSC clouds evaporate, and the chlorine atoms are freed and ozone is rapidly depleted (as discussed in Chapter 2). Destruction is so rapid over the South Pole region in the Southern Hemisphere springtime that it has been termed a "hole in the ozone layer," and it is seen every October (Figure 15.6).

There is a definite year-to-year variation in the development and size of the Antarctic ozone hole. This is demonstrated by plotting the October monthly average ozone amounts over the Antarctic during the period from 1980 to 1991 (Figure 15.7). Notice the reduced ozone amounts in 1987, 1989, 1990, and 1991.

EP/TOMS Monthly average total ozone October 2001

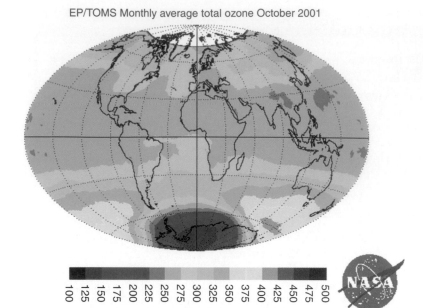

Dobson units

Figure 15.6

The stratospheric ozone hole, as seen by satellite in October 2001. The "hole" is the concentrated region of low ozone values centered over Antarctica.

Why does the ozone hole appear only over the South Pole? Stratospheric clouds composed of ice particles are common over Antarctica but are more rare over the Arctic regions. This is because the Arctic stratosphere does not normally get as cold as the air over Antarctica, due to Rossby waves (Chapter 7) in the stratosphere that prevent the development of an isolated vortex over the North Pole. As a result, while ozone depletion of 15% to 20% has been observed in certain regions of the Arctic stratosphere, the development of a concentrated "ozone hole" has not been observed as of yet. International governmental agreements to limit CFC use should allow the ozone layer to repair itself fully, but perhaps not until the 22nd century.

To summarize, the stratospheric ozone hole is the end result of the innocent release of tiny amounts of chlorine-containing chemicals into the troposphere, which over a period of decades found their way into the stratosphere. Once in the stratosphere, a series of chemical reactions on cloud surfaces led to a catastrophic decrease in the amount of ozone over Antarctica. This indicates that climate change is complicated, and that when studying climate change it is wise to expect the unexpected and to anticipate unforeseen consequences.

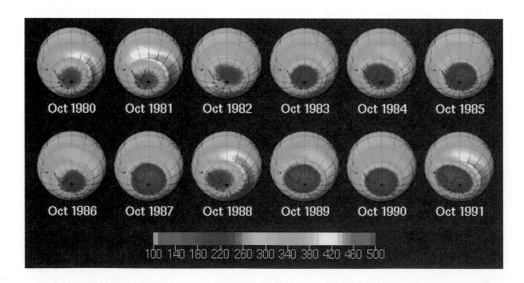

Figure 15.7

October monthly mean ozone amounts over the Antarctic during the period of its discovery in 1980 to 1991, as measured by NASA satellites. Note that each year is different, and that the size of the ozone "hole" increased during this period.

Changing Land Surfaces

As discussed in Chapters 2 and 3, the Earth's surface has a direct effect on regional weather and climate. Large-scale changes in land use, such as urbanization, deforestation, and agriculture, affect the surface of Earth. These changes, in turn, affect the *biosphere*. The biosphere is comprised of all of Earth's living organisms. The biosphere plays an important role in the climate, regulating the carbon cycle, the hydrological cycle, and the heat budget of the planet. Climate changes can be linked to changing land and vegetation patterns, and vice versa.

Desertification

Desertification refers to the spreading of a desert region because of a combination of climate change and human impacts on the land. Practices that leave the land susceptible to desertification include: overgrazing, deforestation without reforestation, diversion of water away from a formerly fertile region, and farming on land with unsuitable terrain or soil.

The consequences of desertification can include the magnification of long-term drought and permanent changes in atmospheric circulation patterns. The human consequences can be severe: a decline in the standard of living, and even famine.

The semi-arid southern fringes of the Sahara Desert are vulnerable to desertification. The Sahel, or Sub-Sahara, is a semi-arid region (climate type Bsh) between 14°N and 18°N with pronounced wet and dry seasons. The year-to-year variation in precipitation depends on the movement of the ITCZ (Chapter 7) and varies considerably.

In the early 1960s, rainfall was plentiful and the nomadic people who live in the Sahel found ample grazing lands for cattle and goats. Herds grew in number and so did the human population. In 1968 the ITCZ did not bring the needed rains as far north as usual, marking the beginning of a severe drought that lasted into the 1980s. The drought, in concert with overgrazing, turned large regions of pasture into a wasteland. The drought peaked in 1973 when rainfall totals were only half the long-term average. The Sahara Desert moved southward into the Sahel and a famine ensued that took the lives of more than 100,000 people and affected more than 2 million. Rainfall eventually returned to the region, but in smaller quantities than were experienced in the 1950s and 1960s. Lake Chad, in West Africa, is now only 5% as large as it was just 35 years ago!

It is possible that this change in regional precipitation is a natural variation. Indeed, the southern boundary of the Sahara shifts position north and south by 100 kilometers (60 miles) in different years. However, the drought may also be enhanced by a biogeophysical positive feedback mechanism. The Sahel lies below the descending branch of the Hadley during its dry season. Sinking air warms and is thus an energy gain for the atmosphere. With a reduction in rainfall there is less vegetation. The reduced vegetation results in an increase in surface albedo, reducing the energy gains of the surface. To make up for reduced energy gains from the surface, the atmosphere subsides, warms, and dries out. This warming and drying associated with the sinking motion of the atmosphere further enhances desert conditions. Thus, the reduced precipitation of a semi-desert region is a positive feedback mechanism in which reduced precipitation leads to changes in the surface budget that result in sinking air, further reducing precipitation.

An even more pronounced example of human-caused desertification has occurred in recent decades in the vicinity of the Aral Sea in central Asia. The government of the Soviet Union diverted the rivers feeding this vast inland lake to provide water for the growing of cotton. Deprived of its sources of water, the

The Shrinking Aral Sea

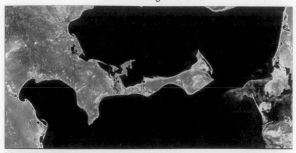

May 29, 1973

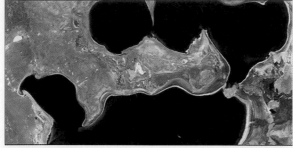

August 19, 1987

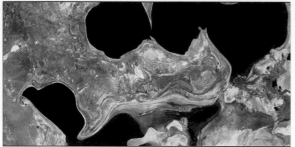

July 29, 2000

NASA

Figure 15.8

The disappearance of the Aral Sea, as seen by satellite during the period from 1973 to 2000. This is a close-up view of a peninsula extending into the lake that progressively grew to replace a sizable portion of the original lake area. Over this period, more than 60% of the lake has vanished, replaced with a dry, dusty, and mostly lifeless plain.

Aral Sea began shrinking rapidly (Figure 15.8). Evaporation left minerals behind, and the shrinking lake exposed gritty, dusty shorelines and sea floors.

As the Aral Sea has shrunk, both the climate and the culture of the region have changed for the worse. Winds picked up this dust and created an Asian "Dust Bowl" (Chapter 12) across a once-flourishing region. The Aral Sea fishing industry collapsed because the lake grew increasingly salty and inhospitable to aquatic life. Meanwhile, the local climate has changed, shortening the growing season and forcing local agriculture to require even more diversion of water from the Aral Sea, further worsening the situation. The lake, the fourth largest in the world in 1960, may dry up completely by the year 2015. In short, the Aral Sea and its surrounding regions are dying, due to desertification caused by the decisions of a now-defunct government.

Urban Heat Islands

The **"urban heat island** effect" refers to the increased temperatures of urban areas compared to a city's rural surroundings. Urban heat islands are a good model with which to explore how changing the energy balance of a region can affect the regional climate.

The reason the city is warmer than the countryside depends on the difference between the energy gains and losses of each region. There are a number of factors that contribute to the relative warmth of cities, such as heat from industrial activity, the thermal properties of buildings, and the evaporation of water. For example, the heat produced by heating and cooling city buildings and running planes, trains, buses, and automobiles contributes to the warmer city temperatures. Heat generated by these objects eventually makes its way into the atmosphere, adding as much as one third of the heat received from solar energy.

The thermal properties of buildings and roads are also important in defining the urban heat island. Asphalt, brick, and concrete retain heat better than natural surfaces. Buildings, roads, and other structures add heat to the air throughout the night and thus reduce the nighttime cooling of the air, so that the maximum temperature difference between the city and surroundings occurs during the night. The canyon shape of the tall buildings and the narrow space between them magnifies the longwave energy gains. During the day, solar energy is trapped by multiple reflections off the many closely spaced, tall buildings, reducing heat losses by longwave radiation (Figure 15.9). Pollution in the city's air also modifies the absorption of longwave and shortwave radiation of the atmosphere.

Evaporation of water may also play a role in defining the magnitude of the urban heat island. During the day, the solar energy absorbed near the ground in rural areas evaporates water from the vegetation and soil. Thus, while there is a net solar energy gain, heating is lessened to some degree by evaporative cooling during evapotranspiration. In cities, where there is less vegetation, the buildings, streets, and sidewalks absorb the majority of solar energy input and warm up rapidly.

The urban heat island is evident in statistical tables of surface air temperatures. The warmer temperatures of urban areas are also apparent in cloud-free satellite

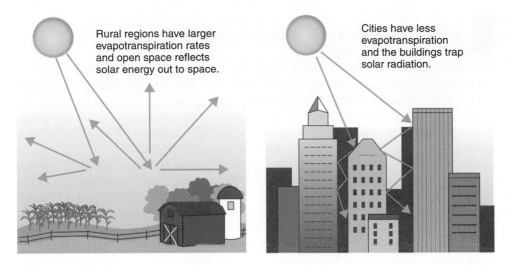

Rural regions have larger evapotranspiration rates and open space reflects solar energy out to space.

Cities have less evapotranspiration and the buildings trap solar radiation.

■ **Figure 15.9**

Concrete roadways and buildings reduce evapotranspiration and absorb more solar radiation than surrounding rural regions. As a result, urban areas are often warmer.

images. Figure 15.10 is a satellite image of infrared radiative energy exiting the atmosphere with a wavelength of 11 microns, which is directly correlated to surface temperature. Urban heat islands appear in this image as dark blemishes that are warmer than the more rural regions around them.

Now that we have investigated several different ways in which humans have changed climate on local and regional scales, we are ready to look at the science of *global* warming.

Urban Heat Island

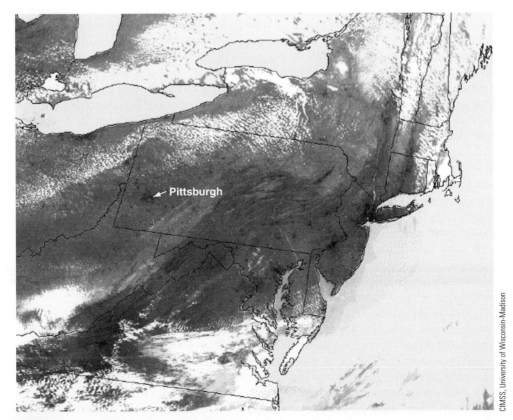

Pittsburgh

■ **Figure 15.10**

An infrared satellite image demonstrating the urban heat island effect. The city of Pittsburgh is marked to help you recognize the heat island feature, a dark "blemish" representing temperatures warmer than in the surrounding rural regions. Other urban heat islands visible in this image include: Cleveland, Ohio; Washington, D.C.; the New York City metropolitan area; Buffalo, Albany and Syracuse, New York; Philadelphia and Harrisburg, Pennsylvania; and Baltimore, Maryland.

CIMSS, University of Wisconsin-Madison

Greenhouse Gases and Global Warming

As we learned in Chapters 1 and 2, the Earth's atmosphere contains gases that are transparent to solar radiation but absorb large amounts of terrestrial infrared radiation. Carbon dioxide, methane (CH_4), water vapor, and CFCs all are "greenhouse gases" that act to trap heat in the Earth's lower atmosphere, just as a greenhouse keeps plants warmer than if they were exposed to the elements. Without the greenhouse effect, the Earth would be too cold for most life to exist.

However, human civilization has significantly changed the amount of greenhouse gases in the atmosphere during the past century. Since the beginning of the Industrial Revolution in the 19th century, concentrations worldwide have increased by 25%. Today, the concentration of carbon dioxide is increasing at a rate of about 0.5% per year. Methane has doubled over the same period because of human activities such as coal mining and agriculture, leaky natural-gas lines, and cattle. CFCs are a human creation and did not even exist in the atmosphere until after 1930.

Based on the radiation concepts from Chapter 2, it would seem obvious that the larger the amount of greenhouse gases, the more energy trapped in the lower atmosphere—and the higher the global temperature. But the global warming debate is not this simple. Climate involves much more than radiation, and includes the interactions of the atmosphere with the land, the oceans, and living things. Because of this, the debate focuses on various possible climate change feedbacks caused by changing atmospheric composition, and scientific conclusions rely heavily on complex computer models of climate that attempt to integrate all of the interactions in predictions of future climate.

Global Warming and Atmospheric Water

Water is crucial to climate, and so it is no surprise that water plays a critical role in the global warming debate. Water vapor and clouds play a vital role in preserving the balance between incoming and outgoing energy in the Earth. Water vapor is a strong greenhouse gas. Water vapor is transparent to solar radiation while being extremely effective at absorbing terrestrial radiation emitted from the Earth's surface.

Human activities add little water vapor to the troposphere, at least directly. Water vapor increases are a result of internal controls of the climate system. As we learned in Chapter 4, the equilibrium between air and water is such that saturation occurs at higher vapor pressures for warmer temperatures than for colder temperatures. Therefore, more atmospheric water vapor may be present in a globally warmed atmosphere. Increased water vapor concentration can warm the atmosphere further, leading to a further increase in atmospheric water vapor in a positive feedback loop. From this perspective, a strong case can be made for global warming resulting from increased greenhouse gases.

However, increased water vapor amounts may also enhance cloud cover that can induce a different global climate change. It is difficult to predict the effect of changes in cloudiness on climate. Different cloud types affect the climate in different ways. High, thin cirrus can lead to a warming, similar to the greenhouse warming. Thick stratocumulus cause a cooling by reflecting large amounts of solar energy back to space. Satellite observations indicate that the average global cloud distribution causes a net cooling of the planet.

Aerosols can also affect climate by causing changes in cloud radiative properties. Aerosols serve as cloud condensation nuclei (see Chapter 4). When more nuclei are present, more droplets will form in the cloud and the droplets will be smaller. This makes the cloud more reflective (Chapter 5), further reducing the solar energy reaching the surface.

Observations of cloud droplets over the Atlantic Ocean downwind of northeast North America indicate that these clouds tend to be composed of drops that are

CIMSS, University of Wisconsin-Madison

Figure 15.11

A visible image from a geostationary satellite of stratus clouds off the coast of California. The ragged lines in the upper potions of the figure are clouds whose reflectance has been enhanced by increased aerosols from ships moving below the clouds. This suggests that human activities can increase the brightness of clouds, and thereby cause a cooling of the planet.

smaller than similar clouds that are in a pristine environment. Evidence of this is seen in satellite images of ship tracks (Figure 15.11). Pollutants from the ship engines rise upward into the cloud and serve as CCN. This increases the number of cloud droplets in the cloud, making the cloud appear brighter. Smaller particles in high concentrations also make it less likely the cloud will yield precipitation. The effect of aerosols on the radiation budget through their impact on clouds is called an *indirect aerosol effect* on climate.

In addition to changing cloud properties, human activities may directly change cloud amount through the creation of contrails (Box 15.1). These airplane-induced clouds may have at least a regional effect on climate. It is currently unknown whether, when all of the interactions between greenhouse gases and water are factored together, the overall impact will be global warming or global cooling. This is a major source of uncertainty in the global warming debate.

Growing a Contrail

Global Warming and the Oceans

The interactions between the atmosphere and the oceans in a globally warming world are also complex and poorly understood. As we learned in Chapter 4, water absorbs energy by warming up and/or changing phase. Since the Earth's surface is 70% water, the oceans can absorb a considerable amount of energy trapped by increasing greenhouse gases. Scientists believe this has slowed the warming of the atmosphere during the past several decades.

But so far we have thought of the oceans as simply a big bathtub that absorbs heat from the atmosphere. On the contrary, the oceans are dynamic and have complex three-dimensional circulations, as we learned in Chapter 8. When this is taken into account, the role of the oceans in climate change becomes far more complicated. For example, vertical motions (convection) in the ocean are very sensitive to temperature. If global warming changes the amount and location of convection, climate change could accelerate rapidly. For example, during the Younger Dryas period about 10,000 years ago (Chapter 14), melting ice sheets apparently interfered

BOX 15.1

Contrails

The white condensation trails left behind jet aircraft are called **contrails** (*con*densation *trails*). The satellite image at right shows multiple contrails over Lake Superior and the upper peninsula of Michigan. Contrails form when hot, humid air from jet exhaust mixes with environmental air of low vapor pressure and low temperature. The mixing is a result of turbulence generated by the engine exhaust.

Cloud formation by a mixing process is similar to the cloud you see when you exhale in cold air and "see your breath." The bottom figure at the right represents how saturation vapor pressure varies as a function of temperature. The blue line is the saturation vapor pressure for ice as a function of temperature (Chapter 4). Air parcels in the region labeled *saturated* will form a cloud. Imagine two parcels of air, A and B, as located on the diagram. Both parcels are unsaturated. If B represents the engine exhaust, then as it mixes with the environment (parcel A) its temperature and corresponding vapor pressure will follow the dashed line. Where this dashed line intersects the blue line, the parcel becomes saturated and a cloud forms.

If you pay attention to contrail formation and duration, you will notice that they sometimes rapidly dissipate, but other times they will spread horizontally into an extensive thin cirrus layer. How long a contrail remains intact depends on the humidity structure and winds of the upper troposphere. If the atmosphere is near saturation, the contrail may exist for several hours. On the other hand, if the atmosphere is dry, the contrail will dissipate as soon as it mixes with the environment.

Contrails are a concern in climate studies because increased jet aircraft traffic may result in an increase in cloud cover over certain regions. It has been estimated that in certain heavy air-traffic corridors such as over Chicago, cloud cover has increased by as much as 20%.

ORBIMAGE

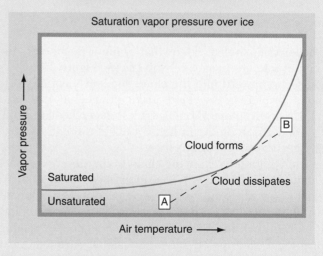

Saturation vapor pressure over ice

with oceanic convection and led to a rapid cooling of the climate of the Northern Hemisphere that lasted hundreds of years. Could atmospheric warming lead to a similar disruption of climate-as-usual? This depends on the details of temperature and precipitation changes caused by global warming, and is an extremely difficult question to answer.

Finally, the oceans are, like the land surface, teeming with life. This underwater biosphere may turn out to hold the crucial clues for global climate change. However, scientists currently understand comparatively little about life under the sea. To improve our global knowledge of life on land and sea, an extensive series of satellite campaigns is in progress and is observing the biosphere in unprecedented detail (Figure 15.12). However, the learning curve can be steep and improved understanding may take years or even decades.

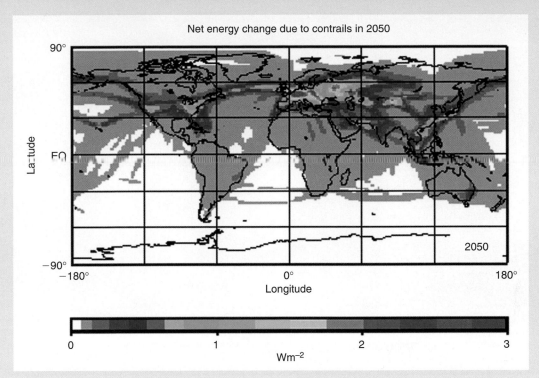

(*Source: Minnis, P., Schumann, U., Doelling, D., Gierens, K., and Fahey, D., "Global Distribution of Contrail Radiative Forcing," Geophys Res Ltrs, 26, July 1, 1999, pp. 1853–1856.*)

An increase in cloud amount changes the region's radiation balance. For example, solar energy reaching the surface may be reduced, resulting in surface cooling. Clouds can also reduce the terrestrial energy losses of the planet by absorbing energy emitted by the surface, resulting in a warming. Recent research at NASA (above) suggests that during the next half-century contrails will help warm some heavily populated parts of the globe by as much as one or two Watts per square meter, an amount comparable to the effect of the "greenhouse gases."

In the days following the September 11, 2001 terrorist attacks, all U.S. commercial jets were grounded and for the first time in decades American skies were briefly contrail-free. This sad period is currently being studied to see if the impact of contrails on weather and climate can be quantified based on their absence during those days.

Climate Modeling

Because of the changes in the Earth's atmosphere, future climate change is a distinct possibility. What will happen? The interactions between the oceans, atmosphere, land, cryosphere and biosphere known today are far too complex for even the best scientists to determine via discussion or pencil and paper. The choices are either to wait and observe what happens to global climate, or to try to predict some details of future climate. Because the consequences of climate change could be wrenching to world civilization (see the next section), future climate prediction has become one of the most active areas in all of the atmospheric sciences in recent decades.

Climate predictions are made using **global climate models (GCMs).** A GCM is a computer program that calculates global climate using mathematical equations

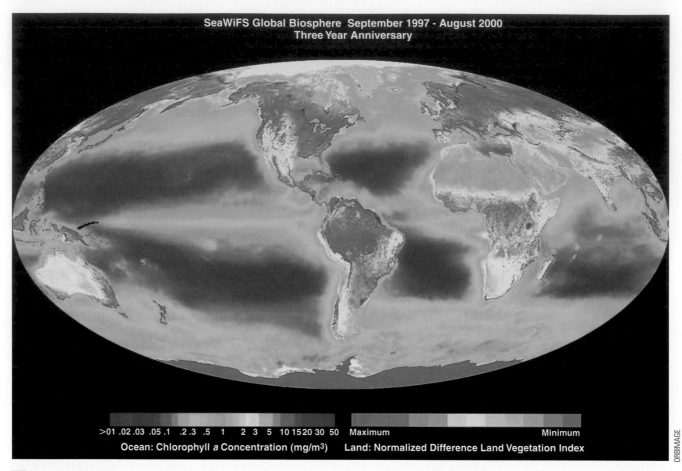

SeaWiFS Global Biosphere September 1997 - August 2000
Three Year Anniversary

>01 .02 .03 .05 .1 .2 .3 .5 1 2 3 5 10 15 20 30 50   Maximum   Minimum
Ocean: Chlorophyll *a* Concentration (mg/m³)   Land: Normalized Difference Land Vegetation Index

ORBIMAGE

■ **Figure 15.12**

A composite satellite image from the NASA SeaWiFS project depicting 3 years of data on land vegetation amount and ocean chlorophyll concentrations, which indicate the locations of plankton in the ocean. Green, yellow, and red areas in the oceans are biologically active.

Daisyworld

derived from physical principles such as the conservation of energy (Chapter 2) and Newton's Laws of Motion (Chapter 6). The roots of today's GCMs can be traced directly to the methods first developed by L.F. Richardson for numerical weather forecasting, which we examined in Chapter 13. There are three big differences, however.

The first difference is that most weather forecast models are primarily atmospheric models. Climate models, however, must include increasingly realistic models of the oceans, biosphere, and cryosphere in order to capture the various interactions discussed in this chapter and illustrated schematically in Figure 15.13.

The second difference between numerical weather forecasting and climate forecasting is in what can and cannot be predicted. Numerical weather forecasting attempts to predict exact values of temperature, pressure etc. for specific locations at specific instants in time. In contrast, long-term climate prediction is limited by chaos theory to statistical estimates of the impact of a change in a climate parameter (for example, carbon dioxide concentration) on world climate. In many ways climate prediction closely resembles the ensemble forecasting approach described at the end of Chapter 13; in fact, climate researchers originally pioneered the ensemble approach. The goal of climate modeling is to determine a *range* of *probable* future climate on large scales, not to intuit the exact temperature at, say, Chicago at an instant in time 50 years from now.

The third difference is that today's numerical weather forecasts can be proved wrong tomorrow and analyzed, and the model can be improved quickly. Most future

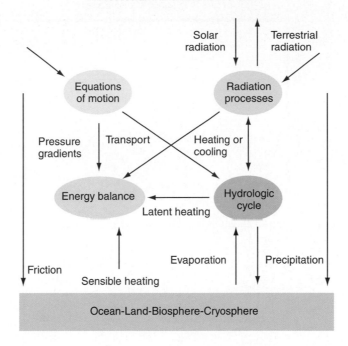

Processes and interactions in a GCM

Figure 15.13

A schematic of the atmospheric processes of a climate model. These interactions are expressed in a global climate model as a series of mathematical expressions in computer code.

climate forecasts, however, will not be proved right or wrong for decades. For this reason, the validity of global climate modeling can be questioned because many of its results cannot be verified immediately. However, the ability of GCMs to account for current climate conditions as well as past climates lends confidence to predictions by GCMs. As with today's weather models, climate models continue to improve and are essential to understanding changes in our atmosphere.

Models are also valuable in identifying the fingerprints of global climate change. The impact of volcanic eruptions on changes in surface temperature was confirmed with observations and model predications of surface temperature after the eruption of Mt. Pinatubo in 1991. Until recently there have been limited observations of climate changes in the polar regions. Climate models have repeatedly indicated that polar regions are very sensitive to a global climate warming. As a result, new efforts have been implemented to study the climate of the poles and document current changes. Models are also very valuable in understanding feedbacks and helping us to separate natural variations from the effects of human activities. Even so, they are only as accurate as the science and the data that are entered into them.

A GCM can be used to determine how sensitive climate is to a given process. For example, if we want to understand how a change in cloud cover impacts climate, we can "force" a change in cloud amount in a climate model and analyze the model's response to this simulation. Or, we can fix the cloud amount and change the vertical distribution of clouds. The climate model is the climate researcher's laboratory to run controlled experiments!

Climate models have been used to assess how increased amounts of carbon dioxide impact climate by allowing us to increase the amount of carbon dioxide in the model atmosphere and have the GCM predict the atmosphere's response. A typical experiment used by climate modelers is to increase the amount of greenhouse gases in the atmosphere to those levels expected by the year 2050 and see how the model responds to these changes. Models predict a warming, though the degree of warming varies with the model used (Figure 15.14). This is because different models make different approximations in order to find a solution to the complex set of equations. So, models tend to differ in their predictions. This variation among the different models can be thought as representing an uncertainty in our understanding of the atmosphere's response to changes. GCMs predict that the global mean

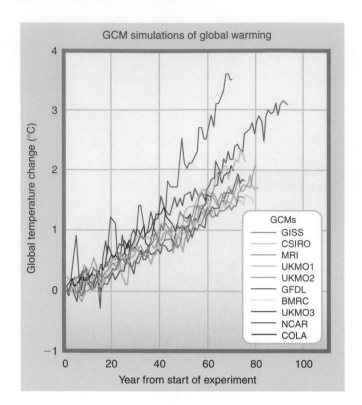

Figure 15.14

A comparison of climate predictions by different climate models shows a warming trend; however, there is uncertainty, as indicated by the differences in the amount of warming predicted. (Source: Climate Change 1995: The Science of Climate Change, *IPCC, p. 300.)*

temperature by the year 2100 will be warmer than today by 1° C (2° F) to 4.5° C (8° F). It was through this type of modeling study that we learned the importance of water vapor as a greenhouse gas. GCMs are also critical in deciphering the impact of aerosols on climate (Figure 15.15).

While variations exist, many GCMs predict consistent qualitative changes in the global distributions of temperature and the hydrological cycle. For example, GCMs

Figure 15.15

The observed and simulated global annual mean warming from 1860 to 1990. Two simulations with a GCM are shown. Both simulations include increasing amounts of greenhouse gases. The second simulation also includes the effects of sulfate aerosols (which tend to cool the planet), and provides a better agreement with the observations. Comparing these two simulations with observations gives scientists an indication of the importance of these processes in climate change. (Source: Climate Change 1995: The Science of Climate Change, *IPCC, p. 300.)*

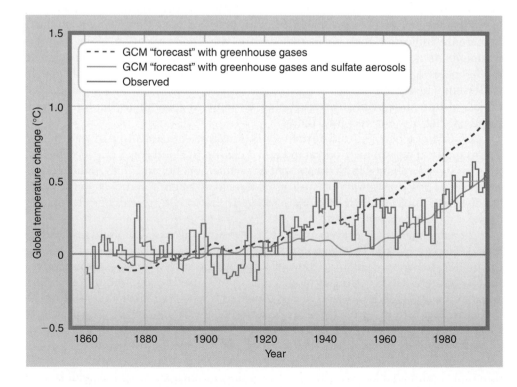

consistently predict that the warming will be greater at the poles than the tropics, and that continents will warm more than oceans. While the troposphere is expected to warm, the stratosphere is predicted to cool. Temperature changes in the troposphere are expected to be greatest in the Northern Hemisphere winter. In addition, the land nighttime air temperatures are expected to rise faster than the daytime temperatures. Recent analysis of temperature observations generally supports these predictions.

GCMs are relatively new, imperfect, and certainly in need of improvement. For example, most GCMs are only now "learning" how to properly include the role of spheres other than the atmosphere. These models are generally designed for global use and do not accurately portray local conditions. Nevertheless, it is likely that future climate change and fluctuations will result from a complex interaction of differing factors—and GCMs will help us unravel the relationship between cause and effect.

Blue Skies

**Atmospheric Chemistry/
Temperature Trends**

Climate Change Assessment

The natural variability of climate over the last few centuries has had a marked effect on human activities, such as travel, health, building structure, art, agriculture, literature, and history. Think about how weather touches your life. Does it influence how you dress or prepare for an outdoor activity? Have you ever been delayed at an airport because of the weather? How often does bad weather close a school? Do you plan your vacations around weather conditions? On longer timescales, a change in climate might also entail a change in lifestyle.

With an increased awareness of the potential impacts on our quality of life, there have been efforts to determine how climate change will affect society. The United Nations Environment Programme and the World Meteorological Organization formed the Intergovernmental Panel on Climate Change (IPCC) in 1988. The IPCC has been directed to assess our confidence in scientists' understanding of climate change and its impact on global and regional scales. We conclude our study of climate change by considering its potential impact on three areas relevant to our society: health, coastal conditions, and forests.

Climate Change and Health

Global warming brings with it the potential for negative health impacts. These include the poleward spread of tropical illnesses and heat waves as well as health issues associated with extreme weather events.

The impact of climate change on health is highly uncertain, but important to explore. With knowledge of the potential impacts and identification of vulnerable regions, public health systems can respond to changing climate conditions and thus help to protect the population from adverse health impacts. Of course, the costs and benefits must be considered.

Increases in heat waves appear likely with the warmer climate of the coming century. As discussed in Chapters 4 and 9, heat waves in urban areas have been associated with increases in mortality. The number of deaths attributed to summer heat waves for six cities during the period from 1964 to 1991 is depicted in Figure 15.16. Note that although Atlanta is hotter than the other cities, it has fewer heat-related deaths. This is because citizens of cities that are normally hotter are accustomed to dealing with extreme heat.

Figure 15.16 also shows a prediction of heat-related mortality rates for the year 2020. This prediction is based on both climate model forecasts and a statistical relationship (derived from historical evidence) between heat waves and increased

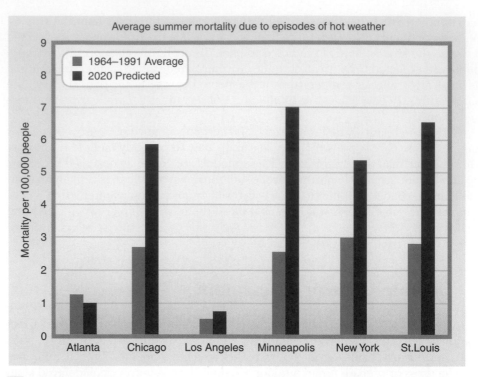

Figure 15.16

Deaths attributed to summertime heat waves in six U.S. cities for the years 1964 to 1991, and a projection of heat-related deaths in 2020 based on climate change estimates. (Source: Adapted from Kalkstein and Green, "An Evaluation of Climate/Mortality Relationships in Large US Cities and the Possible Impacts of Climate Change," Environmental Health Perspectives, *105(1), pp. 84–93.)*

death rates. Thus, this prediction may be extreme, as heat-related illnesses associated with increased heat indices can be prevented through behavioral and technical adaptations as simple as increased air conditioning in the Northeast. Other adaptation methods can include improved heat warnings and improved community-wide emergency plans.

Deaths attributed to weather are normally associated with extreme events such as storms and flooding. Climate changes may change the number of occurrences and the duration of extreme weather events. For example, increases in heavy precipitation have been observed over the past century. Will these increases continue? No one is sure. Climate prediction of precipitation is much less certain than predictions of temperature. This is because precipitation formation is a very complex process (see Chapter 4). Some climate models project a tendency toward increased precipitation in some regions. This tendency poses a risk of floods, the impact of which is difficult to assess. Planning for and responding to natural disasters can be a key to mitigate the risk of death and injury and economic loss due to natural disasters.

Climate Change and Coastal Regions

Over 50% of the total population of the United States lives along coastal regions, and these areas are the fastest-growing regions in the nation in terms of population. Over the next 25 years, the population along the U.S. coasts will increase by about 18 million people.

With this growth, the vulnerability to natural hazards is a constant threat. Currently, economic losses along coastal regions are about $50 billion annually, about ten times greater than in 1970. The potential increase in storms due to climate change may pose a threat to coastal population. This is a particular threat as global warming leads to higher sea levels.

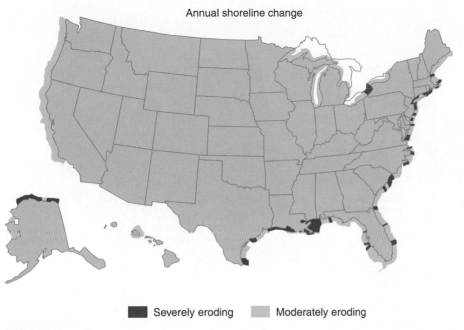

Annual shoreline change

■ Severely eroding   ■ Moderately eroding

■ **Figure 15.17**

A preliminary assessment of the potential of shoreline erosion of the United States. The degree of erosion is associated with changes in sea level, the frequency of storms, and the condition of the shoreline. (Source: http://www.gcrio.org/NationalAssessment/overpdf/overview.html, p. 113)

Global sea level has risen approximately 10 to 20 centimeters (4 to 8 inches) over the last 100 years. Over the next 100 years, the sea-level change is expected to continue. Present estimates of future sea level rises range between 20 and 61 centimeters (8 and 24 inches). These predictions are viewed as reliable because the basic causes—expansion of seawater and melting of ice sheets—are fairly well understood.

An increase in global sea level is expected to have a large impact on coastal erosion. Low-lying coastal regions are most vulnerable to changes in sea level. Figure 15.17 is a preliminary assessment of the shoreline erosion of the United States. The Atlantic and Gulf Coast coastlines are particularly susceptible. These regions have gentle slopes and a rise in sea level results in a large change in the shoreline. A change in the frequency of storms along the coast, such as hurricane storm surges (see Chapter 8), can have a large impact on coastal erosion and flooding, putting lives and property at risk. Undeveloped shorelines can also be quickly eroded during storm surges.

The Pacific Coast may also be affected. For example, beach and cliff erosions increase during El Niño years. During El Niño events sea level rises and more storms hit the Pacific Coast, resulting in more erosion. A long-term rise in sea level coupled with altered storm tracks may enhance this erosion. With the projected increase in coastal populations, climate change is likely to add to any human-caused stresses. Coastal management, including stringent land-use policies, may lessen these stresses.

■ Climate Change and Forests

An increase in atmospheric carbon dioxide levels generally results in higher photosynthesis rates. This is predicted to increase forest growth and productivity. However, the increases may be limited by other conditions, such as the availability of

Dominant Forest Types

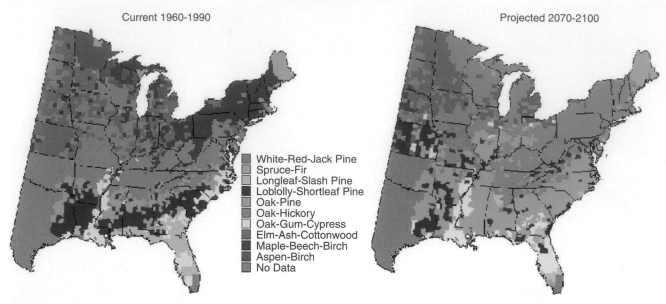

Current 1960-1990 · Projected 2070-2100

- White-Red-Jack Pine
- Spruce-Fir
- Longleaf-Slash Pine
- Loblolly-Shortleaf Pine
- Oak-Pine
- Oak-Hickory
- Oak-Gum-Cypress
- Elm-Ash-Cottonwood
- Maple-Beech-Birch
- Aspen-Birch
- No Data

 Figure 15.18

The current (A) and end-of-21st-century (B) projected forest types for the eastern United States. The distributions are determined by temperature, moisture, and soil conditions. (Source: http://www.gcrio.org/NationalAssessment/overpdf/overview.html, pp. 116–117.)

moisture and nutrients. As discussed in Chapter 14, tree species have adapted to different climate conditions. Trees have also adapted to natural threats, such as fires, insects, droughts, and windstorms. If climate changes, the trees will have to adapt, migrate, or die.

Figure 15.18 depicts today's forest types in the eastern half of the United States. These forests have adapted to their environments' temperature and moisture conditions. In the near future, these forests may face rapid changes in their environment. How will they respond?

In a warmer climate it seems logical that tree species that prefer cooler conditions will shift north. However, trees cannot pull up their roots and migrate. How a given species adapts to environmental changes will depend on its growth rate and its ability to disperse seeds. If the environmental changes are faster than a tree's ability to propagate northward, the species will have difficulty adapting to its new local climate. In addition, the climate change may also impact the frequency of fires, storms, droughts, and pest outbreaks, all of which will impact the rate of transition from one species to another. Figure 15.18 shows one projection of forest distributions at the end of this century. Pine forests decline in the Southeast, while oak forests expand northward. The maple-beech-birch forest types no longer exist in the eastern United States by 2100, according to this scenario.

Are these predictions alarmist? Possibly. However, they are not any more radical than the changes in forests that past climates research (Chapter 14) has revealed for the millennia following the last ice age. For example, ten thousand years ago northern spruce forests were found in the mountains of the Southeast. The difference this time is that humans are influencing the climate change, and the changes may be so rapid that our biosphere is unable to respond.

Closure

We have come to the end of this text, but hopefully it is only the beginning of a lifetime of exploring the atmosphere on your part. Fears of global warming

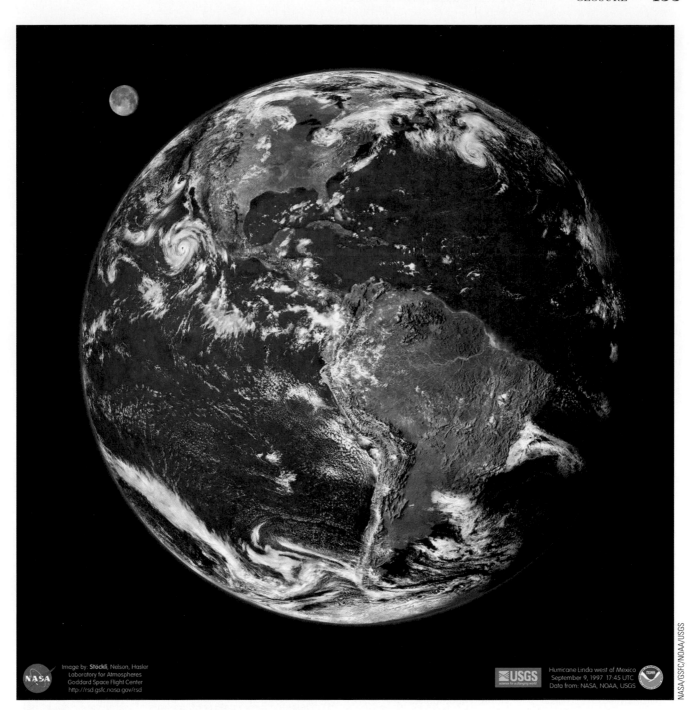

Image by: **Stöckli**, Nelson, Hasler
Laboratory for Atmospheres
Goddard Space Flight Center
http://rsd.gsfc.nasa.gov/rsd

NASA

USGS
science for a changing world

Hurricane Linda west of Mexico
September 9, 1997 17:45 UTC
Data from: NASA, NOAA, USGS

NASA/GSFC/NOAA/USGS

Figure 15.19

This is not an astronaut's photograph of Earth! Instead it is a digital image based on a variety of satellite data from geostationary and polar-orbiting satellites that observe clouds, oceans, and topography. The mountain peaks and valleys have been exaggerated by a factor of 50 so that the topography is clearly visible. Through such tricks, satellite images are even more informative than photographs.

notwithstanding, meteorology is a fascinating subject that is simultaneously mysterious and literally right in front of you. We hope this book has opened your eyes to the many ways we observe the atmosphere (Figure 15.19). Keep looking up at the sky! Finally, we hope that you can use this book to explain what you see and hear and read about in the vast world of weather and climate.

http://info.brookscole.com/ackerman

PUTTING IT ALL TOGETHER

SUMMARY

Recent observations indicate that global surface temperatures are warming. This warming has coincided with rapidly increasing amounts of atmospheric greenhouse gases. These gases play a crucial role in Earth's climate by affecting the energy budget. Human activities are increasing the concentrations of greenhouse gases. Therefore, humans may now be influencing and changing global climate.

Humans have been altering weather and climate on local and regional scales for centuries. Pollution in the form of lead, carbon monoxide, and tropospheric ozone causes health problems and photochemical smog. Oxides of nitrogen and sulfur are primarily to blame for the development of acid rain, which kills trees and aquatic life. The release of chlorofluorocarbons, in combination with stratospheric wind and cloud patterns and chemical reactions, led to the ozone hole over Antarctica. In addition, by diverting rivers and building cities humans have changed local and regional precipitation and temperature patterns markedly. These examples reveal that feedback processes within the atmosphere and connecting the atmosphere with the Earth's surface can cause relatively small initial changes in the atmosphere to amplify into significant climate changes.

The global warming debate is extremely complex. This is largely because scientists are still learning how the atmosphere interacts with water (water vapor, clouds, and the oceans), especially in the situation of rising temperatures due to increasing amounts of carbon dioxide and other "greenhouse gases." Global climate models (GCMs), the very-long–range counterparts to the numerical weather forecast models we studied in Chapter 13, are used to try to predict future climate. Climate simulations using these models indicate that the Earth is warming, and it will continue to warm over at least the next fifty years. Based on the results of different GCMs, the amount of this warming is uncertain. The predictions continue to be refined as we improve our understanding of the atmosphere and its relationship to the oceans and the biosphere. Despite the models' imperfections, the majority of the world's atmospheric scientists tend to concur with the assessment that global warming is, or soon will be, a reality.

Global warming is a pressing topic of international importance because of the potential consequences of significant worldwide warming. Global warming could seriously impair, for example, human health, coastal populations, and forests. For this reason, research on global climate change has become in recent years a dominant part of the field of meteorology.

KEY TERMS

You should understand all of the following terms. Use the glossary and this chapter to improve your understanding of these terms.

Acid deposition
Acid rain
Carbon monoxide (CO)
Contrails
Desertification

Feedback
Global climate models
   (GCMs)
Hydrocarbons

Ice-albedo temperature
   feedback
Lead
Negative feedback
   mechanism
Nitric oxide (NO)
Nitrogen dioxide (NO_2)
Ozone (O_3)
Particulates

Polar stratospheric clouds
   (PSCs)
Positive feedback mechanism
Smog
Sulfur dioxide (SO_2)
Sulfuric acid (H_2SO_4)
Sulfur trioxide (SO_3)
Urban heat island
Volatile organic compounds
   (VOCs)

REVIEW QUESTIONS

1. Why do we use unleaded gasoline in cars today?
2. A Fortune 500 company saw its stock drop rapidly in value because a few worries about its financial health were reported publicly; the drop in stock value generated more worries that were then confirmed, leading to still more investor concern and a massive sell-off of its stock. Is this an example of a positive feedback mechanism or a negative feedback mechanism, and why?
3. How is photochemical smog formed?
4. Using the *Blue Skies* CD-ROM exercise "Atmospheric Chemistry/Smog," examine how the production of tropospheric ozone is related to levels of nitrogen oxides (NO_x, pronounced "Knox") and VOCs. How do concentrations of these pollutants differ from Atlanta to Los Angeles?
5. Overall, is pollution increasing or decreasing across the United States? Why is this change occurring?
6. Do atmospheric scientists expect the ozone hole to get better or worse during the next century? What do they base their predictions on?
7. The city of Los Angeles diverted water from Tulare Lake in California, once the largest freshwater lake west of the Mississippi River, in order to satisfy the city's increasing need for water. The lake dried up completely. How do you think the climate in the vicinity of the lake might have changed as a result?
8. Why are the average temperatures of cities often greater than the surrounding rural region?
9. Can you think of other differences in weather between the city and its rural surroundings?
10. Describe how increased amounts of water vapor could help accelerate global warming. Describe how increased amounts of low clouds could help slow down global warming. How might increased amounts of high clouds affect global warming, and why?
11. Describe how a slight increase in the horizontal extent of glaciers could cause glaciers to expand.
12. What is a global climate model? How is it the same as a numerical weather forecast model? How is it different?
13. Why are climate models used to study future climate change?
14. What is the consensus of global climate models regarding the trend in future global temperature?
15. Using the *Blue Skies* CD-ROM exercise "Atmospheric Chemistry/Temperature Trends," view the movie of global

climate model simulation under the assumption of a doubling of carbon dioxide concentrations. Is global warming actually *global* during the early years of this model simulation, or do some regions cool? In what latitude band does the model show the most consistent trend toward warming? Why do you think this latitude band is so sensitive to greenhouse gas increases?

16. Describe three ways in which a globally warmed world might be different than today's world.

WEB ACTIVITIES

Choose Chapter 15 on the textbook's Web site:
http://info.brookscole.com/ackerman
and select from the following resources:

- Interactive Modules that illustrate and extend your understanding of key topics in this chapter
- Tutorial Quizzes to test your mastery of terms and concepts

For additional readings, go to the InfoTrac College Edition, your online library, at:
http://www.infotrac-college.com

Glossary

Absolutely unstable atmosphere Describes an atmospheric layer with an environmental lapse rate greater than 10° C per kilometer (the dry adiabatic lapse rate). **(3)**

Acceleration The rate at which velocity changes over time. **(6)**

Accretion The growth of an ice particle by collision and coalescence of supercooled water droplets. **(4)**

Acid deposition The falling of acids and acid-forming compounds from the atmosphere to the Earth's surface. **(15)**

Acid rain Acid deposition that occurs in the form of rain. **(15)**

Active sensors Remote sensing instruments that determine atmospheric conditions by sending out energy and measuring the energy that comes back to them. **(5)**

Adiabatic process A process in which a system does not exchange heat with its surroundings. **(2)**

Advection The process by which a property of the atmosphere is transferred by the wind. Usually, meteorologists use advection to refer to horizontal transport of atmospheric properties, such as temperature and moisture. **(2)**

Advection fog A type of fog formed when warm, moist air is blown over a cooler surface and is cooled to the dew point. **(4)**

Aerosols Tiny solid or liquid particles that are suspended in the atmosphere. **(1)**

Aggregation The process of ice-crystal growth by collision and adherence of ice crystals. **(4)**

Air mass A large body of air that has uniform weather characteristics, particularly temperature and humidity. **(9)**

Air-mass modification The process by which the characteristics of an air mass change as it moves away from its region of origin. **(9)**

Air-mass thunderstorm. *See* **Ordinary thunderstorm. (11)**

Albedo The percentage of light reflected by an object when it is illuminated. **(2)**

Alberta Clippers Small, fast-moving extratropical cyclones that originate in western Canada and move south and east across the United States, especially in winter. **(10)**

Altocumulus (Ac) Clouds in the mid-troposphere that are gray or white and that occur as layers or patches. If these clouds have vertical development that resembles towers or turrets, they are called *altocumulus castellanus*. **(4)**

Altostratus (As) Layered clouds in the mid-troposphere that are gray or bluish. **(4)**

Amplitude Half the height from the crest of a wave to its trough. **(2)**

Analog forecast An empirical method of forecasting that uses past weather events that resemble the current conditions to create forecasts for the days ahead. **(13)**

Anemometer An instrument that measures wind speed. **(5)**

Aneroid barometer An instrument that measures atmospheric pressure via changes in the size of a partially evacuated metal box. **(5)**

Angle of inclination The tilt of the Earth with respect to the imaginary plane linking the centers of the Sun and Earth. The term also applies to orbits of other planets and satellites. **(2)**

Angular momentum The product of mass, rotation velocity and perpendicular distance from the axis of rotation. **(7)**

Annual average temperature The average of the monthly mean temperatures. **(3)**

Annual temperature cycle The pattern of change of temperature during the twelve months of the year. **(3)**

Annual temperature range The difference between the values of the warmest and coldest monthly mean temperatures in a year. **(3)**

Anomalies   Departures from typical or normal values. (**3**)

Anthropogenic   Resulting from human activities. (**1**)

Anticyclogenesis   The birth of an anticyclone. (**10**)

Anticyclones   Regions of high pressure. Winds around an anticyclone blow clockwise in the Northern Hemisphere, but counterclockwise in the Southern Hemisphere. (**6**)

Anvil   The upper portion of a mature or dissipating thunderstorm (cumulonimbus) cloud that is flattened on its top side. (**4**)

Aphelion   The location in a planet's orbit when it is farthest away from the Sun. (**2**)

Arctic Oscillation   A sea-surface temperature oscillation of the North Atlantic that affects weather and climate over large parts of the globe. Also called *North Atlantic Oscillation.* (**8**)

ASOS   Automated Surface Observing System. ASOS is the United States' primary network for observing surface weather. (**5**)

Aspect   The compass direction that a land slope faces. (**3**)

Atmospheric window   A narrow range of wavelengths in which the atmosphere absorbs very little of the Earth's emitted infrared energy. The best-known atmospheric window occurs between 10 and 12 microns. (**2**)

AVN   A spectral numerical weather forecast model that is used for short- to medium-range global forecasts. (**13**)

Baroclinic instability   The process through which extratropical cyclones get energy for their growth. Warm air rising and cold air sinking in tilted frontal regions are the source of this energy. (**10**)

Barometer   A device used to measure atmospheric pressure. (**1**)

Barometric pressure   The pressure exerted by column of air as a result of gravitational attraction. (**1**)

Berg wind   A hot, dry wind blowing off the plateau of South Africa. (**12**)

Bergeron–Wegener process   The process by which ice crystals grow more rapidly than water droplets in a cold cloud because of differences in the saturation vapor pressures of ice and liquid water at the same subfreezing temperatures. (**4**)

Bermuda high   A subtropical anticyclone over the North Atlantic Ocean that steers tropical cyclones toward the west and northwest in summer and early fall. (**8**)

Blackbody   A hypothetical object that is a perfect absorber and emitter of radiation. (**2**)

Blocking patterns   Develop when cut-off wind patterns stop or divert the normal eastward progression of weather systems. (**7**)

Blue norther   A fierce north wind behind a fast-racing cold front in western Texas that is accompanied by clear blue skies. (**12**)

Bora   A term used to describe a katabatic wind along the Adriatic coast of Europe. (**12**)

Boreal forest   A forested region, usually a coniferous forest, that adjoins the tundra at the arctic tree line. (**14**)

Boulder windstorms   Downslope winds associated with mountain gravity waves that can cause particularly ferocious windstorms in the vicinity of Boulder, Colorado. (**12**)

Brocken bow   A colored circular region around shadows, which is caused by diffraction by water droplets. (**5**)

Buran   A strong northeast wind in Russia and central Asia. (**12**)

Buys Ballot's Law   Dictates that low pressure lies on the left-hand side of the wind direction in the Northern Hemisphere. Buys Ballot's Law is a consequence of geostrophic balance. (**6**)

Calorie   A unit representing the amount of energy required to increase the temperature of 1 gram of water by $1°$ C. (**2**)

Cape Verde hurricane   An especially dangerous type of tropical cyclone that develops out of an easterly wave near the Cape Verde Islands, just off the western coast of Africa. (**8**)

Capping inversion   A temperature inversion above the Earth's surface that initially forestalls the development of thunderstorms. If the air beneath the inversion is hot and moist enough to rise above the "cap," explosive thunderstorm development results. (**11**)

Carbon dioxide (CO_2)   A colorless, odorless non-toxic trace gas found throughout the atmosphere. It plays a key role in the greenhouse effect, and is increasing in concentration due to the burning of fossil fuels. (**1**)

Carbon monoxide (CO)   A colorless, odorless, toxic trace gas found in the atmosphere at different levels of concentration in connection with air pollution. (**15**)

Ceilometer   A device used to measure the height of cloud bases. (**5**)

Cell   The organizing unit of a thunderstorm, it is a compact region of a cloud that contains a strong updraft. (**11**)

Celsius   A temperature scale in which the freezing point of water at sea level occurs at $0°$ C and the boiling point occurs at $100°$ C. (**2**)

Centrifugal force   An apparent force felt by objects in a turning frame of reference. The force seems to push the objects outward from the center of the turn. (**6**)

Centripetal acceleration   The change of velocity of an object moving in a curved path. This acceleration is directed toward the center of the curved path. (**6**)

Chaos theory   A theory that demonstrates that in a complex system such as the atmosphere, small differences at the beginning of an event can lead to large differences later. **(13)**

Chinook   A dry, warm wind in western North America on the lee side of the Rocky Mountains. The chinook speeds the melting and evaporation of snow. **(12)**

Chlorofluorocarbons (CFCs)   Anthropogenic compounds invented by chemists in 1928 and used as propellants in spray cans, as Styrofoam™ puffing agents, and as coolants for refrigerators and air conditioners. CFCs are primarily responsible for the development of the ozone hole in the stratosphere and also play a role in the enhancing greenhouse effect. **(1)**

Cirrocumulus (Cc)   Thin, rippled convective high clouds. **(4)**

Cirrostratus (Cs)   Layered high clouds that are whitish and usually fibrous, but sometimes smooth. Cirrostratus often produce optical effects such as halos. **(4)**

Cirrus (Ci)   High clouds composed of ice crystals that are whitish and can appear as filaments, patches, or narrow bands. **(4)**

Climate   A representation of a region's weather over a given period of time. A region's climate often is characterized in terms of the average and variation of the climate system over periods of a month or more. **(1)**

Climatology   The description and study of climate. **(1)**

Climatology forecast   A forecast based upon the climatological statistics of a region. **(13)**

Climographs   Graphs that depict climatological information. Climographs usually consist of two climatic elements (e.g., temperature and precipitation) plotted through an annual cycle. **(14)**

Cloud   An ensemble of water drops and/or ice particles in the atmosphere above the Earth's surface. **(1)**

Cloud ceiling   The height of the base of the lowest widespread cloud in the sky. **(5)**

Cloud droplet   A spherical particle of liquid water in a cloud. Cloud droplet sizes range from a few microns to a few tens of microns. **(4)**

Coastal front   The boundary between two air masses along a coastline that acts as a smaller-scale version of a stationary front. **(12)**

Cold-air damming   The stubborn entrenchment of cold air, particularly in and near mountainous regions such as the Appalachian Mountains **(12)**

Cold front   A type of front in which a colder air mass replaces a warmer air mass. **(1)**

Collision-coalescence   The process by which precipitation forms in warm clouds. Drops of different sizes collide and merge, leading to rapid growth into a raindrop. **(4)**

Columns   A type of ice crystal that has hexagonal bases on either end. **(4)**

Comma cloud   A feature of extratropical cyclones observed in satellite pictures in which the cloud system resembles a comma, with a large rounded "head" near the center of the low and a trailing thin "tail" along the cold front. **(10)**

Condensation   The change of phase of water from the vapor to the liquid state. **(1)**

Condensation nuclei   Aerosol particles that assist in forming liquid droplets. **(4)**

Conduction   The transfer of energy from one object to another as a result of the random motions of molecules. Conduction of heat is a consequence of the temperature differences between two objects in physical contact. The transfer always is from warmer to colder objects. **(2)**

Conservation of angular momentum   The principle that the angular momentum of a spinning body (the product of spin and distance from the axis of rotation) must remain constant. Conservation of angular momentum is usually accomplished by increasing (decreasing) spin when the distance from the axis of rotation is decreased (increased). **(7)**

Contact nucleation.   *See* **Nucleation. (4)**

Contrails   Cloudlike streamers that form behind high-flying aircraft. **(15)**

Convection   The transfer of energy by the movements of masses in a liquid or a gas. In meteorology, this term is usually reserved for vertical motions. **(2)**

Convective clouds   Clouds that result from convection. As a result, convective clouds usually exhibit vertical development. **(4)**

Convergence   The horizontal coming-together of air that can lead to lifting of air at the surface. **(4)**

Coriolis force   An apparent force used to describe the observed deflection of moving objects caused by the observer's moving frame of reference. The magnitude of the Coriolis force is proportional to the wind speed and the latitude. It always acts to the right of the wind in the Northern Hemisphere. **(6)**

Corona   A whitish or colored band around the Sun or Moon, which is caused by diffraction of light by water droplets. It differs in appearance from the halo because the corona, unlike a halo, appears to "touch" the Sun or Moon instead of being a thin ring around the light source. **(5)**

cP   The continental polar air mass, which is a cold and dry air mass. **(9)**

Crepuscular rays   Beams of light from the Sun caused by a combination of shadows and scattering. These light and dark bands usually occur at twilight. **(5)**

Critical angle The angle at which refracted rays of light do not escape the substance they are traveling in, but instead move along the surface between two substances.

Crystal habit One of several characteristic shapes of ice crystals. (**4**)

cT The continental tropical air mass, which is in summer a hot and dry air mass. (**9**)

Cumulonimbus (Cb) Precipitation-producing thunderstorm clouds with flattened anvil-shaped tops. (**4**)

Cumulus (Cu) Low convective clouds that develop as individual, detached elements with sharp outlines. A cumulus cloud has a flat base and bulges upward. (**4**)

Cumulus stage The initial stage in the life cycle of a thunderstorm. (**11**)

Cup anemometer An instrument for measuring wind speed that uses small cups to catch the wind. (**5**)

Curvature effect A greater surface curvature increases the rate of evaporation and inhibits cloud droplet growth at the smallest droplet sizes. (**4**)

Cut-off cyclone An aging extratropical cyclone that has become separated from the strong jet-stream winds and from the surface temperature gradients. (**10**)

Cyclogenesis The development of a cyclone. Surface temperature gradients, jet streams, and tall mountain ranges can lead to extratropical cyclogenesis. (**10**)

Cyclones Low-pressure regions around which winds blow counterclockwise in the Northern Hemisphere and clockwise in the Southern Hemisphere. (**6**)

Daily mean temperature The average of the daily maximum and minimum temperatures. (**3**)

Dart leader Cloud-to-ground electrical discharges that follow the initial lightning stroke. (**11**)

Data assimilation The combining of observed data into a numerical forecast model. Tasks of data assimilation include interpolation of scattered data into a regularly spaced pattern and initialization to remove small-scale "bumps" in the data. (**13**)

Data initialization The "smoothing out" of errors and inconsistencies in the initial data that is to be ingested by a numerical weather forecast model, which leads to more accurate weather forecasts. (**13**)

Deep zone The bottom layer of the ocean, which lies below 1000 meters and is uniform in temperature. (**8**)

Dendrites A type of ice crystal that is a flattened hexagon with extended "fingers" or branches at each vertex. Dendrites most closely resemble the classic shape of snowflakes. (**4**)

Dendrochronology The study of tree rings to determine past climates. (**14**)

Density The ratio of the mass of a substance to the volume of space it occupies. (**1**)

Deposition The process by which water vapor in subfreezing air changes into ice. (**2**)

Deposition nucleation The creation of an ice crystal around a ice nucleus that is supersaturated with respect to ice when the water vapor around it changes directly to the ice phase. (**4**)

Deposition nuclei A solid aerosol particle that nucleates an ice crystal from the vapor phase. (**4**)

Derechos Hours-long windstorms associated with a line of severe thunderstorms. Derechos are the result of straight-line winds, not the rotary winds of a tornado. They are often associated with bow-shaped echoes on weather radar. (**12**)

Desertification An increase in the desert conditions of a region. Also applied to the spreading of a desert region as a result of a combination of climate change and human impacts on the land. (**15**)

Dew Liquid water that condenses on objects near the ground when they are cooled to the dew point. (**4**)

Dew point depression The number of degrees difference between the temperature and the dew point temperature. (**4**)

Dew point hygrometer An instrument that measures the dew point temperature by interpreting changes in the reflection of light off of a mirror. (**5**)

Dew point temperature The temperature to which air must be cooled at constant pressure and constant water vapor content to become saturated. (**4**)

Diffluence The spreading-out of horizontal wind direction that often leads to divergence. (**10**)

Diffraction The bending of light around tiny objects that produces patterns of light and shadow, including colored light. (**5**)

Diffuse The action of gas and liquid molecules spreading from areas of higher concentration to areas of lower concentration. (**1**)

Direct methods Observations made by instruments that are in physical contact with the region being measured. (**5**)

Dispersion The process in which white light is separated into its component colors. (**5**)

Dissipating stage The final stage in the life cycle of a thunderstorm. (**11**)

Diurnal Daily, pertaining to actions completed within or that recur every 24 hours. (**2**)

Diurnal temperature cycle The pattern of temperature during the 24-hour period from one midnight to the next. (**3**)

Diurnal temperature range The difference between a day's maximum and minimum temperatures. (**3**)

Divergence The horizontal spreading-out of air that can lead to sinking motions if it occurs at the surface, but which causes rising motion in the lower troposphere if it occurs at jet-stream level. **(4)**

Doldrums Regions of nearly calm winds near the equator. **(7)**

Doppler effect The change in the frequency of a wave pattern caused by the relative motion of the wave emitter and the observer of the wave. **(5)**

Doppler radar A radar that indirectly measures wind speed by detecting the change of frequency that results from the Doppler effect on radar waves that hit moving precipitation particles. **(5)**

Drizzle Liquid precipitation smaller than 0.5 millimeter in diameter. **(4)**

Dry adiabatic lapse rate The change in temperature of a parcel as it rises or descends in the atmosphere, equal to about 10° C per kilometer. **(2)**

Dry climates (B) Climate regions where potential evaporation and transpiration exceeds precipitation. **(14)**

Dryline A type of frontal zone characterized by strong horizontal contrasts in moisture. **(9)**

Dry slot A mostly clear region of an extratropical cyclone that separates the comma cloud head from the comma tail on satellite images. **(10)**

Dust devils Small, vigorous whirlwinds that occur when intense daytime heating helps spin up thin rotating columns of air. **(12)**

Dust storms Weather conditions characterized by strong winds and dust-filled air over an extensive area. **(12)**

Easterly wave A region of clouds and rain associated with a wave-like pattern in tropospheric winds that moves from east to west across tropical regions. A few easterly waves later develop into tropical cyclones. **(8)**

Eccentricity The deviation of an ellipse from a perfect circular shape. **(14)**

ECMWF A spectral numerical weather forecast model that is used for a wide range of forecasts. **(13)**

Eddies Swirls in a fluid, such as air, that interact with the larger-scale wind and help slow it down. **(12)**

Ekman spiral In oceanography, the progressive turning of ocean currents from the surface down to 100 meters because of the combined effect of wind on the sea surface and the Coriolis force. In meteorology, the description of the wind distribution in the atmospheric boundary layer. **(8)**

Ekman transport The effect of the Ekman spiral to move water masses at a right angle to the direction of the surface wind. In the Northern Hemisphere, this movement of water is to the right of the surface wind direction. **(8)**

Electromagnetic energy. *See* **Radiation.**

El Niño An extensive warming of the near-surface water of the equatorial Pacific Ocean between South America and the Date Line. This warming occurs in connection with changes in the atmosphere over the same region, which are known collectively as the *Southern Oscillation.* **(9)**

Energy The capacity to do work. Energy must be conserved, though it can be converted from one form to another, such as from potential to kinetic energy. **(2)**

Enhanced greenhouse effect. *See* **Greenhouse warming.**

Ensemble forecasting A method of weather forecasting that uses the results of chaos theory to assess the amount of confidence that should be placed in a forecast. A forecast model is run repeatedly with slightly different initial conditions. If the resulting forecasts, the "ensemble," agrees closely, then confidence in the forecast is high. If the ensemble exhibits a wide range of different forecasts, then confidence in the forecast is low. **(13)**

Entrainment The mixing of environmental air into an existing air current or cloud. **(11)**

Environmental lapse rate The rate of decrease of temperature with altitude at a particular time and place. **(3)**

Equinoxes The time at which the Sun passes directly overhead at the equator at noon. **(2)**

Eta A gridpoint numerical weather forecast model that is used for short- and medium-range forecasts over the United States. **(13)**

Evaporation The change of phase of water from liquid water to water vapor. **(1)**

Evaporation fog A fog that occurs when water evaporates into cool, moist air and causes saturation. Steam fog and frontal fogs are two common types of evaporation fog. **(4)**

Extratropical cyclone A low-pressure system that forms outside of the tropics and is usually associated with fronts, unlike a tropical cyclone. **(8)**

Eye The usually clear area of lowest pressure at the center of a strong tropical cyclone. **(8)**

Eye wall The circular region of strong thunderstorms immediately surrounding the eye of a strong tropical cyclone. **(8)**

Fahrenheit A temperature scale in which the freezing point of water at sea level occurs at 32° F and the boiling point occurs at 212° F. **(2)**

Fall wind A term used to describe a katabatic wind in Greenland and Antarctica. **(12)**

Fallstreaks Ice particles that fall from a cloud but evaporate above the ground. **(4)**

Fata Morgana A type of superior mirage in which objects near the horizon are distorted vertically into castle-like shapes. **(5)**

Feedback A sequence of interactions in a system where one change leads to some other change, which can act to either reinforce (**positive feedback**) or inhibit (**negative feedback**) the original change. **(15)**

Flash flood A rapidly developing type of flood usually associated with very intense rainfall at or near the site of the flood. **(11)**

Flood A substantial rise in water that covers areas that are usually not under water. **(11)**

Foehn A warm, dry, downslope mountain wind that descends the lee side of the Alps. **(12)**

Fog A stratus cloud that is in contact with the ground. **(4)**

Folklore forecasts Short sayings that attempt to predict the weather based on sky conditions, special days of the calendar, or the behavior of animals. **(13)**

Force The mass of an object multiplied by the change in its speed and/or direction, which is its acceleration. **(2)**

Freezing nucleation. *See* **Nucleation.** (4)

Freezing rain Supercooled rain that freezes on contact with subfreezing objects. **(4)**

Frictional force The resistive force caused by wind blowing over the Earth's surface. It always acts to oppose the wind. **(6)**

Front The transition zone between two air masses of different density. **(1)**

Frontal lifting The forced lifting of warm, less-dense air over colder air in the vicinity of a front. **(4)**

Frontal wave An early stage of the life cycle of an extratropical cyclone that is characterized by an undulation along a previously stationary front, with a developing cold front on its back side and a warm front on its front side. **(10)**

Frost Ice that forms on exposed objects by deposition when an object is cooled to or below its frost point. **(4)**

Frost point The highest temperature at which atmospheric moisture will form frost. It is analogous to the dew point, but is applied when saturation occurs below freezing. **(4)**

Fujita scale A rating system designed to estimate the speed of tornado winds based on the patterns of the damage caused by the tornado. **(11)**

Funnel cloud A tornadic circulation that looks like a narrow cone beneath a thunderstorm but which does not reach the ground. **(11)**

Gap winds Mountain winds that blow through low points, or gaps, in the mountains, creating channels of wind that can easily exceed 160 km/hr (100 mph) in the strongest cases. **(12)**

GEO satellite A satellite in a geostationary earth orbit (GEO), circling the earth once every 24 hours. At an altitude of approximately 36,000 kilometers, the satellite appears stationary over a fixed point at the equator. Most weather satellites are in geostationary orbits. **(5)**

Geostrophic balance An equilibrium achieved when the horizontal pressure gradient and Coriolis forces push equally in opposite directions. **(6)**

Geostrophic wind The horizontal wind created by the balance of the horizontal pressure gradient and Coriolis forces. Although observed winds are rarely exactly geostrophic, the geostrophic wind is often a good first approximation to the direction and speed of observed winds in the midlatitudes. **(6)**

Glacier A mass of perennial ice that originates on land through the accumulation of snow. **(14)**

Global climate models (GCMs) Computer programs that calculate global climate using mathematical equations derived from physical principles such as the conservation of energy and Newton's Laws of Motion. Also called *general circulation models*. **(15)**

Glory Colored rings that appear around the shadow of an object, such as an airplane, and which are caused by diffraction of light by water droplets. **(5)**

Gradient balance A three-way balance of horizontal pressure gradient, Coriolis, and centrifugal forces. **(6)**

Gradient wind The horizontal wind that results from a three-way balance of horizontal pressure gradient, Coriolis, and centrifugal forces. Gradient winds in a curved low-pressure area are slower than the geostrophic wind (subgeostrophic), and are faster than the geostrophic wind (supergeostrophic) in a curved high-pressure area.

Graupel A kind of ice particle formed by accretion inside a cloud. **(4)**

Gravitational force The product of mass and gravitational acceleration. **(6)**

Gravity The mutual attraction between two or more objects. **(1)**

Gravity waves An alternating small-scale pattern of high and low pressure maintained with the help of gravity. Gravity waves are sometimes visible when the rising air in the crests in the waves becomes saturated and forms parallel lines of clouds. **(12)**

Green flash A momentary green light sometimes seen at sunrise or sunset resulting from refraction and dispersion of sunlight. **(5)**

Greenhouse effect The warming of the atmosphere that results from the fact that gases in the atmosphere absorb and emit the Earth's infrared radiation although they are transparent to sunlight. This warming is analogous to the effect of the clear glass in a greenhouse, which keeps the plants inside warmer than they would be otherwise. **(2)**

Greenhouse gases Gases in the atmosphere that are effective absorbers of infrared radiation and ineffective absorbers of solar radiation, such as carbon dioxide and water vapor. **(2)**

Greenhouse warming The possible heating of the planet over and above the natural greenhouse effect as a result of increases in atmospheric carbon dioxide. (**2**)

Grid A set of orderly arranged points on which variables are analyzed or predicted in a numerical weather forecast model. (**13**)

Gridpoint models A class of numerical weather forecast models that divides the atmosphere into grids.

Grid spacing The distance between one grid point and another in a numerical model. (**13**)

Guldberg–Mohn balance A three-way balance of horizontal pressure gradient, Coriolis, and frictional forces. The wind associated with this balance blows at an angle to the isobars from higher toward lower pressure. (**6**)

Gust front The boundary between the outflow of the cold downdraft of a thunderstorm and the warmer, more humid air around it. (**11**)

Gyre An ocean circulation that forms a closed loop that stretches across an entire ocean basin. (**8**)

Haboob A strong dust storm in northern and central Sudan. (**12**)

Hadley Cell A thermally driven circulation pattern comprised of rising air near the equator, poleward flow in the upper troposphere, sinking air over the deserts, and the trade winds. It is an important feature of global-scale winds. (**7**)

Hail Precipitation in the form of lumps of ice produced by thunderstorms. A *hailstone* is a single unit of hail. (**11**)

Hailshaft The region underneath a thunderstorm where hail is falling. (**11**)

Hailswath The path of the hailshaft along the ground which is created as a hail-producing thunderstorm moves. (**11**)

Halo A whitish or colored ring around the Sun or Moon that is produced by refraction of light by ice crystals. (**5**)

Harmattan A dry, dust-bearing wind that develops in winter when cool air from the Sahara Desert moves south and west and displaces warmer, more humid coastal air. (**12**)

Haze A suspension of small particles in the air. Haze reduces visibility by scattering light. (**5**)

Heat A form of energy transferred between systems because of the temperature differences between them. (**2**)

Heat advection The transfer of energy through the horizontal movements of the air. (**2**)

Heterogeneous nucleation. *See* **Nucleation.** (**4**)

Hexagonal plate A type of ice crystal that is a flattened hexagon without elongated branches. (**4**)

Highland climates (H) The climate of regions with high elevations. Highland climate zones are complex and affected by differences in latitude, altitude, and exposure to solar energy. (**14**)

Hira-oroshi A breeze that occurs along the mountainous shoreline of Lake Biwa, west of Tokyo, Japan. (**12**)

Historical climate The climate of the past several thousand years during which humans have kept a record of climate conditions. (**14**)

Homogeneous nucleation. *See* **Nucleation.** (**4**)

Hook echo A curved echo on a weather radar caused by heavy precipitation in a supercell thunderstorm. A hook echo is often an indication of a tornado. (**11**)

Horse latitudes A large region of light winds over the subtropical oceans. (**7**)

Hurricane A strong tropical cyclone found in the Western Hemisphere that has maximum sustained winds exceeding 65 knots (74 mph). *See also* **Tropical cyclone.** (**8**)

Hydrocarbons Compounds made of hydrogen and carbon atoms. (**15**)

Hydrologic cycle A complete description of how water moves between the atmosphere, water surfaces, and land in all three phases. (**1**)

Hydrophobic nuclei. *See* **Nuclei.** (**4**)

Hydrostatic balance An equilibrium achieved when the gravitational and vertical pressure gradient forces push equally in opposite directions. (**6**)

Hygroscopic nuclei. *See* **Nuclei.** (**4**)

Ice ages Periods of global cooling that leads to the creation of vast ice sheets across the Earth's land masses. (**14**)

Ice crystals The geometric organization of water molecules in the solid phase. Hexagonal plates, needles, columns, and dendrites are four different common forms of ice crystals. (**4**)

Ice nuclei. *See* **Nuclei.** (**4**)

Ice-albedo temperature feedback A type of positive climate feedback in which the cooling of the earth leads to the formation of ice sheets that, because of the increased albedo, further cools the planet and supports the continued growth of ice. (**15**)

Immersion nucleation. *See* **Nucleation.** (**4**)

Index cycle The oscillation of Rossby wave patterns between low-amplitude and high-amplitude phases over periods of several weeks. (**7**)

Index of refraction A measure of how optically dense a substance is. It is the ratio of the speed of light in a vacuum to the speed of light in the substance. (**5**)

Indirect methods Observation methods that do not require physical contact between the instrument and the region being measured. Also known as *remote sensing.* (**5**)

Inertia The resistance a body exhibits to any change in its motion.

Inferior mirage. *See* **Mirage.** (**5**)

Infrared satellite image A type of indirect observation that shows the amount of infrared energy emitted by the atmosphere and the Earth's surface as measured by a weather satellite. In an infrared satellite image, white regions are cold and emit less infrared energy than dark regions. **(5)**

Initial conditions The values of atmospheric variables (temperature, dew point, etc.) over a wide area at the starting time of a numerical weather forecast. **(13)**

Insolation The amount of solar radiation reaching the top of Earth's atmosphere. **(3)**

Interpolation A part of the numerical weather forecast process in which data from irregularly spaced observation sites are mathematically adjusted onto a regularly spaced grid for use in the numerical forecast model. **(13)**

Intertropical Convergence Zone (ITCZ) A convective region of thunderstorms separating the northeast and southeast trade winds. **(7)**

Iridescence Small regions of color in the sky, usually in connection with high thin clouds. **(5)**

Isobar A line on a map connecting regions with the same atmospheric pressure. **(1)**

Isobaric charts Maps that depict weather on constant pressure surfaces and include information on the temperature, wind speed and direction, humidity, and the altitude at a given pressure. **(6)**

Isopleth A line on a map connecting locations with the same value of a variable. **(1)**

Isotach A line on a map connecting locations with the same wind speed. **(1)**

Isotherm A line on a map connecting locations with the same temperature. **(1)**

Isotopes Different versions of the same element that have different atomic weights. **(14)**

Jet stream A narrow region of relatively strong winds (i.e., wind speeds greater than 70 knots) usually located in the upper troposphere. **(7)**

Joule A unit used to measure amounts of energy. One Joule equals 0.2389 calories. **(2)**

Katabatic wind A type of strong mountain breeze in which wind gusts can exceed 160 km/hr (100 mph). **(12)**

Kelvin A temperature scale in which the freezing point of water at sea level occurs at 273.16° K and the boiling point occurs at 373.16° K. **(2)**

Khamsin A hot, dry, dusty desert wind in Egypt and the Red Sea. **(12)**

Kinetic energy The energy an object possesses because of its motion. **(2)**

Kirchhoff's Law A law of radiation that states, objects that are good absorbers of radiation are also good emitters of radiation. **(2)**

Köppen scheme The most widely used climate classification scheme. Developed by Vladimir Köppen, it is based on annual and monthly means of precipitation and temperature. **(14)**

Lag The time delay between a cause and an effect. For example, typically the coldest winter temperatures lag several weeks behind the shortest day of the year, partly due to the thermal properties of the Earth and oceans.**(3)**

Lake breezes Winds that blow onshore during the day around large lakes. **(12)**

La Niña An extensive, below-normal cooling of the central and eastern tropical Pacific Ocean. La Niña is roughly the opposite of El Niño. **(9)**

Land breeze A wind that blows offshore from land to water during nighttime in the vicinity of large bodies of water. **(6)**

Lapse rate The rate at which temperature changes with increasing altitude. A positive lapse rate indicates that the temperature is decreasing with height. **(3)**

Latent heat The amount of heat taken in or released by water when it changes phase. **(2)**

Latent heat of condensation The energy per gram released when water vapor turns into liquid water. **(2)**

Latent heat of deposition The energy per gram released when vapor changes directly into ice. **(2)**

Latent heat of fusion The energy per gram released when liquid water freezes. **(2)**

Latent heat of melting The energy per gram absorbed when ice melts. **(2)**

Latent heat of sublimation The energy per gram absorbed when ice changes directly into vapor. **(2)**

Latent heat of vaporization The energy per gram absorbed when liquid water boils. **(2)**

Law of Inertia. *See* **Newton's First Law of Motion.**

Law of Momentum. *See* **Newton's Second Law of Motion.**

Layered clouds Clouds that are much wider than they are tall and form in relatively stable air. **(4)**

Lead A pollutant that is a particulate released into the atmosphere by industry and the burning of leaded gasoline. **(15)**

Leeward The downwind side of an object or region, such as a mountain. **(6)**

Lenticular clouds Clouds in the shape of a lens that sometimes resemble spaceships. They usually form downwind of mountainous regions in connection with small-scale winds above and around mountains. **(12)**

LEO satellite A satellite that orbits the poles in a low-Earth orbit about 850 kilometers above the Earth's surface. **(5)**

LFM An early gridpoint numerical weather forecast model that was used for short-range weather forecasts across North America. **(13)**

Leste A hot, dry, easterly or southeasterly wind that blows from Morocco to the Canary Islands. **(12)**

Levanter A gap wind in the Strait of Gibraltar. **(12)**

Leveche A hot, dry, dusty wind along the southeast coat of Spain. **(12)**

Lifted index A numerical estimate of the severity of thunderstorm activity based on observations near and above the ground. In simplified form, the lifted index is computed by lifting a hypothetical air parcel from the surface to the 500-mb level and then comparing the parcel's temperature to the 500-mb temperature. The more negative the lifted index, the more likely that any thunderstorms that form will be severe. **(11)**

Lifting condensation level (LCL) The level (altitude or pressure) to which air must be lifted dry adiabatically for condensation (or deposition) to occur. **(4)**

Lightning A visible electrical discharge that occurs between the ground and a cloud, between clouds, or within a cloud. **(11)**

Little Ice Age A time period between approximately 1400 and 1850 AD when average global temperatures were lower than at present and there was an expansion of mountain glaciers in the Alps, Alaska, Iceland, and Norway. **(14)**

Longwave radiation Radiant energy characterized by wavelengths primarily between 4 and 100 microns (μm). The Earth emits longwave radiationwith a maximum near 10 μm. **(2)**

Mammatus Pouches that form on the underside of a cloud, usually a thunderstorm anvil. **(4)**

Maunder Minimum The time between the years 1645 and 1715 in which the number of sunspots was dramatically lower than observed before or since. **(14)**

Mature stage The middle and most intense stage in the life cycle of an ordinary thunderstorm. **(11)**

Mercury barometer A device that measures air pressure by the level of mercury that is pushed up into an airless tube. **(5)**

Meridional flow pattern An upper-air flow pattern that occurs when strong Rossby waves cause winds to blow in a pronounced north–south direction. **(7)**

Mesocyclone A vertical column of rotating air within a severe thunderstorm. Tornadoes often form below a mesocyclone. **(11)**

Mesoscale Atmospheric motions with a spatial scale ranging from a few kilometers to several hundred kilometers. Examples are thunderstorms, fronts, tropical cyclones, and sea breezes. **(6)**

Mesoscale convective complex (MCC) A large thunderstorm complex made up of multiple single-cell thunderstorms. An MCC is identifiable on infrared satellite images by its nearly circular shape, which can cover entire states. **(11)**

Mesopause The top of the mesosphere where temperature stops decreasing with altitude. The mesopause is usually at about 80 to 85 kilometers above the Earth's surface. **(1)**

Mesosphere The region of the atmosphere above the stratosphere, between the stratopause and mesopause. The mesosphere lies between approximately 50 and 80 kilometers above the surface of the Earth. **(1)**

Meteorology The study of the atmosphere and its motions, especially on the shorter time scales known as "weather." Meteorology includes the study of the ways weather is affected by interactions among the Earth's land and water surfaces and living things. **(1)**

Methane (CH₄) A trace atmospheric gas that is increasing in concentration, probably due to human activities such as industry and agriculture, which plays a role in greenhouse warming. **(1)**

Microburst A strong, localized downdraft less than 4 kilometers in diameter that sometimes develops underneath a thunderstorm as a result of evaporative cooling. **(12)**

Micron A unit of length equal to one millionth of a meter. Also called a *micrometer* and abbreviated μm. *(1)*

Microscale Atmospheric motions with a spatial scale smaller than 2 kilometers. **(6)**

Midlatitude westerlies The persistent west-to-east winds of the middle latitudes. Also called *westerlies*. **(7)**

Milankovitch cycles (14) Periodic variations of the Earth's orbit that are theorized to have caused the ice ages. **(14)**

Millibars Units of atmospheric pressure. The average atmospheric pressure at sea level is 1013.25 mb, which is equal to 29.92 inches of mercury. **(1)**

Mirage A refracted image of an object caused by differences in air density. In the *inferior mirage*, the refracted image appears below the true object, while in a *superior mirage* it lies above the object. **(5)**

Mistral A term used on the French Riviera to describe a katabatic wind. **(12)**

Mixing ratio The amount of water in the atmosphere in terms of the mass of water vapor per unit mass of dry air. **(4)**

Model A simplified, but relatively accurate approximation of reality. Today's weather forecasts are computed by numerical models that approximate the behavior of the actual atmosphere. **(13)**

Moist adiabatic lapse rate The change of temperature of a rising moist parcel of air. It is quite variable, roughly 6° C per kilometer near the Earth's surface. **(2)**

Moist parcel of air   A parcel of air in which water molecules are changing phase from vapor to liquid or ice. **(2)**

Moist subtropical and midlatitude climates (C)   A climate type characterized by humid and mild winters. At least 8 months of the year have temperatures above 10° C (50° F), with the coolest month below 18.3° C (65° F) and above –3° C (27° F). **(14)**

Molecule   Composed of atoms, a molecule is the smallest unit of a substance that retains the chemical properties of that substance. **(1)**

Momentum   The product of mass and velocity. **(6)**

Monsoon   A circulation pattern characterized by a seasonal reversal of the prevailing winds. **(7)**

Monthly mean temperature   The sum of the daily mean temperatures for a month divided by the number of days in the month. **(3)**

Morning glory   A wind squall of northern Australia that forms on the leading edge of a gravity wave and is often accompanied by a spectacular linear cloud up to 1000 kilometers (620 miles) in length. **(12)**

Mountain breezes   Winds that result from the temperature differences along the slopes of high mountains, which cause pressure gradient forces that drive local wind circulations. **(12)**

mP   The maritime polar air mass, which forms over cool ocean waters and is usually cool and moist. **(9)**

MRF   A type of spectral numerical weather forecast model that is used for medium- and long-range forecasts across the globe. It is the longer-range version of the AVN. **(13)**

mT   The maritime tropical air mass, which forms over warm ocean waters and is warm and moist. **(9)**

Multicell   Thunderstorms which are organized in clusters (such as MCCs) or lines (such as squall lines). **(11)**

Multiple-vortex tornado   An especially damaging tornado that contains smaller spiraling whirlwinds inside the main funnel. **(11)**

Needle   A type of ice crystal that has a hexagonal base and a pointed top. **(4)**

Negative feedback mechanism   *See* **Feedback. (15)**

Newton's First Law of Motion   States that a body resists a change in its motion. This resistance to change is called **inertia. (6)**

Newton's Second Law of Motion   States that force equals mass times acceleration. **(6)**

NGM   A gridpoint numerical weather forecast model that was used to make short-range weather forecasts for the United States. **(13)**

Nimbostratus (Ns)   Precipitation-producing layered clouds, usually found just ahead of a warm front. **(4)**

Nitric oxide (NO)   A pollutant that is a byproduct of high-temperature combustion, such as in automobile engines and electric power generation. **(15)**

Nitrogen dioxide (NO$_2$)   A gas found at all levels of the atmosphere that is emitted by automobile engines. High concentrations of nitrogen dioxide give polluted air its reddish-brown color. **(15)**

Nocturnal inversion   A temperature inversion that develops overnight as a result of strong longwave radiation emissions at ground level. **(3)**

Nocturnal low-level jet   A lower-tropospheric maximum in wind speed that develops at night. This jet supplies moisture and energy to nighttime thunderstorms over the Great Plains. **(11)**

Nor'easters   Extratropical cyclones that affect the northeastern United States and extreme eastern Canada. They are named for the strong northeasterly winds that blow across this region as the path of the low moves northeast along the North American coastline. **(10)**

Normal temperatures   Average temperatures calculated for a location over a long period of time, often 30 years. **(3)**

Norte   A strong, cold northeasterly wind along the Gulf of Mexico. **(12)**

Norwegian cyclone model   The life cycle of the extratropical cyclone as explained by the Bergen School of meteorology. The life cycle progresses from a young frontal wave on a stationary front, to a maturing cyclone with warm and cold fronts, to a weakening occluded cyclone, and finally to a dying cut-off cyclone. **(10)**

Nowcast   A local weather forecast for the next few hours. **(13)**

Nucleation   The initial process of a phase change of water to a liquid droplet or ice crystal. Two basic processes are homogeneous nucleation and heterogeneous nucleation. In *homogeneous nucleation* the process of formation involves only water molecules. In *heterogeneous nucleation* small, nonwater particles serve as sites for particle formation. *Immersion nucleation* occurs when supercooled water freezes around a nucleus. *Contact nucleation* occurs when supercooled water freezes immediately upon touching an ice nucleus, while *freezing nucleation* occurs when supercooled water freezes without a non–water particle present. **(4)**

Numerical modeling   The simulation of fluid motions, such as in the atmosphere, using mathematical approximations of the equations describing the fluid. These complex computations usually require the use of a powerful computer. **(13)**

Obliquity   The angle between the Earth's orbit plane and the plane of the Earth's equator; in other words, the tilt of the Earth. Currently, this value is approximately 23.5°. **(14)**

Occluded cyclone A mature extratropical cyclone that exhibits an occluded front and usually begins to weaken. **(10)**

Occluded front A front that develops as a mature cyclone ages and moves deeper into colder air. **(1)**

Occlusion The process by which a surface low retreats from the tilted areas of strong temperature gradients associated with the warm and cold fronts. **(10)**

Ocean current A narrow, concentrated horizontal flow of water in the ocean that are associated with wind patterns at the ocean surface. Ocean currents are similar to jet streams in the atmosphere, but the currents are much slower than wind speeds in a jet stream. **(8)**

Oceanography The study of the world's oceans. **(8)**

Open wave A young extratropical cyclone with strong warm and cold fronts, but no occluded front. **(10)**

Ordinary cell A thunderstorm cell that is a few kilometers in diameter and has a life cycle of less than an hour. **(11)**

Orographic lifting Rising air flow that is caused by the presence of mountains. **(4)**

Overrunning A condition in which an air mass aloft is moving over an air mass of greater density at the surface. **(9)**

Overshooting top A bubble-like protrusion on the top side of a thunderstorm anvil where the thunderstorm updraft has penetrated into the stratosphere. **(11)**

Ozone (O₃) A chemically active molecule in the atmosphere. Because ozone strongly absorbs ultraviolet light from the Sun, its presence in the stratosphere is a crucial for the existence of life on Earth. In the troposphere, however, it is a toxic pollutant and is the main component of photochemical smog. **(1)**

Ozone layer A region of the stratosphere about 20 to 30 kilometers above the Earth's surface with a maximum of ozone concentration. **(2)**

Ozone hole A depletion of stratospheric ozone that has occurred over the Antarctic continent each spring for the past few decades due to the presence of chlorofluorocarbons (CFCs). **(2)**

Pacific Decadal Oscillation An alternating pattern of sea-surface temperatures in the Pacific Ocean that reverses itself over periods of several decades, affecting weather and climate over large parts of the globe. **(8)**

Paleoclimate Climate of the past. **(14)**

Paleoclimatology The study of climate of the past and the causes its observed variations. **(14)**

Pangaea The supercontinent formed approximately 300 million years ago that subsequently split into the continents of today. **(14)**

Panhandle Hooks Extratropical cyclones that follow curved, hook-shaped paths from the Texas and Oklahoma Panhandles to the Great Lakes, most often in late autumn. **(10)**

Parameterizations Portions of numerical weather prediction models that are devoted to the approximation of phenomena that the model cannot calculate precisely. **(13)**

Parcel of air An arbitrarily defined small-scale "bubble" of air that exhibits uniform conditions within its hypothetical borders. Air parcels are used to simplify explanations of air temperature changes and wind motions. **(2)**

Particulates Airborne solid and liquid aerosols that, when in high concentrations, seriously affect the lives of people and animals, harm plants, or threaten ecosystems. **(15)**

Passive sensors Observational instruments that measure energy received naturally from the region being measured. **(5)**

Perihelion The point on the Earth's orbit when it is closest to the Sun. **(2)**

Permafrost A layer below the surface of the Earth in tundra environments that is perennially frozen. **(14)**

Persistence forecast A forecast that assumes that tomorrow's weather will be the same as today's weather. **(13)**

Photodissociation The destruction of chemical molecules resulting from absorption of high-energy radiation. **(2)**

Pilot leader The initial discharge of negative charges near the cloud base at the beginning of a lightning stroke. **(11)**

Pineapple Express The phrase used to describe the path of extratropical cyclones that approach the West Coast of the United States from the southwest. **(10)**

Planetary scale The largest scale of atmospheric motion, the circulation patterns of which span a hemisphere or even the entire world. **(6)**

Plate tectonics The modern theory of continental movement. Its precursor was Wegener's theory of *Continental Drift*. **(14)**

Polar climates (E) A climate zone poleward of the Arctic and Antarctic circles. Polar climates are extremely cold and have little precipitation. **(14)**

Polar easterlies Low-level air flowing equatorward from the poles that has a strong east-to-west component. **(7)**

Polar front Transition zone separating cold polar air masses from warmer air of the middle latitudes and subtropics. **(7)**

Polar front jet A jet stream found in the middle and upper latitudes in association with the clash of cold and warm air known as the *polar front*. **(7)**

Polar stratospheric clouds (PSCs) Clouds composed of ice and frozen nitrogen particles that form in air temperatures colder than approximately –80° C (–112° F). **(15)**

Positive feedback mechanism.   *See* **Feedback. (15)**

Potential energy   The energy an object has by virtue of its position. **(2)**

Power   The rate of change of energy over time, often expressed in watts. **(2)**

Precession   The wobble of the Earth's axis. Precession alters the relationship of the solstices with the distance from the Earth to the Sun. **(14)**

Precipitation   Any liquid or solid water particle that falls from the atmosphere and reaches the ground. **(4)**

Pressure   Force per unit area. **(1)**

Pressure gradient   A change in pressure across a distance. **(6)**

Pressure gradient force (PGF)   The force that arises from changes in pressure over distance, divided by the air density. The PGF always pushes directly from higher toward lower pressure. **(6)**

Primary rainbow   The brilliantly colored arc of light most commonly seen in the sky, it forms at about a 40-degree angle to sunlight and has blue color on the inside of the arc and red color on the outside of the arc. **(5)**

Primitive equations   The nearly exact physical laws of the atmosphere that, together with the initial atmospheric conditions of a given area, can be used to create a numerical forecast for a later time. **(13)**

Propellers   A method of measuring wind speed in which the blades of a propeller rotate at a rate proportional to the wind speed. **(5)**

Psychrometer   An instrument that measures differences in the temperature of wet- and dry-bulb thermometers as a means of calculating relative humidity. **(5)**

Puelche   A warm, easterly downslope wind along the west slopes of the Andes Mountains in South America. **(12)**

Radar   An instrument used for detecting the presence and distance of objects, such as raindrops, that scatter radio energy. **(5)**

Radar echo   The energy scattered back from a target and detected by the radar receiver.

Radar reflectivity   The amount of energy received by the radar that is scattered back to it, usually by precipitation particles. **(5)**

Radiant energy.   *See* **Radiation. (2)**

Radiation   Energy that moves through space or a medium in the form of a wave with electric and magnetic fields. **(2)**

Radiation fog   A fog that occurs on clear calm nights when air near the surface cools to the dew point. **(4)**

Radiation inversion   A layer of air, usually near the ground, in which the air temperature increases with height as a result of radiational cooling. **(3)**

Radiocarbon dating   A method to determine the age of objects through analysis of elements that undergo radioactive decay in predictable ways. **(14)**

Radiometers   Instruments that measure radiation power. **(5)**

Radiosonde   An instrument package carried upward by weather balloons to measure the vertical profile of atmospheric temperature, relative humidity, and pressure from the surface into the stratosphere. **(5)**

Rain   Liquid precipitation bigger than 0.5 millimeter. **(4)**

Rain gauge   An instrument that measures the amount of rain. **(5)**

Rain shadow   A region downwind of mountains in which precipitation is significantly less than in the upwind regions. **(4)**

Rainbands   Elongated and curved regions of rain that are often associated with tropical cyclones. **(8)**

Rainbow   A bright arc of colored light caused by refraction and reflection of light by raindrops. The primary rainbow is brightest and most commonly seen; the secondary rainbow forms outside the primary rainbow and is fainter. **(5)**

Rawinsonde   An instrument package carried upward with a weather balloon to measure the vertical profile of atmospheric temperature, relative humidity, pressure, and wind direction and speed from the surface into the stratosphere. The wind direction and speed are measured by tracking the weather balloon's motion. **(5)**

Recurvature   A northeastward movement of a previously westward-moving tropical cyclone, caused by midlatitude weather systems. **(8)**

Reflection   The process by which energy incident on the surface is turned back into the medium through which it originated. **(5)**

Refraction   The process by which the direction of energy propagation is changed as a result of spatial variations in properties (e.g. density) of the medium. **(5)**

Relative humidity   The ratio of the observed vapor pressure of the air to the saturation vapor pressure, expressed as a percentage. Relative humidity indicates how close the air is to saturation (i.e., 100% relative humidity). **(4)**

Resistance thermometer   An instrument that observes temperatures by measuring changes in the electrical resistance of a metal. **(5)**

Resolution   The spacing between gridpoints in a numerical weather forecast model. "Fine" resolution results from close spacing of gridpoints; "coarse" resolution results from wide spacing of gridpoints. Fine resolution allows a model to "see" smaller-scale phenomena more accurately than does coarse resolution. **(13)**

Return stroke   The ground-to-cloud electrical current in a lightning stroke that causes the brilliant flash. **(11)**

Right-mover A supercell thunderstorm that moves in a direction that is to the right of other, non-supercell thunderstorms. **(11)**

Rime A white, opaque deposit of ice formed by the rapid freezing of supercooled water drops as they collide with an object at or below freezing. **(4)**

Rossby waves Waves in the midlatitude westerly winds with wavelengths on the order of thousands of kilometers. **(7)**

RUC A gridpoint numerical weather forecast model that is used to make short-term forecasts for the United States. **(13)**

Saffir–Simpson scale The system by which hurricanes are classified on a scale from 1 (minimal hurricane) to 5 (catastrophic hurricane), based on potential wind and sea-water damage and related closely to the winds and lowest central pressure of the tropical cyclone. The Saffir–Simpson scale, unlike the Fujita scale for tornadoes, can be assessed during the lifetime of the storm. **(8)**

Santa Ana winds Warm, dry downslope winds in California that generally blow from the northeast or east. **(12)**

Saturation The condition at which the vapor pressure equals the saturation vapor pressure over a flat surface of water or ice. **(4)**

Saturation vapor pressure The vapor pressure at which the number of molecules leaving a liquid or ice surface equals the number of molecules entering the liquid or ice. It is a function of temperature only. **(4)**

Scattering The process by which light rays change direction of propagation through the interaction with particles, such as molecules, aerosols, and cloud particles. **(5)**

Sea breeze An onshore coastal wind that occurs during the daytime. It is driven by pressure gradients created by the unequal heating of land and ocean. **(6)**

Sea-level pressure The atmospheric pressure a weather station would have if it were located at sea level. **(2)**

Seasonal temperature cycle The pattern of temperature at a location over the course of a year. Also called *annual temperature cycle.* **(3)**

Secondary rainbow A dim second arc of light that sometimes is seen above the primary rainbow at about a 50° angle to sunlight, and which has red color on the inside of its arc and blue color on the outside of the arc. **(5)**

Sensible heat The transfer of energy by warm air without a change of phase of water. **(2)**

Severe midlatitude climates (D) Typically located in the eastern regions of continents, this climate regime has an average temperature of the coldest month that is less than –3° C (27° F). The warmest month has an average temperature exceeding 10° C (50° F). **(14)**

Severe thunderstorm Thunderstorms that produce one or more of the following: a tornado, large hail (diameters greater than 1.9 centimeters [0.75 inch]), or wind gusts of at least 26 m/s (58 mph) **(11)**

Shelf cloud A wedge-shaped cloud that sometimes forms on the leading edge of a thunderstorm gust front. **(11)**

Shortwave radiation Radiant energy with wavelengths between approximately 0.2 and 4 microns. **(2)**

Short-waves Undulations in the large-scale midlatitude tropospheric winds that are smaller and faster-moving than Rossby long-waves. **(7)**

Shower A brief episode of often heavy precipitation. **(4)**

Simoom A strong, dry, and dust-laden desert wind found in the African and Arabian deserts. **(12)**

Sink A process that removes a substance, such as a gas, from its environment, such as the atmosphere. **(1)**

Sirocco A warm south or southeast wind that brings hot, dry conditions to the Mediterranean. **(12)**

Sleet Precipitation in the form of frozen raindrops. Sleet most often falls from nimbostratus clouds ahead of warm fronts, unlike hail which forms via a different process in thunderstorms. **(4)**

Smog Originally used to describe a combination of smoke and fog, this term is now used to describe mixtures of pollutants in the atmosphere. **(15)**

Snow Precipitation in the form of ice crystals or aggregates of crystals. **(4)**

Snowflake An individual ice crystal, or more commonly, an aggregate of ice crystals. **(4)**

Solar constant The amount of solar radiation received at the top of the Earth's atmosphere on a surface perpendicular to the incoming radiation at Earth's mean distance from the Sun. Its value is about 1370 W/m². **(2)**

Solar radiation Electromagnetic radiation from the Sun, which is concentrated in visible wavelengths. **(2)**

Solar zenith angle The angle at the Earth's surface measured between the position of the Sun and an observer's zenith. **(2)**

Solstices The two days during the year when the noon sun is overhead at its northernmost (on or about June 21) or southernmost (on or about December 22) latitudes. On these days, the amount of daylight is at maximum and minimum, respectively, in the Northern Hemisphere. **(2)**

Solute effect A process that enhances the condensational growth of droplets by suppressing evaporation because of elements dissolved in the droplet. **(4)**

Sounding The vertical distribution of temperature, moisture, and wind speed and direction over a location. **(5)**

Source A process that adds a substance, such as a gas, to its environment, such as the atmosphere. **(1)**

Southerly buster A mesoscale cold front that moves across southeastern Australia. **(12)**

Southern oscillation A seesaw in atmospheric pressure between the western and eastern Pacific that is commonly associated with El Niño and La Niña. **(8)**

Specific heat The amount of energy needed to raise the temperature of 1 gram of a substance by 1° C. **(2)**

Spectral models A class of numerical weather forecast models that divides the atmosphere in terms of waves rather than gridpoints. **(13)**

Speed The distance traveled in a given amount of time, equivalent to the magnitude of the velocity. **(6)**

Speed divergence Divergence that occurs due to downwind horizontal wind speed increases (but with no change of wind direction). **(10)**

Split flow pattern An upper air flow pattern in which the jet stream divides into two main streams, and has both west-east (zonal) and north-south (meridional) components. **(7)**

Stepped leaders Cloud-to-ground electrical charges in a lightning stroke that flow downward from the cloud in an attempt to establish a conductive channel all the way from the cloud to the ground. **(11)**

Squall line A line of intense thunderstorms, usually in advance of a cold front. **(11)**

Squamish A gap wind in British Columbia. **(12)**

Stable atmosphere An atmospheric layer with a lapse rate less than 10° C per kilometer. **(3)**

Static stability A measure of the stability of the atmosphere with respect to vertical displacements of air parcels. **(9)**

Station model An efficient method of graphically representing weather conditions at a single location on a weather map. **(1)**

Stationary front A frontal situation in which neither of the air masses on either side of the front are advancing or retreating. As a result, the front remains in approximately the same location. **(1)**

Steam fog An evaporation fog that occurs above a body of water. **(4)**

Stefan–Boltzmann Law A fundamental radiation law that states that the total energy emitted by a blackbody is proportional to the fourth power of its temperature. **(2)**

Storm surge A surge of seawater pushed onshore primarily by winds of a storm, usually a tropical cyclone. **(8)**

Stratocumulus Low clouds that appear in rows or patches and are white or gray. **(4)**

Stratopause The top of the stratosphere where the temperature does not change with height, located at an altitude of 50 kilometers. **(1)**

Stratosphere The region of the atmosphere above the troposphere and below the mesosphere, located between about 15 and 50 kilometers. It is characterized by a temperature inversion. **(1)**

Stratus A cloud layer with a uniform base and a gray color. **(4)**

Subgeostrophic flow A wind that is slower than the geostrophic wind would be in the same horizontal pressure gradient. **(6)**

Sublimation The change of phase of water from ice to vapor. **(2)**

Subtropical highs High pressure regions near the surface observed near 30° latitude. They are semipermanent in their location, shifting with the seasons. **(7)**

Subtropical jet stream A region of strong winds typically found in the upper troposphere between 20° and 40° latitude. **(7)**

Sulfur dioxide (SO_2) An acidic gas found in the atmosphere, released by volcanoes and through the burning of fossil fuels that contain sulfur. **(15)**

Sulfur trioxide (SO_3) A pollutant released into the atmosphere primarily through the burning of fossil fuels that contain sulfur. **(15)**

Sulfuric acid (H_2SO_4) A strong acid that forms in the atmosphere when sulfur is present. Sulfuric acid droplets that form in the stratosphere after strong volcanic eruptions affect climate by reflecting solar radiation. **(15)**

Sundogs Brightly colored regions that flank the sun as a result of refraction of light through ice crystals. Also called *parhelia*. **(5)**

Sun pillar A shaft of light extending vertically from the rising or setting Sun that is caused by reflection of sunlight by ice crystals. **(5)**

Supercell A dangerous type of single-cell thunderstorm that rotates under the influence of vertical wind shear in the wind patterns above and around it. Supercell thunderstorms can produce damaging tornadoes. **(11)**

Supercooled water Liquid water that exists when its temperature is below 0° C. **(4)**

Supergeostrophic flow A wind that is faster than the geostrophic wind would be in the same horizontal pressure gradient. **(6)**

Superior mirage. *See* **Mirage.** **(5)**

Supersaturation A condition in the atmosphere that exists when the relative humidity is greater than 100%. **(4)**

Supertyphoon A severe tropical cyclone in the eastern Pacific Ocean with highest sustained winds in excess of 240 km/hr (150 mph). **(8)**

Surface chart A map of analyzed surface weather conditions. Also called a *sea-level chart*. **(1)**

Surface temperature   The air temperature measured in the shade at 1.5 meters (5 feet) above the ground. (**3**)

Surface zone   The topmost 100 meters of the oceans in which the temperature is uniform due to stirring by winds and waves. Also known as the *mixed layer*. (**8**)

Synoptic scale   Literally "seeing together," this term is used to describe weather on scales that are typically represented on weather maps. (**6**)

Taiga   Forests composed of conifer trees that grow in severe midlatitude climates in Asia. (**14**)

Tehuantepecer   A gap wind on the Pacific Coast of Central America. (**12**)

Temperature   In an ideal gas, the average kinetic energy of its molecules. (**2**)

Temperature inversion   A vertical layer of the atmosphere in which temperature increases with height. (**3**)

Temperature range   The difference between the maximum and minimum temperatures at a location. (**3**)

Terrestrial radiation.   *See* **Longwave radiation.**

Thermal conductivity   The ability of a substance to conduct thermal energy. (**2**)

Thermal lows   Regions in which atmospheric pressure is lowered by heating near the surface due to strong solar heating, which causes lifting of pressure surfaces and thus divergence above the surface. (**7**)

Thermal wind   The relationship between the vertical changes in the geostrophic wind and the horizontal temperature gradient. Where the temperature decreases poleward, the westerly wind above this region increases with increasing altitude. (**6**)

Thermocline   A zone of the ocean in which the temperature decreases rapidly with depth. (**8**)

Thermosphere   The topmost layer of the atmosphere, located above 85 kilometers, in which temperature increases with altitude. (**1**)

Thunderstorm   A cloud or cluster of clouds that produces thunder, lightning, heavy rain, and sometimes hail and/or a tornado. (**11**)

Time zones   Regions of the globe approximately 15 degrees of longitude wide that are assigned the same time. (**1**)

Tornado   A violently spinning vortex of air that extends downward from the bottom of a thunderstorm to the ground. (**11**)

Tornado alley   A region of especially frequent tornado occurrence, often defined to be in the Great Plains from Texas to Kansas. (**11**)

Tornado vortex signature (TVS)   A small-scale red-and-green couplet of inbound and outbound winds on a Doppler radar velocity image that indicates the presence of a spinning vortex inside a thunderstorm. (**11**)

Total internal reflection   The situation in which a light ray traveling from one medium to another makes an angle with the normal that exceeds the critical angle and reflects back into the medium it came from. (**5**)

Trace gas   A gas found in the atmosphere in very small amounts, usually comprising less than 1% of the total atmosphere. (**1**)

Trade winds   Steady winds that occupy most of the tropics and blow outward from the subtropical highs. They are northeasterly in the Northern Hemisphere and southeasterly in the Southern Hemisphere. (**7**)

Transpiration   The process by which plants release water vapor into the air. (**1**)

Trend forecast   A weather forecast that assumes that tomorrow's weather will change as a result of approaching weather systems that, by assumption, are not themselves changing. (**13**)

Tropical cyclone   Circular low-pressure storms with winds of at least 35 knots (39 mph) which are driven by atmosphere–ocean interactions and originate in the tropical oceans. (**8**)

Tropical depression   A tropical disturbance that has developed a weak cyclonic circulation near its center. Winds in a tropical depression are less than 35 knots (39 mph). (**8**)

Tropical disturbance   Disorganized clumps of thunderstorms in the tropics that occasionally develop into tropical depressions, and even more rarely into tropical cyclones. (**8**)

Tropical humid climates (A)   Climates characterized by high monthly mean temperatures, at least 18.3° C (65° F). The range of the annual temperature is small, typically less than 10° C (18° F), and as a result, killing frosts are absent. (**14**)

Tropical storm   A tropical cyclone with highest sustained winds at the center of between 35 and 65 knots (39 to 74 mph). (**8**)

Tropopause   The boundary between the troposphere and the stratosphere where temperature does not change with altitude, normally located between 10 and 15 kilometers above the ground. (**1**)

Troposphere   One of the four main layers of the atmosphere. Most weather occurs in the troposphere, the lowest 10 to 15 kilometers (6.2 to 9.4 miles) of the atmosphere. (**1**)

Turbulence   The irregular, seemingly random pattern of motion in a fluid such as air. (**12**)

Typhoon   Tropical cyclones found in the western Pacific. (**8**)

UKMET   A gridpoint numerical weather forecast model that is used for global short-range and medium-range forecasts. (**13**)

Ultraviolet light   Electromagnetic radiation with wavelengths between approximately 0.2 to 0.4 microns. (**2**)

Updraft   A current of air that has a marked upward vertical motion. Updrafts keep cloud particles suspended in the air. **(4)**

Upslope fog   A fog that forms when air rises and cools to saturation as it climbs over a mountain slope. **(4)**

Upwelling   A wind-driven ocean circulation pattern in which cold, nutrient-rich ocean waters are forced up to the surface as a consequence of Ekman transport. **(8)**

Urban heat island   The increased temperatures of urban areas compared with nearby rural areas due to the radiative properties of cities. **(15)**

UTC   The time standard used around the world by meteorologists, referenced to the time at Greenwich, England; an acronym for *Universel Temps Coordonné* or *Coordinated Universal Time.* **(1)**

Valley breezes   Small-scale winds that result from temperature differences along the slopes of high mountains, which can lead to pressure gradients that drive local wind circulations. An up-valley wind occurs during the day, and a down-valley wind occurs during the night. **(12)**

Vapor pressure   The pressure exerted by water molecules in a given volume of the atmosphere. It is a measure of the contribution of water vapor to the total pressure. **(4)**

Velocity   The change of direction and position of an object with time. **(6)**

Virga   Precipitation that evaporates before reaching the ground. **(4)**

Viscosity   The friction in a fluid, such as air or motor oil. **(12)**

Visibility   The maximum horizontal distance at which objects can be identified. **(5)**

Visible radiation   Electromagnetic radiation with wavelengths between approximately 0.4 and 0.7 microns. Human vision perceives these wavelengths. **(2)**

Visible satellite image   A type of indirect observation which shows the amount of visible light reflected by the atmosphere and the Earth's surface as measured by a weather satellite. In a visible satellite image, white regions such as clouds have a high albedo and are reflecting more light than dark regions. **(5)**

Volatile organic compounds (VOCs)   Pollutants containing carbon that are released into the atmosphere by automobiles. **Hydrocarbons** are a class of VOCs. **(15)**

von Kármán vortex   A cloud vortex generated on the lee of an island in the presence of temperature inversion. **(12)**

Wall cloud   A lowered region of rotating cloud underneath the rear of a severe thunderstorm from which a tornado may form. A wall cloud marks a very strong updraft. **(11)**

Warm front   A front in which cooler air is replaced by warmer air. **(1)**

Warm sector   The wave-shaped region between cold and warm fronts during the early stages of an extratropical cyclone's development. **(10)**

Warning   Issued when hazardous weather is occurring or about to occur. **(1)**

Watch   Issued when the risk of hazardous weather is significant. **(1)**

Waterspout   A type of weak whirlwind that forms underneath cumulus clouds over a large body of water. **(11)**

Water vapor   Water in the vapor phase. **(1)**

Water vapor satellite image   A type of indirect observation which shows the amount of infrared energy emitted by water vapor in the atmosphere as measured by a weather satellite. In a water vapor satellite image, white regions are moister and emit more infrared energy than drier dark regions. **(5)**

Watt   A unit of power or energy per unit time. **(2)**

Wavelength   The distance from one crest of a wave to the next crest. **(2)**

Weak Sun paradox   A puzzle of paleoclimate in which it is known that the Sun's energy output was much weaker billions of years ago, but fossil evidence indicates that the Earth's climate was nevertheless quite warm. **(14)**

Weather   The current state of the atmosphere at a particular location. **(1)**

Weather types   Categories of large-scale weather patterns that were used to classify weather conditions for use in analog forecasts. **(13)**

Wien's Law   A radiation law that describes why hotter objects give off a higher percentage of their radiant energy at shorter wavelengths than do cooler objects. **(2)**

Willy-willy   A dust devil in Australia. **(12)**

Wind-chill   A measure of the increased loss of heat by living organisms due to the movement of the air. **(3)**

Wind-chill equivalent temperature   The no-wind temperature that it "feels" like due to the effect of both temperature and wind on exposed flesh. **(3)**

Wind gust   A very brief and significant increase in wind speed. **(6)**

Wind profiler   A vertically pointing Doppler radar that is used to observe winds at a single location. **(5)**

Wind shear   The change of wind speed and/or direction in the atmosphere along a given direction. **(8)**

Windsock   A tapered, open-ended piece of fabric that is used to estimate wind speed and direction at airports. **(5)**

Wind vane   An instrument used to determine wind direction. **(5)**

Windward   The upwind side of an object or region, such as a mountain. **(6)**

Work   The distance traveled by an object multiplied by the force applied to it in that direction. **(2)**

Younger Dryas   The relatively cold period that existed between 11,000 and 10,000 years ago across portions of the Northern Hemisphere. **(14)**

Zenith   The point directly above an observer. **(2)**

Zonal flow pattern   A situation with weak Rossby waves and strong westerly winds. **(7)**

Zonal index   A numerical measure of the type of large-scale flow pattern. **(7)**

Zonda   A hot wind along the east slopes of the Andes in Argentina. **(12)**

Index

Credits

483

Chapter 11: Figure 11.3: http://k12ocs.ou.edu/teachers/graphic/TstormFreq.gif; **Figure 11.5:** Adapted from Ahrens, C., *Meteorology Today,* 6th ed., Brooks-Cole, 2000, p. 408; **Figure 11.9:** Adapted from Lester, P., *Aviation Weather,* 2nd ed., Jeppesen, 2001, pp. 9–14; **Figure 11.15:** http://k12ocs.ou.edu/teachers/graphic/SupercellSlice.gif; **Figure 11.20:** From American Geophysical Union, *The Tornado,* 1993; **Figure 11.22:** Grazulis, T., *The Tornado: Nature's Ultimate Windstorm,* Oklahoma University Press, 2001. p. 110; **Figure 11.23:** Grazulis, T., *The Tornado: Nature's Ultimate Windstorm,* Oklahoma University Press, 2001. p. 111; **Figure 11.24:** http://k12ocs.ou.edu/teachers/graphic/TornadoFreq.gif; **Figure 11.25:** From Tom Grazulis, *The Tornado: Nature's Ultimate Windstorm,* University of Oklahoma Press, 2001, p. 267; **Figure 11.29:** Orville, R. and Huffines, G., "Cloud-to-Ground Lightning in the United States: NLDN Results in the First Decade, 1989–1999," *Monthly Weather Review,* May 2001, Fig. 3; **Figure 11.32:** http://k12ocs.ou.edu/teachers/graphic/HailstoneLifecycle.gif; **Figure 11.33:** Illinois State Water Survey.

Chapter 12: Figure 12.2: Richardson, L., *Weather Prediction by Numerical Process,* Cambridge University Press, 1922; **Figure 12.5:** Fujita, T., *The Downburst,* Author, 1985, p.108; **Figure 12.7:** www.srh.noaa.gov/bmx/february_22_1998/february_22_1998.html; **Figure 12.8:** www.srh.noaa.gov/bmx/february_22_1998/february_22_1998.html; **Figure 12.9:** Adapted from Kippel, L., Bosart, L., and Keyser, D., "A 25-year Climatology of Large-Amplitude Hourly Surface Pressure Changes over the Conterminous United States," *Monthly Weather Review,* January 2000, page 58; **Figure 12.12:** Adapted from Bentley, M., and Mote, T., "A climatology of Derecho-Producing Mesoscale Convective systems in the Central and Eastern United States, 1986–1995. Part I: Temporal and Spatial Distribution," *Bull Amer Met Soc,* 79, 2527–2540; **Figure 12.13:** Adapted from Ahrens, C., *Meteorology Today,* 6th ed., Brooks/Cole, 2000, p. 261; **Figure 12.14:** http://www.ucar.edu/communications/staffnotes/9902/wind.html; **Figure 12.17:** Adapted from Lester, P., *Turbulence,* Jeppesen, 1994, page 5–34.

Chapter 13: Figure 13.1: Adapted from http://www.srh.noaa.gov/maf/html/wind.htm, Fig. 7; **Figure 13.2:** Adapted from Ahrens, C., *Meteorology Today,* 6th ed., Brooks/Cole, 2000, p. 361; **Figure 13.3:** Adapted from Gedzelman, S.D., *The Science and Wonders of the Atmosphere,* John Wiley and Sons, 1980, p. 14; **Figure 13.4:** Adapted from Krick, I., and Fleming, R., *Sun, Sea and Sky: Weather in Our World and in Our Lives,* Lippincott, 1954, p. 97; **Figure 13.5:** *Some Meteorological Aspects of the D-Day Invasion of Europe,* pp. 165–166. Roger H. Shaw and William Innes, editors, American Meteorological Society. © 1984. Reprinted with permission; **Figure 13.6:** *Richardson's Weather Prediction by Numerical Process,* Cambridge University Press, 1922, p. 184. Copyright © 1922 Cambridge University Press. Reprinted with permission; **Figure 13.7:** Adapted from Daley, R., *Atmospheric Data Analysis,* Cambridge University Press, 1991, pp. 14–15; **Figure 13.12:** Gordon, A., et al., *Dynamic Meteorology: A Basic Course,* Arnold Press, 1998, page 208; **Figure 13.15:** *The Life Cycles of Extra-Tropical Cyclones,* Shapiro and Gronas, eds., American Meteorological Society, 1999. Reprinted with permission; **Figure 13.16:** Adapted from

The Life Cycles of Extra-Tropical Cyclones, Shapiro and Gronas, eds., American Meteorological Society, 1999. Reprinted with permission; **Figure 13.18:** Adapted from *The Life Cycles of Extra-Tropical Cyclones,* Shapiro and Gronas, eds., American Meteorological Society, 1999. Reprinted with permission; **Figure 13.21:** http://deved.meted.ucar.edu/nwp/pcu1/ic1/2_2.htm; **Figure 13.22:** http://www.ecmwf.int/pressroom/newsletter/pdf/88_2.pdf; **Figure 13.23:** Adapted from Kalnay, E., Lord, S., and McPherson, R., "Maturity of Operational Numerical Weather Prediction: Medium Range," *Bulletin of the American Meteorological Society,* vol. 79, no. 12, pp. 2753–2892.

Chapter 14: Figure 14.1: Adapted from Aguado, E. and Burt, J., *Understanding Weather and Climate,* 2nd ed., Prentice-Hall, 2001, p. 414; **Figure 14.3:** Adapted from Trewartha, G., and Horn, L., *An Introduction to Climate,* 5th Edition, McGraw-Hill, 1980, p. 209; **Figure 14.13:** (Adapted from Imbrie and Imbrie, *Ice Ages: Solving the Mystery,* MacMillian, 1979; **Figure 14.16:** From Duvick, D.N. and T.J. Blasing, 1981. "A Dendroclimatic Reconstruction of Annual Precipitation Amounts in Iowa since 1680." *Water Resources Research,* 17: 1183–1189. Reprinted with permission; **Figure 14.17:** Adapted from Graedel, T., and Crutzen, P., *Atmosphere, Climate and Change,* Scientific American Library, 1997, p. 82; **Figure 14.19B:** http://ingrid.ldeo.columbia.edu/SOURCES/.ICE/.CORE/.VOSTOK; **Figure 14.20:** Adapted from Imbrie, J. and Imbrie, J.Z. "Modeling the climatic response to orbital variations," *Science,* vol. 202, pp. 943–953; **Figure 14.21:** From Graedel, T., and Crutzen, P., *Atmosphere, Climate and Change,* Scientific American Library, 1997, p. 94; **Figure 14.23:** Adapted from Graedel, T., and Crutzen, P., *Atmosphere, Climate and Change,* Scientific American Library, 1997, p. 75; **Figure 14.25A:** Adapted from Tarbuck and Lutgens, *The Atmosphere,* 8th ed., Prentice-Hall, 2001, p. 377; **Figure 14.25B:** Adapted from Graedel, T., and Crutzen, P., *Atmosphere, Climate and Change,* Scientific American Library, 1997, p. 77; **Figure 14.26:** Adapted from Ahrens, C., *Meteorology Today,* 6th ed., Brooks/Cole, 2000, p. 512 and : Adapted from Tarbuck and Lutgens, *The Atmosphere,* 8th ed., Prentice-Hall, 2001, p. 373; **Table 14.1:** Adapted from Bradley, R., *Quaternary Paleoclimatology,* Allen and Unwin, 1985, p.2; **Figure 14.28:** (Adapted from Ahrens, C., *Meteorology Today,* 6th ed., Brooks/Cole, 2000, p. 509.

Chapter 15: Figure 15.4: Moran and Morgan, *Essentials of Weather,* Prentice-Hall, 1995, p. 417; **Figure 15.5:** http://toms.gsfc.nasa.gov/multi/min_ozone.jpg; **Figure 15.14:** *Climate Change 1995: The Science of Climate Change,* IPCC, page 300; **Figure 15.15:** *Climate Change 1995: The Science of Climate Change,* IPCC, page 300; **Figure 15.16:** Adapted from Kalkstein and Green, "An Evaluation of Climate/Mortality Relationships in Large US Cities and the Possible Impacts of Climate Change," *Environmental Health Perspectives,* 105(1), pp. 84–93; **Figure 15.17:** http://www.gcrio.org/NationalAssessment/overpdf/overview.html, p. 113; **Figure 15.18:** http://www.gcrio.org/NationalAssessment/overpdf/overview.html, pp. 116–117; **Box 15.1:** (Minnis, P., Schumann, U., Doelling, D., Gierens, K., and Fahey, D., "Global Distribution of Contrail Radiative Forcing," *Geophys Res Ltrs,* 26, July 1, 1999, pp. 1853–1856.

PHOTO CREDITS

Chapter 1: 2: NOAA/NESDIS and Space Science and Engineering Center, University of Wisconsin-Madison; **9:** Space Science and Engineering Center, University of Wisconsin-Madison; **13:** U.S. Department of Commerce, NOAA.

Chapter 2: 40: Courtesy NASA/JPL/Caltech; **41** (left): copyright © Calvin J. Hamilton; **41** (center): NASA; **41:** (right): NASA; **43:** NASA/Goddard Space Flight Center, Scientific Visualization Studio; **49:** NOAA/NESDIS and Space Science and Engineering Center, University of Wisconsin-Madison; **52:** NASA.

Chapter 3: 63: copyright © Steve Ackerman; **68:** Cooperative Institute for Meteorological Satellite Studies, University of Wisconsin-Madison; **70:** NASA; **78:** NASA.

Chapter 4: 85 (bottom left): copyright © Steve Ackerman; **85** (top right): CP Picture Archive (Jaques Boissinot); **92:** copyright © Steve Ackerman; **93:** copyright © Steve Ackerman; **96** copyright © The Photographer's Window; **98** (top): Space Science and Engineering Center and Cooperative Institute for Meteorological Satellite Studies, University of Wisconsin-Madison; **98** (bottom): National Oceanic and Atmospheric Administration/Department of Commerce; **99:** copyright © Ralph Burnett; **102:** copyright © Lil Ackerman; **103:** copyright © Jeffrey Kay; **104** (top): copyright © Steve Ackerman; **104** (bottom): copyright © Elise Collet; **105:** copyright © David W. Martin, Space Science and Engineering Center, University of Wisconsin-Madison; **106:** copyright © Greg Thompson; **107** (top): copyright © Kieth G. Diem, Ph.D.; **107** (bottom): copyright © Lil Ackerman; **108** (top): copyright © Don Lloyd; **108** (bottom): copyright © Kieth G. Diem, Ph.D.; **109:** copyright © Jimmy Deguara/Storm Chaser; **110:** copyright © Greg Thompson; **117:** copyright © Andy Kilgore.

Chapter 5: 127: copyright © Peter Wolf; **128:** copyright © Tom Burgdorf; **129:** NOAA Photo Library, NOAA Central Library; OAR/ERL/National Severe Storms Laboratory (NSSL); **131:** copyright © Lewis Kozlosky, National Weather Service; **133:** National Weather Service; **136:** copyright © Pam Knox; **137:** Jeff Cook/Weatherwise Magazine/Heldref Publications; **138:** copyright © Nola Miller; **141** Space Science and Engineering Center and Cooperative Institute for Meteorological Satellite Studies, University of Wisconsin-Madison; **142** Space Science and Engineering Center and Cooperative Institute for Meteorological Satellite Studies, University of Wisconsin-Madison; **143** Space Science and Engineering Center and Cooperative Institute for Meteorological Satellite Studies, University of Wisconsin-Madison; **144** copyright © Gene Rhoden/Weatherpix; **145** courtesy San Francisco State University Meteorology Program; **147** courtesy San Francisco State University Meteorology Program; **148** (top): College of DuPage; **148** (bottom): Sunset Mirage, Ambosele, Kenya copyright © The Image Bank; **149:** Seiji Miyauchi/Weatherwise Magazine/Heldref Publications; **150:** copyright © Harald Edens; **152:** copyright © Thomas H. Hogan; **153:** copyright © Steve West; **154** copyright © Greg Thompson; **155:** copyright © Steve West.

Chapter 6: 159: AP/Wide World Photos; **164** (top): copyright © 2001 University Corporation for Atmospheric Research; **164** (bottom): copyright © 2001 University Corporation for Atmos-

pheric Research; **169:** copyright © Walter Munk; **179:** copyright © 2001 University Corporation for Atmospheric Research; **180:** copyright © 2001 University Corporation for Atmospheric Research; **184:** Space Science and Engineering Center, University of Wisconsin-Madison.

Chapter 7: 191 (top): data provided by the ASA International Satellite Cloud Climatology Project; **191** (bottom): data provided by the ASA International Satellite Cloud Climatology Project; **194:** Cooperative Institute for Meteorological Satellite Studies, University of Wisconsin-Madison; **195:** National Oceanic and Atmospheric Administration/Department of Commerce; **196:** Cooperative Institute for Meteorological Satellite Studies, University of Wisconsin-Madison.

Chapter 8. 214: copyright © 1997 by the Ocean Remote Sensing Group, Johns Hopkins University Applied Physics Laboratory; **218:** Courtesy of NASA/JPL/Caltech; **223:** NASA; **224:** copyright © Carl Seibert, South Florida Sun-Sentinel, Tribune Publishing; **226** (top): Cooperative Institute for Meteorological Satellite Studies, University of Wisconsin-Madison; **226** (bottom): Cooperative Institute for Meteorological Satellite Studies, University of Wisconsin-Madison; **227:** copyright © Robin Moyer; **230:** National Oceanic and Atmospheric Administration/Department of Commerce; **231:** National Center for Atmospheric Research/University Corporation for Atmospheric Research/National Science Foundation; **234** (top): Cooperative Institute for Meteorological Satellite Studies, University of Wisconsin-Madison; **234** (bottom): National Oceanic and Atmospheric Administration/Department of Commerce; **235:** Courtesy Applied Physics Laboratory, Johns Hopkins University; **239:** South Florida Sun-Sentinel, Tribune Publishing; **240:** courtesy Rosenburg Library, Galveston, Texas; **241:** National Weather Service; **242:** copyright © Dave Saville/FEMA News Photo; **244:** copyright © Bryan Norcross; **245:** National Oceanic and Atmospheric Administration/Department of Commerce.

Chapter 9: 252 (top): Space Science and Engineering Center, University of Wisconsin-Madison; **252** (bottom): Space Science and Engineering Center, University of Wisconsin-Madison; **256** (top): copyright © The Photographer's Window; **256** (bottom): Space Science and Engineering Center, University of Wisconsin-Madison; **257:** copyright © PictureQuest International; **258:** copyright © PictureQuest International; **259:** Cooperative Institute for Meteorological Satellite Studies, University of Wisconsin-Madison; **260:** copyright © Anne Pryor; **262** (top): copyright © The Photographer's Window/American Red Cross; **262** (bottom): Space Science and Engineering Center, University of Wisconsin-Madison.

Chapter 10: 271 (left): American Meteorological Society; **271** (right): John Fitzgerald Kennedy Library; **273** (top). copyright © Ruth Hudson; **273** (bottom): copyright © Paul LaMarre, Jr.; **274:** copyright © National Geographic Society; **275:** Cooperative Institute for Meteorological Satellite Studies, University of Wisconsin-Madison; **276:** National Snow and Ice Data Center, CIRES CB-449, University of Colorado-Boulder; **280:** National Snow and Ice Data Center, CIRES CB-449, University of Colorado-Boulder; **281:** Defense Meteorological Satellite Program; **287:** National Snow and Ice Data Center, CIRES CB-449, University of Colorado-Boulder; **289** (left): Tromso Satellite Station; **289**